ENVIRONMENTAL AERODYNAMICS

MATHEMATICS & ITS APPLICATIONS

Series Editor: Professor G. M. Bell
Chelsea College, University of London

Mathematics and its applications are now awe-inspiring in their scope, variety and depth. Not only is there rapid growth in pure mathematics and its applications to the traditional fields of the physical sciences, engineering and statistics, but new fields of application are emerging in biology, ecology and social organisation. The user of mathematics must assimilate subtle new techniques and also learn to handle the great power of the computer efficiently and economically.

The need of clear, concise and authoritative texts is thus greater than ever and our series will endeavour to supply this need. It aims to be comprehensive and yet flexible. Works surveying recent research will introduce new areas and up-to-date mathematical methods. Undergraduate texts on established topics will stimulate student interest by including applications relevant at the present day. The series will also include selected volumes of lecture notes which will enable certain important topics to be presented earlier than would otherwise be possible.

In all these ways it is hoped to render a valuable service to those who learn, teach, develop and use mathematics.

The Foundation Programme includes:

MODERN INTRODUCTION TO CLASSICAL MECHANICS AND CONTROL
David Burghes, Cranfield Institute of Technology and Angela Downs, University of Sheffield.

VECTOR & TENSOR METHODS
Frank Chorlton, University of Aston, Birmingham.

LECTURE NOTES ON QUEUEING SYSTEMS
Brian Conolly, Chelsea College, London University.

MULTIPLE HYPERGEOMETRIC FUNCTIONS
Harold Exton, The Polytechnic, Preston.

APPLIED LINEAR ALGEBRA
Ray J. Goult, Cranfield Institute of Technology.

GENERALISED FUNCTIONS: Theory, Applications
Roy F. Hoskins, Cranfield Institute of Technology.

MECHANICS OF CONTINUOUS MEDIA
S. C. Hunter, University of Sheffield.

USING COMPUTERS
Brian Meek and Simon Fairthorne, Queen Elizabeth College, University of London.

ENVIRONMENTAL AERODYNAMICS
R. S. Scorer, Imperial College of Science and Technology, University of London.

PHYSICS OF THE LIQUID STATE: A Survey for Scientists and Technologists
H. N. V. Temperley, University College of Swansea, University of Wales and
H. D. Trevena, University of Wales, Aberystwyth.

ENVIRONMENTAL AERODYNAMICS

R. S. SCORER

Professor of Theoretical Mechanics
Department of Mathematics,
Imperial College of Science and Technology,
University of London

ELLIS HORWOOD LIMITED
Publisher Chichester

Halsted Press: a division of
JOHN WILEY & SONS
New York · London · Sydney · Toronto

The publisher's colophon is reproduced from James Gillison's drawing of the ancient Market Cross, Chichester.

First published in 1978 by

ELLIS HORWOOD LIMITED, PUBLISHERS
Coll House, Westergate, Chichester, West Sussex, England.

Distributors:

Australia, New Zealand, South-east Asia:
JOHN WILEY & SONS, AUSTRALASIA PTY LIMITED
1-7 Waterloo Road, North Ryde, N.S.W. Australia

Canada:
JOHN WILEY & SONS LIMITED
22 Worcester Road, Rexdale, Ontario

Europe, Africa:
JOHN WILEY & SONS LIMITED
Baffins Lane, Chichester, Sussex, England

U.S.A., Mexico, S. America and the rest of the world:
HALSTED PRESS a division of
JOHN WILEY & SONS INC.,
605 Third Avenue, New York, N.Y. 10016, U.S.A.

Library of Congress Cataloging in Publication Data

Scorer, Richard Segar, 1919–
Environmental Aerodynamics.

(Mathematics and its applications)
1. Mesometeorology. 2. Dynamic meteorology.
3. Fluid mechanics. 4. Air-Pollution. I. Title.
QC883.5.S36 551.5′153 77-23909

ISBN 0-470-99270-0 (Halsted)
ISBN 85312-063-3 (Ellis Horwood, Publisher: Library Edn.)
ISBN 85312-094-3 (Ellis Horwood, Publisher: Student Edn.)

Set in 10pt Press Roman by Coll House Press.
Printed Offset Litho in Great Britain by Unwin Brothers Ltd, Woking.

Table of Contents

Table of Contents

Table of Contents

Preface

Natural Aerodynamics, the predecessor of this book, was written twenty years ago at a time when air pollution, gliding, and other environmental applications of fluid mechanics were of rapidly growing interest. Since then there has developed even greater interest in mesoscale meteorology, which I like to call out-of-doors hydrodynamics, not only because of its important applications to aviation but because we have begun to regard the environment as something to live in, along with other species, and not merely something crudely to exploit. The physics of the environment has had a fair share of the high school syllabuses: by comparison the beauties of its mechanics have been neglected and this new book will, I hope, enable university teachers to take it much more in their stride.

We are all concerned with man-sized, town-sized, and country-sized phenomena of fluid motion. The medium looks very different from that treated in textbooks on fluid mechanics because they give so little attention to the forces of buoyancy and rotation which are dominant out-of-doors. Because of this the motion has been regarded by engineers and theoreticians as too complicated for them, and accordingly rather avoided. Nevertheless fascinating laboratory experiments and theoretical treatments of a fairly simple kind do exist and there is no reason why they should not be part of standard courses.

The basis of our subject in conventional fluid dynamics is described in the first two chapters of this book. They form a commentary on fluid mechanics which could be specially useful to engineers and physicists who are beginning to grapple with environmental problems. Chapter 3 is a discussion of fluid rotation giving a simpler treatment of secondary flow. This makes it accessible and useful to many who might otherwise have been content with qualitative arguments about pressure and accelerations which often give explanations which are difficult to use with confidence or for prediction.

Chapter 4 discusses many of the subtleties of dynamical meteorology which originate in the Earth's rotation. I hope this survey of wind systems will be helpful to engineers, geographers, and anyone interested in weather.

Chapter 5 and 6 are necessarily somewhat more mathematical because they deal with precisely defined phenomena. Both topics – waves, and billows – have increased in importance for aviation, oceanography, and many branches of physical meteorology and deserve serious mathematics, although even now a great deal remains in the realm of physical inference.

Turbulence is treated in Chapter 7 from the standpoint of mechanism, unlike the more conventional statistical and phenomenological approaches which are less useful out-of-doors. This leads to the next two chapters which are about turbulence as we find it in nature, and contain some remarkably simple but effective methods of analysis, and to Chapter 10 which applies these methods to dispersion of pollution. Chapter 11 surveys mechanical behaviour in which the condensation of clouds is essential. It may also remind us that although the earth is a red and ultramarine planet with a blue atmosphere, it is half covered with white clouds which dominate the motion.

Although the last chapter is short, one may hope that it deals with the part of environmental aerodynamics most ripe for exciting developments in the near future. If this book can inspire a few who might otherwise have thought of themselves as engineers or physicists to enter the biological realm, it will have achieved one of its main aims.

Chapters 3, 5, and 6, with suitable extracts from Chapters 1 and 2, are the heart of one lecture course and Chapters 7–10 of another which have been given in various forms over the last two decades. They contain what does not appear to be assembled elsewhere. Chapters 1, 2, and 4 may be specially useful to atmospheric physicists, who may also find in Chapter 11 many topics which have received little treatment elsewhere. Thus it is to be hoped that the book will be found useful background reading for many environmental specialisms. Some geographers with a feel for the phenomena may find a new way into mathematics.

Courses with an environmental emphasis are being introduced in many disciplines and many participants will have difficulty in extracting for themselves the appropriate parts of more conventional books on fluid mechanics. I hope that they may be helped by this new book to find insight into the joys of the great out-doors.

I acknowledge gratefully the privilege of having many research students to work with me: there is reference to many of their works, and for me their co-operation has always been a source of inspiration. I should also say that my indebtedness to Professors F. H. Ludlam, C-S. Yih and R. R. Long is greater than my references to their writings may imply.

R. S. SCORER
Imperial College, London
February 1977

Author's Introduction

Chance occurrences often open up new realms of research. Gunfire of the Great War in Flanders could be heard in Britain beyond a silent zone: the fall of the Great Siberian Meteorite in 1908 occurred at a time when the first sensitive barographs happened to be in operation for a short time in Britain. These two events aroused interest in the warm layers high in the stratosphere. Half a century later the stimulus given to photochemistry by the smog of Los Angeles quickened the pace in the study of the composition of the stratosphere and particularly of the ozone which is the cause of the warm layers.

Very simple models have recently aroused interest in the possibility that human activities might affect the amount of ozone in the atmosphere, and many unsubstantiated predictions of harmful consequences have set in train hurried programs to measure the minor chemical components of the atmosphere. Inevitably the first obvious message from these observations is that the distribution of all components which vary significantly is subject to those details of the air motion about which information is scarce, especially in the stratosphere. In any age of less frantic and more sober environmentalism the same events might well have received no public attention and the science might have developed slowly in an atmosphere of more modest hypotheses.

Our understanding of the flight behaviour of desert locust swarms showed us that many other airborne life forms are likely to have important aerodynamic behaviour built into their survival mechanisms. These discoveries came about largely because of the interest of glider pilots in air motions on the same scale. At the same time the development of radar gave us the means to observe invisible billows and thermals, and these successes have been followed by aircraft-mounted Doppler radar and computer by which the air trajectory may be determined in flight and small wind discontinuities discovered. These are now known to be regions of concentration of many species of insect which can also be identified by radar. Hence studies of density currents have been greatly furthered by investigation of the insects, while investigation of sea breezes has brought new insight into their life style.

One never knows what will lead to what! We live in an age of great extravagance. Predictions of a dramatic or doom-laden nature are readily publicised, many being made by people unperceptive of the severe limitations of simple models when applied to the infinitely complex great outdoors. This is to be expected because simple models have made possible spectacular advances in laboratory studies of deliberately and skilfully isolated fundamental mechanisms.

It is natural also to enquire into the possibility of disastrous consequences of human activity after an era of great urban smogs, despoiled land and polluted waters. This era has been variously regarded either as one of great economic growth and advance in welfare or as one characterised by a sequence of self-inflicted injuries consequent upon greed. We have certainly far to go in learning how to restrain excess and at the same time protect the freedoms which material progress has given us from rigid regulation at a time of decreasing per capita wealth.

The stratosphere is particularly fertile territory for speculation today when we know just enough to make many a fiction plausible but too little to compel moderation, still less confer certainty. The troposphere is different because we have a good appreciation of its rapid turnover and powerful renewal mechanisms. Nevertheless, in the nineteenth century many cloud forms were 'explained' quite fallaciously on the basis of the then popular science of electrostatics. And little over a decade ago an eminent physicist confidently declared that H-bombs must affect the weather and the public, as always, was ready to believe in man-made causes of what they choose to regard as anomalous climatic events.

Only when a computer was used to 'prove' that cloud droplets could coagulate fairly readily into drizzle and rain did many cloud physicists allow themselves to believe what ordinary meteorologists already knew. From the ground and from aircraft for several decades they observed with their own eyes that rain may fall from a cloud no part of which is at a temperature below freezing.

Typically of our age, it has recently been proposed that the Los Angeles smog could be, and therefore should be, cured by spraying antidotes into the air. Yet the aerodynamics required of such a 'technological fix' probably rule it out as impracticable.

The brazen claims of rainmakers, hail-preventers, and others who distort the realities of science into wishful promises, have had legislators, planners and farmers eating out of their hands. Contracts, and even research grants, could not be refused because of the tempting prospects held out and the absence of certain proof that complex nature is not so easily bent to conform to human whim. The statistics actually demonstrated nothing, but appeared very favourable when suitably selected. The suggestion that the fury of hurricanes could be moderated by cloud seeding was almost as fanciful as an earlier belief that if sufficient silver iodide were distributed in the air it would be possible to stop it raining everywhere for ever. Other 'serious fantasies' include the spreading of carbon black on the artic snow or on the tropical seas, and covering the Sahara desert with photocells to 'solve' the world's energy problem – as if when a snow storm, hurricane, or sand storm had undone our work a raid by ten thousand aircraft could somehow restore the situation.

It now seems that this age of extravagant speculation and beliefs is in retreat. It has taught us much about our world, among other things that it belongs also to other species. One must particularly admire the superb achievements of spectroscopists and engineers who have given us beautiful satellite pictures of the earth in overwhelming detail. The trite dream of universal wealth is fading, and mankind will soon give more attention to living with nature. We may hope they will also try to get to the bottom of things as true scientists have always done, instead of to the top, where the era of growth has encouraged us to think our place to be. The investigations of the air motion described in this book study an important dynamic background to life. Let it not be forgotten, however, that the composition of the atmosphere is almost entirely determined by biological activity of which mankind is only a small part, and that many animal and vegetable forms depend absolutely on air motion for life.

Part 1

FUNDAMENTAL THEORY, VORTICITY, AND WAVES

Chapter 1

FUNDAMENTAL EQUATIONS

1.1 IRROTATIONAL MOTION AND REAL FLUIDS

There are several theorems in the classicial treatment of irrotational motion of incompressible fluids which have great significance for the study of real fluids. These are well described in many good textbooks: here we shall discuss them rather briefly, surveying their significance in a less orthodox manner.

In an inviscid fluid it is assumed that the only forces acting are body forces such as gravity and the fluid pressure which acts perpendicularly to any surface. Viscosity is responsible for any forces which may act along a surface. To calculate the force due to the pressure we consider the total force on the fluid inside a surface S, of which **dS** is an element of area dS, the vector being directed normally outwards from the surface. If **l** is a unit vector in a fixed direction, the force in this direction on the fluid inside S due to the pressure on the element is $-p\mathbf{dS}.\mathbf{l}$, p being the fluid pressure. The total force on the surface is therefore

$$-\int_{\text{surf}} \mathbf{l}.p\mathbf{dS} = -\int_{\text{vol}} \text{div}(\mathbf{l}p)\text{d}\tau = -\mathbf{l}.\int_{\text{vol}} \mathbf{grad}p\ \text{d}\tau \qquad (1.1.1)$$

in which $\text{d}\tau$ is the element of volume, and the integrals are over the whole surface or throughout the enclosed volume. Here we have used the divergence theorem, namely, for any vector field **A**, $\int_{\text{vol}} \text{div}\ \mathbf{A}\ \text{d}\tau = \int_{\text{surf}} \mathbf{A.ds}$, and the identity that $\text{div}(\mathbf{l}p) = \mathbf{1.grad}\ p + p\ \text{div}\ \mathbf{l}$, the last term of which is zero because **l** is a constant vector.

The expression (1.1.1) is the component in the direction **l** and therefore the total vector force on the enclosed fluid is

$$-\int_{\text{Surf}} \text{p}\ \mathbf{dS} = -\int_{\text{Vol}} \mathbf{grad}\ p\ \text{d}\tau \qquad (1.1.2)$$

which shows that the force per unit volume at a point is $-\mathbf{grad}\ p$.

Newton's law of motion, which we take to be absolutely accurate for all purposes in terrestrial fluid mechanics if only because no measurements are practicable for detecting any errors, states that the rate of change of momentum in the direction **l** is equal to the total force in that direction. Accordingly, if ρ is the density of the fluid (mass per unit volume) and **v** is the velocity of the fluid,

$$\int_{\text{Vol}} \frac{D}{Dt}(\mathbf{v}.\mathbf{l})\ \rho d\tau = -\int_{\text{Surf}} \mathbf{l}.p\,d\mathbf{S} + \int_{\text{Vol}} \mathbf{l}.\mathbf{g}\rho d\tau. \qquad (1.1.3)$$

In this D/Dt is the rate of change following the same fluid particles, often called the material rate of change, or the Lagrangian derivative, and $\rho d\tau$ is the element of mass envisaged in the integral. The volume and the surface are thought of as defined by material particles which therefore move with the fluid, and (1.1.3) equates the rate of change of momentum in the direction **l** with the forces in that direction. We have represented the body force as if it were gravity only, which is equal to $\mathbf{g}\rho$ per unit volume. There are body forces, such as electromagnetic ones, which are not necessarily proportional to the mass, but in this book the only body force we shall be concerned with is gravity. Expression (1.1.3) is the **l**-component of

$$\int_{\text{Vol}} \frac{D\mathbf{v}}{Dt}\ \rho d\tau = -\int_{\text{Vol}} \mathbf{grad}\, p\ d\tau + \int_{\text{Vol}} \mathbf{g}\rho\ d\tau$$

and since the volume is unspecified the integrands must be equal, and so

$$\frac{D\mathbf{v}}{Dt} = -\frac{1}{\rho}\ \mathbf{grad}\, p + \mathbf{g}\ . \qquad (1.1.4)$$

From this general form of the equation of motion we deduce the hydrostatic equation. The acceleration is zero, and gravity acts in the negative direction of the gradient of the gravitational potential gz and so (1.1.4) is $\mathbf{grad}\, p = \rho\ \mathbf{grad}(-gz)$, or

$$\frac{\partial p}{\partial z} = -g\rho\ . \qquad (1.1.5)$$

We shall always take the z coordinate as being measured vertically upwards for any situation in which we are concerned with gravity.

In the next section we have set out several mathematical identities of which we shall make use. The identity (1.2.7) may be written in the form

$$\mathrm{div}(\phi\,\mathbf{grad}\,\phi) = \phi\ \mathrm{div}\,\mathbf{grad}\,\phi + \mathbf{grad}\,\phi.\,\mathbf{grad}\,\phi \tag{1.1.6}$$

and this may be integrated, using the divergence theorem, to give

$$\begin{aligned}\int_{\mathrm{Surf}} \phi\,\mathbf{grad}\,\phi.\mathbf{dS} &= \int_{\mathrm{Vol}} \mathrm{div}(\phi\,\mathbf{grad}\,\phi)\,\mathrm{d}\tau \\ &= \int_{\mathrm{Vol}} \phi\nabla^2\phi\,\mathrm{d}\tau + \int_{\mathrm{Vol}} (\mathrm{grad}\,\phi)^2\,\mathrm{d}\tau.\end{aligned} \tag{1.1.7}$$

In order to study the conditions which have to be imposed, or specified, for a field of fluid motion to be determined, we imagine that there are two fields possible in a given situation and denote their velocities by $\mathbf{v}$ and $\mathbf{v}'$. If $\mathbf{u} = \mathbf{v} - \mathbf{v}'$ and if div $\mathbf{v}$ and **curl** $\mathbf{v}$ are both specified at all points in the region of fluid under consideration, so that they are equal to div $\mathbf{v}'$ and **curl** $\mathbf{v}'$ respectively, we must have that

$$\mathrm{div}\,\mathbf{u} = 0 \quad \text{and} \quad \mathbf{curl}\,\mathbf{u} = \mathbf{0} \tag{1.1.8}$$

throughout the region. The second of these equations means that there exists a potential ϕ such that $\mathbf{u} = \mathbf{grad}\,\phi$, and so the first then tell us that

$$\mathrm{div}\,\mathbf{u} = \mathrm{div}\,\mathbf{grad}\,\phi = \nabla^2\phi = 0 \tag{1.1.9}$$

at all points of the region.

The assumption that div $\mathbf{v}$ and **curl** $\mathbf{v}$ are specified everywhere thus gets rid of one of the terms in (1.1.7). If the left hand side were made zero by some conditions imposed at the boundary surface S, the term $\int_{\mathrm{Vol}} (\mathrm{grad}\,\phi)^2\,\mathrm{d}\tau$ would have to be zero, which means that **grad**ϕ would have to be of zero magnitude everywhere so that $\mathbf{v}$ would equal $\mathbf{v}'$. We can therefore set about finding what boundary conditions would achieve this result of making the velocity field unique. We shall discuss why it is useful to assume that div $\mathbf{v}$ and **curl** $\mathbf{v}$ should be zero later.

The integral $\int_{\mathrm{surf}} \phi\,\mathbf{grad}\,\phi.\mathbf{dS} = \int_{\mathrm{surf}} \phi\mathbf{u}.\mathbf{dS}$ can obviously be made zero if the component of $\mathbf{u}$ in the direction of $\mathbf{dS}$, that is normal to the surface, is zero; and this is equivalent to saying that the normal component of $\mathbf{v}$ is specified on the boundary of the fluid. Alternatively, if the boundary consists of a single surface and the tangential component of the velocity $\mathbf{v}$ is everywhere specified on it, the tangential component of $\mathbf{v}'$ is the same, and that of $\mathbf{u}$ is zero. **Grad** ϕ therefore has zero component along the surface so that ϕ is constant on the surface. If the surface consisted of more than one part, ϕ could have different but constant values on the separate parts. In the case of one part

$$\int_{\text{Surf}} \phi\, \mathbf{grad}\, \phi.\mathbf{dS} = \phi \int_{\text{Surf}} \mathbf{grad}\, \phi.\mathbf{dS} = \phi \int_{\text{Vol}} \text{div}\, \mathbf{grad}\, \phi\, d\tau = 0$$

using first the divergence theorem and then (1.1.9). The result is thus as required and we may state **Kelvin's Uniqueness Theorem** as follows: –

If div **v and curl v are specified at all points of a region and either the normal component of v is specified at all points of the boundary, or, when the boundary consists of a single surface, the tangential component of v is specified at all points of it, v is unique.**

If the normal component of **v** is specified at all points on the boundary but div **v** and **curl v** are not specified, we may discuss the energy of the motion to see what the effect is. Let T be the kinetic energy of the motion **v** in the case in which div **v** and **curl v** are zero everywhere, and let **v**′ be the actual motion having kinetic energy T′. If in this case we let $\mathbf{u} = \mathbf{v}' - \mathbf{v}$, and if we assume that the density is uniform throughout the fluid and is taken as unity, we have

$$0 \leq \int_{\text{Vol}} u^2 d\tau = \int_{\text{Vol}} (v'^2 - v^2) d\tau - 2 \int_{\text{Vol}} \mathbf{v.u}\, d\tau$$
$$= 2T' - 2T - 2 \int_{\text{Vol}} \mathbf{grad}\, \phi.\mathbf{u} d\tau$$

where we have written $\mathbf{v} = \mathbf{grad}\, \phi$ because $\mathbf{curl}\, \mathbf{v} = \mathbf{0}$ everywhere. Therefore

$$T' - T \geq \int_{\text{Vol}} \mathbf{grad}\, \phi.\mathbf{u}\, d\tau = \int_{\text{Vol}} \text{div}(\phi \mathbf{u}) d\tau - \int_{\text{Vol}} \phi\, \text{div}\, \mathbf{u}\, d\tau$$
$$= \int_{\text{Surf}} \phi\, \mathbf{u.dS} - \int_{\text{Vol}} \phi\, \text{div}\, \mathbf{u}\, d\tau$$
$$= - \int_{\text{Vol}} \phi\, \text{div}\, \mathbf{v}' d\tau \qquad (1.1.10)$$

because the component of **u** normal to S is zero since **v**′ and **v** have the same normal component on S. If div **v**′ is zero everywhere then the right side of (1.1.10) is zero and $T' \geq T$. The difference between **v** and **v**′ is that **v**′ may have vorticity: if it does not then **curl v**′ is zero and the uniqueness theorem proves that it is the same as **v**. If it is different from **v** somewhere, where **curl v**′ is not zero, then u^2 is positive somewhere and $T' > T$. We can now state **Kelvin's Minimal Theorem** as follows:–

Of all incompressible motions (having div **v** = 0 **everywhere) of which the**

normal component is specified on the boundary, the unique irrotational motion has less kinetic energy than any other motion satisfying the same boundary conditions.

This does not hold for a case in which the density is not uniform because it appears as a variable in the integrand and we cannot arrive at (1.1.10).

If on the hand div $\mathbf{v}'$ is not zero, so that the fluid is compressible, div $\mathbf{v}'$ may take positive or negative values and the integral (1.1.10) may be positive or negative, except in the case in which ϕ is not variable and $\mathbf{v}$ is zero, for in that case the irrotational motion is zero and no other motion can have less kinetic energy. The treatment of the integral in (1.1.10) is further complicated by the fact that the density is not uniform in a compressible fluid. The important point is that, if there are pressure waves being propagated through the fluid, the kinetic energy might be more or it might be less than in the case of the incompressible irrotational motion satisfying the same boundary conditions. In this book we shall only very occasionally be concerned with the case of compressible fluids, and in most cases div $\mathbf{v} = 0$, but we shall be very much concerned with rotational motion in which $\mathbf{curl\ v} \neq \mathbf{0}$.

If we can demonstrate that an inviscid incompressible fluid cannot acquire vorticity, then any motion set up by moving the boundaries must be irrotational (because the initial state of rest is irrotational) and must have the least possible kinetic energy.

First we imagine that the motion is set up by a sudden jerking of the boundary. We shall see in Section 1.4 that the effect of viscosity is to conduct vorticity through the fluid from the boundary in the way heat is conducted from a hot surface. This means that if there is no vorticity in the fluid initially it will take a finite time to produce any by viscosity. When the boundary is jerked suddenly the motion set up is irrotational except for a very thin layer at the boundary. This has the effect of allowing the fluid to slip along the boundary while being subjected to a specified normal component of velocity.

During an infinitesimal time Δt a very large pressure P is created with very large gradients inside the fluid. A finite 'impulsive pressure' Π, equal to $P\Delta t$, is generated and sets up a finite velocity $\mathbf{v}$ during the time Δt. The change in momentum is then equal to the impulse on the surface enclosing the fluid, and so, considering the component in a direction $\mathbf{l}$, we find

$$\begin{aligned}\int_{\mathrm{Vol}} \mathbf{l}.\mathbf{v}\rho\,\mathrm{d}\tau &= -\int_{\mathrm{Surf}} \mathbf{l}\Pi.\mathbf{dS} \\ &= -\int_{\mathrm{Vol}} \mathrm{div}\,(\mathbf{l}\Pi)\mathrm{d}\tau = -\mathbf{l}.\int_{\mathrm{Vol}} \mathbf{grad}\,\Pi\mathrm{d}\tau\end{aligned} \tag{1.1.11}$$

and since this holds for any surface marked out in the fluid and for any direction **l** we must have

$$\rho\mathbf{v} = -\,\mathbf{grad}\,\Pi, \text{ or}$$
$$\mathbf{v} = -\frac{1}{\rho}\,\mathbf{grad}\,\Pi \tag{1.1.12}$$

and this is equal to **grad** ϕ only if the **curl** of it is zero. Using (1.2.8) we see that this would be the case only if

$$\mathbf{grad}\,\Pi \times \mathbf{grad}\,\frac{1}{\rho} = \mathbf{0} \tag{1.1.13}$$

everywhere. This could not be the case unless the surfaces Π = constant happened to be the same as the surfaces ρ = constant; and this would only be the case if it were specially contrived to be so. The motion would therefore be irrotational if ρ were uniform, in which case the velocity and impulsive pressure would be related by

$$\mathbf{v} = \mathbf{grad}\,\phi \quad \text{and } \phi = -\,\Pi/\rho \tag{1.1.14}$$

and motion set up by jerking the boundaries suddenly is therefore irrotational.

We come finally to **Kelvin's Circulation Theorem.** This applies to an inviscid fluid under conservative body forces when the fluid is well mixed. This means that the forces have a potential, of which gravity $g = -\mathbf{grad}\,gz$ is the most important example, and the fluid is homentropic. Being homentropic means the condition of the fluid is not altered if two particles are interchanged. The simplest cases of this are a fluid of uniform density that is incompressible and a compressible fluid in adiabatic motion with all its elements having the same potential temperatures, or entropy. The circulation around a circuit always consisting of the same fluid particles is $\oint\mathbf{v.ds}$, and it changes with time at a rate

$$\frac{D}{Dt}\oint\mathbf{v.ds} = \frac{D\mathbf{v}}{Dt}\cdot\mathbf{ds} + \oint\mathbf{v}.\,\frac{D}{Dt}\,\mathbf{ds} \tag{1.1.15}$$

where **ds** is the vector element of arc of the circuit. Assuming that **v** has no discontinuities in the fluid we have that $\frac{D}{Dt}\mathbf{ds} = \mathbf{dv}$, which is the change in fluid velocity along the infinitesimal arc **ds**, and so the last term in (1.1.15) is $\oint\mathbf{v.dv} = \oint d(\tfrac{1}{2}v^2) = 0$ because **v** is continuous. The first term on the right of (1.1.15) may, by (1.1.4), be written

$$\oint \frac{\mathrm{D}\mathbf{v}}{\mathrm{D}t}\cdot\mathbf{ds} = \oint \mathbf{g.ds} - \oint \frac{1}{\rho}\,\mathbf{grad}\,p.\mathbf{ds} \qquad (1.1.16)$$

of which the first term on the right is equal to $\oint$ **grad** (gz).**ds** $= 0$ because the integral of any gradient vector is zero round a closed circuit, which is the same as saying that **g** is a conservative force field.

The last term of (1.1.16) may be written as

$$\oint \frac{1}{\rho}\,\mathbf{grad}\,p.\mathbf{ds} = \frac{1}{\rho}\oint \mathbf{grad}\,p.\mathbf{ds} = 0$$

in the case where ρ is uniform, or

$$= \oint \mathbf{grad}\left(\int \frac{\mathrm{d}p}{\rho}\right).\mathbf{ds} = 0$$

in the case when the fluid is homentropic, meaning that for all particles the density is the same function of the pressure. Thus we may write $\frac{1}{\rho}\,\mathbf{grad}\,p = \mathbf{grad}\,e$, or $\frac{1}{\rho}\,\mathrm{d}p = \mathrm{d}e$, or $e = \frac{\mathrm{d}p}{\rho}$. In consequence (1.1.15) shows that $\oint$**v.ds** **does not alter with time provided that the circuit is composed always of the same particles. This is the same as saying that the same number of vortex lines always passes through a given material circuit, so that the vortex lines move with the fluid.**

If the fluid is neither hometropic nor of uniform density

$$\frac{\mathrm{D}}{\mathrm{D}t}\oint \mathbf{v.ds} = -\oint \frac{1}{\rho}\,\mathbf{grad}\,p.\mathbf{ds} = \int_{\mathrm{Surf}} \mathbf{curl}\left(\frac{1}{\rho}\,\mathbf{grad}\,p\right).\mathbf{dS} \quad (1.1.17)$$

where the surface of which **dS** is the element is any of which the circuit is the boundary, or edge. By (1.2.8) this is

$$\frac{\mathrm{D}}{\mathrm{D}t}\oint \mathbf{v.ds} = -\int_{\mathrm{Surf}} \mathbf{grad}\,\frac{1}{\rho} \times \mathbf{grad}\,p.\mathbf{dS} \qquad (1.1.18)$$

which is known as **Bjerknes's circulation theorem.** It is related to (1.1.13), and states that the **pressure field in an inviscid fluid can only create circulation when the density is not constant on the surfaces of constant pressure** (i.e. when the isopycnic and isobaric surfaces are not the same). This effect is represented by the term $\mathbf{R} \times \mathbf{f}$ in (1.4.1).

To summarise, therefore, we see that when motion is generated by movement of the boundaries it is irrotational if the fluid is inviscid and of uniform density. Irrotational motion has less kinetic energy than any other possible motion compatible with the same boundary conditions. so that the addition of vorticity always adds energy, and it is clear why turbulence which is a sink of energy is rotational. If the fluid is compressible the boundaries can be moved without setting the whole fluid in motion immediately, so that it can have less energy. On the the other hand there may be pressure waves added to the motion, which thus contains kinetic energy even if the boundaries are not moving, and so a compressible fluid can have more or less kinetic energy than an incompressible fluid satisfying the same boundary conditions when both motions are irrotational.

The motion is determinate only if div **v** and **curl v** are specified in some way at all points of the fluid. **This means that turbulence, which is a state that is too complicated for curl v to be so specified, is indeterminate, and can only be treated by means of additional assumptions or information.**

1.2 NOTATION, COORDINATE SYSTEMS, IDENTITIES

When we express vectors in terms of their components we use rectangular Cartesian coordinates (x, y, z) with z vertically upwards unless otherwise specified. We may use coordinates (x_1, x_2, x_3) when convenient. Thus gravity is

$$\mathbf{g} = (0,0,-g) = -\,\mathbf{grad}\,(gz) = -\nabla(gz) \tag{1.2.1}$$

The total fluid velocity is written as

$$\mathbf{v} = (u, v, w) = (v_1, v_2, v_3) \tag{1.2.2}$$

and its ith component as v_i. The vorticity is

$$\begin{aligned} \omega &= \mathbf{curl\ v} = \nabla \times \mathbf{v} = (\xi, \eta, \zeta) = (\omega_1, \omega_2, \omega_3) \\ \omega_i &= (\mathbf{curl\ v})_i = \epsilon_{ijk}\frac{\partial v_k}{\partial x_j} = \frac{\partial v_k}{\partial x_j} - \frac{\partial v_j}{\partial x_k}\,. \end{aligned} \tag{1.2.3}$$

The divergence of velocity is

$$\text{div}\,\mathbf{v} = \nabla.\mathbf{v} = \frac{\partial u}{\partial x} + \frac{\partial v}{\partial y} + \frac{\partial w}{\partial z} = \frac{\partial v_i}{\partial x_i} \tag{1.2.4}$$

and the material rate of change of the quantity σ associated with a fluid element is

$$\frac{D\sigma}{Dt} = q\frac{D\sigma}{Ds} = \frac{\partial\sigma}{\partial t} + \mathbf{v}.\mathbf{grad}\ \sigma = \frac{\partial\sigma}{\partial t} + v_i\frac{\partial\sigma}{\partial x_i} \tag{1.2.5}$$

where t is time, s is distance along a particle trajectory, and q is the speed which is the magnitude of **v**.

The following identities are noted for reference but are not proved here:

$$\frac{D\mathbf{v}}{Dt} = \frac{\partial\mathbf{v}}{\partial t} + \mathbf{grad}\ (\tfrac{1}{2}q^2) - \mathbf{v} \times \mathbf{curl}\ \mathbf{v}. \tag{1.2.6}$$

$$\text{Div}\ (\sigma\mathbf{v}) = \sigma\ \text{div}\ \mathbf{v} + \mathbf{v}.\mathbf{grad}\ \sigma \tag{1.2.7}$$

$$\mathbf{Curl}\ (\sigma\mathbf{v}) = \sigma\ \mathbf{curl}\ \mathbf{v} + \mathbf{grad}\ \sigma \times \mathbf{v} \tag{1.2.8}$$

$$\mathbf{Curl}\ \mathbf{curl}\ \mathbf{v} = \mathbf{grad}\ \text{div}\ \mathbf{v} - \nabla^2\mathbf{v} \tag{1.2.9}$$

where

$$\nabla^2\mathbf{v} = (\text{div}\ \mathbf{grad})\mathbf{v}.$$

$$\text{Div}(\mathbf{v} \times \boldsymbol{\omega}) = \boldsymbol{\omega}.\mathbf{curl}\ \mathbf{v} - \mathbf{v}.\mathbf{curl}\ \boldsymbol{\omega} \tag{1.2.10}$$

$$\mathbf{Curl}(\mathbf{v} \times \boldsymbol{\omega}) = \mathbf{v}\ \text{div}\ \boldsymbol{\omega} - \boldsymbol{\omega}\ \text{div}\ \mathbf{v} + (\boldsymbol{\omega}.\mathbf{grad})\mathbf{v} - (\mathbf{v}.\mathbf{grad})\boldsymbol{\omega} \tag{1.2.11}$$

These last two identities are true for any two vector fields **v** and $\boldsymbol{\omega}$, although we are mostly concerned with the cases in which $\boldsymbol{\omega} = \mathbf{curl}\ \mathbf{v}$. Also for any scalar σ and the vector $\boldsymbol{\omega} = \mathbf{curl}\ \mathbf{v}$,

$$\text{div}\ \boldsymbol{\omega} = 0 \quad \text{and}\ \mathbf{curl}\ \mathbf{grad}\ \sigma = \mathbf{0}\ \text{always} \tag{1.2.12}$$

1.3 THE NAVIER-STOKES EQUATIONS

The main purpose of this section is to make clear the assumptions on which the **Navier-Stokes** equations are based. They are the most widely used equations in fluid mechanics and it is important not to use them incorrectly in the complex situations which occur in nature.

The most fundamental assumption, without which the equations could not be derived, is that the stresses (forces) in a fluid are related linearly and instantaneously to the rate of strain (distortion). Thus twice the rate of strain produces twice the force, and the fluid has no memory of previous motion: the forces come into existence or vanish immediately when the distorting motion begins or stops. Viscous forces are due to the motion of the molecules and so this assumption imples that the molecular velocities are large compared with the differences of fluid velocity in different parts of the fluid, and this certainly will do for ordinary motions of the atmosphere.

The rate of distortion at a point in the fluid can be represented as the rate of closing up of the angles between three planes of particles which are mutually orthogonal at the moment in question. The quantity $\partial w/\partial y$ is a measure of the rate of rotation about the x-axis of a line of particles on the y-axis. When added to $\partial v/\partial z$, which is the rate of rotation about the x-axis of particles on the z-axis, the sum is the rate of closing up of the angle between lines of particles momentarily located on the axes (see Fig. 1.3.i).

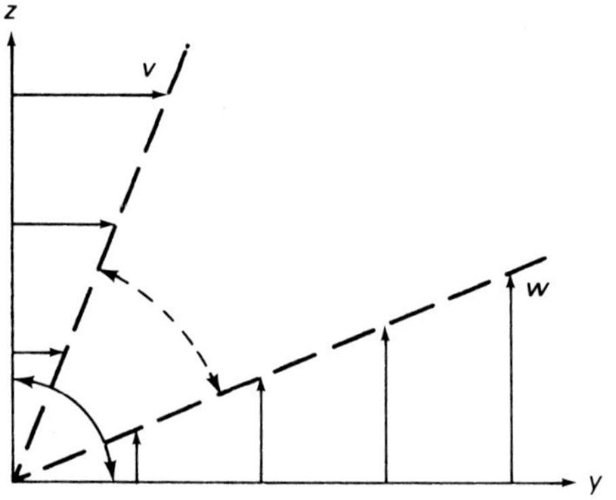

Fig. 1.3.i Rate of fluid distortion measured by gradients of velocity.

This quantity is one of the components of rate of strain. A distortion is also produced if the fluid is stretched along the y-axis or compressed along the z-axis, a rectangular shape becoming a non-rectangular one. The magnitudes of these distortions is represented by the quantities $\partial v/\partial y$ and $\partial w/\partial z$. (Fig. 1.3.ii).

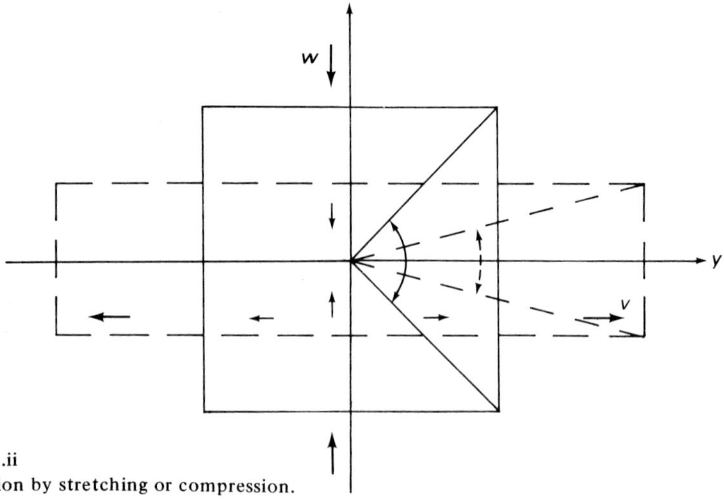

Fig. 1.3.ii
Distortion by stretching or compression.

There are thus six quantities required to define the rate of distortion and which may therefore enter the equations, namely

$$e_{xx} = 2\frac{\partial u}{\partial x}, \quad e_{yy} = 2\frac{\partial v}{\partial y}, \quad e_{zz} = 2\frac{\partial w}{\partial z}$$

$$e_{yz} = e_{zy} = \frac{\partial w}{\partial y} + \frac{\partial v}{\partial z}$$

$$e_{zx} = e_{xz} = \frac{\partial u}{\partial z} + \frac{\partial w}{\partial x} \qquad (1.3.1)$$

$$e_{xy} = e_{yz} = \frac{\partial v}{\partial x} + \frac{\partial u}{\partial y}$$

The vorticity measures the local angular velocity of the fluid and may be thought of as being twice its value. The rotation around the x-axis is obtained by reversing the sign of $\partial v/\partial z$ in e_{yz}, thereby adding the rates of rotation of two lines of particles momentarily at right angles to each other. Thus

$$\xi = \frac{\partial w}{\partial y} - \frac{\partial v}{\partial z}, \quad \eta = \frac{\partial u}{\partial z} - \frac{\partial w}{\partial x}, \quad \zeta = \frac{\partial v}{\partial x} - \frac{\partial u}{\partial y} \qquad (1.3.2)$$

The distortion of Fig. 1.3.ii has no total rotation; also rotations may be accompanied by no distortions. A rigid, or 'solid' rotation Ω produces a velocity in a particle at position $\mathbf{x}$ equal to

$$\mathbf{v} = \mathbf{\Omega} \times \mathbf{x}$$

$$\text{or} \quad v_i = \Omega_j x_k - \Omega_k x_j . \qquad (1.3.3)$$

For rigid material all the e's in (1.3.1) are zero and $\partial w/\partial y = -\partial v/\partial z$, and so

$$\xi = 2\frac{\partial w}{\partial y}, \quad \eta = 2\frac{\partial u}{\partial z}, \quad \zeta = 2\frac{\partial v}{\partial x}. \qquad (1.3.4)$$

Since $\partial u/\partial x = 0$, the velocity of a particle relative to the origin of coordinates has x-component (neglecting higher order terms)

$$u = y\frac{\partial u}{\partial y} + z\frac{\partial u}{\partial z} = \tfrac{1}{2}(y\zeta - z\eta) \qquad (1.3.5)$$

and comparing this with (1.3.3) we see that the velocity is the same as that produced by a rigid rotation equal to $\frac{1}{2}\boldsymbol{\omega}$.

There are nine derivatives of the velocity [set out in (1.3.1)] which define a 3 X 3 matrix whose members a_{ij} are defined by

$$a_{ij} = \partial u_i/\partial x_j \qquad\qquad i,j = 1,2,3.$$

and this may be expressed as the sum of the symmetric matrix

$$e_{ij} = \tfrac{1}{2}(a_{ij} + a_{ji})$$

and the antisymmetric matrix

$$\sigma_{ij} = \tfrac{1}{2}(a_{ij} - a_{ji}).$$

Since the diagonal members of σ_{ij} are zero, the vector $\sigma_{ij}x_j$ is the same as that obtained as the vector product of $\boldsymbol{\Omega}$ and $\mathbf{x}$, where $\Omega_i = \sigma_{jk}$. Thus σ_{ij} represents a solid rotation superimposed in the total motion on the distortion represented by e_{ij}. The solid rotation causes no internal stresses because it does not distort the fluid, and so only the component e_{ij} of the total motion a_{ij} is related to the stresses.

By a suitable transformation of coordinates representing a pure rotational displacement, the matrix e_{ij} may be reduced to a diagonal matrix. This is equivalent to saying that a quadric surface has three principal axes on which the distance from the origin is stationary under rotation, or that the distortion of the fluid elements can be fully described as the sum of extensions or compressions along the three principal axes. This is a property of second degree surfaces in space and is nothing to do with any physical property of the fluid. We shall make use of this fact in a moment.

In specifying the rate of strain and rotation of the fluid relative to the particle at the point by the nine derivatives of the velocity components, we have placed no restrictions on the kind of motion we are discussing other that it is continuous and differentiable. But if we assume that the stresses produced at a point depend on these quantities only, we imply that they are determined by something on a very small scale within the fluid, and not by events taking place a finite distance away. Thus the rates of strain are presumed not to change significantly across the region in which the events determing the strain are taking place. This is equivalent to neglecting any dependence on space derivatives of the velocity higher than the first. This assumption might be wrong if the first derivatives varied significantly over the mean free path of the molecules. The flow around a meteor in the high atmosphere may be incorrectly represented because the meteor may be smaller than the mean free path.

For the equations we are going to derive to be valid, the quantities e_{ij} do not have to be small although in assuming them to be linearly related to the stress we must make similar assumptions about the derivatives of the stress and assume that one component does not influence the relationship between others.

We will also assume that the fluid is isotropic, which means that it has no directional properties. Since we shall be studying stratified fluids we must assume that the shearing motion and the density gradients are too small to affect the linear relationship. This is reasonable if the stresses are produced on the molecular scale because, in a distance many times greater than the mean free path, there is no significant stratification. Where a very hot jet enters a cold environment the temperature gradients may be very large indeed, but here it is assumed that the equations remain valid by allowing for variations of viscosity with temperature without any local anisotropy. The cases in which these assumptions break down are those in which high precision is impossible, because the motion is so turbulent as to make direct application of the equations out of the question.

For the isotropic case some economy of argument is obtained by using the principal axes of rate of strain, denoted by $OXYZ$. For these

$$e_{YZ} = e_{ZX} = e_{XY} = 0 \tag{1.3.6}$$

The rotational displacement effecting the transformation from $Oxyz$ to $OXYZ$ is specified by its three components, which are chosen so that the equations (1.3.6) are satisfied. The result is that, relative to the principal axes, the rate of strain is expressed completely in terms of stretching or compressing motion along the axes. According to the assumptions, then, the stresses p_{ij} are given by equations of the form

$$p_{XX} = -p + ½\lambda\,(e_{XX} + e_{YY} + e_{ZZ}) + \mu e_{XX} \tag{1.3.7}$$

for it must depend on e_{YY} in he same way as on e_{ZZ} because of isotropy. The symbol p_{ij} represents the component in the i-direction of the total stress per unit area across a surface perpendicular to the j-direction. For the principal axes of stress

$$p_{YZ} = p_{ZX} = p_{XY} \tag{1.3.8}$$

and the axes must, from the assumption of isotropy, coincide with the principal axes of rate of strain.

Thus only two coefficients of viscosity are required, λ to define the effect of divergence, for $e_{ii} = 2$ div $\mathbf{v}$, and the other to define the effect of stretching. The coefficient $-p$ appears in (1.3.7) because we know that there is a stress, which we call the **normal pressure**, when there is no motion. The negative sign indicates, by convention, that a stress is positive when it is a tension.

If the axes $Oxyz$ have directions denoted by the unit vectors **l**, **m**, **n** relative to the principal axes $OXYZ$ which vary from point to point (i.e. **l** is a unit vector along O_x) and if **V** is the fluid velocity referred to OXYZ, then

$$\begin{aligned} \tfrac{1}{2}e_{xx} &= \frac{\partial u}{\partial x} = (\mathbf{l}.\mathbf{grad})(\mathbf{l}.\mathbf{V}) \\ &= \left(l_1 \frac{\partial}{\partial X} + l_2 \frac{\partial}{\partial Y} + l_3 \frac{\partial}{\partial Z}\right)(l_1 U + l_2 V + l_3 W) \\ &= l_1^{\,2} \frac{\partial U}{\partial X} + l_2^{\,2} \frac{\partial V}{\partial Y} + l_3^{\,2} \frac{\partial W}{\partial Z} \\ &= \tfrac{1}{2}(l_1^{\,2} e_{XX} + l_2^{\,2} e_{YY} + l_3^{\,2} e_{ZZ}) \end{aligned} \qquad (1.3.9)$$

because

$$\left.\begin{aligned} e_{YZ} &= \frac{\partial V}{\partial Z} + \frac{\partial W}{\partial Y} = 0 \\ \text{and} \qquad e_{XX} &= 2\frac{\partial U}{\partial X}, \text{ etc.} \end{aligned}\right\} \qquad (1.3.10)$$

according to the definition of the principal axes. Because the axes are orthogonal

$$l_1^{\,2} + l_2^{\,2} + l_3^{\,2} = 1 \qquad (1.3.11)$$

and so

$$2 \operatorname{div} \mathbf{v} = e_{xx} + e_{yy} + e_{zz} = e_{XX} + e_{YY} + e_{ZZ} \qquad (1.3.12)$$

and

$$\begin{aligned} e_{yz} &= \frac{\partial w}{\partial y} + \frac{\partial v}{\partial z} = (\mathbf{m}.\mathbf{grad})(\mathbf{n}.\mathbf{V}) + (\mathbf{n}.\mathbf{grad})(\mathbf{m}.\mathbf{V}) \\ &= m_1 n_1 e_{XX} + m_2 n_2 e_{YY} + m_3 n_3 e_{ZZ} \end{aligned} \qquad (1.3.13)$$

because of the relationships (1.3.10). There are, of course, two similar equations for e_{zx} and e_{xy}.

In obtaining all these relationships we put the space derivatives of the l's, m's, and n's equal to zero. This is because, although they vary from point to point with the varations of the principal axes, we are concerned with derivatives of quantities relative to two sets of axes defined at the moment at a point.

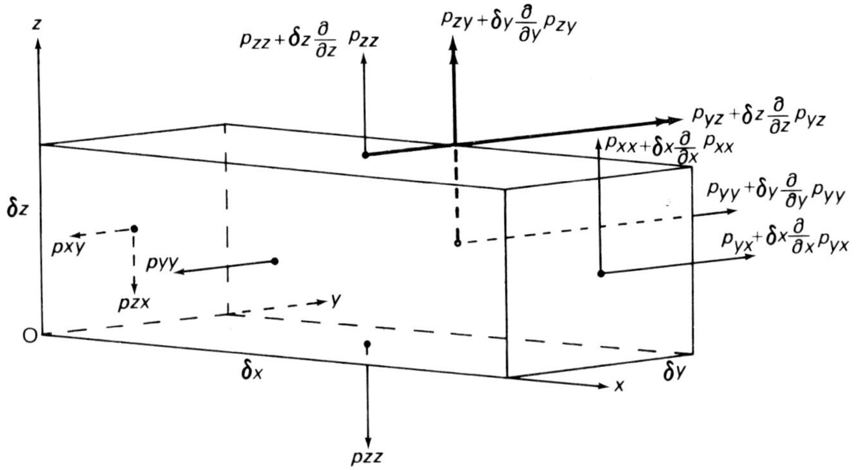

Fig. 1.3.ii Stresses on a box of fluid of dimensions δx, δy, δz, which have a moment about Ox.

In Fig. 1.3.ii we show a rectangular box of sides δx, δy, δz, with one corner at the origin. All the stresses with a non-zero moment about Ox are indicated, and shown acting at the mid points of the appropriate faces of the box. The magnitudes of the moments are calculated as if the forces were uniform over the face on which they act and as if they varied uniformly. Thus the higher order terms in δx etc. which involve higher derivatives than the first are neglected, according to the fundamental principle of calculus. The two forces indicated by the double arrows contribute together a moment about Ox equal to

$$-p_{yz}.\delta x\delta y.\delta z \;+\; p_{zy}.\delta y\delta z.\delta x \qquad (1.3.14)$$

and all the remaining terms in the moment are of higher order. These terms are the products of the forces, the areas on which they act, and their distance from Ox. The sum of all the moments is equal to the moment of intertia multiplied by the angular acceleration of the box. The moment of inertia is of the order of the mass times the square of a linear dimension of the box, and is therefore of the order of $(\delta x)^5$. Since the angular acceleration is necessarily finite and the quantity (1.3.14) is of the order of $(\delta x)^3$, the quantity must be zero; otherwise the acceleration would have to tend to infinity as the size of the box considered tended to zero.

$$\text{Therefore} \quad p_{yz} = p_{zy}$$
$$\text{and similarly} \quad p_{zx} = p_{xz},\; p_{xy} = p_{yx}\;. \qquad (1.3.15)$$

The same result is true for the stresses in an elastic solid as a consequence of the same argument.

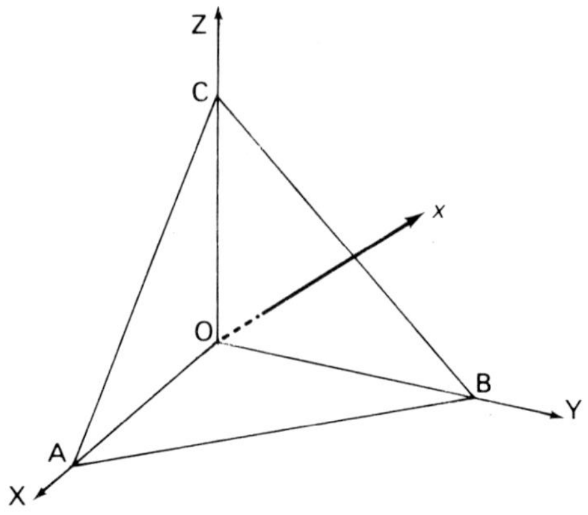

Fig. 1.3.iv Ox is perpendicular to plane ABC.

To relate the stresses to those expressed in terms of the principal axes we consider the forces on a small tetrahedron formed by the origin O and points A, B, C along the axes OXYZ where the plane ABC is perpendicular to the direction Ox. Since the forces are proportional to the areas on which they act, i.e. to the square of the dimension, while the momentum is proportional to the mass, i.e. to the cube of the dimension, the resultant forces must be zero, or otherwise an infinitesimal tetrahedron would experience an arbitrarily large acceleration. Thus the components in the direction Ox are in equilibrium, and so

$$p_{xx}\mathrm{ABC} = l_1 p_{XX}\mathrm{OBC} + l_2 p_{YY}\mathrm{OCA} + l_3 p_{ZZ}\mathrm{OAB}$$

and so since $\mathrm{OBC} = l_1\,\mathrm{ABC}$,

$$p_{xx} = l_1^{\,2} p_{XX} + l_2^{\,2} p_{YY} + l_3^{\,2} p_{ZZ} \tag{1.3.16}$$

and in view of (1.3.11)

$$p_{xx} + p_{yy} + p_{zz} = p_{XX} + p_{YY} + p_{ZZ}. \tag{1.3.17}$$

This last result means that what we may call the **mean pressure**, namely

$$p_m = -\tfrac{1}{3}(p_{xx} + p_{yy} + p_{zz}) \tag{1.3.18}$$

is the same in any set of coordinates and has meaning without reference to the coordinate system.

The forces on the tetrahedron in Fig. 1.3.iv in the direction of Oz are also in equilibrium, and so

$$p_{xz}\text{ABC} = n_1 p_{XX}\text{OBC} + n_2 p_{YY}\text{OCA} + n_3 p_{ZZ}\text{OAB}$$

i.e.

$$p_{xz} = l_1 n_1 p_{XX} + l_2 n_2 p_{YY} + l_3 n_3 p_{ZZ} \tag{1.3.19}$$

The relation (1.3.7) between the stresses and the rate of strain may be written

$$p_{XX} = -p + \lambda \operatorname{div} \mathbf{v} + \mu e_{XX} \tag{1.3.20}$$

and from this, and similar expressions for p_{YY} and p_{ZZ}, we may now obtain the relationship between the stress and the rate of strain in terms of the original coordinates Oxyz. Equations (1.3.16), (1.3.20) and (1.3.9) give

$$p_{xx} = (-p + \lambda \operatorname{div} \mathbf{v})(l_1{}^2 + l_2{}^2 + l_3{}^2) + \mu(l_1 e_{XX} + l_2 e_{YY} + l_3 e_{ZZ})$$

$$= -p + \lambda \operatorname{div} \mathbf{v} + \mu e_{xx}\,. \tag{1.3.21}$$

Since the axes are rectangular $l_i n_i = 0$ and so substituting (1.3.20) into (1.3.19) we find that, using (1.3.13)

$$p_{xz} = \mu e_{xz}\,. \tag{1.3.22}$$

The two equations (1.3.21 and 22) relate the stresses p_{ij} to the rates of strain e_{ij} through two coefficients of viscosity which are physical properties of the fluid and have the dimensions $L^2 T^{-1}$. The mean pressure is now seen to be [from (1.3.18)]

$$p_m = -\tfrac{1}{3}(p_{xx} + p_{yy} + p_{zz})$$

$$= -\tfrac{1}{3}[-3p + 3\lambda \operatorname{div} \mathbf{v} + \mu(e_{xx} + e_{yy} + e_{zz})]$$

$$= p - (\lambda + \tfrac{2}{3}\mu)\operatorname{div} \mathbf{v}. \tag{1.3.23}$$

The quantity p has already been called the **normal pressure** and it has been assumed to be the same for the components of stress in each of the equations (1.3.7) because of the isotropic property of the fluid. When the divergence of velocity is zero and the density is not varying, $p_m = p$. We do not have any practicable means of detecting any difference between p_m and p in any situation in which div **v** was not zero unless the coefficient $\lambda + \tfrac{2}{3}\mu$ were large. Although $p_m = p$ when there is no motion, it does not follow that they are necessarily always the same, and p has not been defined except as the constant in (1.3.7). We now have a definition of it in (1.3.23) and some authors make the error of logic in arguing that p is always the same as p_m because it could not otherwise be defined in such a way as to indicate how it might be measured, except as the same quantity as p_m. We depend either on experiment or additional theory to determine the value of $\lambda + \frac{2}{3}\mu$. Kinetic theory indicates that when the molecules occupy only a negligibly small fraction of the whole volume, $\lambda + \frac{2}{3}\mu$ is zero, and although such an argument is not valid for a liquid this does not matter because then div **v** is very small. No measurement has been devised which gives any magnitude to $\lambda + \frac{2}{3}\mu$, and the success of the assumption that it is zero indicates at least that it has a much smaller magnitude than μ. It is therefore usual to pursue the theory on the assumption that it is zero and that the stress and rate of strain are related by

$$\left.\begin{aligned} p_{xx} &= -p - \tfrac{2}{3}\mu \operatorname{div} \mathbf{v} + \mu e_{xx} \\ p_{yz} &= \mu e_{yz}, \end{aligned}\right\} \tag{1.3.24}$$

or

$$p_{ij} = -\delta_{ij}(p + \tfrac{2}{3}\mu \operatorname{div} \mathbf{v}) + \mu e_{ij}. \tag{1.3.25}$$

We may now obtain the equation of motion for a fluid by equating the rate of change of momentum of unit volume to the sum of the forces on unit volume. Thus

$$\rho \frac{\mathrm{D}v_i}{\mathrm{D}t} = \rho F_i + \frac{\partial}{\partial x_j} p_{ij} \tag{1.3.26}$$

in which **F** is a body force proportional to mass. In our case **F** = **g**.

From this we have by instertion of (1.3.25)

$$\rho \frac{\mathrm{D}v_i}{\mathrm{D}t} = \rho F_i - \frac{\partial p}{\partial x_i} - \tfrac{2}{3}\frac{\partial}{\partial x_i}(\mu \operatorname{div} \mathbf{v}) + \frac{\partial}{\partial x_j}(\mu e_{ij}). \tag{1.3.27}$$

In the case in which μ is the same at all points of the fluid, and is therefore a constant in the last two terms, we may write $\nu = \mu/\rho$ and get

$$\frac{\mathrm{D}v_i}{\mathrm{D}t} = F_i - \frac{1}{\rho}\frac{\partial p}{\partial x_i} - \tfrac{2}{3}\nu \frac{\partial}{\partial x_i}\frac{\partial v_j}{\partial x_j} + \nu \frac{\partial}{\partial x_j}\left(\frac{\partial v_i}{\partial x_j} + \frac{\partial v_j}{\partial x_i}\right) \tag{1.3.28}$$

and ν is called the **kinematic viscosity** because it relates to the motion. μ is called the **dynamic viscosity** and may be defined as the force across unit area in motion with shear of unit magnitude acting in the direction of the shear. In the case of plane flow in which v_i varies in the j-direction the force is $\mu \frac{\partial v_i}{\partial x_j}$ in the i-direction.

If this force varies in the j-direction it produces on unit volume a force in the i-direction equal to $\frac{\partial}{\partial x_j}\mu\frac{\partial v_i}{\partial x_j}$, or $\mu\frac{\partial^2 v_i}{\partial x_j^2}$ when μ is a constant in space. This is the most important effect of viscosity on the motion and it is due to the variation of shear stress from one point of the fluid to another. The force is zero if the velocity gradients are linear.

In turbulent flow the eddies are often thought of as producing an effect similar to that of the molecular viscosity but of greater magnitude. However, if the equation is to be applied on the assumption that μ represents an eddy viscosity, the form (1.3.27) rather than (1.3.28) must be used because usually the coefficient varies from one point to another in the fluid. In particular, the horizontal flow of air over the ground has a level of turbulence and a velocity gradient which both decrease with height, and in some cases the variation of one, and in some cases that of the other, is the more important.

The application of the theory to turbulence also has to be made with great care because turbulence is not usually isotropic, and so the assumptions on which the equation is based fail. This is not serious when the velocity is everywhere in one direction (say x) and varies only in another (say z, i.e. vertically).

The equation (1.3.28) can be written in vector form as follows

$$\frac{\mathrm{D}\mathbf{v}}{\mathrm{D}t} = \mathbf{F} - \frac{1}{\rho}\,\mathbf{grad}\,p + \nu\nabla^2\mathbf{v} + \tfrac{1}{3}\,\upsilon\,\mathbf{grad}\,(\mathrm{div}\,\mathbf{v}) \qquad (1.3.29)$$

which is the **Navier-Stokes equation. It is valid for the motion of a fluid with uniform isotropic viscous properties in which the stress is instantaneously and linearly related to the rate of strain.** Without the viscous terms it is the classical equation for an inviscid fluid, namely (1.1.4). The most important viscous terms in practice are those already mentioned, namely those producing a shearing stress, such as $\nu\frac{\partial^2 v}{\partial x^2}$ and $\nu\frac{\partial^2 w}{\partial x^2}$, which are the forces acting on unit volume of fluid in the y- and z-directions due to non-uniform gradients in the x-direction of the y- and z-components of velocity. These forces act so as to straighten out velocity profiles, and vanish when the profiles are linear.

The final term in (1.3.29) does not concern us much in this book because we consider scarcely any situations in which div $\mathbf{v}$ is not rather small and, still less, has a significant gradient in a case where ν is not negligible or its turbulent equivalent known with any precision.

1.4 THE VORTICITY EQUATION

It will be recalled that Kelvin's minimal theorem (page 20) showed that of all the possible flow patterns which conform to given rigid boundaries of the fluid, the irrotational pattern possesses the least kinetic energy. Moreover this irrotational pattern is unique, and it is the pattern created by a sudden jerking of the boundaries into the motion. If the fluid possesses some vorticity it possesses more kinetic energy, yet the transport of fluid is not increased by the vorticity. The irrotational flow can be thought of as transporting the fluid and the vorticity as additional kinetic energy, often in what we call turbulent form, which the fluid may or may not possess for some cause. Thus if the water in a bucket is stirred with a spoon there must be some vorticity present if it is still moving when the spocn is withdrawn, because the irrotational motion is zero when the bucket is at rest.

This argument applies strictly only to incompressible fluids of uniform density. When the density varies, the motion set up by jerking the boundaries contains vorticity, and if the fluid is compressible, waves are transmitted through it. While they are being transmitted there are internal motions taking place which cannot be discovered from knowledge of how the boundaries are moving at the moment in question. We shall not be concerned with the transmission of sound waves, but turn our attention to the mechanisms which create vorticity and the phenomena which accompany it. The complicated motion set up in a milk bottle by holding it in the hand and rotating the wrist back and forth, or the motion of the air in a car which turns a corner sharply, starts by being irrotational but soon becomes very confused and the milk or air becomes mixed up. The mixing is achieved only by generating vorticity.

The water in a conical flask (or for that matter any fairly steep sided vessel) can be sloshed round by moving the flask quickly in small horizontal circles but without changing its orientation, and the result is to induce rotation in the whole mass of fluid. A wine decanter is very suitable for this experiment. According to Kelvin's circulation theorem, rotation cannot be imparted to the fluid by moving the boundaries, yet there is the rotation to be seen. Such a rapid movement of a flask is a favourite method employed by chemists doing titrations to mix up the liquid when they wish to disperse one more drop of reagent throughout the whole volume. In this case the production of vorticity is achieved by causing the free surface to fold over so that discontinuities of velocity are introduced into the bulk of the fluid.

Vorticity is thus associated with turbulence, overturning, and the production of kinetic energy which does not transport the whole fluid effectively. In machinery it is often wasteful to produce vorticity; on the other hand it may be desirable to generate mixing. It is produced in the flow past obstacles, and also in buoyant convection when a fluid is heated at the bottom or cooled at the top, or when salt is dissolved into the fluid at the top. The vorticity equation, which we shall now derive, describes the way in which vorticity, denoted by **curl v** or $\boldsymbol{\omega}$, can be generated.

The equation of motion (1.3.29) may be written

$$\rho \frac{D\mathbf{v}}{Dt} = \rho F - \mathbf{grad}\, p + \mu(\nabla^2 \mathbf{v} + \tfrac{1}{3}\, \mathbf{grad}\, \mathrm{div}\, \mathbf{v}),$$

and if we consider the case in which μ is a constant (otherwise we would have to take the form (1.3.27)), then by taking the **curl** we eliminate the terms in p and div **v** which occur as gradients. Thus

$$\mathbf{curl}\left(\rho \frac{D\mathbf{v}}{Dt}\right) = \mathbf{curl}\,(\rho\mathbf{g}) + \mu\, \mathbf{curl}\, \nabla^2 \mathbf{v}$$

in the case where the only body force is gravity. By using the identities (1.2.6, 8, 9 and 12) this is found to lead to

$$\rho\, \mathbf{curl}\left(\frac{\partial \mathbf{v}}{\partial t} - \mathbf{v} \times \omega\right) + \mathbf{grad}\, \rho \times \mathbf{f} = \mathbf{grad}\, \rho \times \mathbf{g} - \mu\, \mathbf{curl}\, \mathbf{curl}\, \omega$$

where we have denoted the fluid acceleration, $D\mathbf{v}/Dt$, by **f**. The order of the operators **curl** and $\partial/\partial t$ is interchangeable and so, using (1.2.11)

$$\frac{\partial \omega}{\partial t} + \omega\, \mathrm{div}\, \mathbf{v} - (\omega.\mathbf{grad})\mathbf{v} + (\mathbf{v}.\mathbf{grad})\omega = \mathbf{grad}\, \ln \rho \times (\mathbf{g} - \mathbf{f}) - \nu\, \mathbf{curl}\, \mathbf{curl}\, \omega,$$

from which using (1.2.9) we obtain

$$\frac{D\omega}{Dt} = -\omega\, \mathrm{div}\, \mathbf{v} + (\omega.\mathbf{grad})\mathbf{v} + \mathbf{R} \times (\mathbf{g} - \mathbf{f}) + \nu \nabla^2 \omega. \qquad (1.4.1)$$

Here we have introduced the symbol **R**, to denote $\frac{1}{\rho}\, \mathbf{grad}\, \rho = \mathbf{grad}\, \ln \rho$, which is a representation of the density field. This is the **vorticity equation** and the four terms on the right describe mechanisms by which the vorticity of a parcel of fluid is changed: the left side is equal to this rate of change. We now discuss these mechanisms.

The meaning of the first term on the right is best made clear by transferring it to the left. By means of the equation of continuity, namely

$$\frac{D\rho}{Dt} = -\rho\, \mathrm{div}\, \mathbf{v} \qquad (1.4.2)$$

we then obtain

$$\frac{D}{Dt}\frac{\omega}{\rho} = \left(\frac{\omega}{\rho}\,.\,\mathbf{grad}\right)\mathbf{v} + \frac{1}{\rho}\mathbf{R} \times (\mathbf{g}-\mathbf{f}) + \frac{\nu}{\rho}\nabla^2\omega \qquad (1.4.3)$$

which shows that the second term on the right of (1.4.1) represents a mechanism operating on ω when div **v** is zero which is the same as that operating on ω/ρ when div **v** is not zero. However, our objective is to isolate the terms on the right and elucidate the different vorticity-producing mechanisms which they represent. In the adiabatic motion of a well mixed fluid ρ is proportional to $p^{1/\gamma}$ and so **R**, which is in the direction of **grad** ρ, is also in the direction of **grad** p. If we neglect the effect of viscosity represented by the last term in (1.4.3) for the moment, then the equation of motion is (1.1.4) namely

$$\frac{1}{\rho}\,\mathbf{grad}\;p = \mathbf{g}-\mathbf{f} \qquad (1.4.4)$$

in which case $\mathbf{R} \times (\mathbf{g}-\mathbf{f}) = 0$.

The consequence of this is that we may discuss the meaning of the separate terms in (1.4.1) ignoring the term in div **v** but remembering that when div **v** is not zero the conclusion applies to ω/ρ instead of to ω, provided that the motion is adiabatic and the fluid well mixed.

The clue to the meaning of the term (ω.**grad**) **v** has already been given in the discussion following (1.1.15) on page 22 where it was shown that if the fluid is inviscid and either homentropic (well mixed and adiabatic) or of uniform constant density, the number of vortex lines through a circuit composed always of the same particles is constant. **Thus the vortex lines move with fluid.**

Let it be supposed that we have a set of lines marked out in the fluid and that they represent a vector field **a** which can be deduced from the lines by their direction and the number crossing unit area. Alternatively the lines may be marked along their length so that the magnitude of **a** is represented by the distance between the marks. Thus where the lines are far apart the marks are at short intervals, indicating a small magnitude. This is achieved by marking vortex tubes so as to indicate constant increments of volume along the tube (Fig. 1.4.i) Only if the vector **a** is non-divergent can it be marked out in this way.

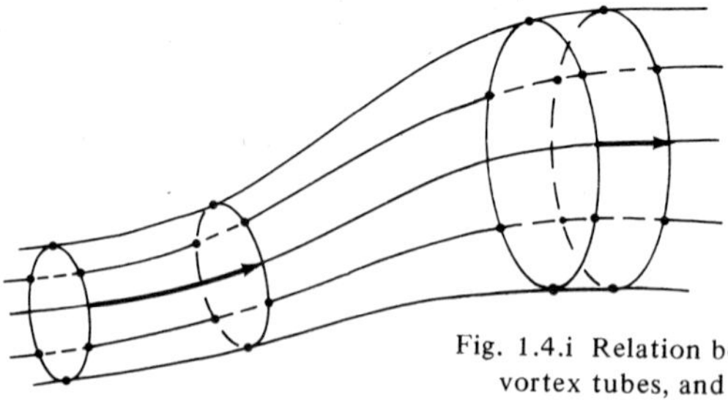

Fig. 1.4.i Relation between vortex lines, vortex tubes, and the vorticity vector

If the motion is also non-divergent, the volume occupied by a given set of particles remains constant, and so the same relationships between the set of lines and the vector field defined by them will persist as the lines are carried by the motion. If the vector **a** is now regarded as a material segment of one of the lines in time Δt, its two ends will be displaced by the velocity field distances $\mathbf{v}\Delta t$ and $(\mathbf{v} + \delta \mathbf{v})\Delta t$ where $\delta \mathbf{v}$ is the difference between the velocities of the particles at the two ends of the vector **a** (see Fig. 1.4.ii).

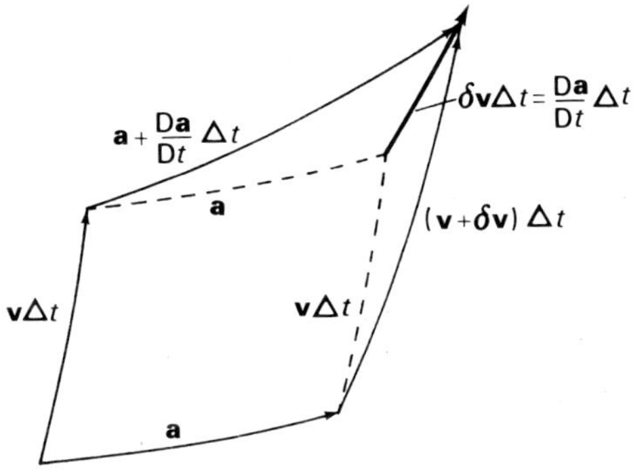

Fig. 1.4.ii Advection of the vorticity vector

Thus
$$\delta \mathbf{v} = (\mathbf{a}.\mathbf{grad})\mathbf{v} \tag{1.4.5}$$

to the first order in the small quantity **a**. At the end of the interval Δt, the material vector has become $\mathbf{a} + \frac{\mathrm{Da}}{\mathrm{D}t}\Delta t$ and so its change is equal to each of the quantities $\frac{\mathrm{Da}}{\mathrm{D}t}\Delta t$ and $(\mathbf{a}.\mathbf{grad})\mathbf{v}\Delta t$. Therefore

$$\frac{\mathrm{Da}}{\mathrm{D}t} = (\mathbf{a}.\mathbf{grad})\mathbf{v} \tag{1.4.6}$$

is the equation giving the change, due to the motion, of a vector field **a** defined by material lines in the fluid carried by the motion. This is the form taken by (1.4.1) for $\boldsymbol{\omega}$ when the fluid is incompressible, of uniform density, and inviscid. For a homentropic fluid the term in **R** in (1.4.3) vanishes and the vorticity must be deduced from the vortex lines carried by the fluid either by the number crossing unit area, as before, or by the length of an intercept of a tube of unit mass, instead of unit volume.

The x-component of $-\boldsymbol{\omega}$ div $\mathbf{v} + (\boldsymbol{\omega}.\mathbf{grad})\mathbf{v}$ is

$$-\xi\left(\frac{\partial v}{\partial y} + \frac{\partial w}{\partial z}\right) + \eta\frac{\partial u}{\partial y} + \zeta\frac{\partial u}{\partial z} \tag{1.4.7}$$

which is therefore a measure of the rate of change of ξ due to the two terms. The first term states that if there is already some ξ, the x-component of vorticity, more will be created if $\partial v/\partial y + \partial w/\partial z$ is negative. This simply refers to the fact that if the fluid is expanded or contracted in planes perpendicular to the x-axis, the moment of intertia about the x-direction is accordingly increased or decreased, and according to the conservation of momentum the rotation about the x-direction is correspondingly decreased or increased. This pattern of distortion is rather special and unlikely to occur by itself: it is more probable that the divergence is zero or nearly so, in which case, since div **v** is zero, the component becomes

$$\xi \frac{\partial u}{\partial x} \tag{1.4.8}$$

which represents the fact that if the fluid is being stretched in the x-direction it is being contracted in planes normal to it. For this reason it is customary to think of **the vortex lines as being intensified by stretching along their length.**

The second and third terms in (1.4.7) are of similar structure, and we therefore consider only

$$\eta \frac{\partial u}{\partial y} \tag{1.4.9}$$

which indicates that if there is already a non-zero y-component of vorticity, η, and the x-velocity, u, varies in the y-direction so as to rotate lines of particles situated along the y-axis, they will acquire a component in the x-direction, and so some ξ will be created (see Fig. 1.4.iii).

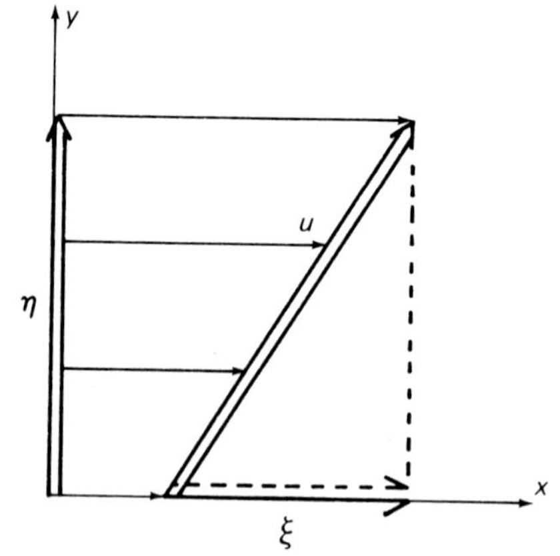

Fig. 1.4.iii Rotation of vortex lines by the fluid motion

The creation of vorticity by the density gradients is represented by

$$\frac{D\boldsymbol{\omega}}{Dt} = \mathbf{R} \times (\mathbf{g} - \mathbf{f}) \tag{1.4.10}$$

This shows that the acceleration has a similar effect to gravity, the inertial and gravitational masses of a piece of matter being the same. Vorticity, or rotation, is created so as to rotate the fluid until the surfaces of constant density are normal to the vector $\mathbf{g} - \mathbf{f}$. When the fluid is at rest $\mathbf{f}$ is zero, and we have the familiar result that rotation of the surfaces of constant density is induced when they are not horizontal so as to bring them into the horizontal.

$\mathbf{R}$ is a vector perpendicular to the surfaces of constant density, and the axis about which the vorticity is generated is horizontal and in the surface, perpendicular to both $\mathbf{g}$ and $\mathbf{R}$. If the fluid undergoes an acceleration, as for instance when it passes round a bend in a pipe, the centrifugal forces are added to the weight. If the acceleration is along the direction of motion, as in a vehicle accelerated in a straight line, the fluid behaves as if $\mathbf{g}$ were in the direction taken by a weight hung on a thread in the vehicle. Water in a partly filled vessel on the vehicle sloshes to the back as the vehicle accelerates forward. In a centrifuge or similarly rotating system $\mathbf{g} \ll \mathbf{f}$, and hence the surfaces of constant density tend to become cylinders whose axis is the axis of rotation.

The creation of vorticity by viscosity is represented by the equation

$$\frac{D\boldsymbol{\omega}}{Dt} = \nu \nabla^2 \boldsymbol{\omega} \tag{1.4.11}$$

when the viscosity is uniform. This is a mechanism which acts on each component of viscosity separately: thus the x-component is

$$\frac{D\xi}{Dt} = \nu \nabla^2 \xi \tag{1.4.12}$$

and this is the same as the equation representing the conduction of heat by a thermal conductivity ν when ξ is the temperature. The effect is to redistribute the vorticity so that its gradient becomes more nearly linear. The usual ultimate effect is to make the vorticity more nearly uniform so that the fluid is made more nearly to rotate as a solid. Where the gradient of ξ increases in the x-direction the flux of ξ conducted decreases (increases negatively) in the x-direction, and so the value of ξ increases.

If ν is not constant, (1.4.12) is correct if the variation in ν is due solely to variations in ρ. If ρ is a constant and μ is variable, the derivation must be made from equation (1.3.27): this is not simple enough to write in vector form and operate upon as we did in the derivation of (1.4.1), even when div $\mathbf{v} = 0$ everywhere. This is important when stresses are produced by eddies because they can vary very substantially across the fluid. Because of its complexity, this situation is best discussed by simplifying the equations as far as possible for the special case under consideration, and this is discussed in Chapter 7.

1.5 THE ISOPYCNIC VORTICITY THEOREM

In the case of an inviscid fluid

$$\frac{D\boldsymbol{\omega}}{Dt} = -\boldsymbol{\omega}\ \text{div}\ \mathbf{v} + (\boldsymbol{\omega}.\mathbf{grad})\mathbf{v} + \mathbf{R} \times (\mathbf{g} - \mathbf{f}). \tag{1.5.1}$$

The first term on the right is parallel to the vorticity. The second term represents the production of changes in the vorticity by moving the vortex lines with the fluid. The third term represents the creation of new vorticity perpendicular to **R**, that is in the surfaces of constant ρ. Consequently, **if at one moment the vortex lines lie in surfaces of constant density**, called isopycnic surfaces, because all new vorticity is created in those surfaces or moves materially with them, **they will remain in them.**

Only four circumstances can alter this situation. Viscosity might produce significant amounts of vorticity not lying in the surfaces or, for some reason, the vorticity might not initially lie in them. The free surface might become folded over so as to produce velocity discontinuities which then diffuse by viscosity, or the density might be altered by heating or some other agency.

The case of special interest to us is that in which both the flow and the isopycnic surfaces are horizontal. If the vorticity is also horizontal initially, it will be found in the surfaces at points downstream. This is a common situation in the atmosphere, for although there is obviously a vertical component of vorticity due to the earth's rotation which represents a turning of the order of one rotation per day, it is usually very small compared with the horizontal component due to the change of wind strength or direction with height.

In the atmosphere away from the bottom boundary the vorticity is, for practical purposes, entirely produced by the action of gravity on density gradients. That means that new vorticity is always horizontal and in the contours of the isopycnic surfaces. The acceleration of the fluid in the ocean and atmosphere is generated as a result of small departues of the density from horizontal uniformity, and is always small compared with **g**, and so the term $\mathbf{R} \times \mathbf{f}$ is usually negligible.

It will be seen in Chapter 5 that the mathematical complexity of waves in the atmosphere is so considerable that almost no progress has been made except in the case to which this theorem is applicable, and even then only in the 2-dimensional case. The theorem shows that in spite of the apparent speciality of this case it is nevertheless relevent to a great many real situations.

It should be noted that airflow over a horizontal surface acquires vorticity due to the drag of the ground, but this does not invalidate the application of the theorem because that vorticity is horizontal and lies in the stream surfaces.

1.6 ADIABATIC MOTION IN THE ATMOSPHERE

If T and p are the temperature and pressure in the air the equation of state is

$$p = R\rho T \tag{1.6.1}$$

where R is the gas constant (not to be confused with the magnitude of $\mathbf{R}$, which is usually denoted by β, or by some authors by S). For adiabatic changes

$$p = k_0 \rho^{\gamma} \tag{1.6.2}$$

where γ is the ratio of the specific heats of air and is about 1.4, and k_0 is a constant for a particle and remains so following the motion, although in a stratified fluid it varies in space.

We define the potential temperature θ, and its reciprocal τ, by

$$\theta = \frac{1}{\tau} = T\left(\frac{p_1}{p}\right)^{\frac{\gamma-1}{\gamma}} \tag{1.6.3}$$

where p_1 is a reference pressure, usually taken as 1000 mb or roughtly mean sea level pressure, in practical meteorology. The 'modified pressure' ϖ is defined by

$$\varpi = \frac{\gamma R}{\gamma - 1}\left(\frac{p}{p_1}\right)^{\frac{\gamma-1}{\gamma}} \tag{1.6.4}$$

and in this way we obtain the equation

$$\frac{1}{\rho}\,\mathbf{grad}\,\mathrm{p} = \frac{1}{\tau}\,\mathbf{grad}\,\varpi. \tag{1.6.5}$$

Consequently, since molecular viscosity is of negligible importance in atmospheric motion except on a very small scale such as around raindrops, cloud droplets, leaves, and so on, the equation of motion (1.3.29) becomes

$$\frac{\mathrm{D}\mathbf{v}}{\mathrm{D}t} = \mathbf{g} - \frac{1}{\tau}\,\mathbf{grad}\,\varpi\ , \tag{1.6.6}$$

and the vorticity equation (1.4.1) is

$$\frac{D\boldsymbol{\omega}}{Dt} = -\boldsymbol{\omega}\,\mathrm{div}\,\mathbf{v} + (\boldsymbol{\omega}.\mathbf{grad})\mathbf{v} + \mathbf{R} \times (\mathbf{g}-\mathbf{f}) \tag{1.6.7}$$

where now $\mathbf{R} = \frac{1}{\tau}\,\mathbf{grad}\,\tau$.

In considering the motion of the air, which is compressible, we can take over very many of the deductions made for an incompressible fluid, including those concerning the vorticity, simply by replacing ρ by τ. In those from which the pressure is not eliminated we may find it necessary to replace p by ϖ. The method fails when there are significant nonadiabatic mechanisms operating. In the cases mentioned above in which viscosity is important we are not usually concerned about compressibility, and so this transformation is not called for anyway. **This simple exchange of symbols takes care of the fact that in vertical motion the air changes its temperature because of changes in pressure due to compression by the air's own weight.** However a modification of the equation of continuity is also required: this is duscussed in Chapter 5, equation (5.7.6).

1.7 BERNOULLI'S EQUATION FOR AN INVISCID FLUID IN STEADY MOTION

For an inviscid fluid the equation of steady motion is

$$\mathbf{grad}\,\tfrac{1}{2}q^2 - \mathbf{v} \times \boldsymbol{\omega} = -\,\mathbf{grad}\,gz - \frac{1}{\rho}\,\mathbf{grad}\,p \tag{1.7.1}$$

the z-axis being vertical. If we now put

$$\frac{1}{\rho}\,\mathbf{grad}\,p = \mathbf{grad}\,\frac{p}{\rho} - p\,\mathbf{grad}\,\frac{1}{\rho}$$

or

$$\frac{1}{\tau}\,\mathbf{grad}\,\varpi = \mathbf{grad}\,\frac{\varpi}{\tau} - \varpi\,\mathbf{grad}\,\frac{1}{\tau} \tag{1.7.2}$$

then

$$\mathbf{grad}\,(\tfrac{1}{2}q^2 + \frac{\varpi}{\tau} + gz) - \varpi\,\mathbf{grad}\,\frac{1}{\tau} = \mathbf{v} \times \boldsymbol{\omega} \tag{1.7.3}$$

with a corresponding equation using p and ρ, which will be taken as read for the rest of this section.

If initially (i.e. somewhere upstream) τ is constant along the streamlines and vortex lines, which means that there exist surfaces containing streamlines and vortex lines which are also isentropic (or, in the case of an incompressible fluid, if the density is constant on streamlines and vortex lines), then $\mathbf{v}.\mathbf{grad}\,\frac{1}{\tau}$ and $\boldsymbol{\omega}.\mathbf{grad}\,\frac{1}{\tau}$ both vanish. Writing

$$gH = \tfrac{1}{2}q^2 + \frac{\omega}{\tau} + gz \qquad (1.7.4)$$

from (1.7.3) we have that

$$\mathbf{v}.\mathbf{grad}(gH) = \boldsymbol{\omega}.\mathbf{grad}(gH) = 0 \qquad (1.7.5)$$

The consequence of (1.7.5) is that gH **is constant in the directions of the vorticity and the velocity and is therefore constant in the surfaces containing these vectors, which we call the Bernoulli surfaces.**

The assumption about the isentropic surfaces was seen in section 1.5 to be relevant to the atmosphere (and for the same reasons to the ocean). In cases where the isopycnic theorem is not applicable so that the vortex lines do not lie in the isentropic surfaces, and never will, the motion is almost certainly unsteady and we cannot obtain any useful result about the constancy of gH.

Bernoulli's theorem is only valid for reversible flows because it is essentially a statement about conservation of energy, and there can be no such result for viscous flows. The classical applications are for a fluid of uniform density for which $\mathbf{grad}\,\tau = \mathbf{grad}\,\rho = \mathbf{0}$. Then

$$\mathbf{grad}(gH) = \mathbf{v} \times \boldsymbol{\omega} \qquad (1.7.6)$$

and so gH is constant along vortex lines and streamlines. Since it is not necessarily constant throughout the fluid Bernoulli surfaces always exist in this case, and each streamline defines a surface, namely the one containing all the vortex lines through it, or vice versa. **Any streamline intersecting one of the vortex lines through a given streamline must intersect them all,** unless there is no vorticity and gH is constant throughout the fluid.

For a well mixed fluid τ is uniform and so the pressure is expressible in terms of ρ [see discussion following (1.1.16)], and in this special case

$$\frac{1}{\rho}\,\mathbf{grad}\,p = \mathbf{grad}\int\frac{dp}{\rho} \qquad (1.7.7)$$

and so

$$gH = \tfrac{1}{2} q^2 + \int \frac{dp}{\rho} + gz = \text{constant on a Bernoulli surface.} \qquad (1.7.8)$$

This is the well known case found in elementary textbooks which is much **used in theory of compressible flow** which is unstratified. **If the motion is also irrotational** then the right side of (1.7.6) vanishes and gH **is constant throughout the fluid.** In addition, if the flow is unsteady but irrotational we can write $\mathbf{v} = \mathbf{grad}\ \phi$ and the equation of motion leads to

$$gH = \tfrac{1}{2} q^2 + \int \frac{dp}{\rho} + gz + \frac{\partial \phi}{\partial t} = \text{constant} \qquad (1.7.9)$$

throughout the fluid. Since the essential feature of stratified or buoyant fluids is that their motion has a significant vorticity, these cases are of little relevance here.

1.8 THE STATIC STABILITY

The potential temperature defined in (1.6.3) may be written

$$\theta = \frac{p}{R\rho}\left(\frac{p_1}{p}\right)^{\frac{\gamma-1}{\gamma}} = k_0^{1/\gamma}\,\frac{p_1^{\frac{\gamma-1}{\gamma}}}{R} \qquad (1.8.1)$$

The static stability is the restoring force due to buoyancy on a unit mass of fluid displaced adiabatically a unit distance vertically. The stability is positive if the force is of opposite sign to the displacement. In an incompressible fluid it is equal to $-\dfrac{g}{\rho}\dfrac{\partial \rho}{\partial z}$ because $\delta\rho$ is the density anomaly and is equal to $\delta z\,\dfrac{\partial \rho}{\partial z}$. Because in a gas the coefficient of expansion is $1/T$, the temperature anomaly may be used in the compressible case, and this is the potential temperature anomaly. Thus the **static stability** is

$$g\beta = -\frac{g}{\tau}\frac{\partial \tau}{\partial z} = \frac{g}{\theta}\frac{\partial \theta}{\partial z} = \frac{g}{\gamma k_0}\frac{\partial k_0}{\partial z} \qquad (1.8.2)$$

the derivatives being obtained in the undistrubed static state. β is positive when the potential temperature increases upwards.

The temperature gradient in which the restoring force is zero and the displaced parcel is undergoing adiabatic changes is the one in which the parcel is indistinguishable from its surroundings, and is called the **adiabatic lapse rate.** It is obtained by combining equations (1.6.1 and 2) with the hydrostatic equation (1.1.5). By eliminating p and ρ and differentiating to obtain $\partial T/\partial z$ we find

$$\frac{\partial T}{\partial z} = -\frac{(\gamma-1)g}{\gamma R} = \Gamma \tag{1.8.3}$$

which is a constant determined by gravity and the physical properties of air. For the earth's atmosphere at sea level it is 9.86°C per km, and for practical purposes may be taken as 1°C per 100 m throughout the atmosphere.

Because the **adiabatic lapse rate** is independent of temperature it would mean that if the atmosphere were in a state of neutral stability with a temperature of 300°K at the ground it would be at the absolute zero at a height of 30 km. Radiative equilibrium between the earth at 300°K and outer space at 0°K would be at $[\frac{1}{2}(300)^4]^{1/4}$ °K, according to a simple application of **Stefan's law**, which is about 250°K, so that it may be deduced that the atmosphere is highly stratified, the stratosphere having a temperature perhaps as high as 250°K at 30 km.

The rate at which a parcel of air is cooled adiabatically if it is displaced in stratified surroundings is very nearly the same as in the neutral, or well-mixed, case. The difference arises from the fact that for (1.8.3) ρ is not the density of the parcel but of the surroundings. But in practice this is of no importance because of other errors. Thus a parcel cannot be displaced very fast in a stratified environment without the temperature difference causing mixing except when waves are the cause of the displacement, in which case the pressure variations due to the waves are of small magnitude. For practical purposes, therefore, in any surroundings we shall not be misled by the assumption that the temperature of an unmixed parcel falls at the rate of 1°C per 100 m when it is displaced vertically.

We shall need later to calculate the density of a displaced parcel. If ρ_0 is its original density, then in adiabatic displacements

$$\frac{\rho}{\rho_0} = \left(\frac{T}{T_0}\right)^{\frac{1}{\gamma-1}} \quad \text{exactly} \tag{1.8.4}$$

$$\text{and } T = T_0 + \Gamma\zeta \text{ approximately} \tag{1.8.5}$$

for a vertical displacement ζ . Therefore

$$\frac{\rho}{\rho_0} = \left(1 + \frac{\Gamma}{T}\zeta\right)^{\frac{1}{\gamma-1}} \simeq 1 + \frac{\Gamma}{(\gamma-1)T}\zeta \tag{1.8.6}$$

$$= 1 - \frac{g}{\gamma RT}\,\zeta = 1 - \frac{g}{c^2}\,\zeta \tag{1.8.6}$$

where $c = (\gamma RT)^{1/2}$ = the velocity of sound = 316 m sec^{-1} approx. (1.8.7)

The approximate formula (1.8.6) is good as long as

$$0.3\,\frac{g}{c^2}\,\zeta \ll 1 \tag{1.8.8}$$

this being the ratio of the first neglected term to the last one taken. The condition (1.8.8) requires that the displacement be small compared with 30 km, and is certainly valid for distances of the order of 3 km, which is greater than those in most practical cases to which we may wish to apply (1.8.6).

The customary concept of static instability is that the buoyancy force is in the direction of the displacement, thereby causing an acceleration in that direction. This is the **displaced parcel theory**. But we shall find it useful to think in terms of a **rotational displacement.** When there is no motion the vorticity equation is

$$\frac{D\boldsymbol{\omega}}{Dt} = \mathbf{R} \times \mathbf{g}, \quad \text{or} \quad \frac{D\xi}{Dt} = g\,R \sin\phi \tag{1.8.9}$$

where ξ is the x-component of vorticity, assumed horizontal, and ϕ is the angular displacement about the x-axis of a local tube of fluid. The tube has angular velocity $\frac{1}{2}\xi$ which means that

$$\frac{D\phi}{Dt} = \tfrac{1}{2}\xi, \quad \text{and} \quad \frac{D\xi}{Dt} = 2\,\frac{D^2\phi}{Dt^2}\,. \tag{1.8.10}$$

In this discussion R denotes the magnitude of $\mathbf{R}$ and is equal to $-\beta\cdot = \frac{1}{\tau}\frac{\partial \tau}{\partial z}$ in the initial equilibrium state, and so (1.8.9) becomes

$$\frac{D^2\phi}{Dt^2} = \frac{g}{2\tau}\frac{\partial \tau}{\partial z}\sin\phi, = -\tfrac{1}{2}g\beta\phi \text{ for small } \phi\,. \tag{1.8.11}$$

This indicates instability, with ϕ growing exponentially with time, if β is negative and the potential temperature decreases with height. This is, of course, the same as the condition derived by the parcel displacement argument.

The interest in this form of the criterion for local static instability is that vorticity is generated by it, and it cannot occur without generating vorticity. The consequence is that mixing takes place and the motion becomes turbulent.

If a parcel of fluid is displaced upwards (Fig. 1.8.i) we can think of this as being the result of vorticity being generated in horizontal rings around it. A parcel which is penetrating through the surrounding fluid (Fig. 1.8.ii) has motion

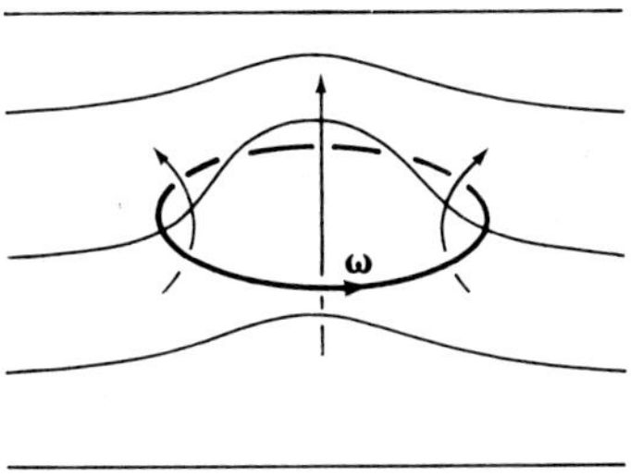

Fig. 1.8.i Generation of vortex rings by the vertical displacement of a stratified fluid.

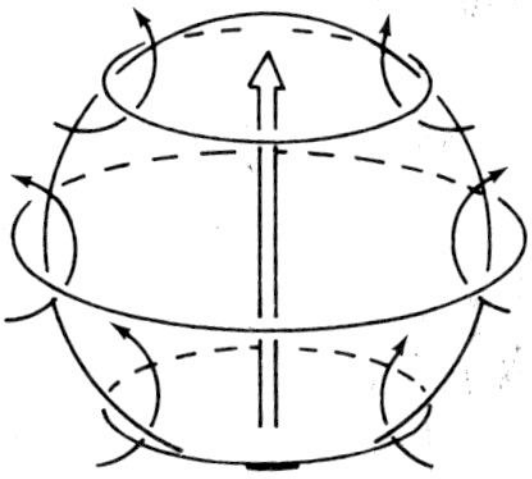

Fig. 1.8.ii Vortex rings generated by static instability transporting a parcel through its surroundings.

which can be thought of as due to the vortex rings which have been generated around it in accordance with (1.8.9). The motion does not remain simple but soon becomes a tangle of small scale vortices which cause the mixing (see Chapter 8): this is because once the surfaces of constant potential temperature become disturbed, the fluid is everywhere subject to the generation of vorticity and irregularities grow anywhere.

Chapter 2

PHENOMENA OF FLUID FLOW

2.1 CLASSICAL INVISCID INCOMPRESSIBLE HYDRODYNAMICS

Flow of an inviscid fluid around a sphere will be used to illustrate the method of classicial mechanics and effects in real fluids. If there is no vorticity **curl v** = **0**, and therefore there exists a potential ϕ defined by

$$\phi_P = \int_0^P \mathbf{v.ds} \tag{2.1.1}$$

such that $\mathbf{v} = \mathbf{grad}\,\phi$. (2.1.2)

The origin 0, from which the integral (2.1.1) is measured is irrelevant, for a change of its position merely represents the addition of a constant to ϕ. If the fluid is incompressible div **v** = 0, and so (2.1.2) becomes

$$\nabla^2\phi = 0, \tag{2.1.3}$$

which is the equation of continuity for an incompressible fluid in irrotational motion.

The equation of motion is employed merely to compute the pressure, and is used in the form (1.7.9), which is Bernoulli's equation.

2.2 SOURCE, DOUBLET, UNIFORM STREAM, AND SPHERE

The potential

$$\phi = -\,m/r \tag{2.2.1}$$

where r is distance from the origin, leads to a velocity m/r^2 radially outwards, which is the motion due to a source of strength (i.e. of total outward volume flux) $4\pi m$. If m is negative we have a sink, into which fluid flows.

With spherical polar coordinates (r, θ, λ), and axial symmetry such that all quantities are independent of λ, and with x measured in the direction $\theta = 0$, the potential

$$\phi = Ux = Ur\cos\theta \tag{2.2.2}$$

represents a uniform stream of speed U in the x-direction.

The potential

$$\phi = -\mu\cos\theta/r^2 \tag{2.2.3}$$

represents the flow due to a doublet. This is made evident by noting that

$$\delta x\,\frac{\partial}{\partial x}\,\frac{m}{r} = -m\,\delta x\,\cos\theta/r^2\,. \tag{2.2.4}$$

The left side of (2.2.4) represents the potential produced by a sink (m/r) together with that due to a negative sink $(-m/r)$ displaced a distance δx away from it in the $-x$ direction. Thus $\mu = m\,\delta x$, which is the moment of the doublet so formed.

If we now add the doublet to the uniform stream, since they are both proportional to $\cos\theta$, we obtain

$$\phi = \left(Ur - \frac{\mu}{r^2}\right)\cos\theta\,, \tag{2.2.5}$$

so that the radial component of velocity is

$$\frac{\partial\phi}{\partial r} = \left(U + \frac{2\mu}{r^3}\right)\cos\theta\,; \tag{2.2.6}$$

and this is zero on $r = a$ if

$$\mu = -Ua^3/2\,. \tag{2.2.7}$$

This, therefore, is the condition for the radial velocity to be zero on the sphere $r = a$, and so

$$\phi = U\cos\theta\left(r + \frac{a^3}{2r^2}\right) \tag{2.2.8}$$

is the potential for the steady flow of the stream U around the sphere of radius a, fixed at the origin. This is shown in the first figure of Fig. 2.2.i.

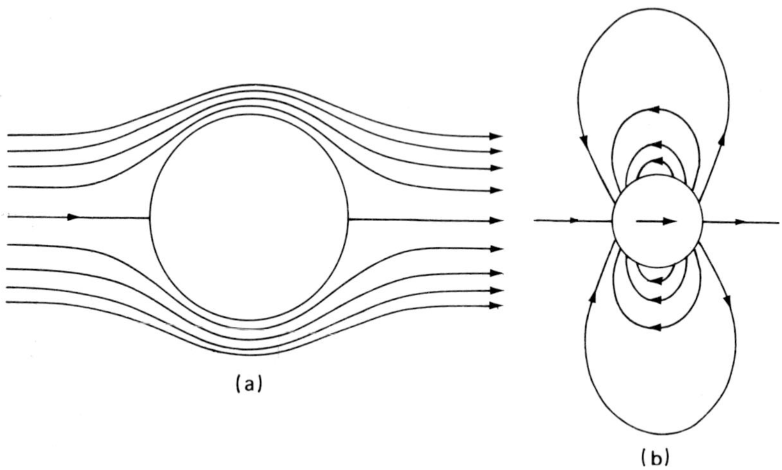

Fig. 2.2.i (a) Streamlines of flow around a sphere shown in a plane through the axis of the flow. (b) Lines of flow due to a sphere moving through otherwise stagnant fluid

2.3 FORCE ON A SPHERE: VIRTUAL MASS

The pressure at any point on the sphere when it is fixed is obtained from (1.7.9) by putting ρ = constant, $\partial/\partial t = 0$, and $q^2 = (\text{grad}\ \phi)^2$. This gives, for $r = a$, when the velocity is constant,

$$p = p_0 - \tfrac{1}{8}\ \rho U^2 (5 - 9\cos^2\theta)\ . \qquad (2.3.1)$$

The pressure on a sphere moving unsteadily in otherwise still fluid is obtained by subtracting the uniform flow (2.2.2) from the potential (2.2.8) to give

$$\phi = Ua^3 \cos\theta/2r^2\ . \qquad (2.3.2)$$

The flowlines are plotted in the second figure in Fig. 2.2.i.

We calculate $\partial\phi/\partial t$ in (1.7.9) by noting that

$$\partial\theta/\partial t = U\sin\theta/r, \quad \partial r/\partial t = -U\cos\theta \qquad (2.3.3)$$

at the fixed point (r, θ) for a coordinate system moving with the sphere with speed $-U$. Then after putting $r = a$, we obtain the pressure on the moving sphere in the form

$$p = p_0 - \tfrac{1}{8}\rho U^2(5 - 9\cos^2\theta) + \tfrac{1}{2}\rho \dot{U} a \cos\theta \,. \qquad (2.3.4)$$

In (2.3.1) and (2.3.4) p_0 is the pressure at an infinite distance in the stream or in the fluid at rest respectively. The pressure produces no resultant force on the sphere if the motion is steady because it is the same on the upstream and downstream sides of the sphere. This is true of any body in an inviscid incompressible irrotational flow of a uniform fluid. But when the sphere has acceleration $\dot{U}$ through a fluid otherwise at rest the total force, F, is obtained by integrating the last term of (2.3.4) over the surface. Thus

$$F = -\int_0^{\pi} p\cos\theta\, 2\pi a^2 \sin\theta\, d\theta = \tfrac{2}{3}\pi\rho a^3 \dot{U} = \tfrac{1}{2}M'\dot{U}\,. \qquad (2.3.5)$$

where M' is the mass of fluid displaced by the sphere.

The quantity $\tfrac{1}{2}M'$ is called the virtual mass of the sphere because it represents the apparent inertia of the fluid which has to be set in motion when the sphere is accelerated.

When a sphere is started in motion from rest the motion is initially irrotational (see 1.1.14), and the force on it is well represented by (2.3.5); but the viscous forces quickly change the flow as described below.

Many other shapes of body may be represented by a distribution of sources and sinks but not any necessarily. Each, however, has its own virtual mass. A flat plate has zero virtual mass for accelerations within its own plane, but a large virtual mass if moved perpendicularly because of the large amount of fluid set in motion. In accelerated or steady irrotational flow in general there is a couple due to the pressure on the surface of any rigid body. There is no couple on a sphere because of its symmetry.

Classical fluid mechanics is concerned with the application of the above methods to a variety of contrived problems, which are described in many excellent textbooks.

2.4 FREE STREAMLINES

The theory of free streamlines is well described in many books (e.g. Lamb, Arts 73-78). A surface is assumed to exist which satisfies the following conditions: that it streams away from a sharp edge on a body the pressure is constant on it and the fluid stagnant behind it. On the other side of this surface, therefore, the velocity is constant, and the free stream surface is assumed to be the boundary of the fluid. It is necessarily a vortex sheet, which is unstable (see Ch. 6) and therefore becomes unsteady downstream, but it is a beginning to the study of wakes provided by classical hydrodynamics.

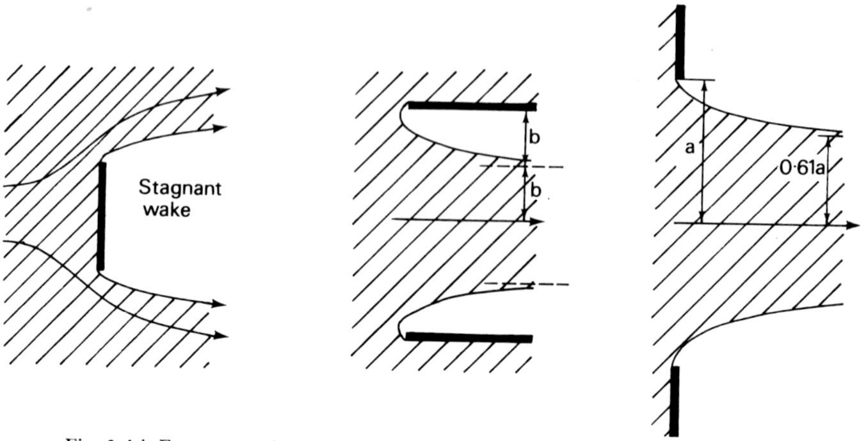

Fig. 2.4.i Free streamlines behind a flat plate, flow into a slit from horizontal boundaries (Borda's mouthpiece), and out of a slit in a vertical wall (vena contracta)

Free streamlines are illustrated in Fig. 2.4.i. The 'vena contracta' is fairly realistic for a jet of water emerging from an orifice: for example the upper half may be taken roughly to represent the flow of water on a horizontal bottom from under a submerged sluice gate. But with significant vertical displacements gravity cannot usually be neglected. The velocity approaches a limit asymptotically, as in the case of 'Borda's mouthpiece'.

A special case of a free streamline may occur where two fluids of different density are in juxtaposition. The potential flow in a corner of angle α is given by

$$\phi = Ur^{2\pi/\alpha} \cos \frac{2\pi\theta}{\alpha} . \tag{2.4.1}$$

When one of the planes is rigid at $\theta = 0$ and the other is the interface with a fluid of different density in which there is no motion and the pressure is hydrostatic, the pressure along the free boundary between the fluids must vary linearly, like gz. In steady flow therefore, $(\partial\phi/\partial r)^2$ must increase linearly along $\theta = \alpha$, according to Bernoulli's equation, because the pressure must be the same on the two sides of the interface. Therefore

$$r^{2\left(\frac{2\pi}{\alpha}-1\right)} \propto r, \quad \text{or } \alpha = 2\pi/3 . \tag{2.4.2}$$

On the basis of this argument the flow of a light fluid over a wedge of stagnant heavier fluid would turn through 60°, Fig. 2.4.ii(a). Hot fluid from a chimney into cool stagnant air would converge initially at this angle, as shown in Fig. 2.4.ii(b). The conclusion only applies close to the corner and is not relevant if friction is important or if the second fluid is not stagnant.

The argument shows, however, that the flow cannot be irrotational on both sides of an interface dividing two fluids resting on a plane unless one is stagnant and $\alpha = 120^\circ$ in the other. Alternatively the fluids must have the same density and speed and the interface must be normal to the boundary as in Fig. 2.4.ii(c). Again, the interface may be horizontal, the two angles the same and ρU^2 the same on both sides as in Fig. 2.4.ii(d), or both may turn through 120° (Fig. 2.4.ii(e)).

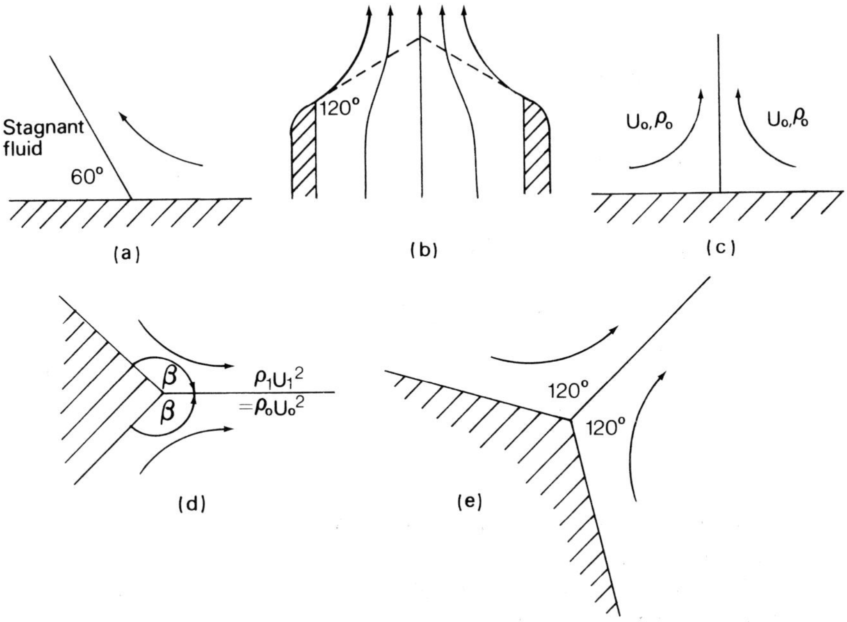

Fig. 2.4.ii (a) Flow of a light fluid up a wedge of stagnant heavy fluid. (b) Emergence of hot gas from a chimney in stagnant air. (c) Front between two fluids perpendicular to boundary. (d) Junction of irrotational streams at a horizontal interface. (e) Junction of irrotational streams at inclined interface.

2.5 SURFACE OF ROTATING FLUIDS

The shape of the free surface of a fluid steadily rotating about a vertical axis may be calculated if the tangential (circumferential) velocity, v, is given. This is because the pressure within the fluid having no vertical acceleration is given by the hydrostatic equation,

$$p = g\rho h + p_0 \ , \tag{2.5.1}$$

where h is the distance below the surface at which the pressure p_0 is taken to be uniform. The fluid acceleration is v^2/r, r being distance from the vertical axis, and this is equal to $\rho^{-1}\,\partial p/\partial r$ according to the equation of motion.

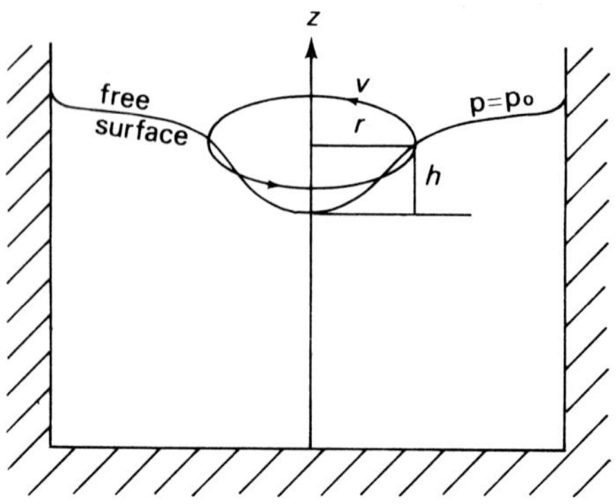

Fig. 2.5.i Coordinates, and shape of surface of rotating fluid

Thus for a fluid in solid rotation Ω

$$\frac{\partial p}{\partial r} = \rho \frac{(r\Omega)^2}{r},$$

so that (2.5.2)

$$h = \frac{\Omega^2}{2g} r^2$$

and the surface is a paraboloid of revolution. h is the height above the centre point where the pressure is p_0.

For fluid in irrotational flow the circulation around the axis is the same at all radii. Consequently $2\pi rv = 2\pi k =$ constant (independent of r) and so $v = k/r$, $\phi = k \log r$. We then find by the same argument that in this case

$$h = h_0 - \frac{k^2}{4gr^4} \qquad (2.5.3)$$

which is the shape of the bathplug vortex, h_0 being the height asymptotically approached at large distance. This is also the shape of the base of a funnel cloud in a tornado if the air is well mixed and has a uniform absolute humidity, for then the cloud base is a surface of constant pressure.

If the tea in a cup is set in rotation, the free surface tends to be paraboloidal near the centre because viscosity makes the rotation nearly solid, while further out the circulation is made nearly uniform by the secondary circulation (see Chapter 3) and so the curvature is convex in vertical sections through the axis, as in (2.5.3). At the wall of the cup surface tension makes it concave again if the surface is wetted.

2.6 LAMINAR VISCOUS FLOW

The stress due to viscosity in a fluid in simple shearing motion with no stretching of the fluid along the streamlines is given by the second of the equations (1.3.24). This says simply that the shear stress is proportional to the viscosity and the shear.

If a fluid of viscosity μ is set in rotation about a vertical axis with tangential velocity v equal to k/r, which is irrotational because it has potential $\phi = k \log r$, the moment of the momentum of a cylindrical shell of given mass and radius r is independent of r. The moment of the tangential stress between such cylinders is also seen to be independent of r, for it is proportional to $\mu 2\pi r.r.\partial v/\partial r$.

Such a velocity distribution is therefore set up around a rigid rotating cylinder in a large vessel. Between two rigid cylinders of radius r_1 and r_2 rotating with angular velocity Ω_1 and Ω_2, the velocity is obtained by adding a solid rotation, which produces no tangential stress within the fluid, to an irrotational motion. Thus, provided the flow is not unstable (see, for example, Chapter 3.10)

$$v = r\Omega + \frac{k}{r} \tag{2.6.1}$$

where

$$k = \frac{r_1{}^2 r_2{}^2(\Omega_1 - \Omega_2)}{r_2{}^2 - r_1{}^2}, \quad \Omega = \frac{r_1{}^2\Omega_1 - r_2{}^2\Omega}{r_1{}^2 - r_2{}^2}. \tag{2.6.2}$$

The steady flow down a uniform circular pipe will produce a shearing stress on a cylinder of fluid of radius r coaxial with the pipe (see Fig. 2.6.i) which is equal to the difference between the pressure force on the two ends.

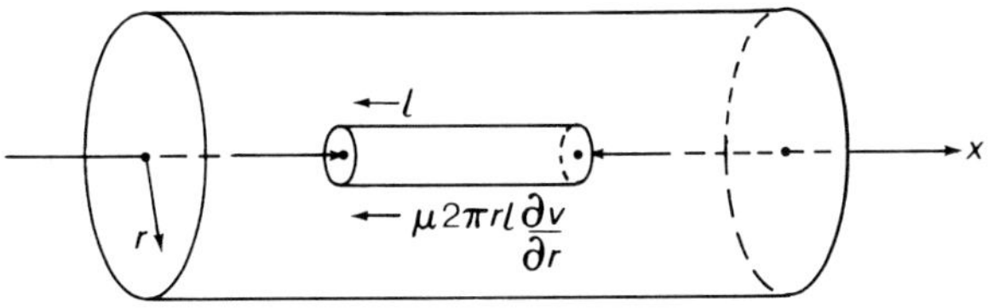

Fig. 2.6.i Forces on a cylinder of fluid in laminar viscous flow in a circular pipe

Thus

$$-\pi r^2 l \frac{\partial p}{\partial x} = \mu 2\pi r l \frac{\partial v}{\partial r}$$

or

$$v = v_0 - \frac{r^2}{4\mu}\frac{\partial p}{\partial x} \tag{2.6.3}$$

where v_0 is the velocity on the axis, $r = 0$. The profile of velocity is therefore parabolic. By similar argument the steady flow between two parallel planes $z =$ constant is also seen to be parabolic, with the formula

$$v = v_0 - \frac{z^2}{\mu}\frac{\partial p}{\partial x} \tag{2.6.4}$$

In this latter case, if the flow is zero on the walls, the velocity in the centre of a gap of depth $2h$ between them is

$$v = -\frac{h^2}{\mu}\frac{\partial p}{\partial x} \tag{2.6.5}$$

and if the gap is narrow so that the flow is very slow, the inertia forces represented by the acceleration in the equation of motion can be neglected so that (2.6.5) may be taken to hold locally everywhere in the form

$$v = -\kappa \frac{\partial p}{\partial x}, \text{ or}$$

$$\mathbf{v} = \mathbf{grad}\,(-\kappa p). \tag{2.6.6}$$

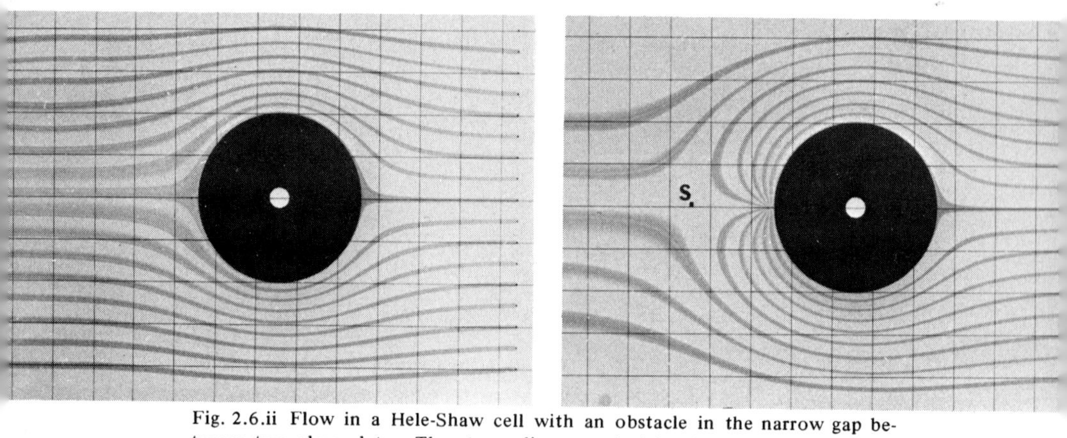

Fig. 2.6.ii Flow in a Hele-Shaw cell with an obstacle in the narrow gap between two glass plates. The stream-lines, marked by dye, show plane irrotational flow. (a) In the left picture the flow passes round a circular obstacle. The dye on the streamlines passing through the low velocity region is spread sideways more than elsewhere. The flow can be seen to be from the right. (b) With suction applied at a sink close to the rear stagnation point of (a) taking 0.6 of the flow. Equally the pattern represents a flow from the left towards a source and a circular cylinder, or flow round a long obstacle delineated by the streamlines from the stagnation point, S. This is the situation envisaged in equation (3.6.1)

The flow in planes parallel to the wall is therefore irrotational. It is the same throughout its depth but of magnitude varying across the gap. Since irrotational flow is unique with given boundaries, the irrotational flow around two-dimensional obstacles may be produced in between parallel plates with the appropriate obstacles inserted in the gap. This apparatus is called a **Hele-Shaw cell**, and streamlines may be made visible by introducing streamers of dye (see Fig. 2.6.ii)

The same principle is embodied in **Darcy's law** for the flow through a porous medium, namely

$$\mathbf{v} = -\sigma \,\mathbf{grad}\, p \,, \tag{2.6.7}$$

which is valid in three dimensions. The flow is assumed to be governed by this law on the ground that the flow is down the pressure gradient, proportional to the porosity σ, and slow enough for inertia forces to be neglected. This is the same as **Ohm's law** for the current in a conducting medium, and **Taylor's electrolytic tank** embodies this result. The equipotentials in a conducting liquid may be mapped out using a probe with an insulated lead. Hence it is possible to discover the irrotational flow around obstacles made from insulating material placed in a tank, with a conducting plate across each end to produce a flow along the tank. Taylor's original application was intended to reveal the flow of air over humps on the ground by making models of the topography out of paraffin wax; but the analogy for this case is poor because there is usually an increase of wind with height so that the flow is not irrotational.

The condition at a rigid boundary of a fluid is that, if the fluid is viscous, the velocity there is zero. The flow around small bubbles of air in water might be thought to follow the pattern of irrotational flow because the shearing stress in the liquid at an interface with a gas is negligible and the boundary condition for an inviscid fluid might be thought to be applicable. However, although surface tension tends to make the surface spherical, impurities introduce resistance to motion on the surface. Because the pressure is greatest, according to (2.3.1), at the front and rear stagnation points on a sphere, the spherical shape due to surface tension is slightly flattened.

One other important factor makes the flow of a viscous fluid near to a free surface different from that of an inviscid fluid, even though there is no shearing stress at the boundary and no restraint on the tangential velocity. This is that the fluid elements are distorted in curved irrotational flow. The same flow in a viscous fluid would therefore have shearing stresses in it, and the flow must be different at the free boundary for this reason. For example, the flow near to the surface of a bath plug vortex in viscous fluid cannot be irrotational with velocity proportional to k/r because it would require a shearing stress to maintain it. Secondary flows are therefore produced by the reduced speed at the surface and there is a flow down the surface into the drain as well as close to the bottom of the bath. (See Chapter 3).

The dissipation of energy by viscous forces in an incompressible fluid may be thought of in terms of the following equation:

Rate at which surface tractions do work + rate of working of normal forces on the surface = rate of increase of kinetic energy + rate at which energy is dissipated internally. (2.6.8)

In the case of steady irrotational flow only the first and last terms are not zero and the rate of dissipation is equal to the input by the surface tractions. Surface tractions are therefore necessary to maintain irrotational flow in a viscous fluid if there is any curvature, because curvature distorts the shape of fluid elements.

If there are no surface tractions and fixed boundaries the rate of dissipation is proportional to the square of the vorticity, and is equal to $\frac{1}{2}\mu\omega^2$ per unit volume, ω being the magnitude of **curl v**. This is the rate at which turbulent energy is dissipated (see Chapter 7).

2.7 BOUNDARY LAYERS

It was shown by equation (1.4.12) that each component of vorticity is conducted through the fluid by the viscosity. If a boundary of a fluid is suddenly set in tangential motion, the fluid at the surface is carried with it and vorticity is created in the fluid close to it, and is therefore gradually conducted into the main body of fluid.

If δ is the distance at which the fluid is moving with, say, 5 per cent of the velocity of the boundary, it is determined by the kinematic viscosity ν (dimensions L^2T^{-1}) and the time t since the moment the boundary was jerked into motion, and by not other factors. Therefore, to have the right dimensions, δ can only take the following mathematical form:

$$\delta \propto (\nu t)^{1/2}. \tag{2.7.1}$$

A vortex sheet suddenly created in the middle of a fluid would diffuse in the same way if it were not unstable (see Chapter 6). Since the same argument applies to the distance at which any multiple of the velocity is achieved, the mathematical form of the velocity profile remains constant, and its dimension expands as in $t^{1/2}$.

By rough analogy, the boundary layer, which is the layer within distance δ of the boundary, grows on a flat plate which moves through a fluid. The time for growth is represented by the distance x from the leading edge of the plate divided by the speed U of the plate through the fluid. Fig. 2.7.i shows how the velocity profile changes downstream on a plate fixed in a stream. If we think of a column of fluid ABCD moving downstream with the layer growing in it, $t = x/U$, and so

$$\delta \propto x^{1/2}\nu^{1/2}U^{-1/2} \tag{2.7.2}$$

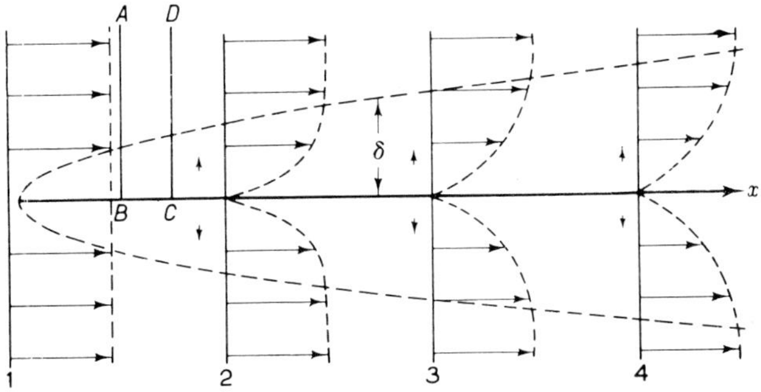

Fig. 2.7.i Change in velocity profile in a stream flowing across a flat plate

At the trailing edge of the plate the vorticity continues to diffuse outwards, but the vorticity of opposite signs on the two sides now begins to cancel out and the curvature of the velocity profile is reversed. In Fig. 2.7.ii the shaded area represents the deficit of momentum flux suffered by the stream, which is equal to the drag force on the plate.

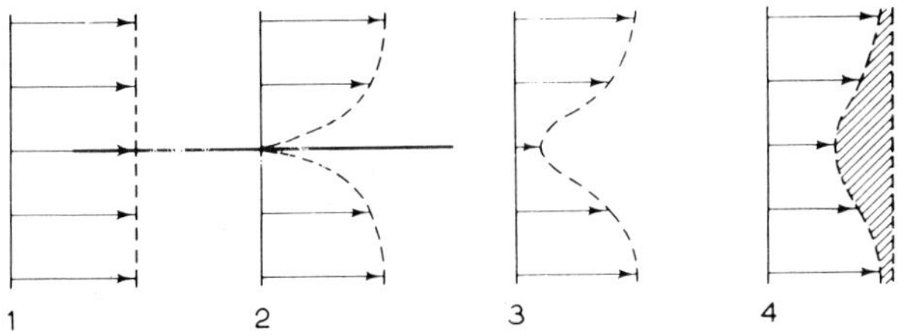

Fig. 2.7.ii Spread of the velocity deficit behind a flat plate

Although the shaded area remains constant downstream it is spread over an ever widening wake. In practice the flow in the wake is not steady unless the **Reynolds number** (see Section 2.8 below) is small.

Boundary layers therefore grow in the downstream direction on all solid surfaces in viscous fluids unless the fluid is accelerated. In Fig. 2.7.i we see that the volume flux in the boundary layer is less than in the same depth of fluid at a point upstream. Therefore continuity of flow requires that there shall be a small velocity away from the surface. This means that the flat plate acts, as far as the fluid outside the boundary layer is concerned, like a body whose thickness increases downstream. If the fluid is being accelerated by motion down a pressure gradient, for example into a converging channel or over the front face of an obstacle where the flow moves away from the forward stagnation point which divides the flow, the thickness of the boundary layer may be reduced. On the other hand, in a diverging channel where the fluid is moving towards higher pressure, the boundary layer is thickened by the decleration, the shear stress is reduced and the fluid is decelerated more quickly than in uniform flow. The pressure somewhere downstream may be equal at a certain point to the stagnation pressure of the slow moving fluid close to the wall, and in that case the

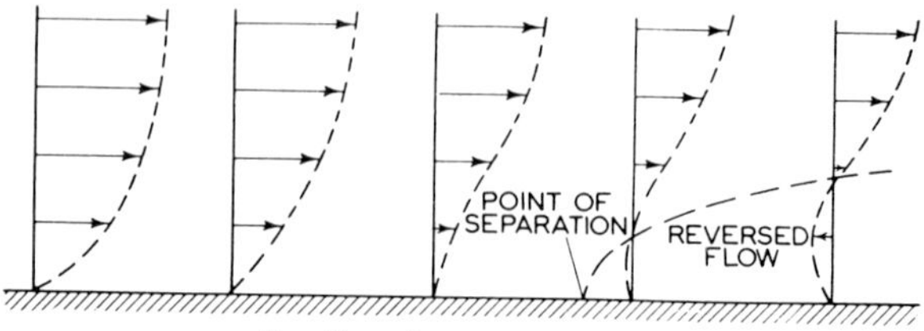

Fig. 2.7.iii Separation and reversal of boundary layer flow in an adverse pressure gradient (e.g. in diverging flow)

flow is reversed beyond that point (Fig. 2.7.iii) and separation takes place. The separation occurs where the forward viscous drag due to the layers further from the boundary is less than the backward force due to the reversed pressure gradient. There is an assumption made in boundary layer theory that, because the acceleration perpendicular to the wall is very small, there is negligible pressue gradient in that direction. The isobars are therefore assumed to extend into the boundary layer from the main flow beyond (see Fig. 2.7.iv).

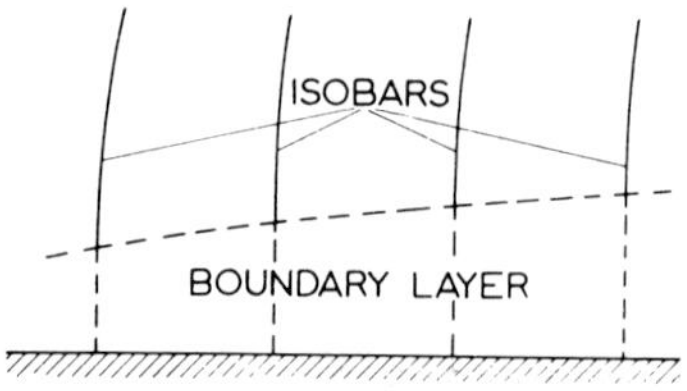

Fig. 2.7.iv External pressure gradient projected through a boundary layer

The separation of the flow in a diverging channel such as is shown in the left diagram of Fig. 2.7.v can be prevented by causing the surface to move in the

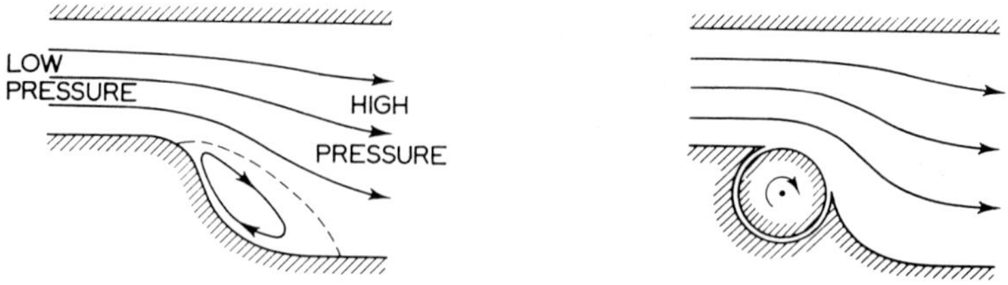

Fig. 2.7.v Left: separation in a diverging channel. Right: separation prevented by motion of the boundary near the point of separation.

direction of flow at the point of separation. Thus a roller may be installed where the curvature is large (Fig. 2.7.v, right).

Another method of preventing boundary layer separation is to introduce a jet of air in the direction of motion, where it would otherwise separate. This is called the **Coanda effect**, and is illustrated for the case of a jet flap on a wing in Fig. 2.7.vi. The flow becomes turbulent when a wing is stalled by separation of the flow from its upper surface. The lift is greatly reduced because the region of low pressure in the high speed flow over the upper surface is lost.

The Coanda effect can be illustrated by blowing a jet of air over a corner of a solid object. If the jet is made to impinge on the wall close to the corner, the

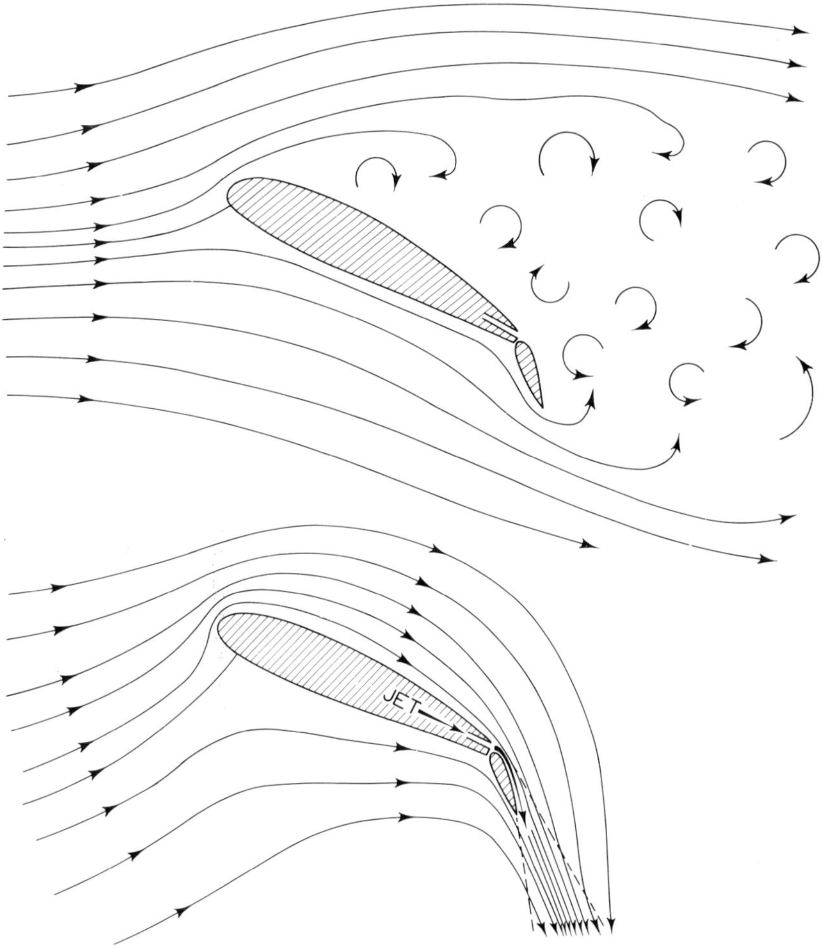

Fig. 2.7.vi The Coanda effect, or jet flap, in which separation from the surface of a wing and stalling is prevented by blowing air over the upper surface of a flap

layer formed in it is too shallow to cause separation. If it is lifted slightly, so that slower moving fluid lies between the velocity maximum and the boundary, separation may occur (Fig. 2.7.vii).

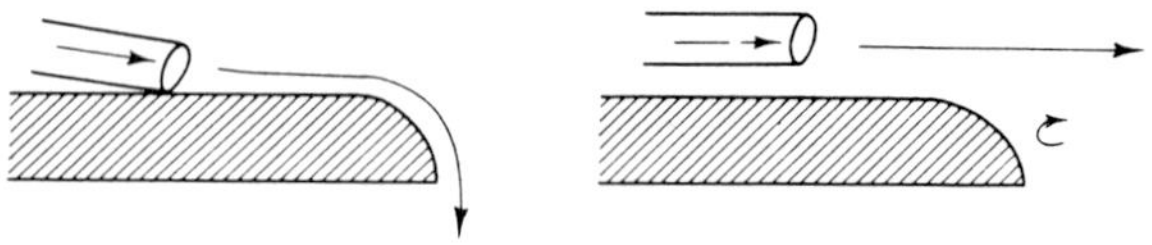

Fig. 2.7.vii A jet blown over an edge close to the surface will adhere to the corner but will separate if slower moving fluid lies between it and the rigid surface

This phenomenon is also illustrated if the jet is of water. The pressure on the outer surface of the thin layer of water is uniform being equal to that of the air, therefore adverse pressure gradients are reduced in magnitude at the surface. It is easy to demonstrate this by holding an upturned pan under the kitchen tap: the water flows without separation over the edge of the base and down the sides of the pan as if it was stuck to them, instead of escaping horizontally, as might be expected at the corner.

The velocity close to the boundary may be increased by placing vortex generators upstream of the separation point. These are small aerofoils whose tip vortices bring faster moving air from outside the boundary layer close to the surface, so as to prevent stagnation there (Fig. 2.7.viii).

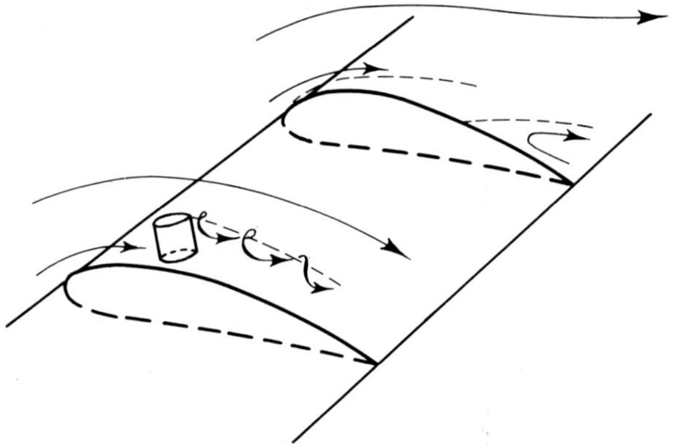

Fig. 2.7.viii Vortex generators prevent separation by bringing faster moving fluid from outside the boundary layer close to the surface

If the air on a sloping hillside is warmed by sunshine heating the ground the buoyant air flows up the slope. This causes separation of flow coming down the slope (Fig. 2.7.ix). On the other hand, if the ground cools at night the boundary layer is drawn by gravity down the slope and separation is stopped.

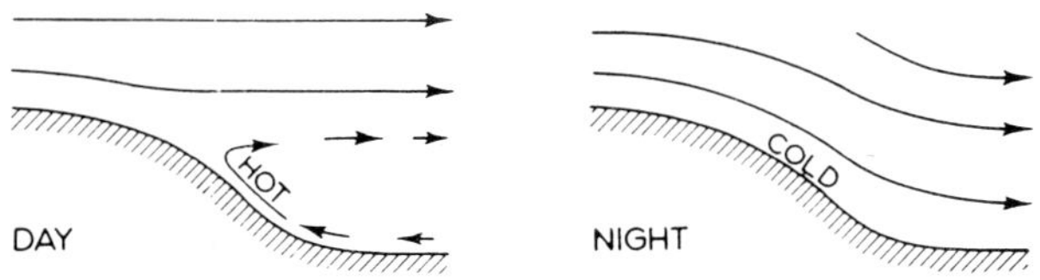

Fig. 2.7.ix The downslope flow of air on a cooled surface can prevent the separation of flow from a hillside and alter the main flow above the cooled layer

Another technique of preventing boundary layer separation is to remove the slow moving fluid by suction through a slot or through a porous surface.

If a trip mechanism, such as a small sharp ridge, is placed on the surface upstream of the point of separation it may cause the boundary layer to become turbulent. This alters the velocity profile by mixing main flow air into the boundary layer and reducing its thickness in the same way as by vortex generators.

Turbulence in channel flow tends to spread the momentum of a stream more uniformly across it, so that the velocity profile is not parabolic, as cal-

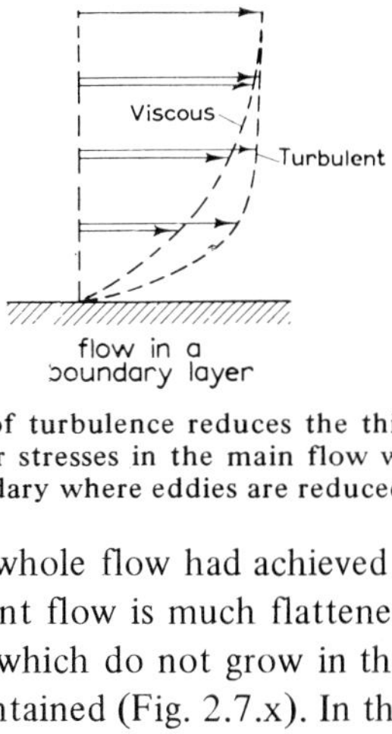

Fig. 2.7.x The presence of turbulence reduces the thickness of the boundary layer by introducing shear stresses in the main flow which are absent close to the boundary where eddies are reduced in size

culated in (2.6.4) where the whole flow had achieved equilibrium under viscous forces. The profile of turbulent flow is much flattened in the middle with thin boundary layers at the walls which do not grow in the downstream direction as long as the turbulence is maintained (Fig. 2.7.x). In this case the boundary layer is made thinner by a reduction of the shearing stress close to the wall where the presence of the wall suppresses large eddies (see Section 7.3). There may be a 'laminar sub-layer' between the turbulent boundary layer and the wall which in effect acts as a lubricant to the flow; but because the velocity gradient in it is greater than in non-turbulent flow the pressure gradient required to maintain this flow is greater when it is turbulent. This is particularly important in the case of pipes.

In a bend of a channel the separation on a wall is not likely to occur on the inside wall because of the reduction of pressure there due to the centrifugal forces in the curved flow (see Section 3.2). However, separation often occurs on the outside of the beginning of the bend where the flow is being decelerated by the high pressure region. Fig. 2.7.x shows how this can be avoided by using a cascade of aerofoils in a rectangular corner of a channel, such as a wind tunnel in which the air is in closed circulation. No large scale eddies are produced when there is a large number of aerofoils. The relative positions of particles originally aligned across the flow is altered and they emerge aligned diagonally. (See also Section 3.3).

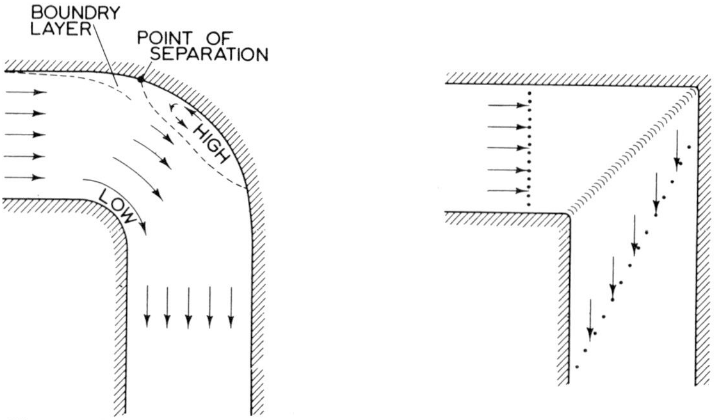

Fig. 2.7.xi The separation in a channel bend occurs on the approach to the high pressure region on the outside. It can be prevented by introducing a 'cascade of aerofoils' into a square corner

There was considerable trouble in the air intakes into the Olympus turbo-prop aero engine. The design reversed the airflow so as to feed it into the back of the engine, and a slow moving boundary layer was formed on the outside of the bend. When the aircraft flew into a cloud containing snow or soft hail, ice was accumulated in this high pressure region. Because the channel was annular, its radius was decreased in the downstream direction and the snow became tightly packed as it was driven downstream. It broke away in large lumps which sometimes extinguished the flame in the combustion chamber. If the bend had been composed of a set of two dimensional bends circular on the outside, arranged radially, the material would have been driven downstream continuously and the flame extinction would not have occurred, provided the wall temperature had been kept above freezing point to lubricate the motion of the ice along it.

2.8 SMALL BODIES MOVING IN A FLUID: REYNOLDS NUMBER

The Reynolds number is a nondimensional number which is the ratio of the inertia forces (mass $\times$ acceleration) to the viscous forces. Thus it is small when the flow is slow. It is the ratio of the term $\mathrm{D}\mathbf{v}/\mathrm{D}t$ to the term $\nu\nabla^2\mathbf{v}$ in equation

(1.3.29). The inertia is represented by density × (a velocity) ÷ (a time), the time being represented by (a distance) ÷ (a velocity). Likewise the viscous force on unit volume is represented by (viscosity) × (gradient of velocity gradient), the gradient being represented by (a length)$^{-1}$. Thus

$$\mathrm{Re} = \frac{\rho U \div [d \div U]}{\mu d^{-1}.d^{-1}U} = \frac{Ud}{\nu} \qquad (2.8.1)$$

where U and d are a representative speed and length appropriate to the situation. Alternatively Re may be regarded as the ratio of a pressure force, represented by $\frac{1}{2}\rho U^2$ through Bernoulli's equation, to a viscous force per unit area represented by viscosity × velocity gradient (i.e. $\mu U/d$).

At small Reynolds numbers the inertia forces become negligible; at large values the viscous forces are negligible, but it should be noted that this is usually because the flow has become turbulent and inertia forces in the form of eddies have become dominant (see Chapter 7). After turbulence has set in the flow pattern is not much altered by further increases in *Re* because the velocities in the eddies are proportional to the mean flow which created them, and so the ratio of their inertia forces to those of the mean flow is not altered by an increase in speed.

At small Reynolds number the flow around a small particle is determined by the viscosity, μ, and the particle dimension, a. The drag force, D, on it when its velocity is U must have the mathematical form

$$D \propto \mu a U \qquad (2.8.2)$$

because no other combination of μ, a, and U generates a quantity with the dimensions of force. The constant of proportionality was first shown by Stokes to be 6π, and is applicable to a solid sphere moving slowly through a liquid. If it moves under gravity and the densities of the sphere and the liquid differ by $\Delta\rho$, the terminal velocity is given by

$$6\pi\mu a U = \tfrac{4}{3}\pi a^3 g \Delta\rho\,. \qquad (2.8.3)$$

For reasons given previously (Sec. 2.6) this formula is not necessarily valid for liquid spheres or bubbles of gas because they allow slip velocities at their surface. However, bubbles very often do obey this equation because of impurities which inhibit motion along the surface.

The difference between high and low Reynolds number drag is well illustrated by a parachute and a dandelion seed. A parachute has a large Reynolds number and is a device for creating intense eddying motion in its wake as it descends through the air. The eddies are generated as the flow separates from the perimeter and from the edge of the hole in its centre, and the drag force, produced by the high pressure on its lower side which is advancing through the air, may be thought of as the force required to produce the eddying motion. All air brakes on aircraft work on this principle: they are called spoilers, because they spoil the laminar airflow over the wing surface and generate eddying energy.

The dandelion seed, by contrast, illustrates the principle that the flow is dominated by viscosity at low Reynolds number. The very small branches present a very large surface over which the air must flow and exert the viscous drag due to the shearing stress at the surface. Eddies are not left behind, but the fluid is dragged downwards with the seed. Because the seed sinks through the air the drag force is always being applied to new air, and so it leaves behind it a steadily lengthening trial of downward moving air. This gradually communicates its downward momentum to a widening volume by viscosity. The energy is dissipated as heat.

2.9 PATHS OF SMALL PARTICLES: EFFICIENCY OF CATCH

When a small particle is 'embedded in' a fluid, which means that its fall speed (terminal velocity) is small compared with the velocity of the stream past some larger obstacle, its acceleration is approximately that of the particles of the fluid. Thus soap bubbles do not collide with a house because they are carried along very approximately by the fluid in which they are embedded. However, they may well collide with a tree because the curvature of the trajectories of the air particles is greater than that in the flow round a house. This causes the bubble, which is heavier than the air it displaces, to travel in a less curved path than the air. Very tiny bubbles have smaller terminal velocities than big ones, and may therefore not collide with the tree trunk but only with small branches as the wind carries them along.

Smoke particles, which have fall speeds of the order of less than one millimetre a second, collide with fine fibres and are therefore filtered out of the air by a cloth, but are not captured on buildings. Indeed, the blackening of buildings is very much affected by the small eddies on sharp corners and protuberances. These produce flows with large curvature which can cause smoke particles to collide with the surface. The broadscale flow around the building does not cause particles to collide with it unless they have a large fall speed: thus heavy raindrops hit the window panes but small drizzle and fog droplets do not. On the other hand, drizzle and fog is very effectively captured on tiny branches and especially on the fine needles of pine trees. Because of this, the rain water falling to the ground is greater under pine trees, which are at a high enough altitude frequently to be in cloud, than in the open in the same neighbourhood. Contrary to popular belief, trees do not attract rain (nor do they, as many economists evidently believe, raise the water table in the ground) and they certainly increase the evaporation to the air of water at depth in the soil: however, in some areas their contribution to ground water through the capture of cloud particles may be very important. They may also cause a more spongy soil to be generated which holds moisture from draining out of the land.

The flow on the upstream side of an obstacle is very approximately potential flow. The paths of particles are indicated in Fig. 2.9.i. Because they have a negative virtual mass, bubbles of gas in a liquid cannot collide with an obstacle because their track is always more curved than the liquid particle trajectories.

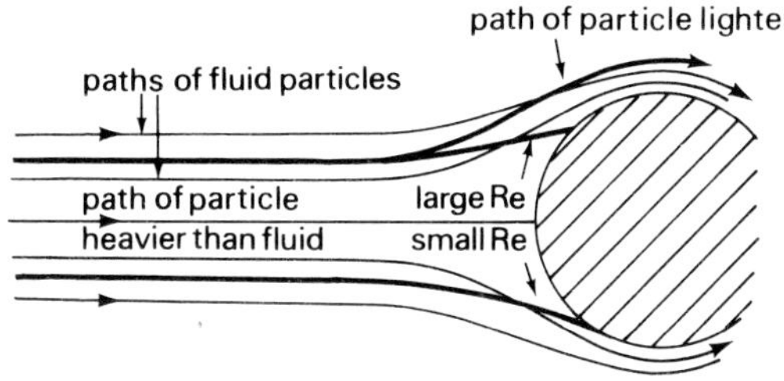

Fig. 2.9.i Paths of particles approaching an obstacle. Particles lighter than the fluid always avoid the obstacle, but heavier particles may collide with it, especially at large Reynolds number flow around the obstacle

If the particle to be captured has a significant size it may be captured even if its centre is not on a collision path. Small cloud or drizzle droplets may be caught in the wake of a raindrop even though they do not collide with the advancing lower surface, because the airflow in the wake is unsteady and may frequently bring cloud particles close to the upper surface of the drop.

The **efficiency of catch** of a body moving through a cloud of particles is defined as the proportion of the particles lying in the cylinder swept out by the geometrical outline of the body which actually collide with and are captured by that body.

The efficiency of catch of bubbles of gas in water is usually very small, but if the body were a wire of smaller dimension than the bubbles it might capture some. The efficiency of catch of a raindrop falling through a cloud depends very much on its size relative to the cloud droplets. If they are very small it is low, but for almost drizzle size it may approach unity. In those circumstances where there is a considerable capture through the wake of the raindrop, it could exceed unity.

The efficiency of collision is not always the same as the efficiency of catch, because it depends on the mechanism of attachment. If a hailstone falls through a cloud of supercooled cloud droplets they stick to it if they freeze on impact. If they do not freeze on impact they may nevertheless wet the surface and be caught that way. On the other hand if they are too plentiful the latent heat of fusion may be enough to raise the temperature of the whole surface of the hailstone to 0°C, and the unsteady motion may cause splashing of water from the surface. Likewise, when a raindrop reaches about 5 mm in diameter the surface tension is no longer sufficient to hold the surface spherical in the presence of the aerodynamic forces, and the raindrop will usually break up if it exceeds this size.

In the early days of the spraying of crops with pesticide, the particles were released from an aircraft and the downward motion of the air in the aircraft downwash together with the gravitational settling of them under their own weight in still air were used to deposit the particles on the crop. Thus large particles were more effectively caught on the crop, but mainly on the upper surfaces only. More recently smaller spray particles have been used and the spray released in a cross-wind: the particles are then captured on the crop as they are carried through it by the wind. This technique is easier to employ because it is not necessary to wait for the air to be calm. It is much less wasteful of pesticide because the large particles used to ensure settling often represent an overdose.

The most efficient way to dose a pest which has an airborne phase is to use a fine spray which sinks very slowly through the air but is composed of particles large enough to be caught on the insect as it flies through the spray cloud. (See Sec. 12.7).

The shapes of obstacles enlarged by capture of material from a fluid stream can vary enormously. Thus a hailstone may be spherical, although often it has a conical face which indicates that it has achieved a preferential orientation as it falls. When supercooled droplets are deposited by freezing onto wires, the growth may be Y shaped, perhaps because sideways oscillation of the wire in the wind produces a stagnation point which oscillates across the front of the wire causing two points of maximum collection. The growth is uneven because a slight protuberance grows more rapidly.

Fig. 2.9.ii Dendritic shapes of ice crystals growing in supersaturated air as they fall. The growth on a protuberance causes the orientation of the falling crystal to expose the opposite point preferentially forward so that the growth is symmetrical. The plates fall down a helical path, rotating as they fall.

In light winds the growth of ice on wires and twigs may be more by condensation (collision of molecules) than by ordinary collision of particles. The rate of sublimation depends on the gradient of concentration of the deposited material, and this will be greater at the tips of small protuberances than on parts of the surface with a larger geometrical dimension. Thus hoar frost often consists of spiky growths, and ice crystals falling through air which is supersaturated with water vapour below 0°C grow with dendritic shapes (Fig. 2.9.ii) at the tips of which the boundary layer is thinner and therefore the concentration gradient greater.

An interesting aspect of falling ice crystal growth is its extraordinary symmetry. As soon as one dendritic (branch) grows larger than the others it tends to be dragged back by viscosity to the rear of the crystal, exposing other dendrites at the front to thinner boundary layers and greater concentration gradients. The growth is also determined by the crystal structure, which imposes hexagonal symmetry.

2.10 DRAG COEFFICIENT: C_D

At large Reynolds number the drag force on a body in a stream U is caused by the creation of kinetic energy of eddies whose velocities are proportional to the stream speed and whose size is determined by the geometry of the body. The drag D is then given by

$$D = C_D \tfrac{1}{2} \rho U^2 d^2 \ , \qquad (2.10.1)$$

where d^2 here represents the frontal area of the body. The Mach number M is U/c where c is the velocity of sound. As U approaches c, the flow over a body is changed because some parts of it become supersonic and the pressure distribution is completely altered by the appearance of shock waves. The pattern of separation is also changed, and a large part of the drag when $U > c$, and $M > 1$, is due to the production of shock waves which travel out as a cone through the air like the wave pattern of a ship on water. The drag is then proportional to a higher power of U, or to put it another way, C_D is decreased as M is increased through unity.

At low Re, i.e. in the Stokes regime, the drag is proportional to U (see, for example, equation 2.8.2). Thus according to the definition of C_D in (2.10.1), C_D must be increased at low Reynolds numbers. Table 2.10.I indicates the drag coefficient of spheres at different Re, and Table 2.10.II indicates more specifically the fall speed of spheres with the density of water falling through air. Liquid spheres disintegrate when the diameter exceeds about 5 mm. For most common objects of laboratory size, except those of deliberate aerodynamic design such as propeller blades and aircraft wings, the relationship (2.10.1) holds over the range of velocities commonly experienced, and C_D is in the

Table 2.10.I
Drag coefficient of spheres

Re	C_D
.002	10,000
.005	5,000
.01	2,400
.05	480
.1	235
.4	64
.8	34
1	27
2	15
5	7
10	4
20	2.6
50	1.6
100	1.1

$$\mathrm{Re} = \frac{2a\,w_t}{\nu}$$

a = radius

w_t = fall speed

$\nu_a = \mu_a/\rho_a$ (for air) = 0.15 cm^2 sec^{-1} at 20°C

ν_w = 0.015 cm^2 sec^{-1} (for water) at 5°C

$$C_D \tfrac{1}{2}\rho_a w_t^{\,2}\pi a^2 = \tfrac{4}{3}\pi a^3 g(\rho_w - \rho_a)$$

$\rho_{a,w}$ = density (air, water)

Table 2.10.II
Fall speed of water drops in air

diameter. mm		Fall speed m sec^{-1}
.001		.0003
.01		.03
0.1		0.27
.2	cloud	.72
.3		1.2
.4		1.6
.5		2.1
.6		2.5
.7		2.9
.8		3.3
.9	drizzle	3.7
1.0		4.0
1.2		4.6
1.4		5.2
1.6		5.7
1.8		6.1
2.0	rain	6.5
2.2		6.9
2.4		7.3
2.6		7.6
2.8		7.8
3.0		8.1
3.2	begins	8.3
3.6	break-up	8.6
4.0		8.8
4.6		9.0
5.2		9.1
5.8	break-up certain	9.2

neighbourhood of unity for $50 < \mathrm{Re} < 10^5$. However, the upper limit varies according to the size of the body because that determines the value of Re at the corresponding Mach number.

It needs to be emphasised that at large Re and small M the degree of turbulence in the fluid stream and the nature of the solid surface upstream of the point of separation can significantly alter the thickness of the boundary layer and also alter the point of separation, as described in Section 2.7. When the separation point is moved back, the wake size is considerably reduced and so not only the drag coefficient but also the drag itself may be reduced by increasing the velocity (and Re). In the great outdoors almost any situation may arise because very large Reynolds numbers are common at low wind speeds.

2.11 SEPARATION AND IRREVERSIBILITY

All viscous flow is irreversible because it converts work into heat by the action of the viscous stresses. On the other hand, most of the energy which makes motion irreversible is generated as eddies. Although the eddies are converted into heat eventually by viscosity, it is the eddy making mechanism which is the primary cause of the irreversibility, and eddies are most often the result of separation of the flow in cases when buoyancy forces are not important.

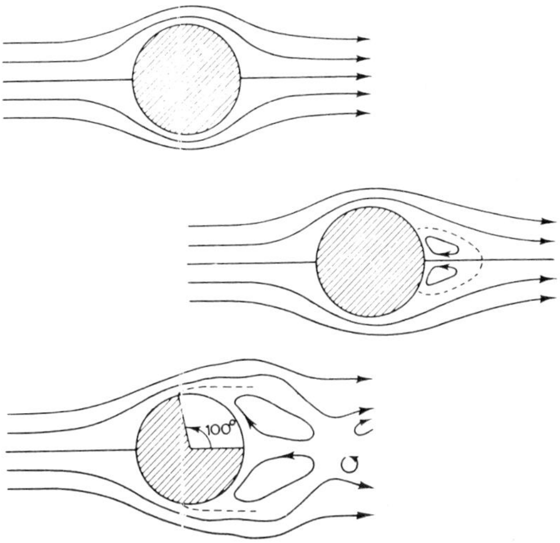

Fig. 2.11.i Initially (or at very low Re) a sphere has no wake, but after a short time the motion separates near the rear stagnation point and the separation point moves forward until at large speeds it is at about 100° from it. The wake is then very turbulent. The point of sepearation may be moved back from the 100° position by turbulence in the airstream or roughness on the sphere causing turbulence in the boundary layer

The two most important examples of eddy-making are wakes and jets. The wake of an obstacle in a fluid stream is initially bounded by the surface of separation, which streams away from the line of separation of a body. In the case of a sphere (Fig. 2.11.i) the separation begins not far from the rear stagnation point S′, but it moves forwards as the flow develops, and for large Reynolds number settles down along a line at about 80° from the forward stagnation point S. In front of this line the flow outside the boundary layer is almost indistinguishable from the calculated potential flow. Behind this line eddies are unsteady and grow in size when they are shed from the body and are carried downstream. The two-dimensional flow behind a cylinder usually takes the form

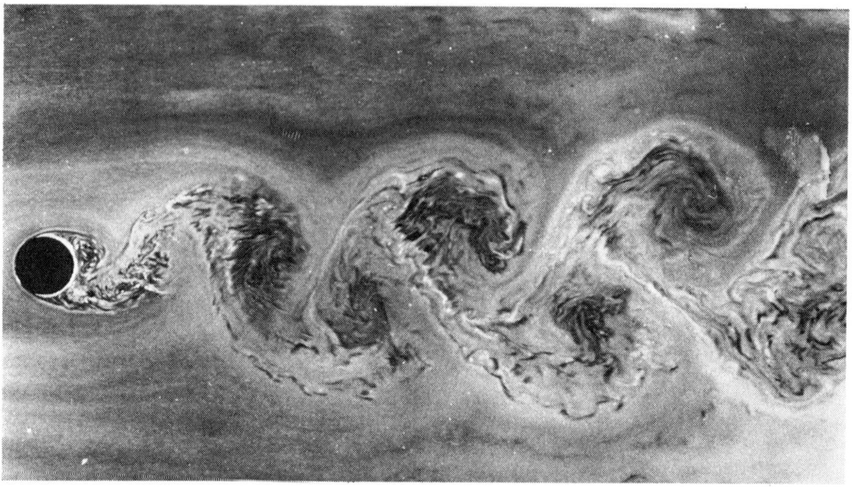

Fig. 2.11.ii Photo of the wake of a circular cylinder made visible by particles floating on the surface of a water channel. In this case the 'particles' were tiny bubbles of froth generated by introducing some detergent and allowing a small amount of air to be sucked into the recirculating pump

of a vortex street (Fig. 2.11.ii, often named after **Von Karman**) in which eddies of opposite rotation are shed alternately: behind a sphere the flow is more chaotic but it sometimes takes the form of a vortex with a helical core.

Behind a cylinder of almost any shape the flow tends to be in the form of a vortex street if the flow settles down. It is then periodic, and this causes wires to oscillate in the wind. The frequency, σ, of oscillation (often called the **Strouhal** frequency) is found to be related to the velocity U of the stream and the obstacle diameter d by

$$\sigma = \mathrm{S}\, U/d, \tag{2.11.1}$$

S being the **Strouhal number,** which is equal to about 0.2 for circular cylinders.

Many experiments performed in wind tunnels at large Reynolds number produce results which are characteristic of the tunnel. This is because of their individual levels of turbulence and fan frequencies, which affect the separation properties of objects, and this particularly applies to artificially smooth laboratory objects such as spheres and circular cylinders. Ordinary objects found in nature usually have rough surfaces and protuberances and are subjected to winds with a considerable level of turbulence (e.g. 0.2, which means that the fluctuating component of the velocity is around a fifth of the wind speed). It should also be remembered that some authors include a factor 2π in the definition of the Strouhal number (2.11.1), while others define it as the reciprocal of the number given there. The island of Jan Mayen (Fig. 2.11.iii) has a Strouhal number of the order of 0.2, even though the Reynolds number is around 10^{10}, and is has an irregular shape.

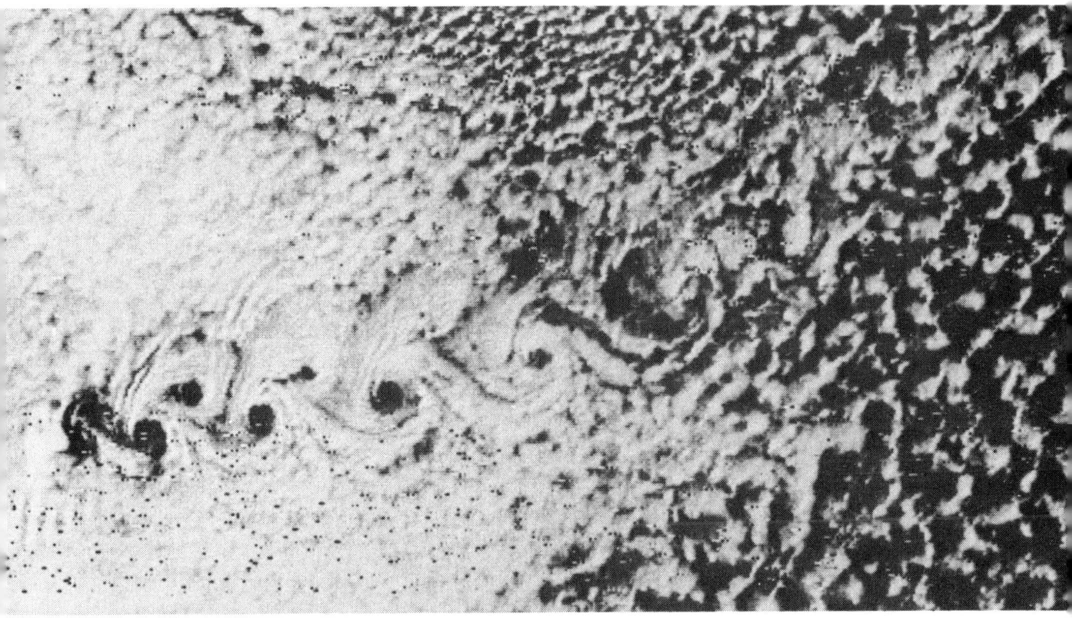

Fig. 2.11.iii A satellite (NOAA 4) photograph of the vortex street in the westerly airflow downstream of Jan Mayen, which has two peaks, 2277m 10 km wide and 769 m 40 km to the SW. The pattern is made visible by the clouds at 1500–2000 m, which lie beneath a stable layer which causes the flow to be nearly two-dimensional when the mountains on the island reach higher than the clouds. The cyclonic vortices are stronger. [Photograph by courtesy of Department of Electrical Engineering, University of Dundee]

Most important in practice as far as separation is concerned, are salient edges. These are edges from which free streamlines may be generated, and the vortex sheets so produced roll up into eddies which are the beginning of the vortex street formation. Although vortex streets are considered theoretically as if they were composed of two parallel rows of alternate vortices, in practice the vortices grow in size as they proceed downstream, and only after about 10 more have been shed does the spacing between neighbours become nearly constant. One special kind of salient edge is the strake, which is a helical protuberance often seen on chimneys. This ensures that the vortices are not shed at the same moment along the whole length of the chimney, and thereby avoids the oscillating force due to the shedding causing the chimney as a whole to vibrate.

Typical salient edges from which separation takes place are the corner of a building, the apex of a roof, the top of a cliff, or wall on a hill. Corners on vehicles have the same effect. They ensure that the flow behind the obstacle is turbulent and produces a drag coefficient of the order of unity. Wakes and separation lines are often made visible by smoke, as illustrated in Fig. 2.11.iv.

Jets are produced when fluid emerges from a hole in a wall or from a nozzle. The fluid on the high pressure side of a hole approaches it very much according

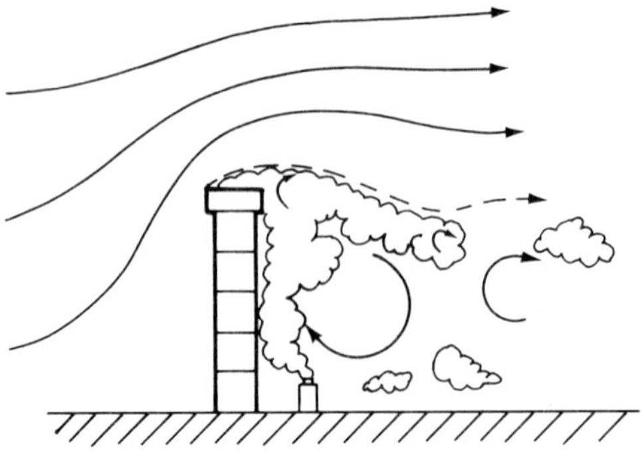

Fig. 2.11.iv The separation on a wall often occurs at the upwind edge of the top. Smoke from a generator behind the wall is carried forward to this line and then is either recirculated in a standing eddy behind the wall or carried away intermittently with moving eddies.

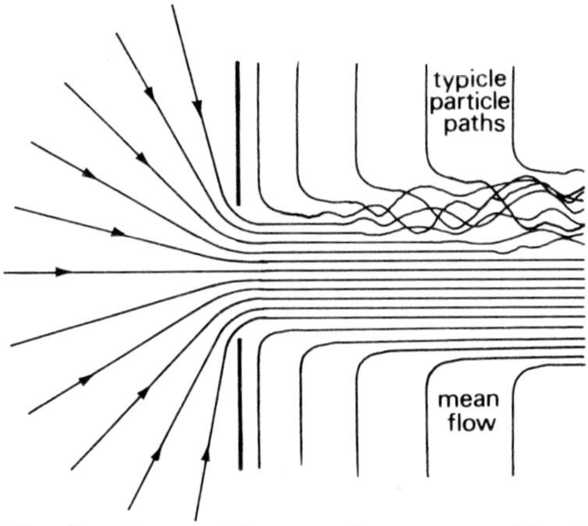

Fig. 2.11.v The flow into a sink or on the upstream side of an obstacle is usually close to potential flow, whereas the flow from an orifice, which is the wake of the obstacle containing the orifice, is turbulent, irreversible, and mixes with and entrains the surroundings.

to a classical theory in which $\mathbf{v} = \mathbf{grad}\ \phi$ and $\nabla^2 \phi = 0$. On the low pressure side (Fig. 2.11.iv) the flow is quite different and consists of a turbulent jet (see Chapter 8). There is often a naivety in the explanation given of the mechanism of entrainment manifested in turbulent jets: it is said in many school textbooks that the ambient fluid flows into the jet because, according to Bernoulli's equation, the large velocity in it is accompanied by a low pressure. This is misleading because in order to set up the jet a large excess of pressure has to be set up on the other side of the wall. When the instability on the edge of the jet mixes some of the ambient fluid into it and carries the fluid with it, by diluting its momentum, a low pressure region is created from where the fluid is entrained. If a laminar constant velocity jet could be set up, the pressure on its boundary would take the ambient value but Bernoulli's equation would still be inapplicable. This is because the fluid inside and outside the jet would have different values of the Bernoulli constant, gH, defined in (1.7.9) since there would be a vortex sheet between them. Thus when the flow becomes unsteady and turbulent Bernoulli's equation is no longer valid.

If flow were reversible it would be most inconvenient for animals, because unless they were moving or there was a wind they would breathe in at each breath the air breathed out from the last one. This would be most unfortunate in sleep! However, the outward breath is a jet and the air breathed in comes from all directions as into a sink. Life is utterly dependent on the processing of fluids within the body, and would be impossible if irreversible motions were not dominant.

In a jet the momentum flux is the same at all distances along it, therefore $R^2 W^2$ is constant, the cross section area being represented by R^2, and the axial component of velocity W being proportional to the volume flux and also to the 'density of momentum'. If W is to decrease then R must increase in proportion, and so the volume flux, which is proportional to $R^2 W$, increases. This means that the jet can only slow down by carrying a greater volume with it, which means that there must be entrainment. and this is initiated by the instability of the vortex sheet. The low pressure is not produced by the high velocity in the jet but only by its mixing with the ambient fluid. Thus a water jet can cause low pressure by entraining air, a fact made use of in the vacuum pump attached to the tap in the chemistry laboratory.

2.12 NON-DIMENSIONAL NUMBERS

A useful way of thinking of the enormous variety of phenomena in fluid flow is to divide them into ranges of the various important non-dimensional numbers. When considering the transition from laminar flow in which inertia forces are negligible (Stokes regime) at low Reynolds number to the high Reynolds number regime, where the viscous forces are negligible and the flow irrotational upstream and fully turbulent downstream of an obstacle in a stream, we imagine the other important non-dimensional numbers to be at the same extreme ends of their range. Thus the Mach number U/c has been assumed to be small throughout, so that effects of compressibility may be ignored.

The most important other numbers which occur in our study are concerned with buoyancy. The **Froude** number is the ratio of buoyancy or gravity forces to inertia forces, or the ratio of gravitational potential energy to kinetic energy. In studies of buoyancy effects in the following chapters this number is in the region of unity, otherwise it is taken as zero. On the other hand, it is not very large. Thus it is usual to make what is called the **Boussinesq approximation**, which is appropriate when density differences are not large. There is a buoyancy number

$$B = \Delta\rho/\rho \tag{2.12.1}$$

which is relevant to convection caused by density anomalies $\Delta\rho$. In most cases $B \ll 1$. The accelerations, according to Archimedes's principle, are gB, which is therefore small compared with g. Boussinesq's approximation is to neglect density anomalies except when combined with g. This is equivalent to neglecting **f** in comparison with **g** in the vorticity equation (1.4.1) on the grounds that **f** is of the order of **g**B, but not, of course, in (1.4.4), which is the equation for **f**.

The buoyancy manifests itself in a stably stratified fluid through the non-dimensional number

$$\beta h \triangleq -\frac{h}{\rho}\,\text{grad}\,\rho \triangleq -h\frac{\delta\rho}{\rho\delta z} \tag{2.12.2}$$

in which h is a typical height in the situation. This is, therefore, the value of B if $\Delta\rho$ is taken as the magnitude of density differences that occur in the situation. The Froude number may be expressed in various forms such as

$$\mathrm{F} = \frac{g\beta h^2}{U^2} \quad \text{or} \quad \frac{gBh}{U^2} \tag{2.12.3}$$

In cases where vorticity is important we find the **Richardson number** to be very relevant. But it differs from the other non-dimensional numbers in that it usually varies across the field and has only a local value, because it is a ratio of gradients. Thus

$$\mathrm{Ri} = \frac{g\beta}{\left(\dfrac{\partial u}{\partial z}\right)^2} \tag{2.12.4}$$

represents the ratio of stabilising effect of the density stratification to the de-stabilising effect of velocity gradients. We shall see in Chapter 6 that simple stability criteria based on a local number are not to be had, and this reminds us that the first form of F given in (2.12.3) contains the gradient β, and is a local number. Many of the solutions to problems containing velocity and density gradients as given are therefore particular and not general at all, for F and Ri are infinitely variable across the field of flow. However in situations in which Re is the relevant number, the flow field is determined by the viscosity before the problem is presented to us.

Vorticity and density stratification therefore have to be studied on the basis of the question, 'what are the significant phenomena?' rather than 'what are the significant non-dimensional numbers?', for there are too many of them.

Chapter 3

SECONDARY VORTICITY

3.1 DEFINITIONS

The **primary flow** in a fluid stream is the motion which transports the fluid. It consists, in steady flow, of the motion along the streamlines. This might seem to be the total flow, but motion in planes normal to the streamlines may exist, namely rotation around them. This is called the **secondary flow** and a measure of it is the magnitude of the **streamwise component of vorticity**, which is called the secondary vorticity. It is important because it represents extra energy, which is often wasteful, has scouring or other effects, and can lead to mixing. We shall discuss the mechanisms whereby it is generated, which are of particular interest because the vorticity generated at a boundary by viscosity is perpendicular to the direction of flow over the boundary. It occurs in pipes and rivers and in the corners of the cross section of channels. In all these cases it occurs as a result of a component of vorticity existing normal to streamlines, and is generated usually as a result of curvature of the primary flow.

The distance measured along the path of a particle is denoted by s and the fluid speed by q. The curvature of the path is κ, a is the tangential component of $\mathbf{f}$, the acceleration, and τ is the torsion of the path.

We use a system of moving axes, as employed in elementary differential geometry. The unit vector directed along the tangent to the particle path is denoted by $\mathbf{t}$, which is called the **unit tangent**.

By definition
$$\mathbf{t} = \mathbf{v}/q \ . \tag{3.1.1}$$

Since $\mathbf{t}.\mathbf{t} = 1$, by differentiation with respect to s we obtain

$$\mathbf{t}.\frac{\partial \mathbf{t}}{\partial s} = 0 \tag{3.1.2}$$

which means that the change in $\mathbf{t}$ is always in a direction normal to $\mathbf{t}$, because it is of fixed magnitude. $\mathbf{t}$ and $\partial\mathbf{t}/\partial s$ define a plane, called the **osculating plane**, which is the plane of closest contact with the particle path. This may be thought of as defined by three successive points on the curve or two successive tangents, just as a tangent is defined by two successive points. The unit vector in the direction of $\partial \mathbf{t}/\partial s$ is denoted by **n** and is called the **unit principal normal.** The curvature, κ, is given by

$$\frac{\partial \mathbf{t}}{\partial s} = \kappa \mathbf{n} \tag{3.1.3}$$

In the case of circular motion it is seen that κ^{-1} is the radius and so this is the radius of curvature in other cases. The magnitude of the curvature is

$$\kappa = \mathbf{n}.\frac{\partial \mathbf{t}}{\partial s} = \left|\frac{\partial \mathbf{t}}{\partial s}\right| \tag{3.1.4}$$

Since $\mathbf{t}.\mathbf{n} = 0$,

$$\mathbf{n}.\frac{\partial \mathbf{t}}{\partial s} = -\,\mathbf{t}.\frac{\partial \mathbf{n}}{\partial s} = \kappa. \tag{3.1.5}$$

The **unit binormal,** denoted by **b**, is defined by

$$\mathbf{b} = \mathbf{t} \times \mathbf{n} \tag{3.1.6}$$

and with **n** defines the plane normal to the curve. **t, n, b** form a right handed system of axes.

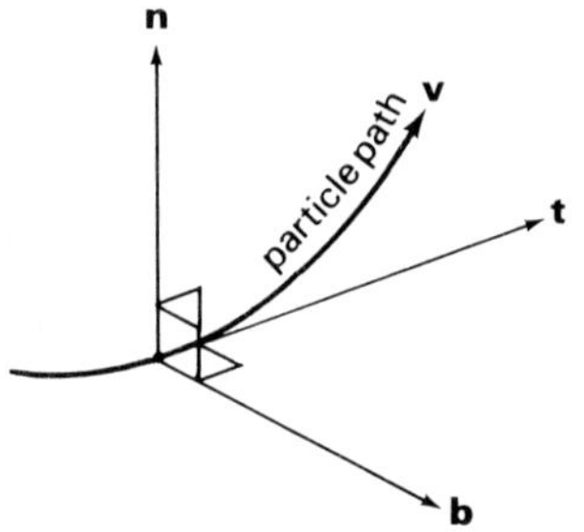

Fig. 3.1.i The unit tangent, normal, and binormal to a particle path

As we proceed along the trajectory **t** is tipping up at rate κ in the direction of **n**. At the same time **n** and **b** are rotating around **t** at rate τ, which defines the torsion. It is the rate at which the osculating plane rotates around **t** as the point moves along the curve. For a plane curve $\tau = 0$.

As **t** tips in the direction of **n**, **n** tips backwards in the direction of $-\mathbf{t}$ at rate κ, and rotates towards **b** at rate τ. **b** rotates in the direction of $-$**n** at rate τ, and is in a fixed direction normal to the plane if the curve is plane. The acceleration of a particle has components in the direction of **t** and **n** only. We now give mathematical expression to these ideas.

Since **n** is a unit vector, $\partial\mathbf{n}/\partial s$ has no component along **n** and so

$$\frac{\partial \mathbf{n}}{\partial s} = -\kappa\mathbf{t} + \tau\mathbf{b}. \tag{3.1.7}$$

The coefficient of **t** must be κ so as to conform with (3.1.5) when the scalar product of (3.1.7) with **t** is taken, and (3.1.7) may be taken as defining τ. Since $\mathbf{t}.\mathbf{b} = 0$

$$\mathbf{t}.\frac{\partial \mathbf{b}}{\partial s} = -\mathbf{b}.\frac{\partial \mathbf{t}}{\partial s} = 0 \quad \text{by (3.1.3)}$$

and so $\partial\mathbf{b}/\partial s$ has a component only in the **n** direction of magnitude τ. This may be obtained by differentiating $\mathbf{b}.\mathbf{n} = 0$, and using (3.1.3). Thus

$$\frac{\partial \mathbf{b}}{\partial s} = -\tau\mathbf{n}\,. \tag{3.1.8}$$

Equations (3.1.3., 7, 8) are called **Frenet's formulas.**

Since the acceleration is in the plane of **t** and **n**, in steady flow by (1.2.6)

$$\mathbf{f} = \frac{\mathrm{D}\mathbf{v}}{\mathrm{D}t} = (\mathbf{v}.\mathbf{grad})\mathbf{v} = \mathbf{grad}\,\tfrac{1}{2}q^2 - \mathbf{v}\times\boldsymbol{\omega} \tag{3.1.9}$$

$$= (\mathbf{f}.\mathbf{t})\mathbf{t} + (\mathbf{f}.\mathbf{n})\mathbf{n} \tag{3.1.10}$$

$$= q(\mathbf{t}.\mathbf{grad})q\mathbf{t} \qquad \text{by (3.1.1)}$$

$$= q^2(\mathbf{t}.\mathbf{grad})\mathbf{t} + q\mathbf{t}(\mathbf{t}.\mathbf{grad}\,q) \tag{3.1.11}$$

$$= \kappa q^2\mathbf{n} + a\mathbf{t} \tag{3.1.12}$$

because $\mathbf{t}.\mathbf{grad} = \partial/\partial s$, and $a = \mathbf{v}.\mathbf{grad}\,q$ is the tangential acceleration. Thus,

$$\mathbf{f}.\mathbf{t} = \mathbf{t}.\mathbf{grad}\,\tfrac{1}{2}q^2 = a, \tag{3.1.13}$$

$$\mathbf{f}.\mathbf{n} = \kappa q^2 = \text{the centrifugal force on unit mass,} \tag{3.1.14}$$

$$\mathbf{f.b} = 0 \text{ by definition of } \mathbf{b} \tag{3.1.15}$$

$$\mathbf{f} \times \mathbf{t} = (\mathbf{f.n})(\mathbf{n} \times \mathbf{t}) = \kappa q^2 \mathbf{b} \tag{3.1.16}$$

$$\begin{aligned} \boldsymbol{\omega}.\mathbf{f} &= \boldsymbol{\omega}.\mathbf{grad}\ \tfrac{1}{2}q^2 \quad \text{(by (3.1.9)} \\ &= \boldsymbol{\omega}.\mathbf{grad}\ (\tfrac{1}{2}\mathbf{v.v}) \qquad (3.1.17) \\ &= \mathbf{v}.(\boldsymbol{\omega}.\mathbf{grad})\mathbf{v}. \end{aligned}$$

The secondary vorticity is

$$\omega_S = \boldsymbol{\omega}.\mathbf{t} = \frac{1}{q}\ \boldsymbol{\omega}.\mathbf{v} \tag{3.1.18}$$

and represents twice the rotation around the direction **t** in unit time. In studying secondary vorticity a better representation is often the amount of rotation in unit distance, especially if we relate it to the geometry of a pipe shape or channel which causes curvature, or position in relation to a given obstacle. Thus the relevant quantity is ω_s/q.

3.2 THE SECONDARY VORTICITY EQUATION

Although it might seem to reduce the generality of the discussion, we shall consider only steady flow. It must be recognised that secondary flow loses its precision of definition in unsteady flow, and motion in three dimensions is complicated enough. We shall show that instabilities can be studied even in steady flow, because if small disturbances can be shown to grow in the downstream direction that is equivalent to instability following the fluid. The fact that the disturbance is not propagated upstream simply means that we have fixed our coordinates to the upstream end of the disturbance, and our main interest in instability is how a steady stream may become turbulent.

We take the vorticity equation in the form

$$(\mathbf{v.grad})\boldsymbol{\omega} = -\boldsymbol{\omega}\ \text{div}\ \mathbf{v} + (\boldsymbol{\omega}.\mathbf{grad})\mathbf{v} + \mathbf{R} \times (\mathbf{g} - \mathbf{f}) - \nu\nabla^2\boldsymbol{\omega}. \tag{3.2.1}$$

Then, using the appropriate results of the last section, we have

$$(\mathbf{v}.\mathbf{grad})\,\frac{\boldsymbol{\omega}.\mathbf{t}}{q} = \boldsymbol{\omega}.\left[\frac{1}{q}(\mathbf{v}.\mathbf{grad})\mathbf{t} + \mathbf{t}\left(\mathbf{v}.\mathbf{grad}\,\frac{1}{q}\right)\right] + \frac{\mathbf{t}}{q}\,.\,(\mathbf{v}.\mathbf{grad})\boldsymbol{\omega}$$

$$= \boldsymbol{\omega}.\left[\kappa\mathbf{n} - \frac{1}{q^3}\,\mathbf{t}(\mathbf{v}.\mathbf{grad}\,\tfrac{1}{2}q^2)\right] + \frac{\mathbf{t}}{q}.[-\boldsymbol{\omega}\,\mathrm{div}\,\mathbf{v} + (\boldsymbol{\omega}.\mathbf{grad})\mathbf{v} + \mathbf{R}\times(\mathbf{g}-\mathbf{f}) + \nu\nabla^2\boldsymbol{\omega}]$$

$$= \boldsymbol{\omega}.\left[\kappa\mathbf{n} - \frac{1}{q^2}\,\mathbf{t}(\mathbf{f}.\mathbf{t})\right] - \frac{\boldsymbol{\omega}.\mathbf{t}}{q}\,\mathrm{div}\,\mathbf{v} + \frac{1}{q^2}\,\boldsymbol{\omega}.\mathbf{f} + \frac{\mathbf{t}}{q}\,.\,[\mathbf{R}\times(\mathbf{g}-\mathbf{f}) + \nu\nabla^2\boldsymbol{\omega}]$$

(3,1.10)

$$= \boldsymbol{\omega}.\left[\kappa\mathbf{n} + \frac{1}{q^2}\mathbf{n}(\mathbf{f}.\mathbf{n})\right] - \frac{\boldsymbol{\omega}.\mathbf{t}}{q}\,\mathrm{div}\,\mathbf{v} - \frac{1}{q}\,\mathbf{R}.\mathbf{f}\times\mathbf{t} + \frac{\mathbf{t}}{q}\,\mathbf{R}\times\mathbf{g}.\mathbf{t} + \nu\,\frac{\mathbf{t}}{q}\,.\,\nabla^2\boldsymbol{\omega}$$

(3.1.14)

$$= 2\kappa\boldsymbol{\omega}.\mathbf{n} + \kappa q\mathbf{R}.\mathbf{b} + \frac{1}{q}\,\mathbf{R}\times\mathbf{g}.\mathbf{t} - \frac{\boldsymbol{\omega}.\mathbf{t}}{q}\,\mathrm{div}\,\mathbf{v} + \nu\frac{1}{q}\,.\nabla^2\boldsymbol{\omega}\,. \tag{3.2.2}$$

The term in this equation in div **v** on the right has the same physical meaning as in the original vorticity equation and is of very small importance. Likewise the viscous term has no new significance. We shall give no further attention to these two terms but will concentrate on the effects of curvature and gravity. For these purposes the equation is strictly for an **incompressible inviscid fluid**, and takes the form

$$\mathbf{v}.\mathbf{grad}\,\frac{\omega_s}{q} = 2\kappa\boldsymbol{\omega}.\mathbf{n} + \kappa q\mathbf{R}.\mathbf{b} + \frac{1}{q}\,\mathbf{R}\times\mathbf{g}.\mathbf{t}\,. \tag{3.2.3}$$

Since $\quad \boldsymbol{\omega}.\mathbf{n} = \boldsymbol{\omega}.\mathbf{b}\times\mathbf{t} = \mathbf{t}\times\boldsymbol{\omega}.\mathbf{b} = \dfrac{1}{q}(\mathbf{v}\times\boldsymbol{\omega}).\mathbf{b}$

$$= \frac{1}{q}\,(\mathbf{grad}\,\tfrac{1}{2}q^2 - \mathbf{f}).\mathbf{b} = \frac{1}{q}\,\mathbf{grad}\,\tfrac{1}{2}q^2.\mathbf{b} \tag{3.2.4}$$

we have that

$$2\kappa\boldsymbol{\omega}.\mathbf{n} + \kappa q\mathbf{R}.\mathbf{b} = \kappa q\left[\frac{2}{q^2\rho}\,\rho\,\mathbf{grad}\,\tfrac{1}{2}q^2 + \frac{2}{q^2\rho}\,\tfrac{1}{2}q^2\ \mathbf{grad}\ \rho\right].\mathbf{b}$$

$$= \kappa q\ \mathbf{grad}\ \ln(\tfrac{1}{2}\rho q^2).\mathbf{b}\ .$$

If for the moment we ignore the gravity term, as would be justified when $\mathbf{f} \gg \mathbf{g}$

which for example could be the case in a curved pipe, then (3.2.2) becomes

$$\mathbf{v}.\mathbf{grad}\ \frac{\omega_s}{q} = \kappa q\ \mathbf{grad}\ L.\mathbf{b} \tag{3.2.5}$$

where

$$L = \ln(\tfrac{1}{2}\rho q^2) \tag{3.2.6}$$

In the case where there are no density gradients we can use (3.2.4) to represent $\boldsymbol{\omega}.\mathbf{n}$ in (3.2.3), but in general we can see that the gradients of ρ and gradients of q have a similar effect in producing secondary vorticity when there is curvature. If the flow is horizontally stratified the term in **g** vanishes initially (upstream) and so the application of (3.2.5) is not restricted to the case in which $\mathbf{f} \ll \mathbf{g}$. In thinking of the forces acting on the fluid we see that the inertial effects are represented by **grad** L.

Returning to (3.2.3) we notice that, in the derivation, the term $2\kappa\boldsymbol{\omega}.\mathbf{n}$ originated from two different parts of the vorticity equation. First, it comes from **(v.grad)t**. This means that as the direction of **t** changes along the path of the particle, vorticity which was directed along the principal normal now finds itself having a component along the tangent, so creating secondary vorticity.

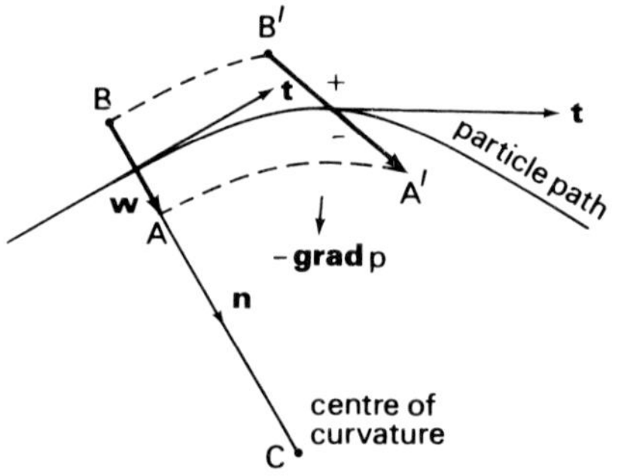

Fig. 3.2.i Streamwise vorticity is created out of vortex lines across the stream by the rotation of the direction of the stream and by the consequent rotation of the vortex lines in the opposite direction

The second half comes from the term $(\boldsymbol{\omega}.\mathbf{grad})\mathbf{v}$, which refers to the advection of the vortex lines with the fluid. On the inside of the bend the velocity is greater than on the outside, so that a line of particles AB which was initially perpendicular to **t** is rotated to A′B′ so as no longer to be normal to the streamlines. Thus $\boldsymbol{\omega}$ acquires a component along **t** if it had a component along **n** initially.

The acceleration of the flow on the inside of a bend is a property of irrotational flow. It can be thought of as arising because, in order to pass round the bend, the fluid must be subjected to a pressure gradient force towards the centre of curvature. The particles on the inside therefore move towards lower pressure and are accelerated, while those on the outside are decelerated. Alternatively we may consider the case of irrotational motion: since there is no vorticity in the plane of the bend the rotation of two lines of particles at right angles to one another must be equal and opposite. Thus the tangent, represented by a line of particles along the streamline, rotates in the opposite direction to a line along the normal. Consequently, a component of vorticity along the normal, which introduces no vorticity into the plane of the motion (which may therefore be considered irrotational), is rotated towards the tangent and the angle between them closed up at twice the rotation of either. From this we see why the two contributions are equal.

The effect of the secondary vorticitiy is seen by considering the case of horizontal flow passing round a bend in a horizontal plane. The vector **g** is downwards, in the direction of **b**. This is, to a first approximation, in a fixed direction, although the swirl introduced by the secondary vorticity may cause the streamlines not to be plane. In Fig. 3.2.ii we have illustrated the case of flow entering the bend with velocity increasing upwards, and having vorticity in the direction of $-\mathbf{n}$. The Bernoulli surfaces containing the streamlines and vortex lines are initially horizontal planes parallel to the plane of the motion. The **Bernoulli vector**, **B**, defined as the normal to the Bernoulli surface at a point, is given by

$$\mathbf{B} = \mathbf{t} \times \boldsymbol{\omega}. \tag{3.2.7}$$

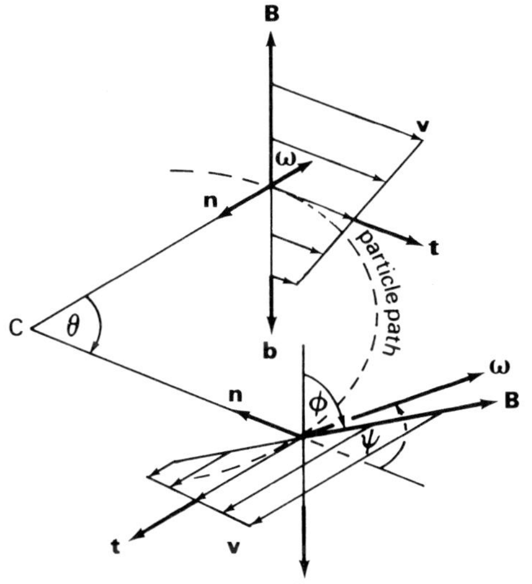

Fig. 3.2.ii The Bernoulli vector which moves with the fluid, is rotated around the streamline through angle ϕ when the flow turns through angle θ, because of the streamwise component of ω which is produced by the curvature

The effect of the bend is to rotate the vector **B** around **t** through an angle ϕ while the vorticity vector $\boldsymbol{\omega}$ is rotated in the plane of the motion through an angle ψ relative to **t** and **n**. θ is the angle of the bend. The fast moving fluid is displaced towards the outside of the bend.

The rotation of the vector **B** will be the consequence of the creation of secondary vorticity when the flow is confined in a circular pipe, as in Hawthorne's experiments to be described shortly. But in a wide shallow river the flow near the middle will be almost horizontal. This results in the displacement of the faster moving surface fluid towards the outside of the bend only producing vertical motion near the sides of the river, so that in the middle **B** will remain vertical.

The meaning of the term $\kappa q\mathbf{R}.\mathbf{b}$ is now evident. Its effect is to bring the densest fluid to the outside of the bend: it represents the effect of density gradients in L in (3.2.5), and a density gradient in the direction of **b** produces secondary flow in the same sense as a vorticity vector in the direction of **n** (which is a gradient of speed in the direction of **b**).

The effect of gravity is to modify this effect by tending to move the densest fluid to the bottom of the stream. In this case it is represented by $\frac{1}{q}\mathbf{R}\times\mathbf{t}.g\mathbf{b}$ because g is parallel to **b**: this is equal to $-\frac{g}{q}\mathbf{R}.\mathbf{n}$, which tends to rotate the isopycnic surfaces until they contain **n**, which is horizontal in this case.

3.3 SQUIRE'S FORMULA

If the fluid is inviscid and of uniform density and in steady motion the secondary vorticity equation is simply

$$\frac{\mathrm{D}}{\mathrm{D}t}\frac{\boldsymbol{\omega}.\mathbf{t}}{q} = 2\kappa\boldsymbol{\omega}.\mathbf{n} \tag{3.3.1}$$

or

$$q\frac{\partial}{\partial s}\frac{\omega_s}{q} = 2\kappa\omega_n \tag{3.3.2}$$

where suffixes s and n refer to components along the streamline and normal. For a small change in stream direction $\mathrm{d}\theta$, since $\mathrm{d}\theta = \kappa \mathrm{d}s$,

$$\mathrm{d}\frac{\omega_s}{q} = 2\frac{\omega_n}{q}\kappa \mathrm{d}s = 2\frac{\omega_n}{q}\mathrm{d}\theta$$

Approximately, and perhaps exactly, the speed may remain constant as the streamline turns through $d\theta$, and Squire's version of the formula is (Squire and Winter, 1951)

$$d\omega_s = 2\omega_n d\theta \tag{3.3.3}$$

which expresses the way in which vorticity normal to the flow becomes streamwise, the factor 2 being of particular significance as explained in the previous section.

The particular application with which Squire was concerned was the passage of boundary layer air through a cascade of aerofoils in the corner of a closed circuit wind tunnel. The flow is made to turn through 90° in a horizontal plane by passing between the aerofoils shown in horizontal section in Fig. 3.3.i. In this way the flow is guided round the corner as a set of separate flat layers of air and the consequences of the large velocity gradients avoided because the

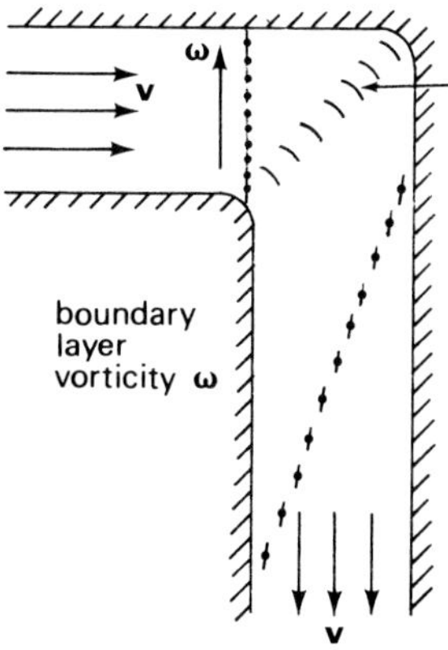

Fig. 3.3.i In a wind tunnel the boundary layer vorticity is across the stream. In traversing a bend through a cascade of aerofoils a line of particles becomes broken into several sections forming a diagonal line, and in each section of the cascade secondary flow occurs. But this produces a small displacement compared with that generated in the longer bend that would be necessary without the cascade.

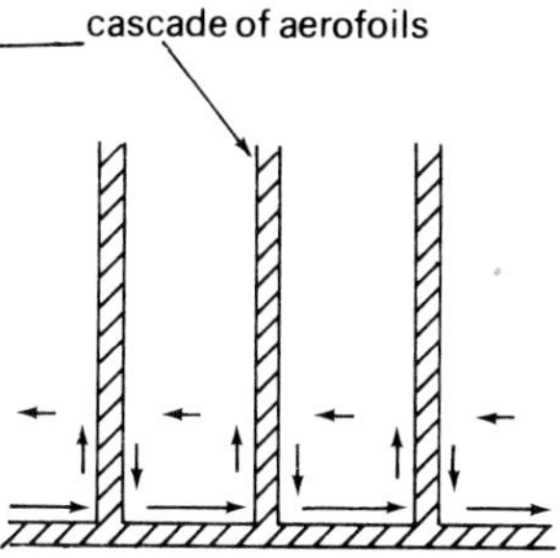

Fig. 3.3.ii The flow between the cascades is mainly observable as boundary layer flow towards the inside of the bend in each section of the cascade. The diagram shows the secondary motion near the bottom looking downstream.

bend for each particle is short. A marked line of particles before the bend becomes broken into a series of pieces each of which is rotated, and emerges as a nearly continuous line with only very small scale turbulence, of the order of the thickness of the aerofoils.

At the top and the bottom of the tunnel, however, there is a boundary layer with vorticity directed across the stream (Fig. 3.3.ii). In each bend this causes secondary flow to be generated with a component of velocity across the floor and ceiling towards the inside of the bend, with a small outward flow in the body of the tunnel to compensate. Squire's formula shows that the amount of secondary flow emerging from the cascade is independent of the dimensions of the cascade and dependent mainly on the angle turned through, but its duration is less in a short bend and the up- and down-motions on the aerofoils quickly cancel one another after the bend.

3.4 FLOW IN RIVER BENDS

The water close to the bed of a river is retarded by the frictional drag of the bed and by the drag due to the wakes of obstacles and protrusions on the bed. There is also drag due to solid particles carried with the stream which are intermittently slowed down or brought to rest on the bed. Since rivers are usually much wider than they are deep we can think of them as having slow moving fluid at the bottom with the main body of the water above moving faster (Fig. 3.4.i).

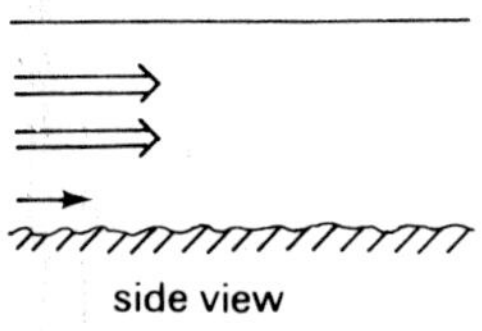

Fig. 3.4.i Velocity profile in a river

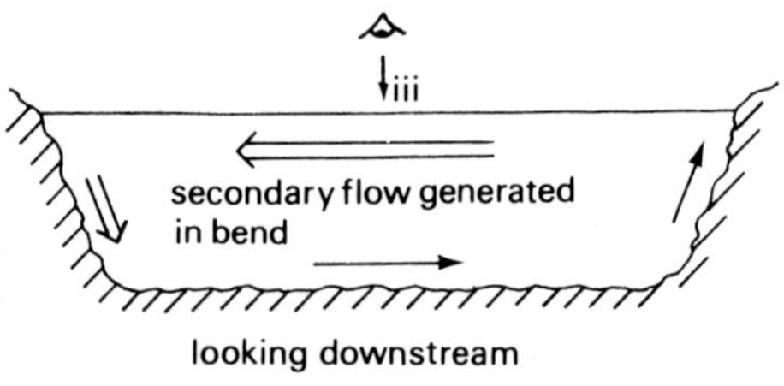

Fig. 3.4.ii Secondary motion generated in a bend to the right

After the bend the vorticity vector which was initially directed across the stream will lie at an angle to it and a component ω_s will be generated. The slow moving bottom water will be turned through a much greater angle than the faster moving surface water. We may think of them both being subjected to the same pressure gradient towards the centre of curvature which is determined by the slope of the free surface with the vertical acceleration being negligible, so that the hydrostatic equation applies throughout the depth.

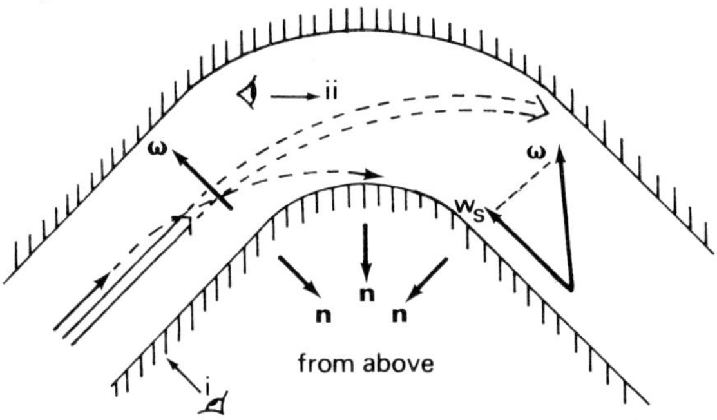

Fig. 3.4.iii Paths of particles in the mainstream and the boundary layer of a river in a right hand bend. An upstream component ω_S is generated.

The effect is that fast moving water is moved to the outside bank (Figs. 3.4.ii and iii). This causes people to observe that the theoretical idea that the flow is decelerated on the outside and accelerated on the inside is not realised, but the flow is three-dimensional and not plane. The outer bank is therefore eroded, while slow moving fluid from which silt is readily deposited rises on the inside. The channel is therefore moved bodily outwards on a bend, which causes meandering to be increased with time (Fig. 3.4.iv).

We have argued on the assumption that all the vorticity we are concerned with is in the fluid as it enters the region of consideration, and that no more is produced by the walls in that region. In practice new vorticity is continuously being produced. If this did not happen the secondary flow in a second bend in the opposite direction to the first would be very much influenced by (a) how far the stream had been rotated, so that the Bernoulli surfaces had taken up new positions, (b) by the magnitude of the secondary vorticity and the angular momentum associated with it on leaving the first bend. In rivers the vorticity is generated fairly rapidly in the bottom fluid, so that in practice the flow in a second bend may be very little influenced by what happened in the first.

In other situations, as soon as the velocity field has been significantly altered by the secondary flow, the kind of first approximation we have been discussing is no longer applicable except in a qualitative way. In many cases the geometry of the situation is so complicated that there is no question of a detailed calculation anyway. The growth of meanders* often leads to the formation of the ox-bow lakes of geography book fame.

*Named after the river Menderes in Caria, Asia Minor; named *Meander* in Latin, *Maiandros* in Greek, and described by Xenophon ca 375 BC.

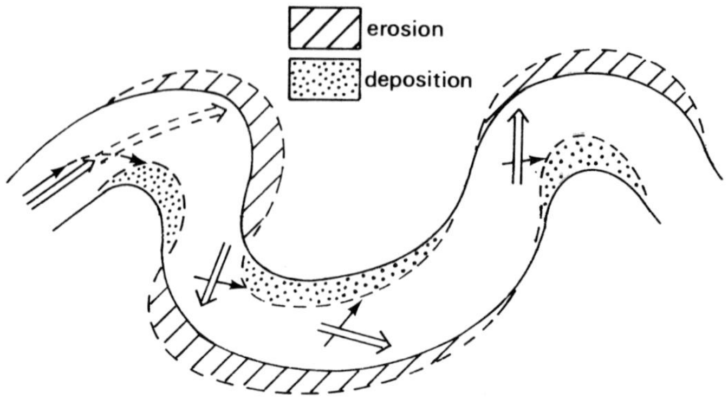

Fig. 3.4.iv Because of the erosion of the outer bank of a bend by the faster moving component of the stream, and the deposition of silt on the inside where the slow moving bottom water rises, the meandering of a river is increased with time

One of the most remarkable cases of secondary flow occurs where a side channel draws water from a straight-flowing river. The flow on the bed can be thought of as determined by the drag of the flow above together with the acceleration due to the pressure field imposed by the shape of the free surface. This pressure field is determined by the flow of the bulk of the fluid above the boundary layer on the bed, and this is indicated in the diagram for a case in which a side channel draws off one third of the flow of the main stream. The

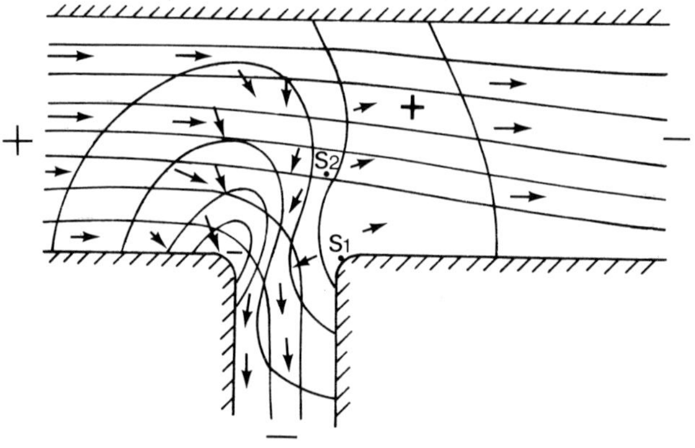

Fig. 3.4.v The contours indicate the topography (+ and −) of the free surface and the streamlines of the main flow where a side channel draws off about a third of the flow of a straight stream. The arrows indicate the bottom flow and S_1 and S_2 the stagnation points in it.

pressure field then consists of a small pressure gradient in the downstream direction, which balances the drag, together with larger gradients which produce the decelerations in the main stream and the curved flow into the side channel.

The main flow is of almost constant depth and is approximately non-divergent, and can therefore be represented by streamlines whose spacing is inversely proportional to the speed (Fig. 3.4.v). The flow at the bottom, indicated by short arrows in the diagram, is not non-divergent, and the stagnation points in particular are places of downflow. Their precise position is very much dependent upon the details of the case. Low and high pressure is indicated by negative and positive signs.

The most important features are the rising pressure into which the main-stream flow is decelerated and the low pressure on the inside of the curved flow into the side stream. The consequence is that at least two thirds of the bottom flow goes into the side channel. Configurations can be designed to make all the bottom flow go that way. A simple explanation often given for the collapse of Mesopotamian civilisation is that those in charge of the irrigation on which the agriculture and the civilisation depended, did not understand secondary flow. Consequently the channels were continuously blocked with silt because they insisted on drawing them off at right angles from the main stream. Every effort to dredge them led to the raising of the average level of the irrigated areas so that eventually they could not be irrigated by gravity, and pumping on the scale required became impossible.

There is some residual secondary flow in the side channel and so the maximum accumulation is on the upstream side of it. There is one stagnation point, S_1, beneath the pressure maximum where the main flow stagnates on the downstream corner of the sidestream entrance. There is another, S_2, at the point where the pressure gradient force and the force due to the drag of the upper layers are equal and opposite. At this point the main flow line crosses the isobar at right angles.

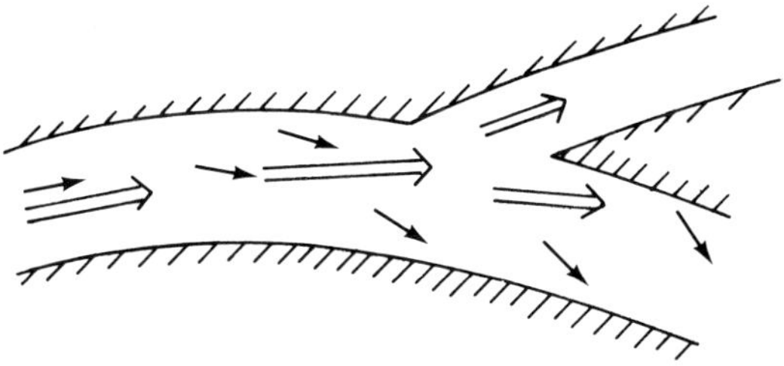

Fig. 3.4.vi By drawing water from the outside of a bend in the main channel almost no bottom flow and silt enter the side channel

To avoid drawing silt into a channel it should be taken from the outside of a bend in the main stream (Fig. 3.4.vi) where the bottom flow is away from the outside bank. Or the amount of silt taken from the main channel can be reduced to less than its proportionate amount by building a wall which causes the side channel flow to be decelerated as it enters, before the flow is curved into the side channel direction. In this way the bottom flow in the main stream is directed away from the side stream entrance to a slight extent (Fig. 3.4.vii).

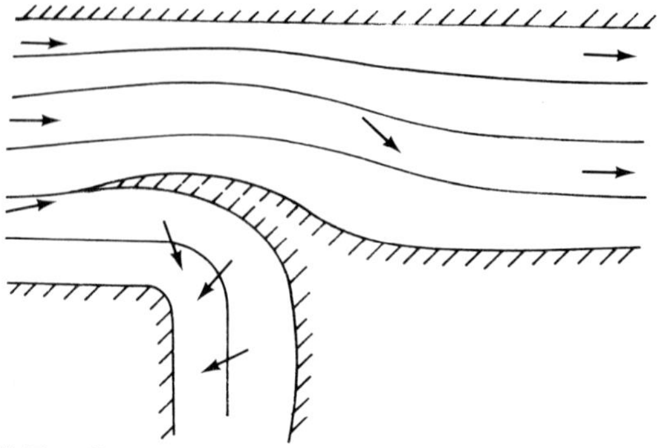

Fig. 3.4.vii The silt carried into a side channel can be reduced if the flow into it is decelerated before entering the bend

We now turn to discuss the residual secondary flow remaining downstream of a bend in the flow.

3.5 OSCILLATIONS IN A PIPE HEAD

In this discussion we ignore the effect of density gradients except to remark upon their effect in a qualitiative way at the end. Gravity may therefore be ignored, and for convenience we speak of the motion taking place in a horizontal bend (Fig. 3.5.i) of a pipe of circular cross section. The basic equation (3.2.5) may then be written

$$q \frac{\partial}{\partial s} \frac{\omega_s}{q} = \kappa q \, \mathbf{grad} \, (\ln \tfrac{1}{2} \rho q^2).\mathbf{b} \qquad (3.5.1)$$

In the notation of 3.2 illustrated in Fig. 3.5.ii, ϕ is the angle through which the Bernoulli surfaces have been rotated around the streamline. If, for the purposes of discussing the main mechanism under study, we contrive a case in which the Bernoulli surfaces have the same rate of secondary rotation with distance downstream across the whole section of the pipe,

we may write
$$\omega_s = 2 \frac{\mathrm{D}\phi}{\mathrm{D}t} = 2q \frac{\partial \phi}{\partial s} . \qquad (3.5.2)$$

If we also assume that all the flow has the same curvature as the centre line of the pipe (which is equivalent to assuming both that the diameter of the pipe is small compared with the radius of curvature of the bend and that change in curvature due to the helical motion of the particles is small compared with the curvature) then $\mathbf{b}$ may be regarded as a constant vertical vector and κ a constant scalar. The significance of this assumption is illustrated in Fig. 3.5.iii. If $\phi = 0$ when the Bernoulli surfaces contain vertical generators with the fast moving fluid on the outside of the bend, (3.5.1) becomes

$$2q\,\frac{\partial^2 \phi}{\partial s^2} \;=\; \kappa q\;\mathrm{grad}(\ln \tfrac{1}{2}\,\rho q^2)\,\sin\phi, \tag{3.5.3}$$

and if θ is the angle turned through by the central (undisturbed) streamline

$$\kappa\,\mathrm{d}s \;=\; \mathrm{d}\theta$$

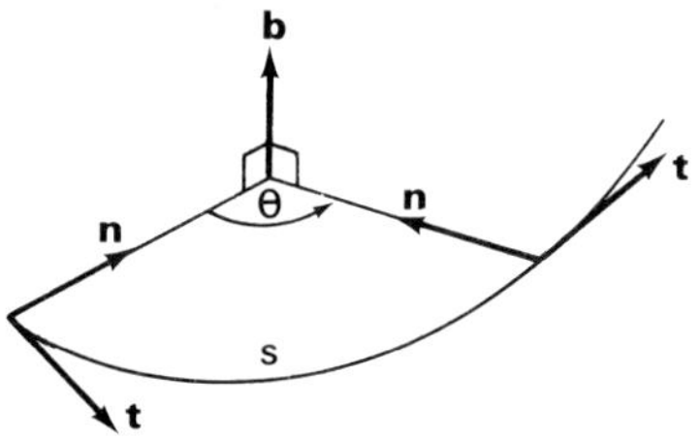

Fig. 3.5.i Coordinates for the motion around a horizontal bend with **b** vertical

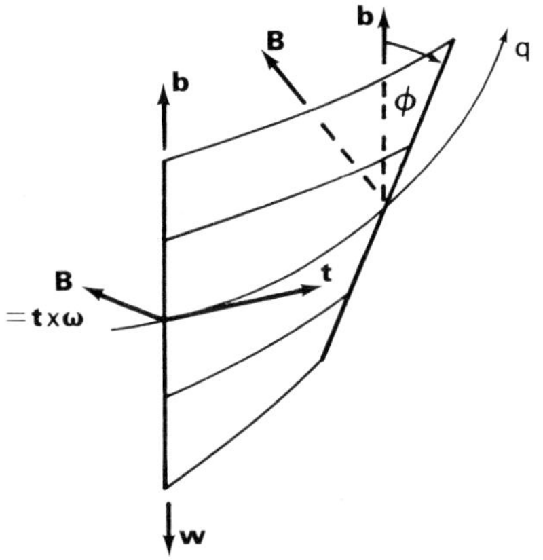

Fig. 3.5.ii The rotation of the Bernoulli surface, to which the vector **B** is perpendicular, through angle ϕ as it transverses a bend. $\phi = 0$ in the equilibrium position in which **B** is parallel to **n**.

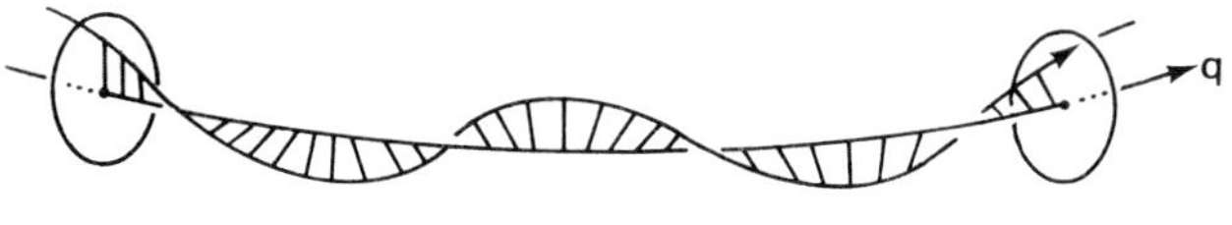

Fig. 3.5.iii When there is secondary motion the paths of the particles are helical. The assumptions made in the simple theory assume that the helix is of small radius and has long pitch compared with the radius of the bend. The effects of curvature are calculated as if the particles passed along the axis of the pipe

and so
$$\frac{\partial^2 \phi}{\partial \theta^2} = \frac{1}{2\kappa} \operatorname{grad}(\ln \tfrac{1}{2} \rho q^2) \sin \phi \tag{3.5.4}$$

where the gradient is measured across the Bernoulli surfaces in the negative direction of **B**, which is parallel to **n** when $\phi = 0$.

This equation represents a system which executes stable oscillations when the gradient of q is negative, and the period of oscillation in θ for small values of ϕ, is (by analogy with the pendulum equation)

$$2\pi(-2\kappa/\operatorname{grad} \ln \tfrac{1}{2} \rho q^2)^{1/2} \tag{3.5.5}$$

The unstable situation in which q increases inwards ($\phi > 90°$) is discussed in Section 3.8.

In a real case it is unlikely that the period indicated by (3.5.5) will be the same for all streamlines. For it to be the same, if y is the coordinate in the direction of **B**, it would be required that

$$\frac{\partial}{\partial y} \quad \frac{1}{\kappa}\frac{\partial}{\partial y} \ln \tfrac{1}{2} \rho q^2 \quad = \text{const (indep. of } y) \tag{3.5.6}$$

or, ignoring the variations in κ,

$$\rho q^2 \propto e^{-\alpha y} \tag{3.5.7}$$

where α is a constant and the zero of y may be put at any convenient distance. With any other form of profile a more complicated motion results.

Hawthorne (1951) demonstrated the reality of these secondary oscillations by blowing air through a pipe of circular cross section of radius a. The stream was given a velocity gradient in the vertical direction and passed horizontally round a bend. A straight section ended at station 0, where the position of the Bernoulli surfaces was identitifed in the cross section by measuring the contours of constant stagnation pressure p_0 where

$$p_0 = \rho g H = \tfrac{1}{2} \rho q^2 + p + \rho g z ,$$

or, since in this case the effect of gravity can be ignored and the hydrostatic equation subtracted out, we may write, with $\rho g z$ added to p (or subtracted from p_0)

$$p_0 = p + \tfrac{1}{2} \rho q^2 . \tag{3.5.8}$$

These contours were measured across the open end of the pipe by a pitot tube, consequently there might be some very slight errors due to the influence of the pressure field in the diverging and turbulent flow outside. A section of curved pipe was then attached, and the position of the Bernoulli surfaces measured in the same way at the new open end. In order to determine the magnitude of the secondary vorticity at this position, which we call station 1, a section of straight tube of length b was attached and the position of the Bernoulli surfaces measured at its downstream open end, station 1b.

The expected behaviour is indicated in diagram 3.5.iv, where the Bernoulli vector **B**, perpendicular to the surfaces, is used to denote their orientation. The line AB is the intersection of the central Bernoulli surface with plane of the cross section, and is normal to **B**.

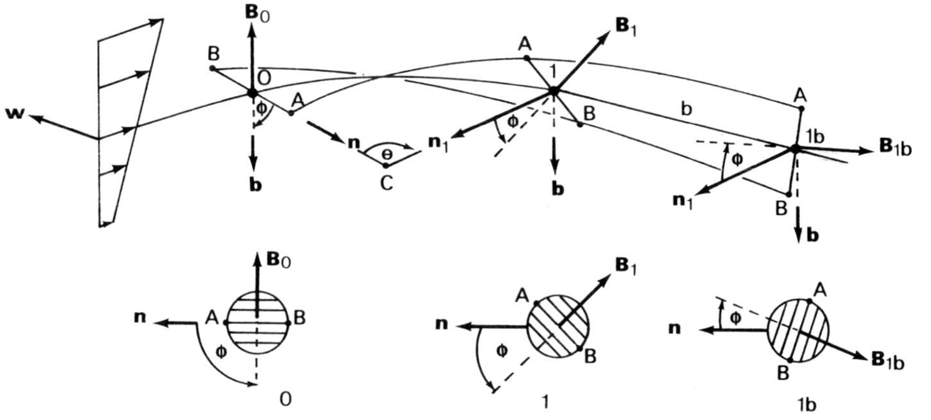

Fig. 3.5.iv On the left is shown the velocity profile and vorticity vector in the stream approaching station O, which is at the beginning of the pipe bend. At station O the various vectors and the position of the line of particles AB which lie on a Bernoulli surface are shown at the beginning of the bend and at station 1 at the end of the bend. Finally at station 1b they are shown at the end of a straight section of pipe of length b. Below the diagrams show the positions of the surfaces in the pipe cross section as seen looking upstream.

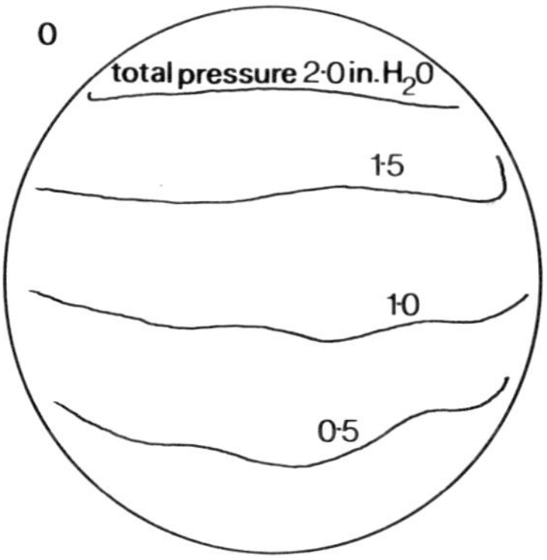

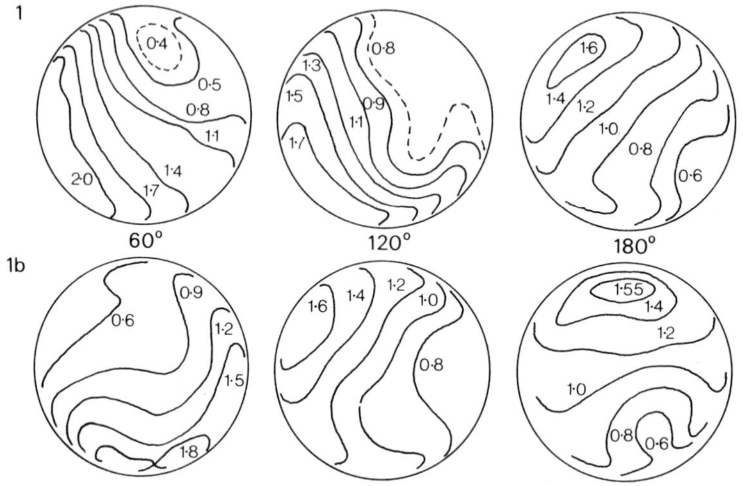

Fig. 3.5.v The disposition of the Bernoulli surfaces in the pipe cross section on page 95 at station O, and on this page at stations 1 and 1b after bends of 60°, 120° and 180°, as measured in Hawthorne's experiments.

ϕ is the angle between the plane of the bend given by **n** and the negative direction of **B**. The change in ϕ in the length b represents the secondary motion persisting as it existed at station 1. The cross sections are viewed facing upstream.

The straight section b was then removed and another section of the bend added, to increase θ, and the same procedure repeated. The diagrams in Fig. 3.5.v show the results obtained by Hawthorne. The pipe had an internal diameter of 6 in and the radius of the bend was 30 in. Measurements were taken at intervals of 30° of bend and the linear extension tube had length b equal to 25 in. The upper diagrams show the configuration of the stagnation pressure at each value of θ, and the lower ones the configuration at the end of the straight extension. The dynamic pressures are measured in inches of water, and the maximum shows a slow decline due to friction as the length of pipe is increased, while the minimum is increased.

In Hawthorne's experiment the velocity profile was linear, with zero velocity at one extreme. Thus the condition (3.5.7) was not satisfied and the different Bernoulli surfaces rotated through different angles. Nevertheless the results expected from the simple theory are clearly apparent and may be summarised in Fig. 3.5.vi which indicates the angle through which the Bernoulli's vector rotated. The oscillations appear as considerably damped, for when $\theta = 180°$, = one period in this experiment roughly, the secondary motion is very small but $\mathbf{B}_1$ has not returned to the initial position.

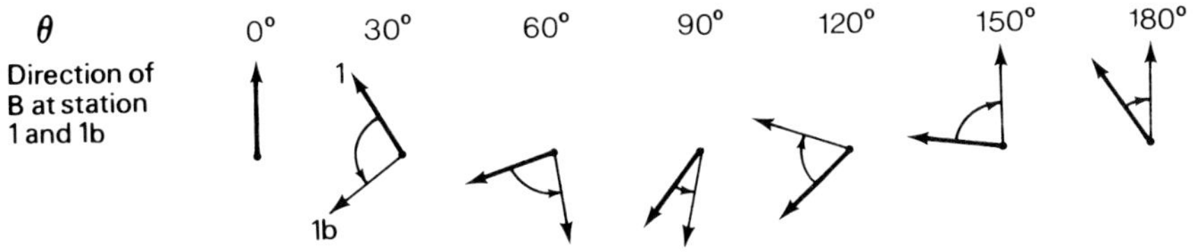

Fig. 3.5.vi Summary of Hawthorne's results in the bend of a circular pipe. θ is the angle of the bend. The direction of **B** at station 1 shows the magnitude of the rotation achieved at station 1 and the magnitude of the secondary rotation is indicated by the difference between $\mathbf{B}_1$ and $\mathbf{B}_{1b}$.

In another experiment Hawthorne used a pipe of internal diameter 6⅜ in with a radius of curvature of the central line of 9$\frac{9}{16}$ in. He turned the flow through 90° and then added increasing lengths of straight pipe to study the fate of the Bernoulli surfaces as friction and other influences began to operate. The ultimate state would, of course, be with the maximum velocity in the middle but his results, shown below in Fig. 3.5.vii, clearly indicate the tendency of the low stagnation pressure fluid to move towards the middle. This result will be discussed further in Section 3.8.

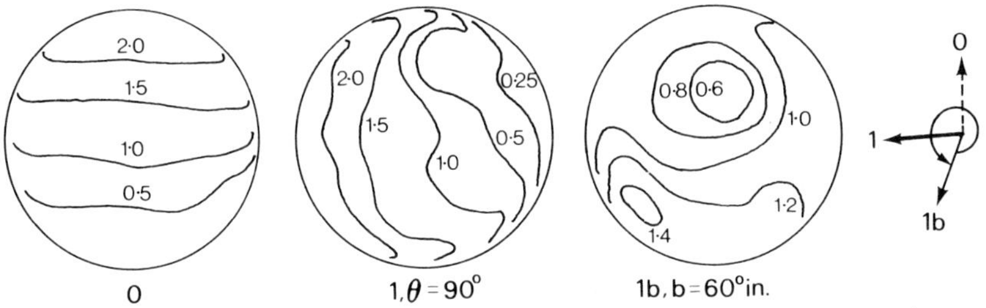

Fig. 3.5.vii The flow with Bernoulli surfaces depicted at station O was turned through 90° and the distribution was measured at increasing lengths of straight pipe. At $b = 60$in the tendency for the low stagnation fluid to occupy the centre of the channel is clear

The flow in a ventilating duct of rectangular cross section was also studied by Hawthorne, where the tendency of the high stagnation pressure fluid to move to the outside of the bend, with oscillations around that position, was clearly revealed.

Dean (1927) studied the case of a very long bend, as in a closely coiled pipe, where an equilibrium between the inertial and viscous forces is established. The high velocity fluid is produced by the pressure gradient in the middle of the pipe, and low velocity fluid at the walls by friction. The high speed fluid moves to the outside of the bend, and is decelerated by friction, while the low speed fluid moving inwards to take its place is accelerated. The velocity configurations arrived at are shown in Fig. 3.5.viii. These are taken from a paper by McConalogue and Srivastava (1968) who extended Dean's work. The contours show the downstream velocity and the streamlines of the secondary flow. In the case shown in the upper half diagram the flux down the pipe was about one quarter of that in the case shown in the lower half, and the secondary flow was about one seventh. Reference to the original paper should be made for details.

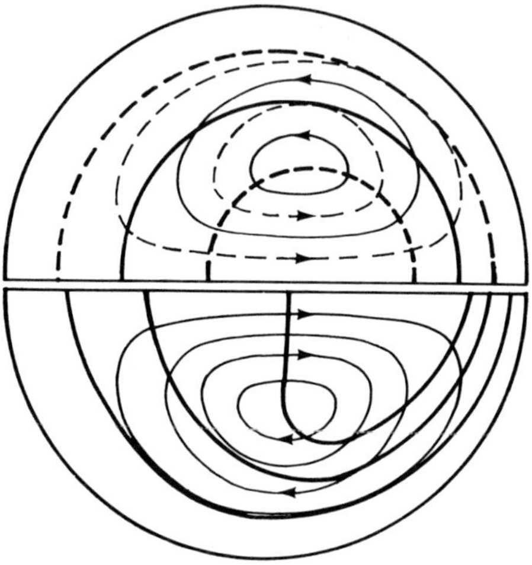

Fig. 3.5.viii Steady state flow of a viscous fluid in a long pipe bend shown in cross section. The upper and lower halves of this diagram represent different cases, the flux down the pipe being about four times greater in the lower half. The contours show the Bernoulli surfaces, and the streamlines the secondary flow, according to calculation by McConologue and Srivastava.

The pressure drop in a pipe bend is a combination of the loss due to friction and that due to the setting up of the secondary flow. If the flow emerges from the bend just at the extreme point of an oscillation when the secondary flow is zero, the loss will be less than if, for that bend, the secondary flow has large magnitude at that point. The loss in some pipe bends of 90° can be nearly doubled by the generation of secondary circulation. Indeed, it has long been known that this loss, expressed as a proportionate increase in the loss that would be experienced due to friction in a straight pipe of the same length as the bend, does not increase monotonically with the angle, θ and correspondingly with the length of the bend. Such a result is to be expected according to this simple theory, and the graph in Fig. 3.5.ix shows the kind of relationship to be expected with a given size of pipe and radius of bend. If a given θ is chosen and κa varied, a similar periodic variation would be found.

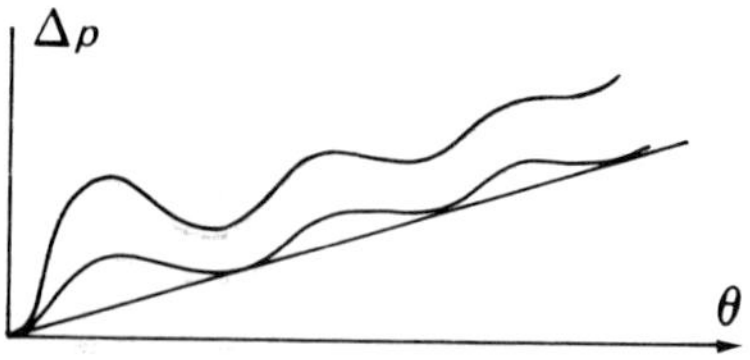

Fig. 3.5.ix The pressure drop to be expected in a pipe bend increases linearly with the length (angle) of the bend but in addition a further loss occurs because of the secondary flow set up. The middle curve shows how this would vary if the oscillations had the same period over the whole cross section, but in practice the loss is more like the upper curve.

The straight line indicates the pressure drop, Δp, in the same length of straight pipe, and the lower curve the minimum loss in a bend of angle θ. The upper curve shows the loss to be expected if all the Bernoulli surfaces did not rotate together, so that at no point in the bend is the secondary velocity zero everywhere across the section. This problem was first studied experimentally by Eustice (1925).

3.6 SECONDARY MOTION BEHIND OBSTACLES; HAWTHORNE'S HOT HUMP

The flow behind obstacles is usually turbulent, and so it is not realistic to pretend to much detail in a theory. The very simple theory which follows is intended really to give an idea of the secondary vorticity which has been created when the flow reaches the widest point of the obstacle. Beyond there, separation can be a dominant factor in determining the flow, and this varies very much with the obstacle shape.

To discuss general principles, we need to have a simple representation of an obstacle. This is done by placing a doublet in a stream, the strength being related to the size of the obstacle. A source may be added to represent the inflow in the wake of the obstacle towards it relative to the main stream. A more complicated obstacle could be represented by a system of doublets, or even a distribution of sources and sinks separated by a finite distance but producing closed circulations around which the main stream flows. Its wake could be represented by a system of sources rather than one alone, but that is not a profitable line of discussion unless we have a particular pressing problem on hand.

A uniform stream is represented by a stream function increasing linearly in the perpendicular direction. A simple two-dimensional obstacle in it could be represented by the classicial stream function

$$\psi = Uy + m\theta + \frac{\mu}{r}\sin\theta \tag{3.6.1}$$

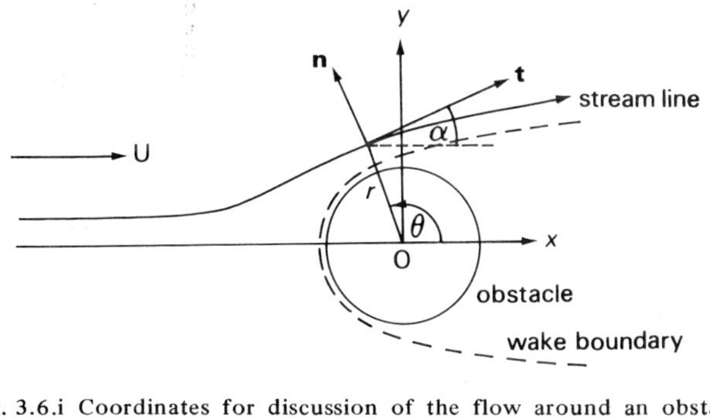

Fig. 3.6.i Coordinates for discussion of the flow around an obstacle with a wake

where (x,y) and (r,θ) are rectangular and polar coordinates in the plane normal to the obstacle, which in this case is a circular cylinder of radius $(\mu/U)^{1/2}$. The wake size is determined by the source strength m. The formula (3.6.1) enables us to calculate the position of any streamline, and in principle we could do the same with any more complicated formula representing a more complicated case. The situation is illustrated in Fig. 3.6.i and Fig. 2.6.ii(b).

We now imagine that the actual flow past the cylinder varies in the direction of the z-axis. Thus we might have a velocity stratification which would be accompanied by a corresponding variation in the values of m and μ so as to retain the same geometry of the flow, together with a density stratification, represented by

$$\frac{\partial U}{\partial z} = \lambda U, \quad \frac{\partial m}{\partial z} = \lambda m, \quad \frac{\partial \mu}{\partial z} \lambda \mu, \quad \frac{\partial \rho}{\partial z} = -\beta \rho \tag{3.6.2}$$

Clearly (3.6.1) would not actually represent the flow at every level but would be a fairly good first approximation.

The gradient of L in (3.2.6) is in the direction of z and may be calculated at any point of the flow. Since it is proportional to ρq^2 the gradient of its logarithm will be proportional to $\beta\lambda^2$. Equation (3.2.5) or (3.5.1), which neglects the effect of gravity, is now

$$q \frac{\partial}{\partial s} \frac{\omega_s}{q} = \kappa q \frac{\partial}{\partial z} \ln \tfrac{1}{2} \rho q^2 \tag{3.6.3}$$

which, when integrated with respect to s along the streamline is

$$\frac{\omega_s}{q} = \int_{x=-\infty} \frac{\partial \alpha}{\partial s} \frac{\partial}{\partial z} \ln \tfrac{1}{2} \rho q^2 \, ds$$

$$= \int_{x=-\infty} \frac{\partial}{\partial z} \ln \tfrac{1}{2} \rho q^2 \, d\alpha \tag{3.6.4}$$

This is an easy quantity to evaluate in principle, but very laborious in practice by numerical step-by-step integration. The formula looks simple enough but depends on following the variables along the particular streamline from far upstream in order to evaluate the secondary vorticity at a point in the field. However it is a relatively easy calculation to program for a computer, and the secondary vorticity is thereby obtained in one of the planes of the flow, say the plane normal to the stream through the obstacle at its widest point.

The configuration of the Bernoulli surfaces is not readily obtained from the distribution of ω_s because, except when it represents a solid rotation of a section of the flow, it cannot be assumed to be precisely equal to $q\partial\phi/\partial s$, as in the last section. To map out these surfaces it would be necessary to obtain the field of secondary vorticity in each plane of a series of steps in the x-direction, to calculate the value of div **v** in each plane, and then by integration obtain the velocity distribution in each plane. From this the increment of displacement of the Bernoulli surfaces could be calculated. The calculation in this exercise would be an order of magnitude greater than needed to obtain the first approximation to the value of ω_s alone, and if it were undertaken it would be just as well to proceed from knowledge of the displaced positions to calculate a second approximation to the value of **grad** (ln ½ ρq^2) before proceeding to the next step. However, such an exercise is scarcely justified in view of the other errors inherent in the crudity of the original approximation used, (3.6.1), to obtain the streamlines.

The problem requires specification of the three non-dimensional numbers

$$\frac{U}{\lambda \mathrm{m}}, \quad \frac{U}{\lambda^2 \mu}, \quad \frac{\lambda}{\beta} \tag{3.6.5}$$

so that there is a three-dimensional family of cases even for the simple case of a cylinder in a flow with exponential functions of z for velocity and density in the undisturbed stream. In any real case there must be ends to the cylinder and the functions are not exponential, and so only very broad results are likely to emerge from any study if they are to be simple enough to comprehend, unless the calculation is undertaken because of a special interest in some definite configuration.

An axisymmetric obstacle can be represented using Stokes's stream function, thus

$$\psi = \tfrac{1}{2} U r^2 \sin\theta + \mathrm{m}\cos\theta + \frac{\mu}{r^2}\sin^2\theta \tag{3.6.6}$$

This case is more complex, and a decision has to be made as to the definition of the streamlines to be used in the calculation of (3.6.4). Hawthorne and Martin (1955) performed this calculation for the flow over a hemisphere assuming the flow to be approximately the same as the potential flow round a sphere, each Bernoulli surface being horizontal far upstream. They also included the effect of gravity and retained the term $\frac{1}{q}\mathbf{R}\times\mathbf{g.t}$ from (3.2.3) but, because variation in ρ was negligible in comparison with variations in q^2 and $g >> f$, they wrote ½q^2 for ½ρq^2 in L. They expressed their calculations in the form of maps of the magnitude of ω_s/q far downstream. However, as we have already pointed out it is more realistic to express them as values in the plane where the obstacle is widest. Indeed they carried out an experiment in which the rotation was made visible by trails of smoke issuing from holes on the hemisphere. A faring was placed in the wake of the hemisphere to reduce separation and turbulence behind it so that the secondary vorticity would be visible, and the trails were seen to be helical, showing a very complicated overturning of the Bernoulli surfaces.

The series of experiments described by Hawthorne and Martin show the flow over Hawthorne's Hot Hump. This consisted of a hemisphere placed on a boundary. The viscous drag of the boundary upstream produced a velocity gradient and the surface was heated (or cooled) to produce a density gradient by conduction. When the hot boundary was at the bottom of the fluid the density gradient was unstable, but it was stable when the hemisphere was placed on the roof of the wind tunnel. The sense of the secondary vorticity was observed by the helical motion of smoke trails downstream of the median plane.

With forward shear and a cold surface on the bottom of the tunnel the two effects are as shown in Fig. 3.6.ii.

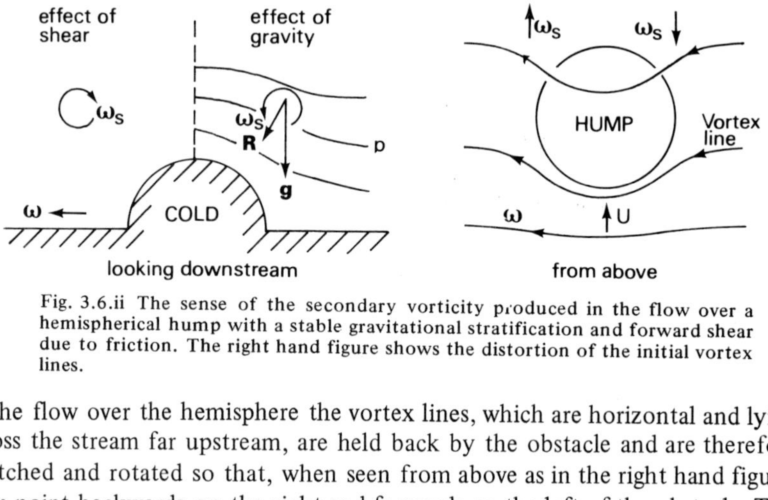

Fig. 3.6.ii The sense of the secondary vorticity produced in the flow over a hemispherical hump with a stable gravitational stratification and forward shear due to friction. The right hand figure shows the distortion of the initial vortex lines.

In the flow over the hemisphere the vortex lines, which are horizontal and lying across the stream far upstream, are held back by the obstacle and are therefore stretched and rotated so that, when seen from above as in the right hand figure, they point backwards on the right and forwards on the left of the obstacle. This introduces a secondary component which produces downward motion behind the obstacle, as indicated in the left half of the left diagram (Fig. 3.6.ii). The effect of gravity on the density gradient is to produce a motion tending to restore the isopycnic surfaces to the horizontal because they are humped up over the obstacle. This brings **R** back into coincidence with **g**.

Over a hot hump the secondary vorticity generated by gravity is in the opposite direction. Close to the floor where the vertical displacement is negligible the effect is very small, whereas the effect of the holding back of the vortex lines is a maximum. Consequently there is on each side a surface below which

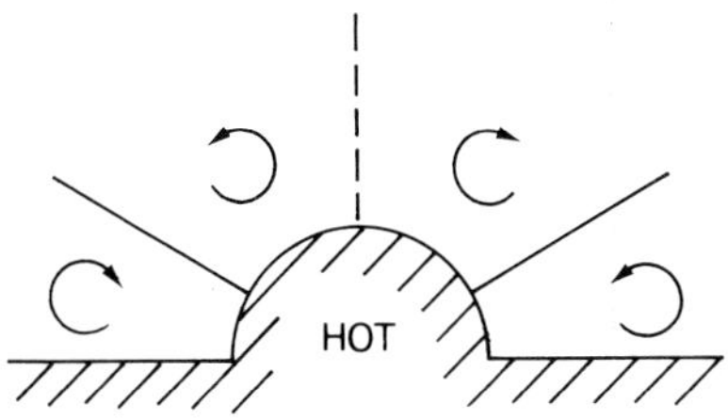

Fig. 3.6.iii The secondary circulation set up over a hot hump viewed in the downstream direction.

the vorticity was dominantly produced from the shear and above which it was dominantly produced by gravity. There is therefore an inflow towards the axis of the flow along these surfaces. Both effects fall to zero on the vertical plane of the flow.

No account is taken in this theory of the possibility of the propagation of gravity waves through the fluid, and we shall see in Chapter 5 that this method is not appropriate for that problem. In the gravitationally stable case of the cold hump there would be a wave pattern similar to that produced by a ship on the surface of water (see 5.15). In the other, unstable, case the presence of the hump would set off buoyant convection. This is discussed in Chapters 9, 10, and 11.

An obvious conclusion from these considerations is that when the density gradients are small and can be neglected the secondary motion generated by an obstacle will produce a downward flow behind it towards the ground. This is the case on many occasions when the low level air is stirred up by motion over rough ground, since there is usually shear generated by ground drag.

3.7 SECONDARY FLOW IN FRONT OF OBSTACLES

The distortion of the vortex lines by the presence of an obstacle in a stream with shear is responsible for some interesting phenomena on the upstream side of the obstacle. This is well illustrated by the case of a cylindrical pillar standing up from the bed of a river. The main body of the flow is fairly uniform and so the flow on the upstream side of the cylinder is very similar to the potential flow pattern with a line of stagnation points down the front. On this line the pressure is a maximum for the neighbourhood and increases away from the bottom, and this feature of the pressure distribution is communicated to the boundary layer on the bed of the stream because the vertical accelerations are small compared with gravity. Consequently we would expect a flow outwards from the bottom of the column (Fig. 3.7.ii) in the upstream direction close to the bed. An alternative argument leading to this conclusion is that the vortex lines of the oncoming flow are all wrapped around the obstacle (Fig. 3.7.i) and they are stretched so as to intensify the vorticity, which is the same as saying that they are continuously accumulated by this wrapping around the obstacle at its foot (Fig. 3.7.i).

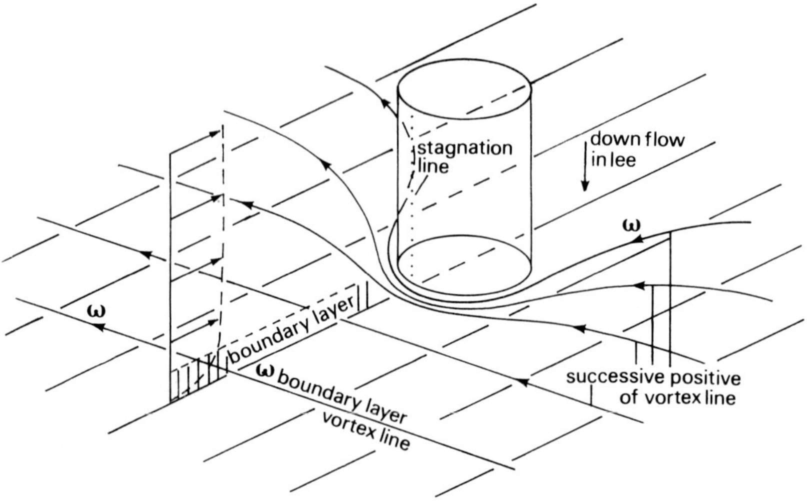

Fig. 3.7.i The vortex lines in the boundary layer are wrapped round an obstacle protruding from the boundary and produce downflow along the front stagnation line, outflow at the base of the obstacle on the upstream side, and downflow behind it.

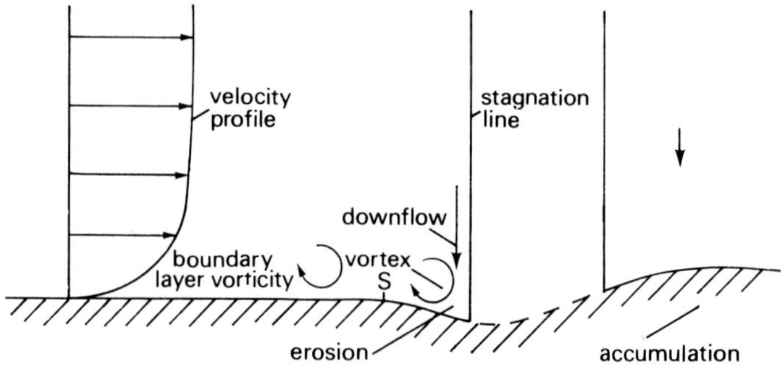

Fig. 3.7.ii The central plane of 3.7.i showing where the bed of the flow is eroded at the foot of an obstacle.

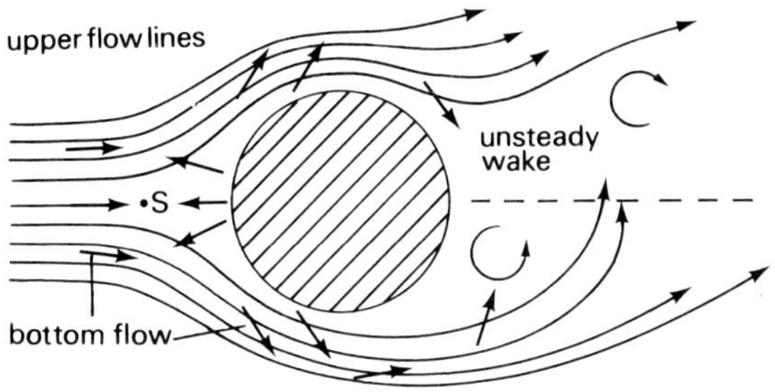

Fig. 3.7.iii Main flow, shown by flow lines, and flow in the boundary layer close to the surface around a cylindrical pillar standing in a stream, shown by heavy arrows.

The photograph below shows streamers of water paint drawn out along the streamlines at the bottom boundary in a water channel. The paint begins to leave the boundary in the unsteady part of the flow.

Outside the boundary layer the flow upstream is very similar to potential flow (see Fig. 2.6.ii) and contains a vortex street in the wake of the cylinder.

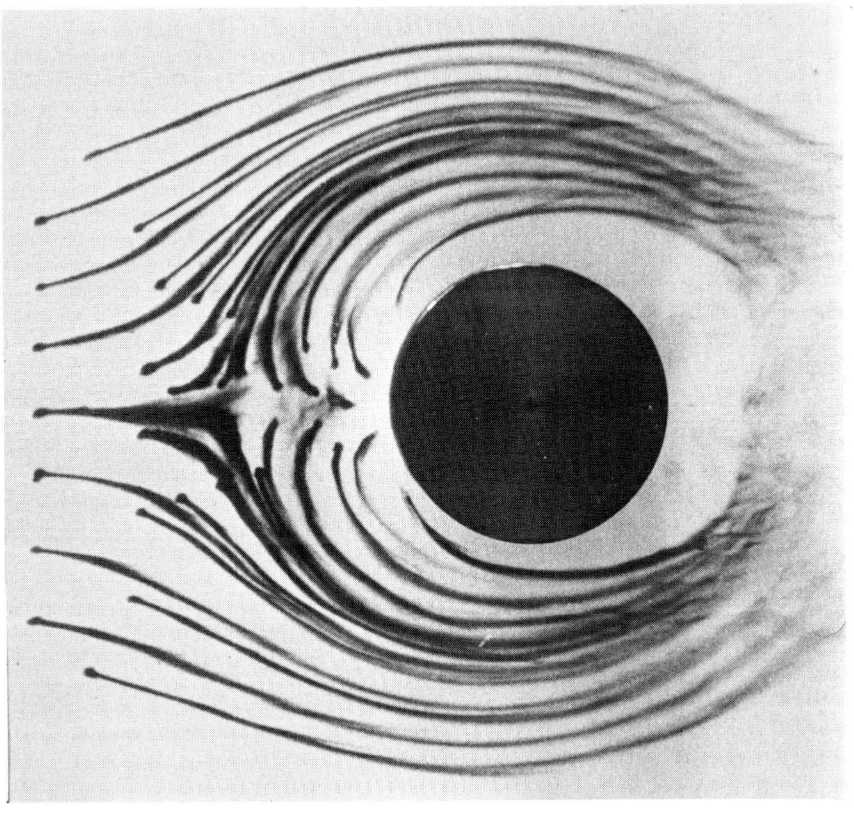

The outflow at the base occurs all round the front of the obstacle (Fig. 3.7.iii) as far as the point where the flow separates from the sides. In the wake two influences are at work. A general downflow is induced by the secondary vorticity; but close to the floor the flow may be into the wake. This is because the vortices in the wake, which may or may not be moving downstream, have low pressure in their centres towards which there is an inflow in the boundary layer. The consequence is that there is an accumulation in the wake of some of the material eroded from the bed at the foot of the pillar.

The erosion at the foot has often been the cause of the undermining of bridge piers. The most effective way to avoid this is to build the pier with a sharp upstream edge which cuts the stream into two without any significant region of stagnation. The vortex lines are cut by, rather than wrapped around, the obstacle, and as a consequence no significant secondary flow is set up (Fig. 3.7.iv).

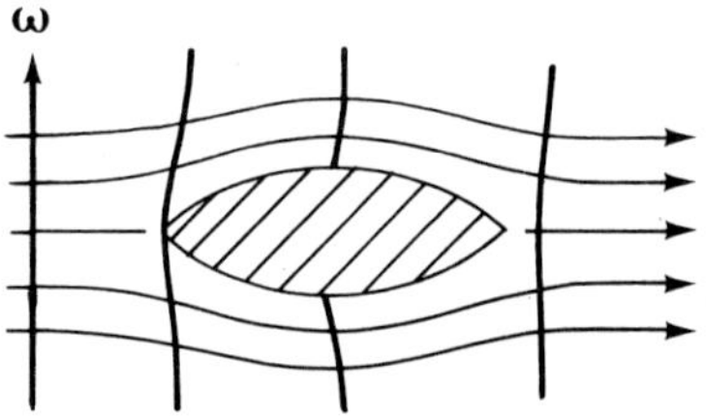

Fig. 3.7.iv Vortex lines cut by a sharp upstream edge of an obstacle produce little secondary flow.

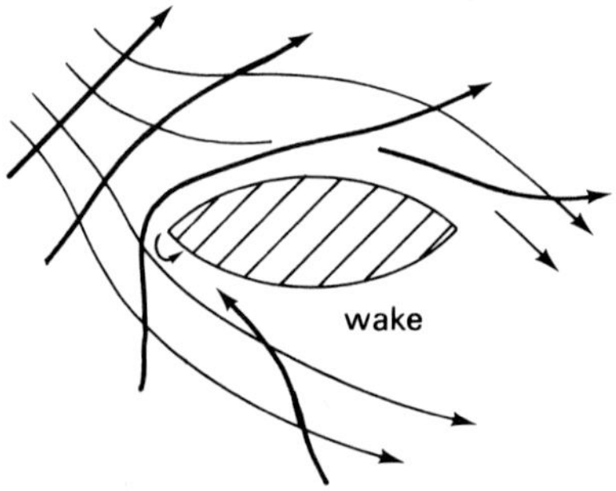

Fig. 3.7.v If the direction of flow becomes oblique to a sharp edged pier the vortex lines may be wrapped around it and cause disastrous erosion at its base.

Such a design can only be used with success if the stream direction is known for certain to be more or less unvarying. In the case of a wide shallow, or braided, river, the direction of flow relative to a bridge pier may change after or during a period of flood and the erosion in that case, due to the wrapping of vortex lines around one side, may be disastrous (Fig. 3.7.v).

In a wind stream past an obstacle such as a large cooling tower or tall building, the shear may extend to the top of the building and beyond, although it may be more intense in a shallow boundary layer close to the ground. In that case the downflow induced by the wrapping of vortex lines is present above the boundary layer, and this may cause pollution emitted from the top to be drawn down in the wake. This phenomenon must not be confused with the downwash due to vortices shed from the obstacle (see Chapter 10).

3.8 MOTION IN A TEA CUP

The configuration of the vortex lines in a tea cup stirred so as to give it vorticity about a vertical axis is very complex. A circulation in vertical planes, with horizontal circles as vortex lines, develops in the boundary layer at the bottom. This causes a circulation of the main flow with a slow outward flow above the boundary layer (Fig. 3.8.i). This in turn produces a vortex with a vertical axis with the same circulation at all radii and a decrease of tangential velocity almost like r^{-1}. Near the axis of symmetry the viscosity prevents the

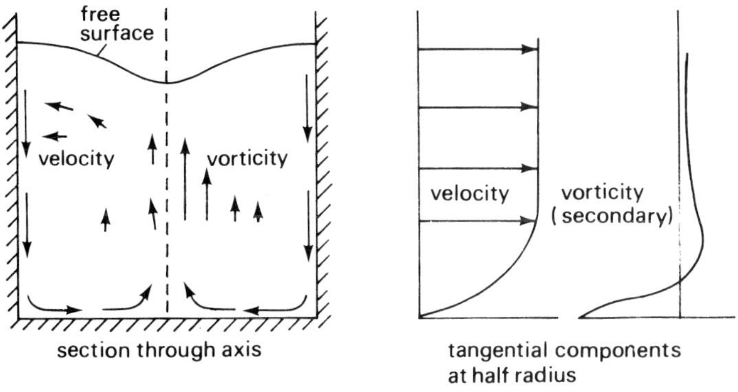

Fig. 3.8.i The velocity and vorticity components in a vertical plane through the axis of a cup with fluid rotating in it, and the velocity and vorticity profiles of the flow.

large shears that this velocity distribution would require. Consequently the rotation becomes more nearly solid, with tangential velocity varying like r, and the surface concave near the middle (see Section 2.5).

The inflow in the boundary layer at the bottom produces the slow upward and outward flow in the main body of the fluid and also causes tea leaves to accumulate in the middle of the bottom of the cup even though they are heavier than water and therefore might be expected to be centrifuged to the outside.

On the outer vertical wall the flow is centrifugally unstable (see Section 3.10) and the fluid whose tangential velocity has been reduced there descends the wall and feeds the inflow in the bottom boundary layer with fluid of smaller circulation than the main body of fluid.

If a cup of liquid is rotated about a vertical axis of symmetry with the fluid initially at rest, a secondary flow is set up with the fluid in the bottom boundary layer being centrifuged outwards and forced up the outer wall. This is different from the reverse of the flow just described in a few details. The flow on the outer wall is initially turbulent because of the large shear there but the instability is gradually suppressed by the stable stratification due to the increase of circulation outwards (see Section 3.10).

3.9 FLOW IN THE CORNERS OF STRAIGHT CHANNELS

It is observed that there is a secondary circulation with flow from the centre of a channel towards a corner, and outwards from the corner along the walls. This is most noticeable in channels of triangular section, having acute angled corners.

A ready explanation is provided as follows: the vortex lines generated in the boundary layer of a uniform stream are parallel to the walls and perpendicular to the stream (Fig. 3.9.i). In a corner the vortex lines turn to follow the general shape of the section, but since the flow in a corner is retarded more than near the centre of the wall, the vortex lines are held back there. This produces a streamwise component of vorticity with a secondary flow as described. This explanation is likely to be correct, but it must be emphasised that it does not amount to a proof that this kind of secondary flow must occur. This is because, on the face of it, the vortex lines in the corner could well coincide with the lines of constant velocity so that they would not actually be distorted.

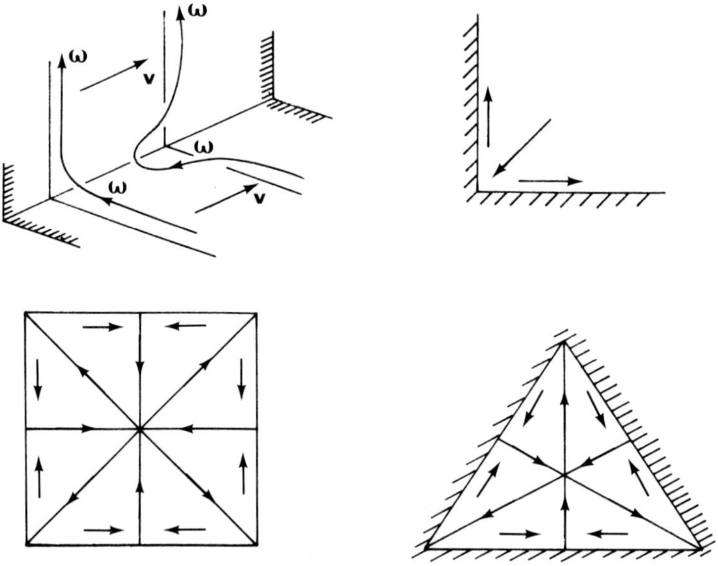

Fig. 3.9.i Distortion of the boundary layer vortex lines in a corner of a channel, and the resultant secondary flow in a cross section of a square and of a triangular channel.

A corollary of this argument is that the flow along a convex edge will generate a secondary flow with flow towards the edge along the solid surface and away from it elsewhere because the vortex lines will be carried forward by the main stream more rapidly near the edge (Fig. 3.9.ii). This phenomenon is not known to be common.

Fig. 3.9.ii Distortion of vortex lines in flow along an angular edge, and the consequent secondary flow.

It is also possible that, if it is observed, it might be due to the upflow towards the edge induced in corners below.

The argument, based on qualitative inference, given above, is very weak, and is given because it is easy to remember. But any vortex line moving under its own influence is retarded where the curvature is large because of the opposing influences of the nearby elements of vorticity, whereas in places where the curvature is small it is moving under the influence of the rest of the vortex line and all the elements contribute in the same direction. If a closed vortex line is bulged outwards in the neighbourhood of one point on it, the solid angle subtended by it at points in the neighbourhood of a point on the bulge is reduced. Accordingly the rate of change of solid angle in unit displacement, which is proportional to the velocity in the direction of the displacement, is reduced.

3.10 LOCAL INSTABILITY IN CURVED FLOW

In Section 1.8 it was shown that static instability may be thought of either in terms of a displaced parcel which finds itself subject to a force in the direction of the displacement, or in terms of the generation of vorticity in a parcel in the same direction as a rotational displacement. This view of instability is local, and is appropriate when the disturbance is small compared with the extent of the fluid, especially when the disturbance occurs far from solid boundaries.

A local treatment of instability is not appropriate when only disturbances of a special kind are unstable. Such is the case when wavy disturbances in only a small range of wavelengths are unstable, the shorter wavelengths usually being damped by viscosity and the larger ones being inhibited by the presence of rigid boundaries. In such cases the physical dimensions and physical properties with dimensions (usually viscosity, sometimes thermal conductivity) play a significant role in determining the instability characteristics and a global, or overall, analysis is necessary to elucidate them. Global analyses of particular cases are given in many good textbooks. The local analysis given here is not described elsewhere, except by Scorer (1967), where the more complex analysis adds confusion to a few misprints: some may enjoy the issues raised thereby.

This type of local analysis ignores viscosity: this is justified on the ground that the instability is induced by curvature of the flow which may be applied suddenly to a parcel as it moves in a larger flow pattern, and may develop quickly so that viscosity has not time to affect the flow, even though the vorticity in the stream was itself originally produced by viscosity. The linearly displaced parcel method tells us nothing about the kind of unstable motion that will be generated except in the case of instability in a fluid at rest. Taylor studied such a case by causing the interface between oil and water to be changed from stable, with the oil lying above the water in the field of gravity, to unstable when the whole fluid was accelerated downwards with an acceleration in excess of g. He observed that any small distrubance imposed on the interface grew: this is a property of local instability, that it makes no dimensional reference to the kind of disturbances it treats.

Since a parcel of fluid may enter a region in which there is curvature for a short time only, the instability produced may be switched on and off. The present analysis is therefore appropriate because it follows the fate of a parcel: by contrast a global analysis studies disturbances by reference to a fixed frame determined by the rigid boundaries of the fluid.

This study is based on the assumption that the flow is steady. The instability is thus manifested by a downstream growth of the disturbance. The disturbance must therefore be in the form of the generation of streamwise (secondary) vorticity, for otherwise the streamlines would not be fixed in space. Chapter 6 discusses the growth of billows, a form of instability which cannot be treated by the present analysis because it cannot very conveniently be represented by steady flow.

It is clear that if the sign of the curvature κ is reversed in (3.5.4) the oscillations which it represents are unstable, and this is physically equivalent to the reversal of gravity in a static fluid, or to Taylor's experiment. For the purposes of our discussion gravity will not be included, and its effect on the argument can easily be seen in any case under consideration by retaining the terms in g. Since gravity is a constant it has an important effect on the flow in a long straight path, and it has a modifying effect on the flow in a bend if there is a large density gradient, but it is of no particular interest in this context. Thus the relevant equation for steady flow is, from (3.2.2),

$$\mathbf{v}.\mathbf{grad}\ \frac{\boldsymbol{\omega}.\mathbf{t}}{q} = 2\kappa\boldsymbol{\omega}.\mathbf{n} + \kappa q\ \mathbf{R}.\mathbf{b} \tag{3.10.1}$$

for studying the effect of curvature.

The flow may now be thought of as being locally the same as that of a helical vortex in which coaxial cylinders of fluid have different speeds of rotation around the axis and linearly along the axis. In this situation the streamlines and vortex lines are helices on the cylinders and the particle trajectories have curvature and torsion, which may be specified, and which are constant for each particle in the undisturbed state. The specification of curvature and torsion of one streamline and of the velocity and vorticity (velocity gradients) at a point on it determines the motion field in the neighbourhood. The radius of the cylinder on which the streamline lies could be specified instead of the torsion. We do not consider any effects due to tangential acceleration, for these were discussed in Section 1.4 (page 41) and do not directly produce secondary motion.

We now imagine that the fluid receives a small rotational displacement around the streamline, taking its density and vortex lines with it. The density is assumed to be initially constant on each of the coaxial cylinders of the flow, otherwise the acceleration is finite in the absence of a disturbance.

The vector $\mathbf{R}$ is initially along the principal normal, and we imagine it to be directed positively outwards from the axis, because that is the more stable state. The vorticity $\boldsymbol{\omega}$, the tangent $\mathbf{t}$, and the binormal $\mathbf{b}$, all lie in the plane perpendicular to $\mathbf{R}$ and $\mathbf{n}$. The positions of $\mathbf{R}$ and $\boldsymbol{\omega}$ after the displacement are represented by $\mathbf{R}_1$ and $\boldsymbol{\omega}_1$.

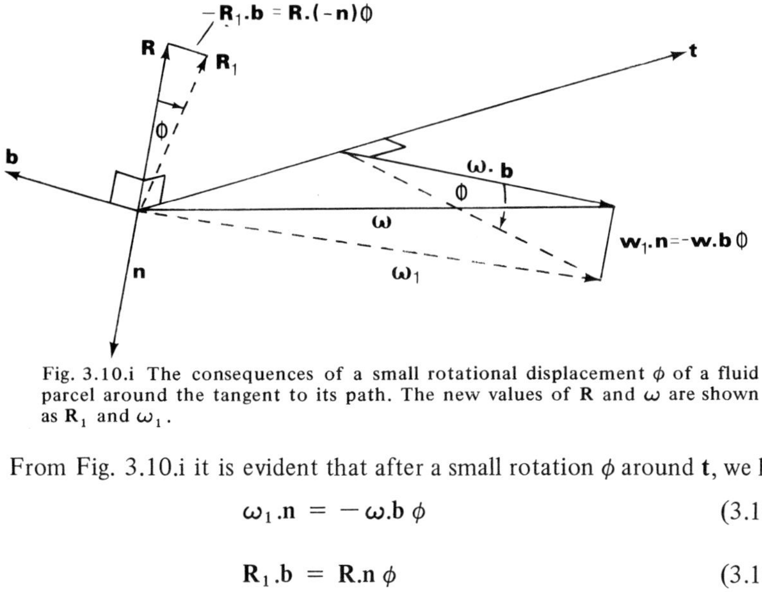

Fig. 3.10.i The consequences of a small rotational displacement ϕ of a fluid parcel around the tangent to its path. The new values of $\mathbf{R}$ and ω are shown as $\mathbf{R}_1$ and ω_1.

From Fig. 3.10.i it is evident that after a small rotation ϕ around $\mathbf{t}$, we have

$$\omega_1.\mathbf{n} = -\omega.\mathbf{b}\,\phi \tag{3.10.2}$$

$$\mathbf{R}_1.\mathbf{b} = \mathbf{R}.\mathbf{n}\,\phi \tag{3.10.3}$$

and so writing

$$\mathbf{v}.\mathbf{grad}\,\frac{\omega.\mathbf{t}}{q} = q\frac{\partial}{\partial s}2\frac{\partial\phi}{\partial s} \tag{3.10.4}$$

equation (3.10.1) becomes

$$\frac{\mathrm{D}^2\phi}{\mathrm{D}s^2} = \left(-\frac{\kappa}{q}\,\omega.\mathbf{b} + \frac{\kappa}{2}\mathbf{R}.\mathbf{n}\right)\phi = \frac{\kappa}{q}\mu^2\phi. \tag{3.10.5}$$

This represents an unstable situation if the coefficient of ϕ is positive, i.e. if

$$2\mu^2 = -2\boldsymbol{\omega}.\mathbf{b} + q\mathbf{R}.\mathbf{n} > 0 \qquad (3.10.6)$$

because κ is always defined to be positive by the choice of the direction of **n** towards the centre of curvature.

The meaning of the term $q\mathbf{R}.\mathbf{n}$ is simple, for when there is no vorticity the condition (3.10.6) implies that if the density increases inwards towards the centre of curvature the centrifugal force is destablising. As in the case of the linear displacement of a parcel, there is an assumption inherent in this treatment that the pressue field is unaltered by the displacement and that it is determined by the motion of the undisplaced fluid surrounding the displaced parcel. Any accelerations which arise are due to the centrifugal forces no longer being in equilibrium with the pressure field. To analyse the matter further we need to set out the elementary properties of a helix.

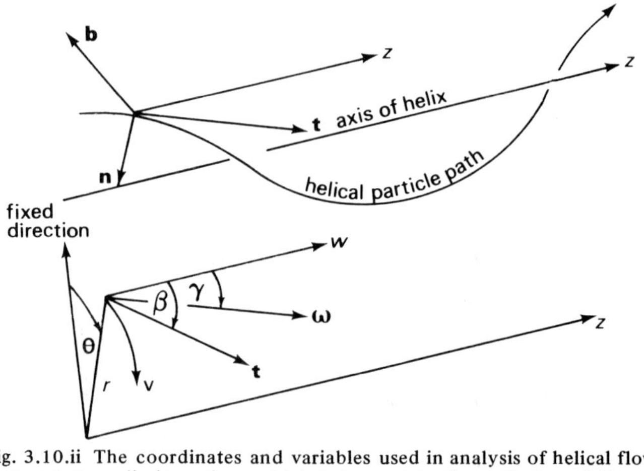

Fig. 3.10.ii The coordinates and variables used in analysis of helical flow. The vectors **t**, **b**, $\boldsymbol{\omega}$ lie in a plane which is tangent to the cylinders of the motion, the z-direction being along the generator. v is the circumferential and w the axial velocity component.

The particle is at a point whose cylindrical polar coordinates (Fig. 3.10.ii) are (r,θ,z). It has velocity (u,v,w) in these directions, z being along the axis of the helix. In the helical motion

$$u = 0, \qquad v = v(r), \qquad w = w(r), \qquad \beta = \beta(r) \qquad (3.10.7)$$

where β is the angle between the tangent and the direction of the axis along which the velocity component is w. The velocity component v is called circumferential. γ is the angle between the vorticity vector and the direction of the axis. The position vector **x** is

$$\mathbf{x} = (r,\theta,z) \text{ in cylindrical polar coordinates} \tag{3.10.8}$$

$$= (r\cos\theta,\ r\sin\theta,\ z) \text{ in fixed cartesian coordinates.} \tag{3.10.9}$$

We have to perform all differentiations with respect to s in cartesian coordinates because the direction of r is varying,

so that
$$\mathbf{t} = \frac{\partial \mathbf{x}}{\partial s} = \left(-r\sin\theta\,\frac{\partial\theta}{\partial s},\ r\cos\theta\,\frac{\partial\theta}{\partial s},\ \frac{\partial z}{\partial s}\right). \tag{3.10.10}$$

But
$$\mathbf{v} = q\mathbf{t} = q(0, \sin\beta, \cos\beta) \text{ in cyl. coords.} \tag{3.10.11}$$

$$= q(-\sin\beta\sin\theta, \sin\beta\cos\theta, \cos\beta) \text{ in cartesian coords,} \tag{3.10.12}$$

and so, from (3.10.10), we have

$$r\frac{\partial\theta}{\partial s} = \sin\beta,\quad \frac{\partial z}{\partial s} = \cos\beta. \tag{3.10.13}$$

Differentiating (3.10.12) with respect to s we have

$$\kappa\mathbf{n} = \frac{\partial \mathbf{t}}{\partial s} = \left(-\sin\beta\cos\theta\,\frac{\partial\theta}{\partial s},\ -\sin\beta\sin\theta\,\frac{\partial\theta}{\partial s},\ 0\right) \tag{3.10.14}$$

$$= \frac{1}{r}\sin^2\beta\,(-\cos\theta, -\sin\theta, 0) \text{ in cartesian coords.} \tag{3.10.15}$$

$$= \frac{1}{r}\sin^2\beta\,(-1, 0, 0) \text{ in cyl. coords.} \tag{3.10.16}$$

Since $\mathbf{n}$ is a unit vector
$$\kappa = \frac{1}{r}\sin^2\beta = \frac{v^2}{q^2 r}. \tag{3.10.17}$$

Likewise the torsion in calculated by differentiating $\mathbf{b}$: thus

$$\mathbf{b} = (0, -\cos\beta, \sin\beta) \text{ in cyl. coords.} \tag{3.10.18}$$

$$= (\cos\beta\sin\theta, -\cos\beta\cos\theta, \sin\beta) \text{ in cartesian coords.} \tag{3.10.19}$$

$$\frac{\partial \mathbf{b}}{\partial s} = \left(\cos\beta\cos\theta\,\frac{\partial\theta}{\partial s},\ \cos\beta\sin\theta\,\frac{\partial\theta}{\partial s},\ 0\right)$$

$$= \frac{1}{r}\sin\beta\cos\beta\,(\cos\theta, \sin\theta, 0) = -\tau\mathbf{n}. \tag{3.10.20}$$

Therefore

$$\tau = \frac{1}{r}\sin\beta\cos\beta = \kappa\cot\beta = \frac{vw}{q^2 r} \tag{3.10.21}$$

In this case of axial symmetry in which the quantities are functions of r only, the **curl** of any vector **a** is

$$\textbf{curl a} = \left(0, -\frac{\partial a_z}{\partial r}, \frac{1}{r}\frac{\partial}{\partial r}(ra_\theta)\right) \text{ in cyl. coord.} \tag{3.10.22}$$

and so

$$\boldsymbol{\omega} = \textbf{curl v} = \left(0, -\frac{\partial w}{\partial r}, \frac{1}{r}\frac{\partial}{\partial r}(vr)\right) \text{ in cyl. coords.} \tag{3.10.23}$$

and from (3.10.18), noting that $v = q\sin\beta,\ w = q\cos\beta$, we get

$$\begin{aligned}\boldsymbol{\omega}.\mathbf{b} &= \frac{\partial w}{\partial r}\cos\beta + \frac{1}{r}\sin\beta\frac{\partial}{\partial r}(vr)\\ &= \frac{\partial w}{\partial r}\cos\beta + \frac{\partial v}{\partial r}\sin\beta + \frac{v}{r}\sin\beta\\ &= \frac{\partial q}{\partial r} + \frac{v^2}{qr}\end{aligned}$$

The coefficient μ^2 in (3.10.5), whose sign determines the stability, is

$$\mu^2 = -\frac{\partial q}{\partial r} - \frac{v^2}{qr} - \frac{1}{2}\frac{q}{\rho}\frac{\partial\rho}{\partial r} \tag{3.10.25}$$

for an incompressible fluid of density ρ which is initially a function of r only. But in the equilibrium flow the equation of motion is

$$\frac{v^2}{r} = \frac{1}{\rho}\frac{\partial p}{\partial r}, \tag{3.10.26}$$

whence

$$\mu^2 = -\frac{1}{\rho q}\frac{\partial}{\partial r}(p + \tfrac{1}{2}\rho q^2) = -\frac{1}{\rho q}\frac{\partial p_0}{\partial r} \tag{3.10.27}$$

where p_0 is the stagnation pressure measured by a pitot tube fixed to the coordinate system and directed into the flow, along $-\mathbf{t}$.

The implication of this last result is that flow in which the stagnation pressure decreases outwards is unstable for small rotational displacements of fluid parcels around the streamline. Therefore a stable configuaration is one in which the lowest stagnation pressure fluid is closest to the axis. We have already noted in Section 3.5 that Hawthorne observed the tendency of the fluid to move into this configuaration in a straight pipe (see Fig. 3.5.vii).

The consequences of this result are much further reaching than appears at first sight because there are cases in which the sign of μ^2 depends on the choice of the coordinate system, although this obviously has nothing to do with the mechanics. Thus

$$p_0 = p + \tfrac{1}{2}\rho q^2 = p + \tfrac{1}{2}\rho\,(v^2 + w^2) \tag{3.10.28}$$

and

$$\frac{\partial p_0}{\partial r} = \frac{\partial p}{\partial r} + \frac{1}{2}\frac{\partial \rho}{\partial r}\,(v^2 + w^2) + \frac{1}{2}\rho\left(2v\frac{\partial v}{\partial r} + 2w\frac{\partial w}{\partial r}\right) \tag{3.10.29}$$

In the helical motion there is an absolute value for v, the circumferential velocity, which must be measured relative to a non-rotating coordinate system, angular velocities having absolute values. But there is no absolute value for w, to which we may add a constant w_0 by moving the coordinate system with speed w_0 in the negative direction of z. In the case in which $\partial\rho/\partial r$ is zero this alters only the term $\rho w \partial w/\partial r$ into $\rho(w + w_0)\partial w/\partial r$, and its magnitude and sign may be chosen freely provided $\partial w/\partial r$ is not zero. In that case the sign of $\partial p_0/\partial r$ depends on the zero chosen for w, and so does the sign of μ^2, which means that **there are always some disturbances which are unstable when $\partial w/\partial z$ is non-zero.**

The effect of adding w_0 to w is to alter β, the pitch of the helix. The dynamics must, of course, remain unaltered by the change in the velocity of the coordinate system, but by altering the direction of $\mathbf{t}$ which is determined by the ratio v/w, we are merely studying a different perturbation of the flow.

Clearly the motion may be stabilised or de-stabilised by changing the sign and magnitude of $\partial\rho/\partial r$, instability being made more likely by having the density decrease outwards, for then the term $\dfrac{1}{2}\dfrac{\partial \rho}{\partial r}\,(u^2 + w^2)$ in (3.10.29) is negative. If we now ignore the effect of density gradients we may revert to the equation (3.10.5) in the form

$$\frac{\partial^2 \phi}{\partial s^2} = -\frac{\kappa}{q}\,\boldsymbol{\omega}.\mathbf{b}\,\phi\ . \tag{3.10.30}$$

The sign of $\boldsymbol{\omega}.\mathbf{b}$ is determined by the position of $\mathbf{t}$ in relation to $\boldsymbol{\omega}$ in the plane which is tangent to the cylinder of the flow. If ω lies between **t** and **b** the scalar product ω.**b** is positive and the motion is stable. If $\beta < \gamma$ the motion is unstable (see Fig. 3.10.iii).

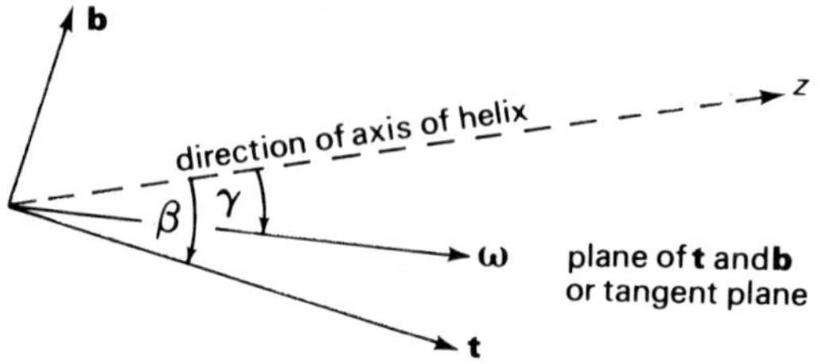

Fig. 3.10.iii The positions in the tangent plane of ω and **t** relative to the direction of the axis of the local helix. The motion is unstable if the vector **t** about which the perturbation rotates the fluid lies between the vorticity and the direction of the axis.

The motion is thus unstable for rotations around an axis in the tangent plane (normal to **n**) which lies between the vorticity vector and the direction of the axis of the helix appropriate to the local motion (see Fig. 3.10.iv). In principle this helical axis may be found by noting the values of κ and τ from which β may be obtained from (3.10.21) and the direction is at angle β to **t** in the plane normal to **n**. The quantities κ, τ, **t** and **n** can be found at any point in the neighbourhood of which the trajectory is defined using only first derivatives of the velocity field.

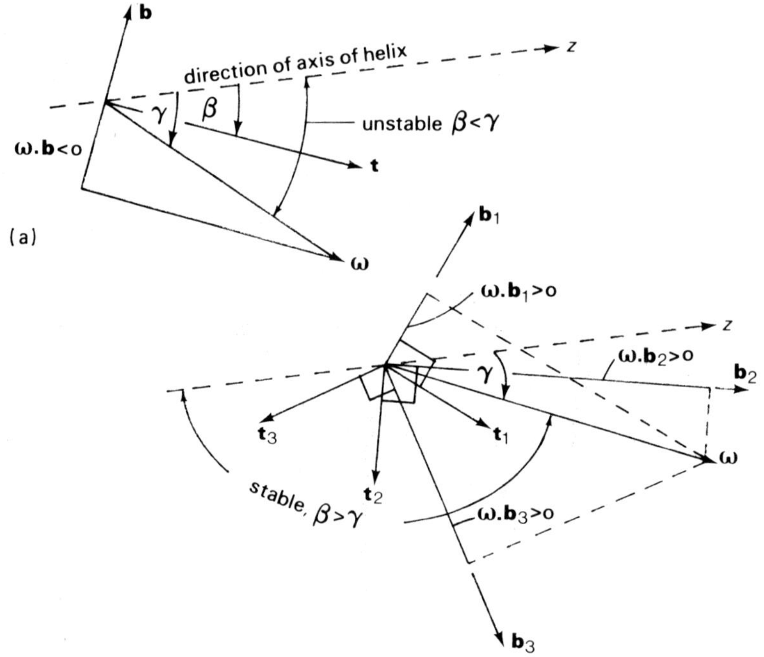

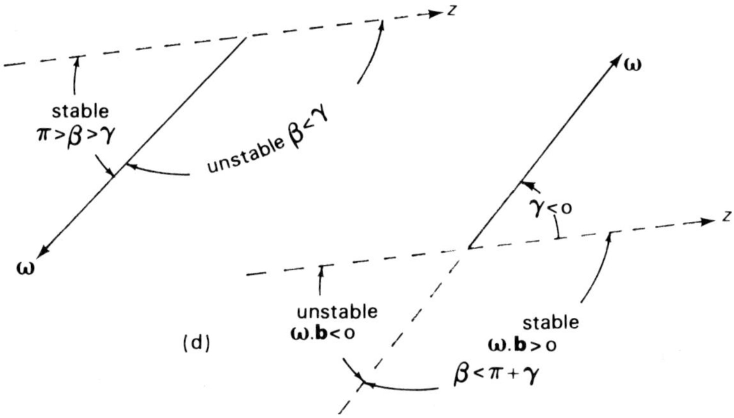

Fig. 3.10.iv The various directions about which rotations are unstable for a given flow (given helical axis and given vorticity vector). In (a) we have instability, and for the three direction $\mathbf{t}_1$, $\mathbf{t}_2$, $\mathbf{t}_3$ in (b) we have stability. In (c) and (d) the results are shown for backward and left-handed vorticity.

We have pursued this argument assuming the helix to be right handed, and β positive, as in the diagram above. The diagrams corresponding to those above when β or γ is negative are simply mirror images of those shown which indicate the directions about which rotations are unstable and stable for various positions of ω relative to the axis of the helix. When $\mathbf{t}$ is pointing backwards along the axis it simply means that w is negative.

To interpret these results in terms of a general disturbance we imagine helical flow on coaxial cylinders disturbed by the formation of a bump, or blister, on one of them. Figure 3.10.v shows the contours of such a blister. The blister is formed as a hump on one of the Bernoulli surfaces by rotation

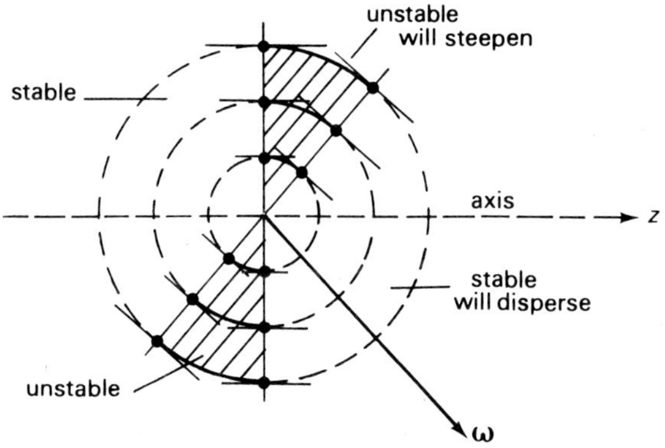

Fig. 3.10.v The parts of the blister on a Bernoulli surface in the unstable region will steepen.

of the fluid around the direction of the tangent to the contours so that this direction is unstable where it lies between the direction of the axis and the direction of the vorticity vector. The part of the blister marked unstable will steepen, while the stable part of the disturbance will disperse as stable waves. It is to be expected that the blister would lengthen in the direction of greatest instability.

When there is no density gradient

$$\frac{\kappa}{q}\mu^2 = -\frac{\kappa}{q}\,\boldsymbol{\omega}.\mathbf{b} = \frac{\sin^3\beta}{\nu r}\,\omega\sin(\gamma-\beta)\ , \tag{3.10.31}$$

and so because $dz = ds \cos\beta$, and $q dt = ds$, for constant q equation (3.10.5) may be represented in the following three forms

$$\frac{D^2\phi}{Ds^2} = \frac{\omega}{\nu r}\sin^3\beta\sin(\gamma-\beta)\,\phi = \mu_s{}^2\,\phi\ , \tag{3.10.32}$$

$$\frac{D^2\phi}{Dt^2} = \frac{\nu\omega}{r}\sin\beta\sin(\gamma-\beta)\phi = \mu_t{}^2\,\phi\ , \tag{3.10.33}$$

$$\frac{D^2\phi}{Dz^2} = \frac{\omega}{\nu r}\,\frac{\sin^3\beta}{\cos^2\beta}\sin(\gamma-\beta)\phi = \mu_z{}^2\,\phi\ . \tag{3.10.34}$$

These represent the growth of a disturbance following a parcel of fluid along its trajectory s, in time t, or along the axis z. If we have identified an unstable disturbance by adding a velocity w_0 to the coordinate system, it is possible that in any practical case this would mean that the growth as z or s increases might appear to be large. However, because the coordinate system has a large velocity relative to the apparatus, the parcel could well be a long way from the region of interest before any significant growth had occurred. This is most likely to be the case when γ, the angle between the vorticity vector and the axis, is small, and a large velocity w_0 has to be imposed in order to make **t** lie between them.

In practice, therefore, it is most sensible to fix the coordinate system to any apparatus we may have and use (3.10.33) to find out which circumstances create instability. Perturbations caused by the presence of the apparatus will cause secondary rotations around the tangent to the flow relative to the apparatus. Thus, if there is swirling flow inside a straight circular pipe, an imperfection on the pipe surface is likely to set off instability if the tangent lies at a smaller angle to the axis then the vorticity vector.

The vortex lines are, in the situation being considered, helices, and the angle γ is increased for a given distribution of circumferential velocity (or swirl) $v(r)$, by increasing the magnitue of $-\dfrac{\partial w}{\partial r}$, the circumferential component of vorticity (see 3.10.23). In the case of a swirling flow passing down a circular pipe, with no boundary layer on the outer wall at the entrance, the flow rapidly becomes unstable because the retardation of the fluid at the outer wall produces low stagnation pressure fluid there, and this will tend to move towards the centre of the pipe as mentioned in Section 3.5 and in remarks following (3.10.27). The vorticity vector may be directed along the axis if the swirl is a

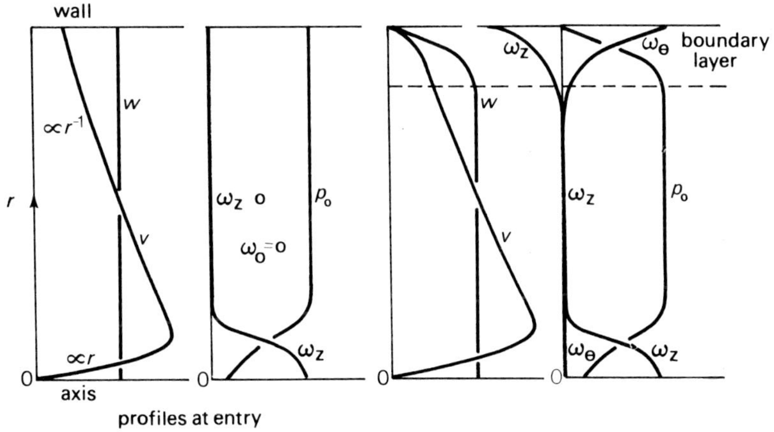

Fig. 3.10.vi Profiles of velocity components, stagnation pressure and vorticity at entry and downstream when irrotational swirling flow enters a pipe.

solid rotation (as it must be in the centre of the flow), or there may be no vorticity, if $v \propto r^{-1}$ and w is independent of r (see Fig. 3.10.vi). Further down the pipe w decreases towards the boundary which gives $\boldsymbol{\omega}$ a right handed circumferential component, and v decreases towards the boundary giving $\boldsymbol{\omega}$ a backward pointing component. Thus γ exceeds 90°, and whatever the direction of $\mathbf{t}$, it must lie between 0° and 90° in forward moving right handed swirling flow. Consequently the boundary fluid moves inwards, and in the case where the original swirling flow is irrotational except for a small solidly rotating core, this fluid will very rapidly penetrate to the core. The vortex will then appear to break down or explode according to the visualisation employed.

This kind of breakdown will not occur if the vorticity vector has large magnitude and is directed forwards along the axis initially because the instability will then be confined to the boundary layer. In the case of the main body of the fluid being in the neutral, irrotational state to begin with (except for a small rotational core), the onset of breakdown will be delayed until the boundary layer is thick enough for the instability to overcome the stabilising effect of viscosity which damps small disturbances, and the global effect of the rigid outer boundary which damps large ones. When the instability sets in therefore, there is quite a large mass of fluid ready to penetrate to the core as soon as a disturbance begins to grow, and the breakdown appears like an explosive global phenomenon involving the whole mass rather than the boundary layer alone.

To find the most unstable disturbance we find the maximum of μ_t by differentiating (2.10.33) with respect to β and setting $\partial\mu_t/\partial\beta = 0$. This gives $\beta = \frac{1}{2}\gamma$, and from the equation itself it is obvious that $\mu_t = 0$ when $\beta = 0,\gamma$. Thus **the most unstable disturbances are those which are a rotation of the fluid around lines bisecting the angle between the axis and the vorticity vector**. This means that the flow is most unstable if γ is contrived to be 2β where β is the angle made by the flow relative to a coordinate system at rest relative to the disturbing influence.

To make the flow as stable as possible it can be seen from the instability diagrams (Fig. 3.10.iv) that in right handed swirl it is best to have vortex lines which have a left handed screw, i.e. $\gamma < 0$. This is achieved by having w increasing in magnitude with radius, which ensures that the fluid on the outside has the largest stagnation pressure. However this flow would be upset by a disturbing body such as a bullet or a fish or an obstacle drawn along by a wire which was overtaking the flow, for relative to it w would be increasing in magnitude towards the core of the flow.

An interesting application of equation (3.10.34) can be made to the case of swirling flow with initially uniform axial velocity w, and a large vorticity directed along the axis. This is not explosively unstable as in the case just discussed of small initial vorticity in the main bulk of the fluid, and we may imagine that γ increases from zero to a small positive value as the boundary layer is formed. The most rapidly growing disturbance has been shown to be with $\beta = \frac{1}{2}\gamma$, the one which achieves the greatest amplitude at the point of exit from a given apparatus is the one having the largest value of μ_z.

$$\frac{\partial}{\partial\beta}\,\mu_z{}^2 = \frac{\omega}{2\nu r}\,\frac{\sin^2\beta}{\cos^3\beta}\,[5\,\sin(\beta-\gamma)+\sin(3\beta-\gamma)] \qquad (3.10.35)$$

which is zero where $\beta = 0$ (minimum) and where

$$5(\gamma-\beta) \triangleq 3\beta-\gamma$$

when the angles are small. This is where

$$\beta = \tfrac{3}{4}\gamma \qquad (3.10.36)$$

and this is the feature we would expect to see in the largest disturbance at a given value of z. It is no easy task to determine this in an experiment where w is changing in the boundary layer, but clearly γ must exceed β by a significant amount in any case.

3.11 INSTABILITY OF PLANE ROTATING FLOW

Two particular cases of swirling flow are of special interest, namely those in which there is no motion along the z-axis initially, where the motion is in circles around a common axis. In the first case (Fig. 3.11.i) the vorticity is directed in the same direction as the rotation, in which case the circulation increases with radius, and v decreases more slowly than r^{-1}, or increases with r,

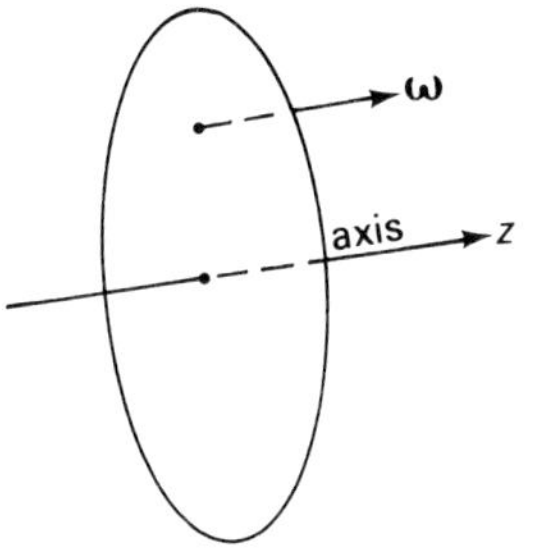

Fig. 3.11.i Plane flow in circles with circulation increasing outwards is stable to toroidal disturbances.

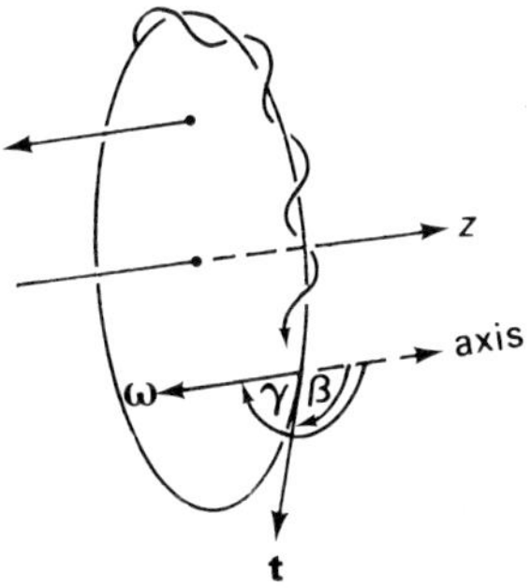

Fig. 3.11.ii Toroidal disturbances are the most unstable when the circulation of plane flow decreases outwards.

and $\omega_z = \dfrac{1}{r}\dfrac{\partial}{\partial r}(vr) > 0$. Here $\gamma = 0$ and there are no unstable disturbances (Fig. 3.11.i). If an axial component w which varies with r is introduced, immediately some unstable disturbances exist. If $\omega_z = 0$, then $v \propto r^{-1}$, and if an axial component of velocity which varies with r is introduced ω has only one component, namely ω_θ, and the most unstable disturbances appear to be rotations around helices with $\beta = 45°$; but this is not a correct conclusion when $v = 0$ because $\kappa = 0$ and the growth rate is zero.

The second case is that in which ω is directed in the negative direction of z (assuming v to have a right handed rotation). In this case $\gamma = 180°$ and it so happens that if $w = 0$ (in the most obvious coordinate system) then $\beta = 90°$, and the most unstable disturbances are toroidal, namely rotations around the circular streamlines (Fig. 3.11.ii). The circulation and stagnation pressure decrease outwards in this case, and v decreases more rapidly than r^{-1}.

This instability is well known, and has been explained by linearly displaced parcel theory. According to that argument, in the unstable situation the moment of momentum about the axis decreases outwards, so that if a parcel is given a radial displacement by means of a radial force its moment of momentum is conserved and it finds itself with a greater circumferential velocity than its new neighbours. In that case the pressure field which exerts a radial force on the fluid exactly balancing the centrifugal force is too small to move the displaced parcel in the same circular path as its new surroundings. It will therefore be accelerated outwards and the disturbance is unstable. An inward displacement puts the parcel in a pressure field which exerts a force towards the axis in excess of the centrifugal force.

This argument is a good explanation of the phenomenon, but it does not enable us to deduce that the most unstable disturbance is the toroidal one, nor does it tell us the effects of variations of w with r.

An interesting corollary of this result is that if a stream with forward shear (velocity increasing away from the boundary) is set into wave motion by the passage of the stream over an uneven bottom, the motion is unstable to toroidal disturbances in the troughs of the waves where the centre of curvature of the flow lies in the direction of the increasing stream velocity (Fig. 3.11.iii). The growing disturbance takes the form of corrugations of the Bernoulli surfaces along the direction of the flow. The instability lasts while the fluid passes through the wave trough, and is stabilised in the following crest. However the dispersion of the stable waves in the crest provides an ideal disturbance for the further growth of the waves in the following trough.

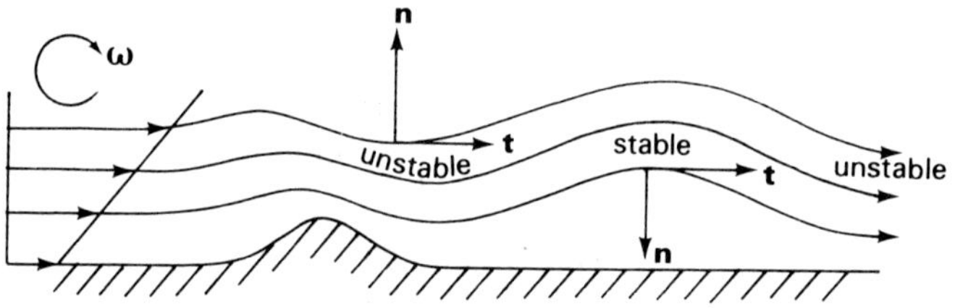

Fig. 3.11.iii Regions of instability in waves in a stream with shear.

In the troughs **b** is directed into the paper so that $\boldsymbol{\omega}.\mathbf{b} > 0$, but in the wave crests the directions of **n** and **b** are reversed and **b** is outwards from the paper. Since $\boldsymbol{\omega}$ is unchanged, $\boldsymbol{\omega}.\mathbf{b} < 0$ in the crests.

This type of instability is sometimes called **Görtler instability** because he vividly described how it occurs in the boundary layer of a fluid travelling over a concave boundary, e.g. Görtler (1959). Scorer and Wilson (1963) showed that it might occur in standing waves in the atmosphere where a shear layer is contained in the undisturbed stream. In this connection it should be noted that the shear (vorticity) produced in a wave trough is so as to stabilise the flow and prevent the occurrence of Görtler instability; it is therefore necessary that the flow should contain the shear where it is moving horizontally if the instability is to occur further downstream. Billow instability is, in fact, much more likely to occur in standing waves, and is indeed very common (see Chapter 6).

It may be noted that equation (3.10.34) shows that μ_z becomes infinite in the case $\beta = 90°$, which means simply that because there is no motion in the z-direction, an unstable disturbance will grow indefinitely within the apparatus.

3.12 ROTATIONS AROUND THE PRINCIPAL NORMAL AND BINORMAL

Rotations around the binormal are the same as rotations around the tangent with an appropriate axial velocity w_0 added. The equation for the growth of the binormal component of vorticity is (see Scorer, 1967)

$$\frac{D}{Dt}\frac{\boldsymbol{\omega}.\mathbf{b}}{q} = -2\tau\boldsymbol{\omega}.\mathbf{n} - \kappa q\mathbf{R}.\mathbf{t} \tag{3.12.1}$$

and this can be transformed directly into (3.10.1) for steady flow by imagining that an appropriate w_0 is added and the new coordinates are denoted by a prime (Fig. 3.12.i).

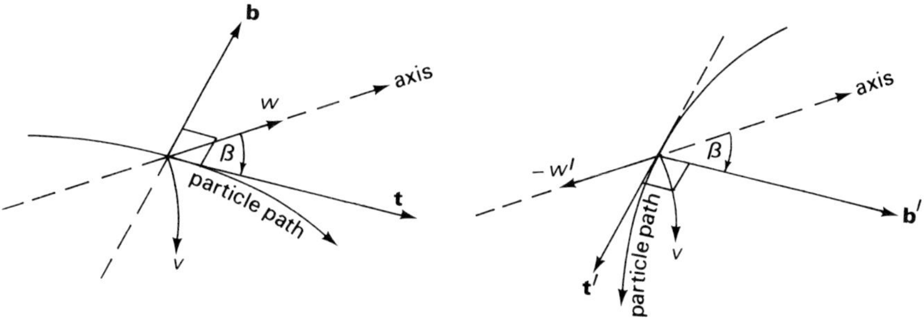

Fig. 3.12.i Coordinates changed so that rotation around **b** becomes a rotation around **t**′, by a suitable addition of axial velocity.

Then $\mathbf{t}' = -\mathbf{b}, \quad \mathbf{b}' = \mathbf{t}, \quad \beta' = \beta + 90°$,

$$\mathbf{v}' = \mathbf{v}, \quad \frac{v}{w} = \tan\beta, \quad \frac{v'}{w'} = \tan\beta' = -\cot\beta\,,$$

$$w' = -v\tan\beta, \quad q' = \frac{v}{\cos\beta} = q\tan\beta\,,$$

$$\kappa' = \frac{1}{r'}\sin^2\beta' = \frac{1}{r}\cos^2\beta = \kappa\cot^2\beta = \tau\cot\beta\,,$$

$$\boldsymbol{\omega} = \boldsymbol{\omega}', \quad \mathbf{R} = \mathbf{R}', \quad \mathbf{n} = \mathbf{n}'\,,$$

$$\tau' = \kappa'\cot\beta' = -\kappa\cot^2\beta\tan\beta = -\kappa\cot\beta\,.$$

Equation (3.12.1) applied to the new coordinates is

$$\left(\frac{\mathrm{D}}{\mathrm{D}t}\right)' \frac{\boldsymbol{\omega}'.\mathbf{b}'}{q'} = -2\tau'\boldsymbol{\omega}'.\mathbf{n}' - \kappa' q' \mathbf{R}'.\mathbf{t}'$$

$$= 2\kappa \cot\beta\, \boldsymbol{\omega}.\mathbf{n} - \tau \cot\beta\, q \tan\beta\, \mathbf{R}.(-\mathbf{b})$$

$$= \cot\beta\,(2\kappa\boldsymbol{\omega}.\mathbf{n} + \kappa q \mathbf{R}.\mathbf{b})$$

$$= \frac{q}{q'} \frac{\mathrm{D}}{\mathrm{D}t} \frac{\boldsymbol{\omega}.\mathbf{t}}{q} \tag{3.12.2}$$

In steady flow when there is no tangential acceleration q is a constant and we can say that

$$\left(\frac{\mathrm{D}}{\mathrm{D}t}\right)' \boldsymbol{\omega}'.\mathbf{b}' = \frac{\mathrm{D}}{\mathrm{D}t}\boldsymbol{\omega}.\mathbf{t} \tag{3.12.3}$$

and (3.12.1) gives the same information as (3.10.1), which must be the case because it is concerned with the same component of vorticity seen in a different coordinate system.

The equation for the normal component of vorticity is readily shown to be (Scorer, 1967)

$$\frac{\mathrm{D}}{\mathrm{D}t} \frac{\boldsymbol{\omega}.\mathbf{n}}{q} = 0 \tag{3.12.4}$$

to the first degree in ϕ, and so the helical swirling flow is neutral to rotations around the principal normal, i.e. rotations in which the position of the Bernoulli surfaces remains unchanged.

We have thus demonstrated that the discussion of the local instability in curved flow is completely covered by the equation for rotational displacements around the tangent, namely (3.10.1).

Chapter 4

THE ROTATING EARTH

4.1 FORMAL MATHEMATICAL TREATMENT

When Ruskin wrote "These questions about movement lead partly far away into high mathematics" he was expressing a cry of anguish which has often been echoed throughout the world of geographers and others who are very much concerned with what goes on in the great outdoors. Engineers and navigators learn the mathematical rules because they have an instinctive and habitual faith in their correctness; but many of them even then can only give fallacious physical explanations of the formulas they use. Mathematicians tend not to care about physical explanations. In all these categories of people there are exceptions, of which the reader is no doubt one; and for them we shall try to produce both mathematical and physical arguments which explain the phenomena we observe. We shall consider the mathematical argument first because that will establish the precise formula which requires physical explanation.

We shall not consider the effects of the gravitational fields of the sun and moon because they only produce small tidal effects, or effects which are important over periods comparable with a month, or year, and these are so small that they do not produce effects on the motion in the troposphere which are as large as the normal errors of measurement of the motion. For a similar reason we do not expect the effects of the earth's rotation to have effects in times which are small when compared with a day, and this is demonstrated in detail in Section 5.8 for the case of gravity waves.

Newton's law of motion applies to motion relative to a frame of reference which is not accelerating. The motion of our practical frame of reference (Fig. 4.1.i), which is fixed to the surface of the earth at the point 0, which we occupy, consists of a rotation around the axis of the earth once a day, with an angular velocity denoted by Ω. This is the sum of a motion in an easterly direction of speed $\Omega R \sin \theta$, where R is the radius of the earth and θ is the co-latitude, of a rotation about an axis through 0 parallel to the axis of the earth, and of an acceleration towards the axis of the earth of magnitude $\Omega^2 R \sin \theta$. Of these

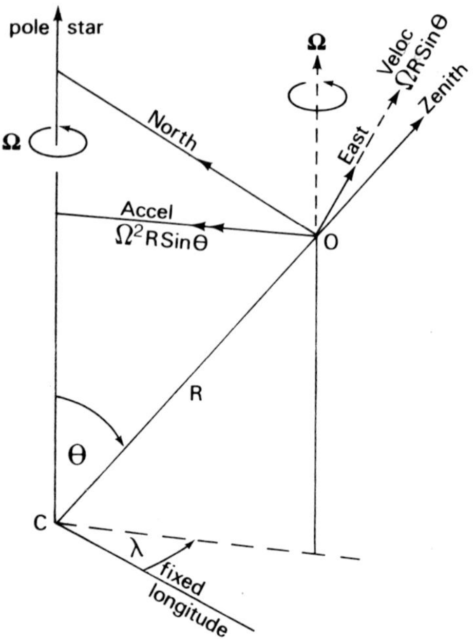

Fig. 4.1.i The velocity, acceleration, and rotation of a coordinate system OENZ at a point on the earth's surface relative to a non-rotating frame at the centre of the earth.

the rotation and the acceleration will affect the motion relative to a frame of reference which has no acceleration. The mathematical statement of this situation is as follows.

A vector which has three components in the direction of the axes, such as a position vector, a displacement, or a time derivative of one of these, and which has magnitude r and rate of change $\mathbf{dr}/\mathrm{d}t$ relative to the axes, has an additional rate of change $-\boldsymbol{\omega} \times \mathbf{r}$ relative to a set of axes, denoted by a suffix 1, which has the same origin and is rotating relative to the original set with angular velocity ω. Therefore

$$\left(\frac{\mathbf{dr}}{\mathrm{d}t}\right)_1 = \frac{\mathbf{dr}}{\mathrm{d}t} - \boldsymbol{\omega} \times \mathbf{r}\,. \tag{4.1.1}$$

This is simply a statement of the fact that a point whose position vector is $\mathbf{r}$ in both coordinate frames at the instant in question and which is fixed in the non-rotating frame, has velocity $-\boldsymbol{\omega} \times \mathbf{r}$ in the rotating one, because $\boldsymbol{\omega} \times \mathbf{r}$ is a vector perpendicular to $\boldsymbol{\omega}$ and $\mathbf{r}$ and of magnitude equal to $\boldsymbol{\omega} \times$ (the projection of r on the normal to $\boldsymbol{\omega}$) (see Fig. 4.1.ii).

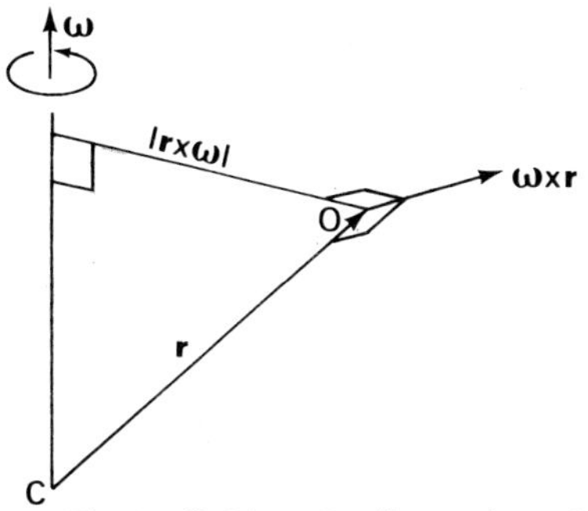

Fig. 4.1.ii Diagram illustrating the basis of the argument for equation (4.1.1).

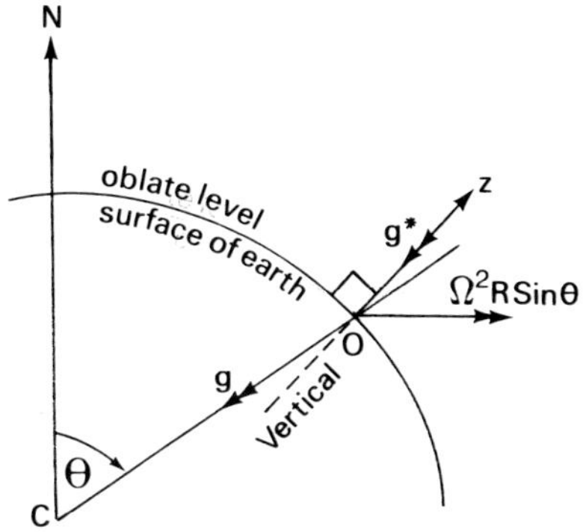

Fig. 4.1.iii Apparent gravity is perpendicular to the surface of the earth, which is an oblate sphere. The force of gravity is towards the centre of the earth, and the difference is the centrifugal force of the earth's rotation.

Therefore
$$\frac{d\mathbf{r}}{dt} = \left(\frac{d\mathbf{r}}{dt}\right)_1 + \omega \times \mathbf{r} = \left[\left(\frac{d}{dt}\right)_1 + \omega \times\right]\mathbf{r} \tag{4.1.2}$$

Evidently this equation relates the time derivatives of a vector relative to the two frames, and we may apply it a second time, or apply it directly to $d\mathbf{r}/dt$ instead of to $\mathbf{r}$. Thus

$$\frac{d^2\mathbf{r}}{dt^2} = \left[\left(\frac{d}{dt}\right)_1 + \omega \times\right]\left[\left(\frac{d}{dt}\right)_1 + \omega \times\right]\mathbf{r}$$

$$= \left(\frac{d^2\mathbf{r}}{dt^2}\right)_1 + \left(\frac{d}{dt}\right)_1 \omega \times \mathbf{r} + \omega \times \left(\frac{d\mathbf{r}}{dt}\right)_1 + \omega \times (\omega \times \mathbf{r})$$

$$= \left(\frac{d^2\mathbf{r}}{dt^2}\right)_1 + 2\omega \times \left(\frac{d\mathbf{r}}{dt}\right)_1 + \omega \times (\omega \times \mathbf{r}) \tag{4.1.3}$$

or
$$\frac{d\mathbf{V}}{dt} = \left(\frac{d\mathbf{v}_1}{dt}\right)_1 + 2\omega \times \mathbf{v}_1 + \omega \times (\omega \times \mathbf{r}) \tag{4.1.4}$$

where $\mathbf{V}$ is the velocity relative to the fixed frame, and $\mathbf{v}_1$ relative to the rotating frame. ω is assumed to be constant.

This gives the acceleration 'in space' in terms of the position, velocity, and acceleration relative to a rotating frame with origin at the centre of the earth, C. If we now transfer to a rotating frame with origin at the point 0 on the earth's surface we are at, the velocity $\mathbf{v}_1$ is the same as the velocity relative to our local frame fixed to the earth, which we call $\mathbf{v}$.

Therefore
$$\frac{\mathrm{d}\mathbf{V}}{\mathrm{d}t} = \left(\frac{\mathrm{d}\mathbf{v}}{\mathrm{d}t}\right)_0 + 2\boldsymbol{\omega} \times \mathbf{v} + \boldsymbol{\omega} \times (\boldsymbol{\omega} \times \mathbf{r}) \qquad (4.1.5)$$

where suffix 0 indicates the value in the frame of reference with which we are moving at 0, and $\mathbf{r}$ is the radius vector from the centre of the earth.

An important particular case of acceleration is that due to gravity. The weight vector of a body of unit mass as normally measured is the mass multiplied by 'apparent gravity'. This is defined as the acceleration of a freely falling particle, instantaneously at rest, relative to our frame moving with the earth. If we call this quantity $\mathbf{g}^*$, then since $\mathbf{g}$ is the actual acceleration we have that, since $\mathbf{v} = \mathbf{o}$

$$\mathbf{g} = \mathbf{g}^* + \boldsymbol{\omega} \times (\boldsymbol{\omega} \times \mathbf{r}) \qquad (4.1.6)$$

This relates the attraction of the earth towards its centre C to the apparent vertical, down which $\mathbf{g}^*$ is directed, the difference being due to the centrifugal force represented by $\boldsymbol{\omega} \times (\boldsymbol{\omega} \times \mathbf{r})$, whose magnitude $\Omega^2 R \sin\theta$ we have already remarked upon. The earth is an oblate spheroid and the vertical does not point to the centre of the earth except at the poles and on the equator (see Fig. 4.1.iii).

The equation of motion for fluid relative to our local axes fixed to the earth is therefore derived from the basic equation

$$\frac{\mathrm{d}\mathbf{V}}{\mathrm{d}t} = \mathbf{F} + \mathbf{g}\ , \qquad (4.1.7)$$

where $\mathbf{F}$ represents *all* the forces other than gravity, and is

$$\left(\frac{\mathrm{d}\mathbf{v}}{\mathrm{d}t}\right)_0 + 2\boldsymbol{\omega} \times \mathbf{v} = \mathbf{F} + \mathbf{g}^* \qquad (4.1.8)$$

From now on we shall write this in the conventional form

$$\frac{\mathrm{D}\mathbf{v}}{\mathrm{D}t} = -\frac{1}{\rho}\,\mathbf{grad}\ p + \mathbf{v} \times \mathbf{f} + \mathbf{g} + \mathbf{F}\ . \qquad (4.1.9)$$

$\mathbf{F}$ now includes any viscous or other forces *not* due to the pressue gradient or gravity, and $\mathbf{g}$ is identified with observed, or 'apparent', gravity, namely

$$\mathbf{g} = (0, 0, -g). \qquad (4.1.10)$$

The quantity

$$\mathbf{f} = 2\Omega = (0, 2\,\Omega \sin\theta, 2\,\Omega\cos\theta) = (0, f_2, f_3) \tag{4.1.11}$$

is called the **Coriolis parameter**, the components being in the Eastward, Northward and upward vertical directions.

The symbol f is conventional for the Coriolis parameter in discussions of our topic. The vector **f** is used elsewhere, particularly in Chapter 3, to represent the acceleration of the fluid, but there is no cause for confusion.

4.2 PHYSICAL INTERPRETATION

The quantity $\mathbf{v} \times \mathbf{f}$ appears to have been added to the forces on a fluid which produce its acceleration. It is usually called the 'deviating force', or 'Coriolis force' after the French engineer who died in 1843. The horizontal component of this force, which is due to the vertical component of **f**, is an acceleration to the right of magnitude vf_3, where v is the horizontal speed relative to the ground.

In this chapter the discussion will view the fluid FROM ABOVE in the NORTHERN HEMISPHERE. To interpret it for the southern hemisphere it may be viewed as if from below, i.e. from inside the earth, or left and right may be interchanged. CYCLONIC means COUNTERCLOCKWISE WHEN VIEWED FROM ABOVE in the northern hemisphere, but clockwise in the southern hemisphere. Clockwise is a northern hemisphere convention: clocks turn with the sun, which is opposite to the earth's rotation. CYCLONIC means rotating WITH THE EARTH.

Considering therefore the vertical component of the earth's rotation, it causes the ground to rotate cyclonically beneath us. If we travel from O to P at a uniform speed v in time t (Fig. 4.2.i) the rotation of the earth will have

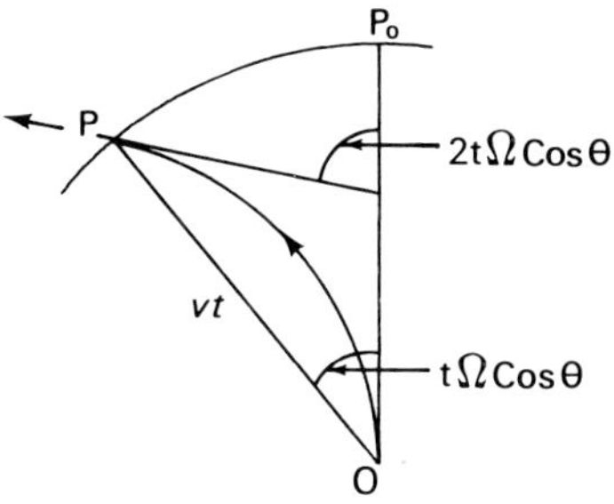

Fig. 4.2.i A straight journey from O to P on the earth follows a curved path in space.

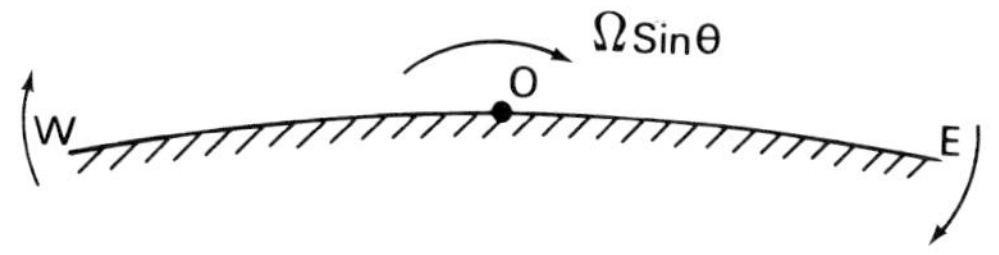

Fig. 4.2.ii The horizontal component of the earth's rotation causes the western horizon to tip upwards.

caused our end point also to have traversed the circular arc P_oP along which it has travelled from its original position P_o, and we shall in space have traversed the circular arc from O to P. Since the vertical component of the earth's rotation is $\Omega \cos \theta$, the angle P_oOP is $t\,\Omega \cos \theta$ and the arc P_oP traversed by P is $vt \times t\,\Omega \cos \theta$. In time t with acceleration a from rest a particle traverses a distance $\frac{1}{2}at^2$. The acceleration in this case therefore is $2v\,\Omega \cos \theta$.

The factor 2 is a source of mystery to many people, because it appears that in getting to P one has merely travelled along OP in a straight line across the ground. However, the track is along the circular arc and the direction of the motion 'in space' (i.e. relative to fixed axis) has turned through an angle $2t\,\Omega \cos \theta$ by the time it arrives at P: it therefore behaves as if the rotation were twice the imagined value.

The horizontal component of the earth's rotation is $\Omega \sin \theta$ and is around a horizontal axis pointing northward. The eastern horizon is sinking and the western horizon is rising (Fig. 4.2.ii), so that the weight of a particle is decreased or increased according as it moves eastward or westward by an amount $2v\,\Omega \sin \theta$ per unit mass. With a typical velocity of 10 m sec^{-1} and Ω equal to $2\pi/(24 \times 60 \times 60)$ rad sec^{-1}, this acceleration is of the order of 10^{-3} m sec^{-2}, which is about $10^{-4} \times$ gravity. It is therefore not productive of any discernible effects because small changes in temperature produce larger changes in the weight of the air. If the whole depth of the atmosphere were moving at 10 m sec^{-1} the pressure at the ground would be changed by about 0.1 mb, which is about the weight of the bottom metre of air. The pressure at the ground varies normally by a few mb in the time between successive weather charts and across the area represented on the chart, and the velocity does not change enough to affect the last significant figure reported in a pressure measurement, namely 0.1 mb.

Ferrel's law (after the American meteorologist, d. 1891) states that the air is deviated to the right as as a result of the earth's rotation; Buys-Ballot's law (after the Dutch meteorologist, d. 1890) states that if the motion is straight there is lower pressure on the left. These are merely particular applications of the principle.

4.3 THE RELATIONSHIP BETWEEN PRESSURE GRADIENT AND WIND

There can be no general relationship between pressure gradient and wind because it is related to the acceleration. Decelerating air is most often moving towards high pressure. It is not correct to say that but for the earth's rotation the air would move towards low pressure; however, we may say, that but for friction it would move equally often towards high pressure. The fact that it moves nearly along the isobars is a measure of the small influence of friction on the motion, because it means that on the whole the pressure field is not being required to do much work on the air.

The facts that the pressure is the easiest parameter to measure accurately about the air and that the air motion is nearly horizontal makes it easy to get a useful relationship. As a first approximation the vertical component of (4.1.9) is

$$0 = -\frac{1}{\rho}\frac{\partial p}{\partial z} - g \tag{4.3.1}$$

because the vertical acceleration is small compared with g the vertical velocity being small, and the deviating and frictional forces are also negligible. This is the well known **hydrostatic equation**, and deviations from it in the atmosphere are usually negligible except in waves of the kind described in Chapter 5, and in some violent storm downdraughts or funnels.

There are several types of horizontal motion, according to the relative importance of the terms in the horizontal component of (4.1.9), namely

$$\frac{\mathrm{D}\mathbf{v}_h}{\mathrm{D}t} = -\frac{1}{\rho}\,\mathbf{grad}_h\, p + \mathbf{v}_h \times \mathbf{f} + \mathbf{F}_h \tag{4.3.2}$$

the horizontal component of the velocity $\mathbf{v}$, of $\mathbf{F}$ and of the gradient of p being taken, and the vertical component of $\mathbf{f}$.

(1) The geostrophic wind

Very often the air travels large distances with only a small change in speed and direction during a day. The acceleration is nearly zero. Also the frictional force is small compared with the remaining terms and all that is then left of (4.3.2) is

$$0 = -\frac{1}{\rho}\frac{\partial p}{\partial x} + vf_3 + 0 \tag{4.3.3}$$

in which the x direction is taken across the direction of motion and v is the speed. The wind is then in geostrophic (earth-turning) balance and

$$v_g = \frac{1}{\rho f}\frac{\partial p}{\partial x}, \tag{4.3.4}$$

where f is written for f_3 and v_g is the geostrophic wind speed. It is inversely proportional to the spacing of the isobars, with low pressure on the left.

(2) The ageostrophic wind

This is the departure of the actual wind from geostrophic, denoted by v_a. Thus

$$\mathbf{v} = \mathbf{v}_g + \mathbf{v}_a\,. \tag{4.3.5}$$

In the absence of friction, therefore, but with a significant acceleration, we find, on subtracting (4.3.3) from the appropriate component of (4.3.2), that

that $$\frac{D\mathbf{v}}{Dt} = \mathbf{v}_a \times \mathbf{f} \tag{4.3.6}$$

The ageostrophic wind is thus proportional to the acceleration and directed at right angles to the left of it.

(3) Antitriptic winds

These were defined by Jeffreys to be those in which the frictional force balanced the pressure gradient force. The concept is only useful for winds close to the ground where the frictional force is opposed to the velocity. The motion is then directed towards low pressure. If we imagine that the drag is proportional to kv^2 we would obtain

$$kv^2 = -\frac{1}{\rho}\frac{\partial p}{\partial y} \tag{4.3.7}$$

where the y-direction is the direction of motion and v the speed.

A wind in a ravine or canyon, a sea breeze, a katabatic or anabatic wind, usually has a significant antitriptic component.

(4) Shearing stresses

If the wind varies with height, from causes to be discussed below, and there is some mixing of superposed layers, momentum is transferred to the slower moving layer.

In the case of drag at the surface, work must be done to overcome it and so there is a component of velocity towards low pressure (Fig. 4.3.i). On the other

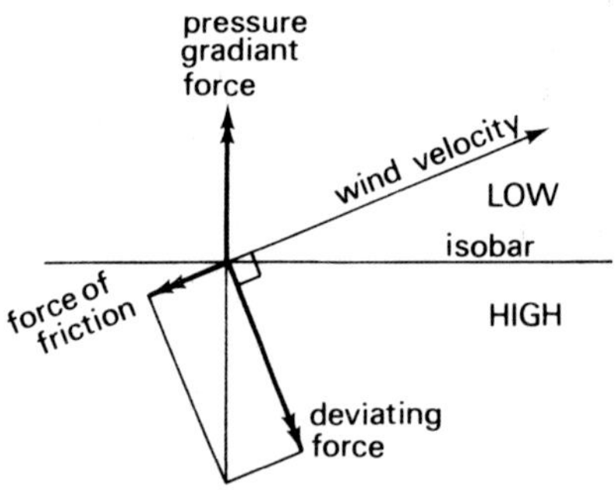

Fig. 4.3.i Deviation of the wind from the isobar on account of surface friction.

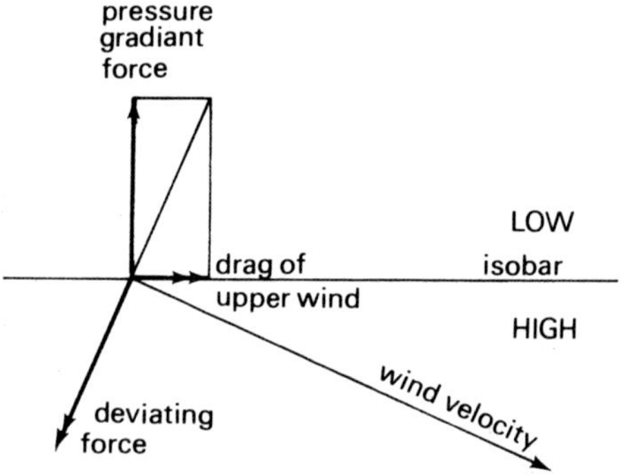

Fig. 4.3.ii Deviation of the wind from the isobar due to drag of upper wind.

hand the friction at the surface may be small, as it often is over the sea, while the wind speed increases with height enough for the lower layers to be dragged forward by the upper layers, whose direction may also differ. The wind then has a component towards high pressure (Fig. 4.3.ii). It is not correct to say that because of friction the surface wind always has a component towards low pressure: it may have one towards high pressure, and if it has one towards low pressure this could be on account of a retarding stress due to layers above moving more slowly or in a different direction. In the Trade winds the flow across the isobars is primarily due to the slower moving upper layers, a change in direction with height, or an equatorward drift. The matter is discussed further in Section 4.11.

(5) The gradient wind: curved flow

When the motion is circular we may readily represent the acceleration by the quantity v^2/r, where r is the radius of curvature of the air path. If only the pressure gradient force and deviating force are also important we have the following equation, balancing the radial acceleration with the radial forces:

$$\mp \frac{v^2}{r} = \mp \frac{1}{\rho} \frac{\partial p}{\partial r} + vf \qquad (4.3.7)$$

The upper sign is appropriate to cyclonic motion in which the pressure gradient force is radially inwards and the deviating force outwards. They are reversed for an anticyclone (Fig. 4.3.iii).

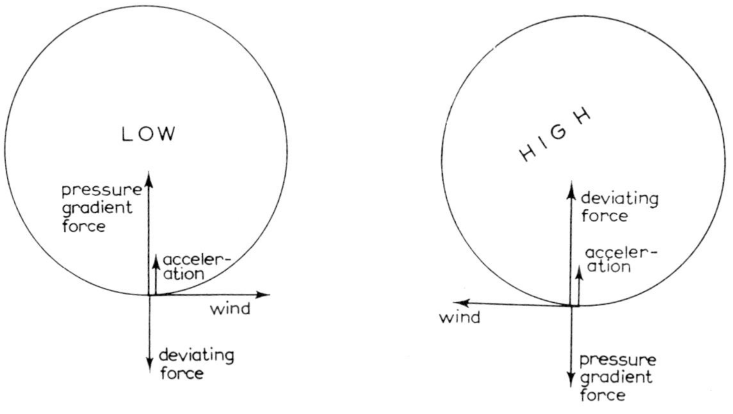

Fig. 4.3.iii Balance of forces in curved frictionless flow.

The gradient wind equation, (4.3.7) is a good approximation to reality when the flow is curved without much tangential acceleration.

(6) Changing pressure field: isallobaric wind

The isallobaric wind is that component of the wind due to the changing pressure field. If, during the change, the wind is roughly geostrophic we may differentiate the geostrophic wind equation with respect to time and obtain

$$0 = -\frac{\partial}{\partial t}\left(\frac{1}{\rho}\,\mathbf{grad}\,p\right) + \frac{\partial}{\partial t}\,(\mathbf{v} \times \mathbf{f}) \tag{4.3.8}$$

or

$$\mathbf{f} \times \frac{\partial \mathbf{v}}{\partial t} = -\frac{1}{\rho}\,\mathbf{grad}\,\frac{\partial p}{\partial t} \tag{4.3.9}$$

If we now assume that there is very little variation in the acceleration from place to place we may equate Dv/Dt to $\partial v/\partial t$. This is by no means always justified in practical cases, but pursuing this approximation we have

$$\mathbf{f} \times \frac{D\mathbf{v}}{Dt} = -\frac{1}{\rho}\,\mathbf{grad}\,\frac{\partial p}{\partial t} = \mathbf{f} \times \mathbf{v}_i \times \mathbf{f} \tag{4.3.10}$$

because in this case the isallobaric wind is the ageostrophic wind given by (4.3.6). The isallobaric wind is therefore directed at right angles to the isallobars (lines of constant $\partial p/\partial t$) towards the region of greatest rate of fall of pressure. It has magnitude given by

$$\mathbf{v}_i = \frac{1}{\rho f^2}\,\frac{\partial p}{\partial x} \tag{4.3.11}$$

because $\mathbf{v}_i$ and $\mathbf{f}$ are approximately perpendicular.

This classical piece of theory is a very unsatisfactory explanation of why the wind blows across the isobars towards a region of maximum pressure fall when the gradient is increased very rapidly. For if the fall is rapid, geostrophic flow does not have time to be approximately balanced and the inflow merely represents this fact. (4.3.11) is then quite inaccurate because in a small fraction of a day the Coriolis effect cannot be dominant.

(7) The geostrophic acceleration: confluence and diffluence

When the flow is almost straight and the pressure distribution nearly steady so that curvature and isallobaric effects can be neglected, the ageostrophic component may be deduced from the rate of change of the geostrophic wind along the streamlines, that is along the isobars. Figure 4.3.iv indicates the way in which the air flows across the isobars so as to move towards high pressure

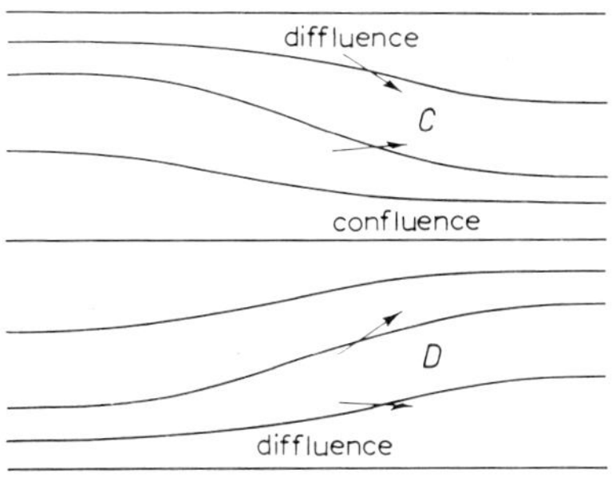

Fig. 4.3.iv Ageostrophic motion due to confluence and diffluence. An important consequence is convergence (C) and divergence (D).

where it is being decelerated and towards low pressure where it is being accelerated.

The flow is called confluent of diffluent where it is accelerating or decelerating respectively. If it were geostrophic it would be non-divergent because the velocity would everywhere provide the same flux between two isobars. But where there is diffluence on the left and confluence on the right there is convergence (marked C in the diagram), and divergence occurs in the opposite case. Since the acceleration is approximately $(\mathbf{v}_g . \,\mathbf{grad})\, \mathbf{v}_g$, according to (4.3.6) the magnitude of the ageostrophic wind is given by

$$v_a = \frac{1}{f}\,(\mathbf{v}_g . \,\mathbf{grad})\mathbf{v}_g \,, \tag{4.3.12}$$

the direction being across the isobars towards low pressure for an acceleration.

4.4 MOTION IN THE UPPER AIR; ISOBARIC CONTOURS

It is not convenient to draw maps of the pressure distribution except at the surface. Instead we draw contours of an isobaric surface. Figure 4.4.i shows a vertical cross section of an isobaric surface perpendicular to the isobars at height h. If Δp is the difference between the pressure at A and at B, both at height h, since the pressure is the same at A as at A′ a height Δh above B, we must have that $\Delta p = g\rho\Delta h$. Therefore

$$\frac{1}{\rho}\frac{\partial p}{\partial x} = g\,\frac{\partial h}{\partial x} \tag{4.4.1}$$

and the isobars at height h may be represented by the contours of the isobaric surface through the point. The geostrophic wind equation (4.3.4) may be written

$$v_g = \frac{g}{f}\frac{\partial h}{\partial x} \tag{4.4.2}$$

and a map of contours of an isobaric surface looks very like a map of isobars at a constant height. The advantage of the former is that the factor g/f is independent of the density, and therefore of height, and the same scale factor may be used for contour maps at any height.

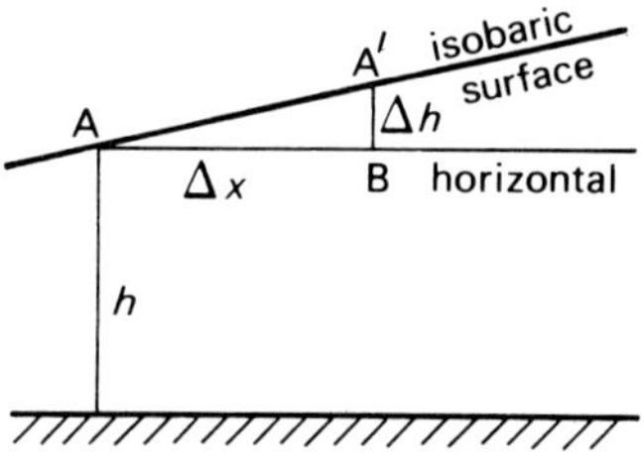

Fig. 4.4.i Relationship between isobars at constant height and contours of an isobaric surface.

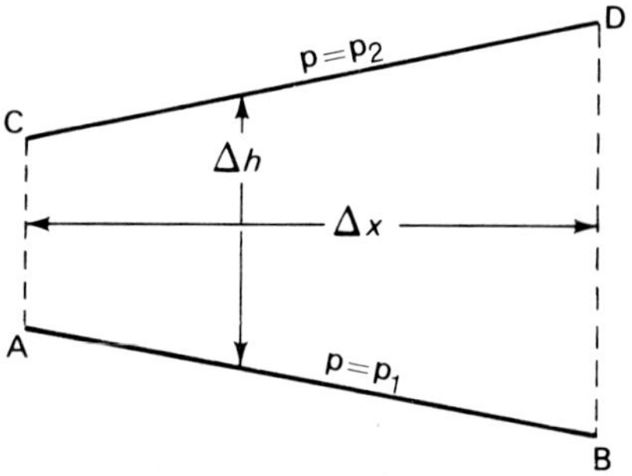

Fig. 4.5.i Relationship between thermal wind and isobaric thickness.

4.5 THE THERMAL WIND

If there is a horizontal gradient of temperature the horizontal pressure gradient must change with altitude because two neighbouring columns of air do not have the same weight. Figure 4.5.i represents a vertical cross section of the air perpendicular to the horizontal isotherms in constant pressure surfaces p_1 and p_2. Since (4.4.2) holds at the two levels denoted by suffixes 1 and 2, the difference in geostrophic wind Δv_g is given by

$$\Delta v_g = \frac{g}{f}\Delta\frac{\partial h}{\partial x} = \frac{g}{f}\frac{\partial}{\partial x}\Delta h \tag{4.5.1}$$

where Δh is the thickness of the layer between the two isobaric surfaces. Since the height of a column of air of given horizontal section and given weight is proportional to temperature, the thickness is proportional to temperature, at a given pressure. This means that

$$\frac{\partial}{\partial x}\Delta h = \frac{\Delta h}{T}\left(\frac{\partial T}{\partial x}\right)_{p=\text{const}}, \tag{4.5.2}$$

and so
$$\frac{\partial v_g}{\partial z} \simeq \frac{\Delta v_g}{\Delta h} = \frac{g}{fT}\left(\frac{\partial T}{\partial x}\right)_{p=\text{const}} \tag{4.5.3}$$

At a level where the isobaric surfaces are horizontal, the vertical gradient of geostrophic wind blows along the isotherms with low temperature on the left, and is therefore called the thermal wind. Where the isobaric surfaces are not horizontal a correction is required, and so in practice the isotherms are not drawn on charts, but instead contours of the thickness between two isobaric surfaces are drawn, and these represent the difference between the winds at the two levels. This equivalent to using (4.5.1) rather than (4.5.3). The name thermal wind persists because the wind shear is attributable to horizontal temperature gradients.

4.6 ESTIMATING THE VERTICAL VELOCITY

The problem facing a weather forecaster is as follows: all cloud formation and rain is due to upward motion of the air, and the disappearance of cloud is due to subsidence. But the vertical velocity of the air over a large area is too difficult to measure because of its small magnitude (10^{-3} up to 1 m sec^{-1}). Furthermore, because of the sparsity of observations and their lack of accuracy, it is not possible to measure the horizontal wind so that the horizontal gradients of u and v can reveal w by means of the simplified version of the equation of continuity, namely

$$w = \int \left(\frac{\partial u}{\partial x} + \frac{\partial v}{\partial y}\right) \mathrm{dz}\ . \tag{4.6.1}$$

To add to the difficulty, if we were to measure the pressure accurately, and were to assume the horizontal wind to be geostrophic to determine its value, we would find it to be non-divergent, and leading to zero vertical velocity. Mathematically we note that the components of the geostrophic wind are

$$u_g = -\frac{1}{\rho f}\frac{\partial p}{\partial y}, \quad v_g = \frac{1}{\rho f}\frac{\partial p}{\partial x} \tag{4.6.2}$$

so that
$$\frac{\partial u_g}{\partial x} + \frac{\partial v_g}{\partial y} = 0 \tag{4.6.3}$$

except for horizontal gradients of ρf which are relatively too small to lead to a significant value of w; and in any case variations in ρf are not a significant cause of vertical motions.

In other words, vertical motion is associated with the ageostrophic component of the wind. We are not concerned here with the mathematical techniques used in numerical weather forecasting, but with understanding the basic mechanisms whereby upward motion can be associated with measurable and, preferably, easily observable phenomena. For some people these ideas are very helpful in understanding what can be seen in the sky and on a weather chart and in making one's own forecast or making the best use of public ones.

If the wind always blew geostrophically there could be no vertical motion, and so vertical motion is determined by the ageostrophic component of the wind. When vertical motion occurs there must be a stretching of vertical columns of fluid because the vertical velocity is zero at the ground. Almost the whole atmosphere possesses vorticity because it is rotating with the earth, and even in anticyclones the absolute vorticity is positive, i.e. in the same direction as the earth's rotation, even though it is less than that of the earth. Consequently vertical stretching of the vortex lines intensifies the vorticity, at least near the ground. At some higher altitude the vertical column of fluid must be subject to vertical compression and horizontal divergence of the flow, and this results in a decrease in the vertical component of vorticity, probably to a value less than that of the earth. In consequence, relative to the earth, the circulation appears anticyclonic.

Near the bottom of the atmosphere, therefore, there must be an increase of cyclonic vorticity when air rises. This means an increase in circulation round a circuit of given size near the bottom, which requires an increase in the pressure gradient, and therefore a fall in pressure at the centre beneath the rising air. This is the logic behind the close association of falling pressure and rising air, which in turn is associated with cloud and rain. Thus it is not so much low pressure as falling pressure that is associated with rising air: low pressure has been thought of as the cause of rain and the association has sometimes been explained by asserting that friction causes the air to flow in towards low pressure in the manner we have discussed in Chapter 3. It should be pointed out that the rain is associated mainly with troughs of low pressure, that the air behind these is very often subsiding and clear of cloud, and that the vertical motion is mainly due to the lifting of warm air by the often subsiding cold air, and has little to do with friction. It is true that friction is a cause of inflow, but this often has the effect of causing the low pressure centre to fill up because the friction decreases the circulation.

In practice therefore, a forecaster estimates the amount of vertical motion as much by keeping a watch on the vorticity of the geostrophic wind as by any other means. This can be determined directly from the pressure field and is equal to $\frac{1}{\rho f}\nabla^2 p$. This vorticity is increased by convergence at low levels, and low pressure centres are formed or deepened by convergence. The amount of convergence that is produced in this way far exceeds that due to friction in those latitudes where the earth's rotation is important. Close to the equator, where the vertical component of the earth's rotation is negligible, large gradients of pressure are not generated.

4.7 SUTCLIFFE'S DEVELOPMENT THEORY

Development is the name given to the formation of low pressure centres and the vertical motion and weather phenomena which accompany it. We concentrate our attention on the horizontal acceleration field which is the cause of all the ageostrophic motion in the absence of friction.

If suffixes 1 and 2 denote the wind vector at two levels, $\mathbf{grad}_h$ denotes the horizontal component of the gradient, and a prime denotes the difference between the two levels, then

$$\mathbf{v}' = \mathbf{v}_2 - \mathbf{v}_1 \tag{4.7.1}$$

and the acceleration at the upper level is

$$\frac{D\mathbf{v}_2}{Dt} = \frac{\partial \mathbf{v}_2}{\partial t} + (\mathbf{v}_2 \,.\, \mathbf{grad}_h)\,\mathbf{v}_2$$

$$= \frac{\partial \mathbf{v}_2}{\partial t} + (\mathbf{v}_1 \,.\, \mathbf{grad}_h)\,\mathbf{v}_1 + (\mathbf{v}' \,.\, \mathbf{grad}_h)\mathbf{v}_1 + (\mathbf{v}_2 \,.\, \mathbf{grad}_h)\mathbf{v}' \,. \tag{4.7.2}$$

Consequently

$$\frac{D\mathbf{v}_2}{Dt} - \frac{D\mathbf{v}_1}{Dt} = (\mathbf{v}' \,.\, \mathbf{grad}_h)\mathbf{v}_1 + \left[\frac{\partial}{\partial t} + \mathbf{v}_2 \,.\, \mathbf{grad}_h\right]\mathbf{v}' \tag{4.7.3}$$

According to (4.3.6) the ageostrophic motion is entirely produced by the acceleration. Since it is also responsible for the convergence or divergence of the horizontal component of the motion it is responsible for the vertical motion. Upward motion is associated with convergence below and divergence above, and also with increasing cyclonic circulation and falling pressure in the lower layers. Equation (4.7.3) gives the difference between the accelerations above and below and expresses this difference, which determines the amount of cyclogenesis, in terms of the wind field below ($\mathbf{v}_1$), the advection by the thermal wind ($\mathbf{v}' \,.\, \mathbf{grad}_h$), and the rate of change of the thermal wind field ($\mathbf{v}'$) as seen at the upper level (where $\mathbf{v} = \mathbf{v}_2$). We now consider the various mechanisms involved.

(1) Thermal steering of the surface pressure field

As a practical approximation to the wind field we may take the geostrophic wind, and for $(\mathbf{v}'.\,\mathbf{grad}_h)\mathbf{v}_1$ we write $(\mathbf{v}_g'.\,\mathbf{grad}_h)\mathbf{v}_g$. This is a vector in the direction in which the surface isobars are turned as we move along with the thermal wind, i.e. along the isotherms with cold air on the left. This moves the surface pressure pattern in the direction of the thermal wind, and is called **thermal steering.**

To exemplify the argument leading to this conclusion we imagine a closed cyclonic circulation in the surface isobars, with straight isotherms of the thickness pattern across the top of it (Fig. 4.7.i). The cold air is indicated to lie to the north so that this is a simplified version of a typical mid-latitude situation of a cyclone in the westerlies. The low level geostrophic wind $\mathbf{v}_{g1}$, is circulating

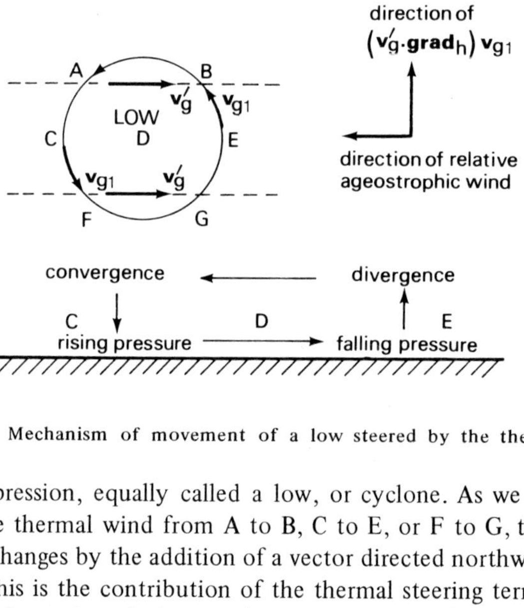

Fig. 4.7.i Mechanism of movement of a low steered by the thermal wind.

around the depression, equally called a low, or cyclone. As we pass along the direction of the thermal wind from A to B, C to E, or F to G, the surface geostrophic wind changes by the addition of a vector directed northwards (i.e. along FA, or GB). This is the contribution of the thermal steering term (first on the right hand side) to the relative acceleration given by (4.7.3). According to (4.3.6) the ageostrophic wind is directed at right angles to this to the left, namely westwards (to the left in Fig. 4.7.i). This means that the upper air is undergoing a displacement as shown in the lower diagram, namely to the left relative to the lower layers, and this implies upward motion with cyclonic development at the surface ahead of the depression and anticyclonic development as a consequence of downward motion behind it. Thus the depression is moved in the direction of the thermal wind.

A second example is provided by a pattern of diffluence, as shown in Fig. 4.7.ii. Here the geostrophic wind is decreasing in the downstream direction

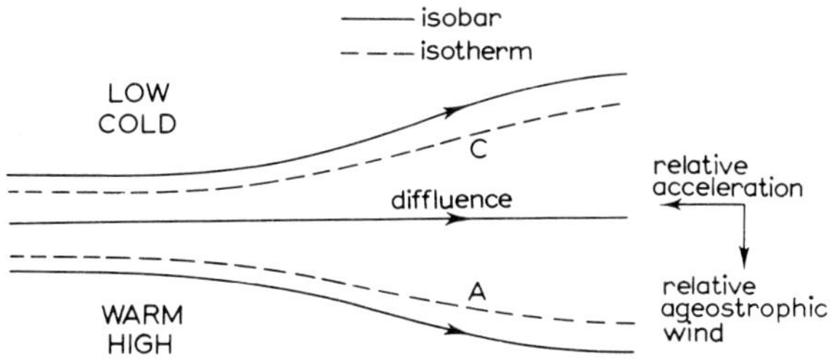

Fig. 4.7.ii Movement of a diffluence pattern in the direction of the thermal wind.

which is also taken to be the direction of the thermal wind; and so the relative acceleration is a vector directed in the upstream direction, opposite to the wind in this case. The relative ageostrophic wind is therefore at right angles to this, indicating low level convergence on the left and divergence on the right. Therefore there is cyclogenesis (C) or falling pressure on the left and anticyclogenesis (A), or rising pressure on the right so that the pressure gradient is increased, and the pressure pattern is moved in the direction of the thermal wind. It is interesting to compare this with the ageostrophic motion indicated in Fig. 4.3.iv, which is deduced without reference to the upper air and the thermal wind pattern, and indicates a quite different effect.

Equation (4.7.3) could equally be written with the suffixes 1 and 2 interchanged and the sign of $\mathbf{v}'$ changed. Then we would be looking at how the thermal wind advected the upper air pressure field. But similar conclusions would not be valid because there is not necessarily the same relationship between cyclogenesis and vertical motion as exists at the surface where there is a rigid horizontal boundary to give rise to it. One could argue that of course the pressure pattern is moved along with the upper wind because it is due to the weight of the upper air: but that would be erroneous because the pressure field at the surface is not actually due to the upper air which moves along with the thermal wind. The isotherms in the case illustrated in Fig. 4.7.i do not indicate warm air over the cyclone. In fact 'cold lows' are common in the atmosphere, hurricanes being the extreme case. The low pressure is produced by warm air in the stratosphere, and the low pressure usually indicates that the tropopause is lower than average: that warm air is not moved along with the cyclone, but that a wave travels along the tropopause and is induced by the cyclone below.

Thus although we have indicated by two examples how the pressure pattern is moved along by the thermal wind we have only indicated the argument that can be used in any particular case. To prove this result of Sutcliffe's in a general way would require a formal representation of the surface boundary condition and some other restraint on the motion in the neighbourhood which would make cyclogenesis the necessary consequence of the ageostrophic motion deduced from the acceleration, because obviously we have assumed that the compensating vertical motions occur in a particular way. Nevertheless this may be regarded as a theorem for the circumstances supposed, which are like the real atmosphere.

This is a superb example of deducing all the aspects of the motion that are important from the geostrophic wind field which cannot actually produce them. It represents a very considerable breakthrough in dynamical thinking.

(2) The formation of lows by heating

The term $\partial \mathbf{v}'/\partial t$ on the right hand side of (4.7.3) represents the effect of local changes in the temperature field. We illustrate this by imagining the air to be warmed over a region, such as Britain or Iberia on a summer's day. The thermal wind will blow with warm air on the right, that is anticyclonically around the heated region (Fig. 4.7.iii): as it develops the local rate of change is in that same direction, and by Equation (4.7.3) this is also the direction of the relative

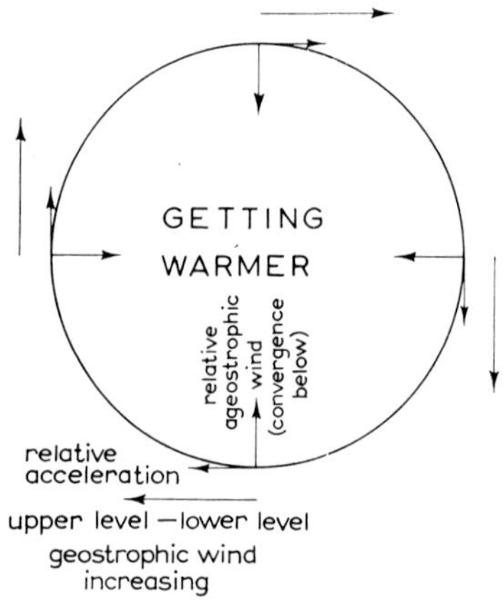

Fig. 4.7.iii The generation of a surface low by an anticyclonic acceleration aloft around warmed air.

acceleration. The ageostrophic component is at right angles to this to the left, that is outwards from the heated region. This produces divergence aloft and convergence below, and therefore upward motion in the middle, an increase in low level vorticity, and the establishment of a heat low.

We conclude that heating produces centres of low pressure. This may seem rather an obvious result to a reader well schooled in the idea that hot air rises after producing low pressure beneath it because it is less dense. He is probably familiar with this in the case of the establishment of sea breezes and similar phenomena; but in many such cases the earth's rotation is negligible and friction important, and the winds may even be antitriptic. In the present case the earth's rotation is the dominant influence and friction negligible. It illustrates the principle that the earth's rotation is a cause of the development of low pressure by dynamical mechanisms. The hurricane is the extreme example which develops because there is a heat source, but the lighter air that has been warmed by the heat source is not the cause of the low pressure at the centre.

The low pressure is intensified by the inflow at the bottom, because it accompanies the vertical stretching of the vortex lines, and the increase in the circulation. Indeed the Coriolis force, which makes the inflowing air deviate to the right, may be said to produce the low pressure. By contrast, if the low pressure is produced in the absence of the earth's rotation the inflow causes the low to fill up, and friction retards this filling a little. In our case the effect of friction is to facilitate inflow without increasing the circulation, thereby hastening the filling up of the low.

(3) The intensification of temperature gradients

A horizontal temperature gradient may be intensified at an already existing zone of large temperature gradient by the wind blowing across the isotherms towards a region where it blows along them, for example in a frontal zone. In Fig. 4.7.iv the flow depicted in the left diagram will bring warm and cold air closer together

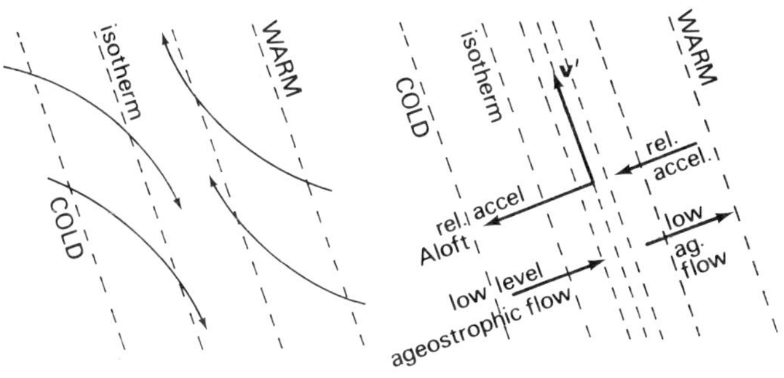

Fig. 4.7.iv Intensification of thermal gradient.

and intensify the temperature gradient by advection in the upper air, represented by the term $(\mathbf{v}_2 \,.\, \mathbf{grad}_h)\,\mathbf{v}'$ in (4.7.3). The thermal wind, which is along the isotherms, is being accentuated by advection and so the relative ageostrophic wind is at right angles to the vector $\mathbf{v}'$ which is being generated, producing a flow away from the zone of increasing temperature gradient aloft and motion towards it below. The zone is therefore a region of falling pressure. In this case it is a trough of low pressure, not a centre, and the development is typical of a cold front.

The same result can be deduced for a warm front, which is illustrated in Fig. 4.7.v.

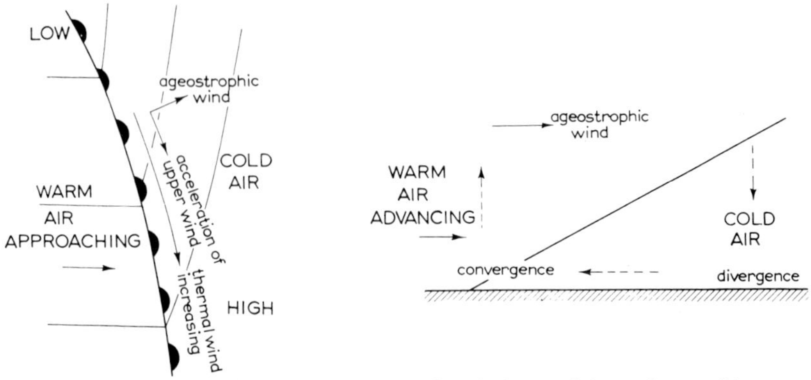

Fig. 4.7.v The left diagram shows the surface isobars and the surface position of a warm front. On the right are shown the consequences in a vertical section of the ageostrophic wind caused by the approach of the warm air.

When a cold front approaches an anticyclone, and the thermal wind increases, the pressure falls ahead of it and rises behind it so as to displace the front through the anticyclone, which reforms behind it. The warm front approaching a cold air mass may produce the opposite result, depicted in Fig. 4.7.v, where the divergence ahead of the advancing warm air is accompanied by rising pressure, and the front is halted. In general the intensification of a temperature gradient produces rising pressure in the cold air, and falling pressure in the warm air, but it must be remembered that it is only one of the mechanisms we have been concerned with and any actual case might produce a more complex result.

It cannot be too much emphasised that whatever part friction may play in atmospheric motion it is not one of the major causes of significant development on the scale of cyclones and anticyclones.

4.8 FRONTS

It will be evident in Chapter 7 that gradients of temperature and velocity are not smoothed out by the bulk motion of a fluid, but only by the molecular motion. The idea that eddies serve to smooth out gradients is derived from the view of turbulence in which small scale detail is defocussed. Actual motion almost invariably intensifies gradients of temperature and velocity, otherwise the mechanism of molecular diffusion would never be significant. These points are not affected by the presence or otherwise of the earth's rotation, but since large gradients of velocity are vorticity and gradients of density generate vorticity, we would expect to find the same mechanism of gradient intensification taking place on a large scale in an atmosphere which is all possessed of vorticity because it is on a rotating earth.

The classical picture of a front, formulated by Margules, demonstrates the importance of rotation. It is a special case of the thermal wind mechanism producing a discontinuity of geostrophic wind at a discontinuity of temperature. If a temperature discontinuity ΔT is inclined at an angle α to the horizontal, the horizontal pressure gradient in the vertical plane containing the steepest slope in the air below the front exceeds that above (see Fig. 4.8.i) by $\Delta\rho g \sin\alpha$.

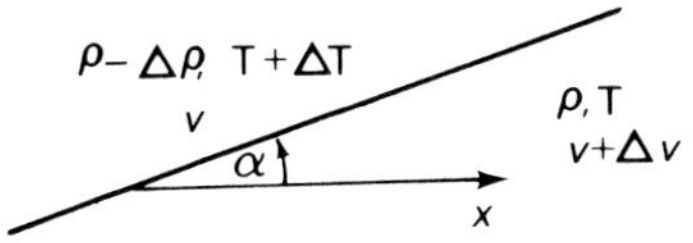

Fig. 4.8.i A discontinuity of density $\Delta\rho$ and velocity Δv inclined at angle α to the horizontal.

If v and $v + \Delta v$ are the geostrophic winds directed normal to this plane, then

$$v = \frac{1}{(\rho - \Delta\rho)f}\frac{\partial p}{\partial x}, \quad v + \Delta v = \frac{1}{\rho f}\left(\frac{\partial p}{\partial x} + \Delta g \sin\alpha\right); \tag{4.8.1}$$

from which we deduce that, since $\Delta\rho/\rho = -\Delta T/T$

$$\Delta v = \frac{\Delta T}{T}\left(\frac{g}{f}\sin\alpha - v\right). \tag{4.8.2}$$

The term $v\Delta T/T$ represents the effect of the difference in density in the basic geostrophic wind equation (4.3.3) and is negligible in practical cases. Thus in CGS units g/f is of the order of 10^7, while v is of the order of 10^3: and so if $\sin\alpha$ is of the order of 1:100 we may write (effectively using the Boussinesq approximation) that

$$\Delta v_g = \frac{\Delta T}{T}\frac{g}{f}\sin\alpha \tag{4.8.3}$$

Since the geostrophic wind is discontinuous at fronts, so also is the ageostrophic wind, and there may be up and downslope motions in either air mass, and these may increase the temperature discontinuity. In Figure 4.8.ii motions

Fig. 4.8.ii Intensification of temperature contrast at a front.

which typically occur in the vertical plane normal to a front are shown. If air from A and B come into juxtaposition at C the temperature gradient can be sharpened, with the consequence that the cold air advances undercutting the warm air according to the mechanism described in Section 4.7 (3).

In some cases the motion leads to a decrease in the temperature gradient, as often happens at a cold front of type (a), called a katafront, in which the cold air is warmed by subsidence. The consequence is low level ageostrophic flow away from the front and sinking motion in the warm air also. In that case the cloud systems produced when fronts cause cyclogenesis are rapidly dispersed.

Figure 4.8.iii depicts the motion in three dimensions at a warm front of a typical cyclone, where the motion is like that in Fig. 4.8.ii(a). The cold front

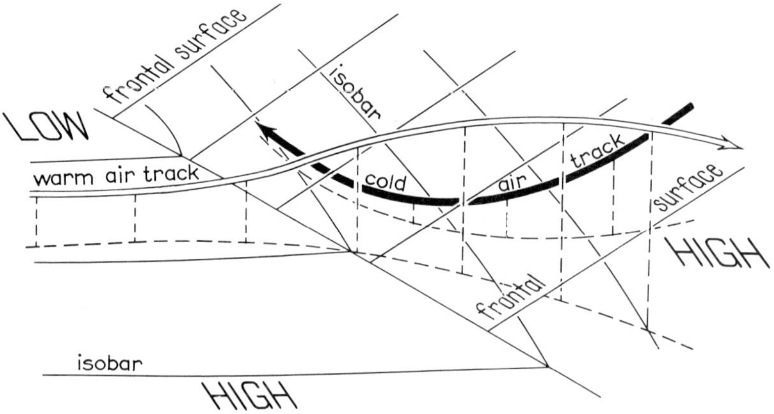

Fig. 4.8.iii The isobars at ground level, the frontal surface at a warm front, and the typical tracks of warm air above the surface and cold air ahead of and below it.

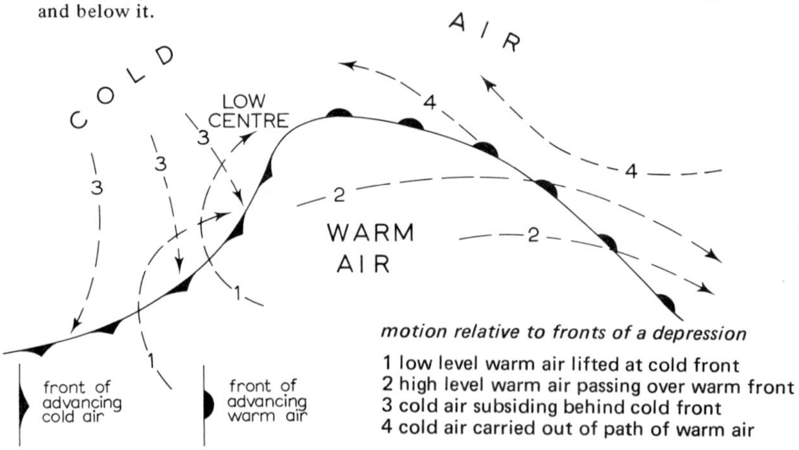

Fig. 4.8.iv Typical trajectories of air in a warm sector cyclone. The position of the fronts is shown as their position at ground level. The warm sector is larger at greater height.

is a similar motion in reverse, and the tracks of different parts of the warm and cold air masses are shown in Fig. 4.8.iv.

4.9 JET STREAMS

The warm air above a front is subject to the thermal wind and therefore usually blows almost along the front with the cold air on the left. A typical cross section of a frontal zone is shown in Fig. 4.9.i, in which the vertical scale is exaggerated. It shows the order of magnitude of the wind maximum and its horizontal and vertical extent, but obviously individual cases differ considerably. The jet stream is the name given to the wind maximum. Such concentrations of

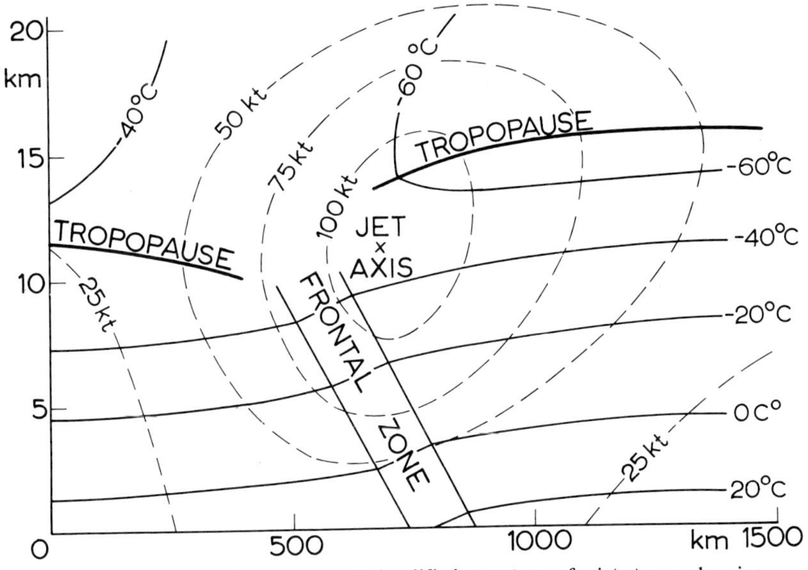

Fig. 4.9.i Typical dimensions and simplified structure of a jet stream showing isotherms in the section and wind speed blowing across it (into the paper).

velocity are easily reproduced in rotating masses of fluid in which temperature gradients are generated. In Fig. 4.9.i the front is depicted as a zone of large gradient rather than a discontinuity, and this is nearer to reality, but the effects are the same.

Above a jet stream the temperature gradient is reversed, usually in the stratosphere, and the wind decreases with height. The strength of a jet stream varies along a front, and has a maximum to the west of the centre of a cyclone which is active on a wave of a front. Fig. 4.9.ii shows a typical distribution of the jet stream maximum in relation to the wave, which is moving much more slowly, perhaps at 30 to 50 kt. The warm air in the jet stream therefore overtakes the cyclone, and is accelerated into the maximum at the jet entrance at A, and decelerated as it emerges at the jet exit at B.

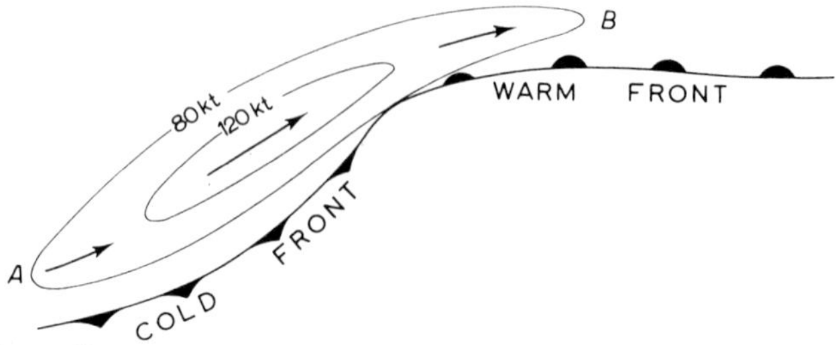

Fig. 4.9.ii Typical position and speed distribution of jet stream maximum in relation to surface position of fronts in a warm sector cyclone. This diagram is best interpreted with Figs. 4.8.iv and v and 4.9.i in mind.

The energy required to produce the acceleration is derived from a direct circulation in which the cold air sinks and the warm air rises over it, thereby decreasing the potential energy of the air. The motion in this circulation is ageostrophic, and is largely accounted for by the mechanism of the acceleration of the geostrophic wind described in Section 4.7 (3). At a jet exit the deceleration is effected by an indirect circulation in which the warm air sinks and moves ageostrophically towards higher pressure: kinetic energy is converted into potential energy. The jet maximum, or *axis* of the jet, is usually about 1 km higher at the entrance than at the exit because the maximum is further downwind at the lower levels.

In spite of their name, depressions are the object of the most delightful mechanisms in the atmosphere. Their general anatomy, with the direction of motion of the warm air revealed by cirrus and altocumulus clouds, was known to mariners long before aviators could explore them, before balloon soundings were in regular use, and before scientists had made any significant analysis of them.

The condensation of clouds increases the time scale of depressions because rising warm air may be continuously warmed by the condensation and release of rain. Thus there are probably no cyclones of comparable scale on Mars. Venus and Jupiter have deeper atmospheres with clouds of quite different chemical composition. The readiness of water to form rain and its very large latent heat probably make the earth's atmosphere beautiful in a way which is rather rare in the universe. Certainly no other planet in the solar system has such rapidly varying and partly clouded skies.

4.10 HURRICANES

In Sections 7.8 and 7.10 mechanisms whereby vortices may be generated as a result of convection clouds are described. The part played by clouds is discussed in more detail in Section 11.10. The starting point is in an air mass possessing some absolute rotation as a result of being on the rotating earth. The feature of hurricanes which makes them different from larger cyclones is that they depend on a continuous supply of heat to maintain the circulation. This heat source is the rain, whose latent heat of condensation is given to the air in which the rain condenses. This heat source exists only because the sea level pressure at the centre is lower than in the surrounding air: thus the air temperature is lower by between 3° and 5°C at the centre because the air expands adiabatically as it moves inwards. In consequence much more violent convection is supported there.

Because of the strong winds and rough sea, the air spiralling inwards at the sea surface is subject to considerable retardation due to friction. In the absence of friction the tangential velocity relative to fixed axes would increase like r^{-1}, r being distance from the centre: instead, it is observed to increase inwards roughtly like $r^{-1/2}$. Near the centre the rotation of axes fixed to the earth is a small contribution which can be neglected. The air does not penetrate to the centre of the storm but only to the outer edge of the eye, which is a region free of rain clouds composed of air sucked down from the stratosphere by the low pressure in the vortex. The stratospheric air in the eye is like the air in a bath plug vortex, and is very warm due to its descent from the stratosphere. It is very dry also, but near the surface it mixes with other air and sometimes contains cloud. Some stratospheric air is mixed with the clouds surrounding the eye, and by evaporating the clouds it is cooled. Therefore there is a violent cascade of cold air down the inside of the wall of cloud (see Fig. 11.10.ii).

The air rises in the ring of clouds around the eye and spreads outwards beneath the tropopause. Because there is little friction in this outflow the tangential velocity relative to fixed axes decreases like r^{-1} and at three or four times the radius of the eye the rotation of the spreading canopy of the cloud is anticyclonic relative to the earth. This is discussed further in Section 11.10.

In the large horizontal gradient of wind speed due to the rapid increase in tangential velocity as the air approaches the centre, the convection clouds are drawn out into spiral lines. The air is cooled as it moves, increasing convection, and this is intensified by the ageostrophic motion producing convergence and upward motion of the air in the area towards which the centre is moving. The sky clears more rapidly in the quarter away from which the centre is moving.

If a hurricane passes over land it rapidly declines in intensity because the heat source is discontinued. It is often alleged that it is the increase in surface friction that is the more significant cause of the filling up of the low pressure centre; but this is belied by the fact that if a cold front is drawn into the circulation of a hurricane as it crosses from sea to land it may continue longer before the pressure rises in the centre. This illustrates the important fact that it is the action of lifting the warm air (either by the cold air behind the cold front or by the surrounding air which has not been similarly warmed by the sea) that causes both the convergence and intense rotation. The low pressure is produced by the lowering of the stratosphere in the eye so as to provide warmer air overhead.

Hurricanes may become mid-latitude cyclones if they move to a high enough latitude (say 37°) and are regenerated by the inflow of a cold front. The sea is not warm enough to provide the heat source needed to maintain a hurricane except in low latitudes in the autumn. The heat flows from the sea largely as latent heat in the form of evaporated water.

Hurricanes are called typhoons in the Pacific and Indian Oceans. The general name is tropical cyclone. They occur in the Arabian sea, in the Bay of Bengal, in the Western Pacific and in the Atlantic oceans in low latitudes. Occasionally they cross the Malay Peninsula into the Indian Ocean and migrate towards Bengal. They are known in Northern Australia, and the China Sea. Their movement is from the East in the Trade Winds at first but they curve along the East side of the continents, sometimes entering the Westerlies of the Eastern US or Japan. They occasionally cross Central America into the Eastern Pacific from the Caribbean Sea, and pass from the Arabian sea into southern Arabia, where they have been responsible for outbreaks of plagues of desert locusts.

The Figures 11.10.i and ii show a typical cross section of a mature hurricane and a radar picture of one near the coast of Florida. The dimensions of a hurricane can vary from case to case over a wide range, but typically the circulation and cloud systems extend over 100–300 km diameter; the eye may be from 5 to 50 km wide; and the speed of its centre from very slow to 15–20 kt. For further descriptions of these dramatic storms see *Clouds of the World* (Scorer, 1972) and *Hurricanes* by I. R. Tannehill.

Apart from the strong wind, the most severe effect is the raising of a tidal wave should the occurrence of a natural high tide coincide with a strong on-shore wind in the right hand advancing quarter of the storm. If the speed of the storm is such as to raise a wave in its centre of about 1 m above normal sea level, this wave is increased in height as it moves into shallow coastal water, and ships of up to 5 000 tons have been carried as much as 400 m inland on occasion.

4.11 THE EKMAN SPIRAL

We consider the situation in which the air velocity is horizontal and constant in direction at each height over a large area. Thus there are no accelerations and the equation of motion (4.3.2) is then

$$0 = -\frac{1}{\rho}\mathbf{grad}p + \mathbf{v}\times\mathbf{f} + \frac{\partial}{\partial z}K\frac{\partial v}{\partial z}. \qquad (4.11.1)$$

We have expressed the stress, represented in (1.3.25), by the pressure p and an eddy viscosity, K, multiplying a velocity gradient according to the conventional mathematical treatment to represent the Reynolds stress, which is described in Section 7.3, as if it were a viscous stress. The effect of variations in ρ are negligible except in so far as they produce a thermal wind which will be represented by the vertical variation of the horizontal pressure gradient. The pressure gradient may be represented by the geostrophic wind (u_g, v_g) according to (4.3.4) and so the two components of (4.11.1) are

$$\begin{aligned} (u - u_g)f &= \frac{\partial}{\partial z} K \frac{\partial v}{\partial z}, \\ (v - v_g)f &= -\frac{\partial}{\partial z} K \frac{\partial u}{\partial z}. \end{aligned} \tag{4.11.2}$$

G. I. Taylor applied Ekman's original theory concerning wind driven motions in the ocean to the atmosphere. That work is well summarised by Brunt (1939) p. 253. If we take the classicial assumptions, that there is no thermal wind so that (u_g, v_g) is a constant, and that K is independent of z, the solution of (4.11.2) is readily seen to be

$$u + iv - (u_g + iv_g) = C\, e^{-(1+i)mz + i\gamma} \tag{4.11.3}$$

in which C and γ are real constants determined by the boundary condition at the bottom of the atmosphere. Here

$$m^2 = f/2K \tag{4.11.4}$$

with m taken to be positive. The negative real exponent is chosen in (4.11.3) on the grounds that the wind becomes equal to the geostrophic wind at sufficient height. Taylor's theory was designed to obtain an estimate of the magnitude of K, at a time when its order was not known, by measuring the deflection of the surface wind from the isobars. He applied the boundary condition that the stress was in the direction of the wind at the surface. This is expressed by saying that $u + iv$ is parallel to $\partial u/\partial z + i\partial v/\partial z$ at $z = 0$. The reader is referred to Brunt for details.

Equation (4.11.3) shows that the horizontal wind vector approaches the geostropic value with increasing z along a spiral. In the ocean a positive exponent is used for wind driven motions which fade out downwards, and Ekman assumed the ocean surface to be level so that the geostrophic flow was zero. This is simply to say that the pressure gradient in the air is cancelled by an immeasurably small tilt of the ocean surface. In that case the Equations (4.11.2) become

$$U + iV = -i\frac{K}{f}\frac{\partial^2}{\partial z^2}(U + iV)\,, \tag{4.11.5}$$

of which the solution is

$$U + iV = e^{(1+i)mz}\,(U_0 + iV_0) \tag{4.11.6}$$

where K and m now have values appropriate to the water, and $U_0 + iV_0$ is the current at the surface. The velocity vector thus follows out a spiral of magnitude decreasing with depth.

The total momentum of the wind-driven motion, which is in the direction of what may be called the 'surface ocean current', is perpendicular to the stress applied by the wind. Assuming the stress to be along the wind direction, this is perpendicular to the wind itself. The surface stress is equal and opposite to the Coriolis force on the current. An important consequence is that an anticyclone generates convergence of the surface ocean flow, even though there may well be a significant outflow of the air near the surface. This mechanism is illustrated in Fig. 4.11.i and is the classicial explanation of the Sargasso Sea under the Bermuda high pressure area. For the effect to be significant the anticyclone must

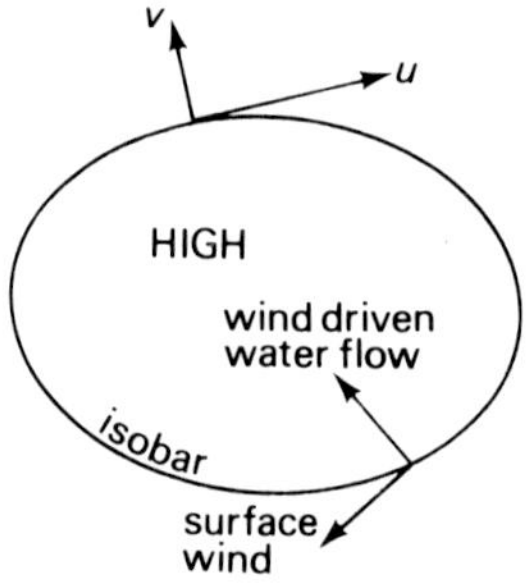

Fig. 4.11.i u is the wind component along the isobar, and v the outward flowing component from an anticyclone; the surface water is driven in towards the centre of the high.

be present in the same area for long periods of time, so that the corresponding geostrophic flow in the ocean, which is comparatively slow, can produce a noticeable effect.

The cross-isobar flow of the wind due to drag at the surface may be estimated in the case of no thermal wind ($u_g + iv_g$ = const) by integrating Equations (4.11.2) from the surface up to infinite z. Thus if we choose a coordinate system such that $v_g = 0$, and estimate the total transverse flow v, we have

$$\int_0^\infty v\,\mathrm{d}z = -\frac{1}{f}\left[K\frac{\partial u}{\partial z}\right]_0^\infty = \tau_{xz}/\rho f\,, \tag{4.11.7}$$

where τ_{xz} is the stress at the surface in the x direction. This result does not depend on any assumptions about K. Equation (4.11.7) suggests that the surface stress could be determined by observations of ν, but attempts to do this by Sheppard and others revealed the importance of the thermal winds. The winds did not follow out a spiral according to (4.11.3), but revealed that the geostrophic wind varied very significantly with height. This was so even in the area of the Windward Islands, where horizontal homogeneity had been expected of the large geographical extent of the Trade winds.

The Ekman spiral in the ocean and the Taylor spiral in the atmosphere are of considerable theoretical interest because of the influence of the earth's rotation on motion systems, but the details of any actual case are much more complicated, and no precise conclusions of the simple theory can be made use of quantitiatively.

Chapter 5

WAVES IN A STRATIFIED FLUID

5.1 THE PURPOSE OF A THEORY; ASSUMPTIONS, APPROXIMATIONS, NOTATION

Obviously we try to solve as many of our technical problems by mathematics as we can: in every case we have to make assumptions and approximations in order to make the analysis tractable, but we must insist that the most important aspects of the mechanisms of reality are retained in our model.

There are many original papers using very sophisticated mathematics designed to resolve certain mathematical issues, but seldom do they take us any nearer to matching our theory to the observational (out of doors) and experimental (laboratory) facts known. For example, one way to determine the precise nature of a steady flow pattern is to imagine the flow started up from rest and see what pattern the motion tends towards after a long time: this method has been used to discover what the flow is like at great heights and close to an obstacle. There do not exist, nor are there likely to exist, any observations from nature which would confirm that theory. Moreover the theory cannot be used in any discriminating way in practice to help understand observations, nor does it make any useful difference to predictions we might make of the flow over any actual mountain because other factors dominate any particular case. Yet the issue of the flow at great height has been given more attention by theorists than almost any other. Indeed many regard that aspect of the problem as solved, whereas in practice the real atmosphere never behaves like the models used; in fact almost all actual cases have features such as a level at which the wind speed is zero, which make them insoluble by the mathematics available today.

One very great difficulty in practice is that actual mountain shapes are very complicated, and the air flow very variable. It therefore seems at first sight desirable to develop a theory by which we can calculate the flow over any given shape in any given stream. However, to do this for the two-dimensional case of a simple stream is difficult enough, and most mountains are not two-dimensional while even the best methods are not applicable in the three-dimensional case. Consequently they are not really useful in the sense that they can be employed to predict the airflow today or tomorrow. However many of the mechanisms which dominate what happens in actual cases can be understood by studying special theoretical cases in which they are dominant.

If more complicated calculations for everyday use could be done we would still get very little further than we would without them. This is because the differences between reality and theory lie in the considerable unsteadiness of actual flows, and in the fact that the flow often separates from the surface, particularly in the lee of a steep obstacle where the predictions would be most useful if they could be made. Buoyancy produced by clouds and changes in the wind and temperature structure during the day, together with a variety of mathematically quite intractable situations leading to turbulence of various kinds, all militate against quantitiative precision.

In such circumstances it is gratifying that a simple theory does enable us to gain considerable insight into the mechanisms, and it is these simpler mathematical aspects which will occupy our attention. It must also be appreciated that great precision in prediction is not necessarily of much use nor worth obtaining. We go through life managing most of the time by having a large safety factor which is scarcely ever invoked, and we have other things to do and enjoy besides giving attention to the quantitiative aspects of our experiences which will be soon over. The observational and computational effort needed to make accurate calculations about all we do and experience would simply be wasted, and for that reason much that is theoretically possible will never be done. The final criterion is whether we can understand the world we live in, and find our way through it, not whether we can predict it and control it.

Except in those paragraphs where it is stated otherwise we shall be studying steady, adiabatic or incompressible, two-dimensional ($\partial/\partial y = 0$), inviscid flow of a stream that is disturbed from straight horizontal flow over level ground by passage over an obstacle. The undisturbed, or initial, state is represented by the suffix 0. Thus u_0 is the velocity possesed by a particle far upstream along the streamline. In the adiabatic (or incompressible) case τ (or ρ) is constant along a streamline and equal to τ_0 (or ρ_0).

The velocity and vorticity are assumed to have the following forms

$$\mathbf{v} = (u,v,w), \ \mathbf{v}_0 = (u_0,v_0,0); \ \boldsymbol{\omega} = (\xi,\eta,\zeta), \ \boldsymbol{\omega}_0 = \left(-\frac{\partial v_0}{\partial z_0}, \frac{\partial u_0}{\partial z_0}, 0\right). \quad (5.1.1)$$

In motion which does not vary in the y-direction there are no pressure gradients or other forces (except the coriolis force in specified cases) in that direction and so

$$v = v_0 \text{ everywhere,} \quad (5.1.2)$$

which means that each Bernoulli surface has its own velocity in the y-direction. Since this does not change at all it has no effect on other components of the motion. Thus v_0 will not appear in our equations representing the dynamics and $v_0(z)$ may be added to any result we obtain when $\partial/\partial y = 0$. Likewise ξ and ζ, the vorticity components,will not appear in the analysis.

The continuity of mass flow between two Bernoulli surfaces requires that (see Fig. 5.1.i)

$$\rho_0 u_0 \mathrm{d}z_0 \quad = \quad \rho u \mathrm{d}z \quad = \quad -\rho w \mathrm{d}x \,, \tag{5.1.3}$$

which can be written as follows

$$u \quad = \quad \frac{\rho_0}{\rho} u_0 \frac{\partial z_0}{\partial z}, \qquad w \quad = \quad -\frac{\rho_0}{\rho} u_0 \frac{\partial z_0}{\partial x} \,. \tag{5.1.4}$$

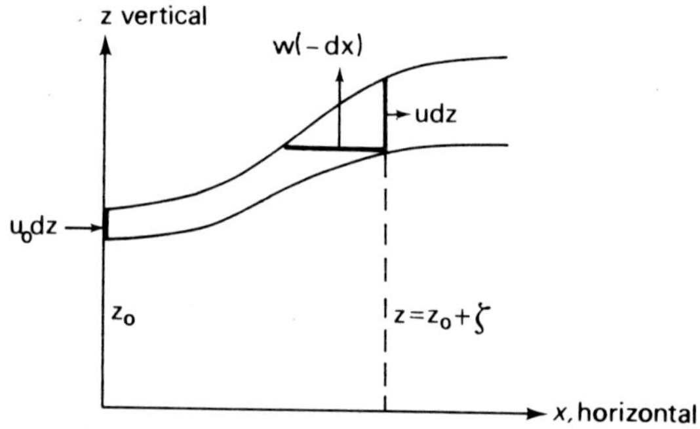

Fig. 5.1.i Coordinate system for the study of steady two-dimensional disturbances for which $\partial/\partial y = 0$. The mass flux between two given flow surfaces is constant.

Thus

$$\mathbf{grad}\, z_0 \; = \left(\frac{\partial z_0}{\partial x}, 0, \frac{\partial z_0}{\partial z}\right) \; = \; \frac{\rho}{\rho_0 u_0}(-w,0,u) \; = \; \frac{\rho}{\rho_0 u_0}\, \mathbf{v} \times \mathbf{j}, \tag{5.1.5}$$

where

$$\mathbf{j} \; = \; (0,1,0). \tag{5.1.6}$$

If n is length measured normal to the Bernoulli surfaces, on which z_0 is constant,

$$\frac{\partial}{\partial z_0} \; = \; \frac{\partial n}{\partial z_0}\frac{\partial}{\partial n} \; = \; \frac{1}{\mathrm{grad}\, z_0}\frac{\mathbf{grad}\, z_0 . \mathbf{grad}}{\mathrm{grad}\, z_0}, \quad \text{where } \mathrm{grad}\, z_0 \; = \; |\mathbf{grad}\, z_0|,$$

$$= \; \frac{1}{(\mathrm{grad}\, z_0)^2}\, \mathbf{grad}\, z_0 . \mathbf{grad} \tag{5.1.7}$$

$$= \frac{\rho_0 u_0}{\rho q_1^{\ 2}} \mathbf{v} \times \mathbf{j}.\mathbf{grad}. \tag{5.1.8}$$

The form (5.1.7) is applicable in any steady motion; while (5.1.8) is appropriate to the two-dimensional case, in which the speed is

$$q_1^{\ 2} = u^2 + w^2, \tag{5.1.9}$$

and $\text{grad}q_1^{\ 2} = \text{grad}q^2$.

5.2 THE EQUATION FOR STEADY MOTION

For inviscid motion the Vorticity Equation (1.4.1) takes the form

$$\frac{\mathrm{D}\boldsymbol{\omega}}{\mathrm{D}t} = -\boldsymbol{\omega} \,\mathrm{div}\,\mathbf{v} + (\boldsymbol{\omega}.\mathbf{grad})\mathbf{v} + \mathbf{R} \times (\mathbf{g} - \mathbf{f}), \tag{5.2.1}$$

and the equation of continuity for steady motion is

$$\rho \,\mathrm{div}\mathbf{v} = -\mathbf{v}.\mathbf{grad}\rho. \tag{5.2.2}$$

In the kind of two-dimensional motion under consideration the y-component of velocity, v, has a gradient only in the direction normal to the flow surfaces. These contain $\boldsymbol{\omega}$, and so the second term on the right of (5.2.1) has no y-component. Using (5.2.2) the y-component of (5.2.1) in steady motion is

$$\mathbf{v}.\mathbf{grad}\,\eta - \frac{\eta}{\rho}\mathbf{v}.\mathbf{grad}\rho = [\mathbf{R} \times (\mathbf{g} - \mathbf{grad}\,\tfrac{1}{2}q^2) + \mathbf{v} \times \boldsymbol{\omega}]_y. \tag{5.2.3}$$

But
$$\mathbf{R} = \frac{1}{\tau}\frac{\mathrm{d}\tau}{\mathrm{d}z_0}\mathbf{grad}\,z_0 = -\beta\,\mathbf{grad}\,z_0\ , \tag{5.2.4}$$

and this is parallel to $\mathbf{v} \times \boldsymbol{\omega}$. (There is an exactly similar equation for the incompressible case with ρ written for τ). Since $\mathbf{v}.\mathbf{grad}\rho_0 = 0$,

$$\mathbf{v}.\mathbf{grad}\left(\frac{\rho_0}{\rho}\eta\right) = \frac{\eta\rho_0}{\rho^2}\mathbf{v}.\mathbf{grad}\rho + \frac{\rho_0}{\rho}\mathbf{v}.\mathbf{grad}\,\eta$$

$$= \left[\frac{\rho_0}{\rho}\beta\,\mathbf{grad}\,z_0 \times \mathbf{grad}(gz + \tfrac{1}{2}q^2)\right]_y$$

$$= \left[\frac{\beta}{u_0}(\mathbf{v}\times\mathbf{j})\times\mathbf{grad}(gz + \tfrac{1}{2}q^2)\right]_y \quad \text{by (5.1.5)},$$

$$= \frac{\beta}{u_0}\mathbf{v.grad}(gz + \tfrac{1}{2}q^2)$$

$$= \mathbf{v.grad}\left[\frac{\beta}{u_0}(gz + \tfrac{1}{2}q^2)\right] \tag{5.2.5}$$

because β/u_0 is constant along a streamline and can therefore be taken inside the operator **v.grad**. Consequently

$$\frac{\rho_0}{\rho}\eta - \frac{\beta}{u_0}(gz + \tfrac{1}{2}q^2) = \text{constant along a streamline.} \tag{5.2.6}$$

In the incompressible case $\rho_0 = \rho$ and so the equation is slightly simpler.

The value of the constant in (5.2.6) may readily be obtained by an alternative derivation, as follows:
As in (1.7.4)

$$gH = \tfrac{1}{2}q^2 + \frac{\varpi}{\tau} + gz = \text{constant along a streamline,} \tag{5.2.7}$$

and the equation of motion is $$\mathbf{grad}(gH) = \varpi\,\mathbf{grad}\frac{1}{\tau} + \mathbf{v}\times\boldsymbol{\omega}. \tag{5.2.8}$$

By (5.1.7) $$\frac{\mathrm{d}}{\mathrm{d}z_0}(gH) = \frac{1}{(\mathrm{grad}\,z_0)^2}\mathbf{grad}\,z_0.\mathbf{grad}(gH)$$

$$= \frac{\varpi}{\tau}\tau\frac{\mathrm{d}}{\mathrm{d}z_0}\frac{1}{\tau} + \frac{\rho_0 u_0}{\rho q^2}\mathbf{v}\times\mathbf{j}.\mathbf{v}\times\boldsymbol{\omega}$$

$$= \frac{\varpi}{\tau}\beta + \frac{\rho_0 u_0}{\rho}\eta \tag{5.2.9}$$

because the component of $\mathbf{v}\times\boldsymbol{\omega}$ which is normal to **v** and to **j** is $\mathbf{v}\times\eta\mathbf{j}$, and so

$$\mathbf{v}\times\mathbf{j}.\mathbf{v}\times\eta\mathbf{j} = q^2\eta. \tag{5.2.10}$$

Thus, substituting for ϖ/τ from (5.2.7) into (5.2.9), which eliminates the pressure and is therefore equivalent to using the vorticity equation, we have

$$\frac{d}{dz_0}(gH) = \beta(gH - \tfrac{1}{2}q^2 - gz) + \frac{\rho_0 u_0}{\rho}\eta, \qquad \text{and so}$$

$$\frac{\rho_0}{\rho}\eta - \frac{\beta}{u_0}(\tfrac{1}{2}q^2 + gz) = \frac{1}{u_0}\left[\frac{d}{dz_0}(gH) - \beta gH\right] = \frac{1}{u_0\tau}\frac{1}{dz_0}(gH\tau), \qquad (5.2.11)$$

which is constant on a streamline because it is completely determined by values far upstream ($\beta = \beta_0, \tau = \tau_0$). The corresponding equation for the incompressible case has $\rho = \rho_0$ on the left and ρ for τ on the right; thus it is

$$\eta - \frac{\beta}{u_0}(\tfrac{1}{2}q^2 + gz) = \frac{1}{u_0\rho}\frac{d}{dz_0}(\rho gH). \qquad (5.2.12)$$

In (5.2.11) H and τ [and in (5.2.12) H and ρ] are constant along a streamline.

We have thus obtained a value for the constant in (5.2.6). These results depend fundamentally on the isopycnic vorticity theorem (1.5) and are only true if somewhere in the flow the vorticity vector lies in the isopycnic surfaces; otherwise we could not write $\eta\mathbf{j}$ for $\boldsymbol{\omega}$ to obtain (5.2.10).

The result (5.2.6) is possible only in two-dimensional motion because only in that case is the velocity field defined by the spacing of the Bernoulli surfaces, which also defines the density field.

5.3 THE EQUATION FOR MOTION IN THREE DIMENSIONS

Instead of (5.2.9) we have

$$\frac{d}{dz_0}(gH) = \frac{d}{dz_0}\frac{1}{\tau} + \frac{1}{(\mathbf{grad}\,z_0)^2}\,\mathbf{grad}\,z_0 . \mathbf{v} \times \boldsymbol{\omega}$$

$$= \beta(gH - \tfrac{1}{2}q^2 - gz) + \frac{|\mathbf{v} \times \boldsymbol{\omega}|}{|\mathbf{grad}\,z_0|} \qquad (5.3.11)$$

because $\mathbf{v} \times \boldsymbol{\omega}$ is parallel to $\mathbf{grad}\,z_0$ everywhere if it is somewhere on each streamline, by the theorem of (1.5). Whence

$$\frac{|\mathbf{v} \times \boldsymbol{\omega}|}{|\,\mathrm{grad}\,z_0\,|} - \beta(\tfrac{1}{2}q^2 + gz) = \frac{1}{\tau}\frac{d}{dz_0}(gH\tau)$$

$$= \text{constant on a stream (Bernoulli) surface} \qquad (5.3.2)$$

This form of the equation is deceptively simple. To use it the equation of continuity must be used simultaneously, which is very difficult to do, and it is unlikely that any attempt to do this will be made. This is because of the difficulty of specifying boundary conditions for a medium capable of transmitting waves similar to the motion under study.

5.4 THE WAVE EQUATION FOR AN INCOMPRESSIBLE STRATIFIED STREAM IN STEADY TWO-DIMENSIONAL MOTION

In this case $\rho = \rho_0$ and we may use (5.1.4) to express the velocity and vorticity in terms of the field of z_0 as follows:

$$\eta = \frac{\partial u}{\partial z} - \frac{\partial w}{\partial x} = \frac{\partial}{\partial z}\left(u_0 \frac{\partial z_0}{\partial z}\right) + \frac{\partial}{\partial x}\left(u_0 \frac{\partial z_0}{\partial x}\right)$$

$$= \frac{\mathrm{d}u_0}{\mathrm{d}z_0}(\nabla z_0)^2 + u_0 \nabla^2 z_0. \tag{5.4.1}$$

$$q^2 = u_0^{\,2}(\nabla z_0)^2, \tag{5.4.2}$$

and so (4.2.6) becomes

$$\frac{\eta}{u_0} - \frac{\beta}{u_0^{\,2}}(\tfrac{1}{2}q^2 + gz) = \text{constant along a streamline}$$

$$= \nabla^2 z_0 + \frac{1}{u_0}\frac{\mathrm{d}u_0}{\mathrm{d}z_0}(\nabla z_0)^2 - \frac{\beta}{u_0^{\,2}}[\tfrac{1}{2}u_0^{\,2}(\nabla z_0)^2 + gz]$$

$$= \nabla^2 z_0 + (\alpha - \tfrac{1}{2}\beta)(\nabla z_0)^2 - \frac{g\beta z}{u_0^{\,2}}. \tag{5.4.3}$$

Here

$$\alpha = \frac{1}{u_0}\frac{\mathrm{d}u_0}{\mathrm{d}z_0} \tag{5.4.4}$$

represents the vorticity in the undisturbed stream, so that

$$\alpha - \tfrac{1}{2}\beta = \frac{\mathrm{d}}{\mathrm{d}z_0}(\ln u_0 + \tfrac{1}{2}\ln\rho) = \frac{\mathrm{d}}{\mathrm{d}z_0}\ln(\rho^{1/2}u_0) = S \tag{5.4.5}$$

represents the total stratification of the stream due to shear and density gradients.

More insight is gained by expressing this equation in terms of the displacement of the fluid, ζ, given by

$$z_0 = z - \zeta. \tag{5.4.6}$$

Then
$$\nabla z_0 = \nabla z - \nabla \zeta = \left(-\frac{\partial \zeta}{\partial x}, 1 - \frac{\partial \zeta}{\partial z}\right)$$

$$(\nabla z_0)^2 = 1 - 2\frac{\partial \zeta}{\partial z} + (\nabla \zeta)^2 \tag{5.4.7}$$

$$\nabla^2 z_0 = -\nabla^2 \zeta,$$

and (4.4.3) becomes

$$\nabla^2 \zeta - (\alpha - \tfrac{1}{2}\beta)\left[1 - 2\frac{\partial \zeta}{\partial z} + (\nabla \zeta)^2\right] + \frac{g\beta}{{u_0}^2}\zeta = \text{const on a streamline.} \tag{5.4.8}$$

In particular

$$\nabla^2 \zeta + (\alpha - \tfrac{1}{2}\beta)\left[2\frac{\partial \zeta}{\partial z} - (\nabla \zeta)^2\right] + \frac{g\beta}{{u_0}^2}\zeta = 0 \tag{5.4.9}$$

in the case where the undisturbed stream is horizontal with ζ and its gradients equal to zero, the constant in (5.4.8) then being $-\alpha + \frac{1}{2}\beta$. Lamb (see §235 Eq. 12) gave this equation in linearsied from with $\alpha = 0$.

5.5 PRACTICAL APPROXIMATIONS

In motion which is essentially determined by the action of gravity on the density stratification, the first and last terms in (5.4.9) are necessarily important. For example, we could have a case in which $\alpha = \frac{1}{2}\beta$, in which waves are obviously possible. We now consider the importance of the term in $\frac{1}{2}\beta$, noting that there could obviously be cases in which α is large enough to be important because its magnitude can be made as large as we choose by having a large enough velocity gradient.

When a streamline has a vertical tangent, $\partial\zeta/\partial z$ is equal to unity and $(\nabla\zeta)^2$ is the same order of magnitude as $\zeta\nabla^2\zeta$. If the wavelength is $2\pi/k$ then comparing the first and last terms in (5.4.9), which are both involved in an essential way for gravity waves, we see that k^2 is the same order of magnitude as $g\beta/{u_0}^2$. The term in $\frac{1}{2}\beta$ is comparable only if β, or $\frac{1}{2}\beta\zeta^2 k^2$, is comparable with $g\beta\zeta/{u_0}^2$. This would require ${u_0}^2$ to be comparable with $g\zeta$, which is unlikely for waves of large enough amplitude for the flow to be nearly vertical somewhere. Or, comparing the first term in (5.4.9) with $\beta(\nabla\zeta)^2$, we see that the second is only important if $\beta\zeta$ is comparable with unity. This means that the displacement would have to be comparable with the height in which the denisty (or potential temperature) is decreased (increased) by a factor e^{-1}, often called a scale height. In an actual case $\beta\zeta$ is roughly equal to $\Delta\rho/\rho$, and the density anomaly is usually a small fraction of the density.

In the atmospheric case β is of the order of 10^{-2} km^{-1}, and so for the term in β to be important the displacement would have to be of the order of 100 km.

In neglecting this term we are making the Boussinesq approximation (see Chapter 2.12) because the terms in β not containing g originate in the term $\mathbf{R} \times \mathbf{f}$ in the vorticity equation. As a result we are saying that the accelerations, since they are produced by buoyancy forces created by density anomalies in which $\Delta\rho \ll \rho$, are small compared with gravity. The density variations are taken to have a small effect on the inertial mass of the fluid, but a significant effect through gravity.

Another approach is to assume that the coefficients in (5.4.9) are constants and that the displacements are small enough for the non-linear term $(\nabla\zeta)^2$ to be negligible. In that case we may write

$$\xi = e^{Sz}\zeta \tag{5.5.1}$$

where

$$S = \alpha - \tfrac{1}{2}\beta, \tag{5.5.2}$$

and Equation (5.4.9) then becomes

$$\nabla^2\xi + \left(\frac{g\beta}{{u_0}^2} - S^2\right)\xi = 0. \tag{5.5.3}$$

When $\alpha = 0$ the term in β without g is important only if $\tfrac{1}{4}\beta$ is comparable with $g/{u_0}^2$. This condition is never fulfilled in practice because the scale height is always much more than the height from which free fall under gravity produces a speed u_0.

In practical cases velocity gradients, represented by α, can be much more important than density gradients, represented by β, except that density discontinuities can be realised and kept sharp much more easily than discontinuities of velocity. Density discontinuities are treated in Section 5.8.

If S is neglected in (5.5.3) it might as well be neglected in (5.5.1), which is equivalent to neglecting β in (5.4.9). The substitution (5.5.1) has sometimes been used in order to obtain an equation of the form (5.5.3), which is easier to handle than (5.4.9). This is only possible if S is a constant throughout the whole depth of the medium under consideration. If a solution of (5.4.9) is sought in such a case, of the form $\exp i(Sz + kx)$, it is found that for a range of values of k there is no solution which tends to zero for large z. This is on account of the term in $\beta\partial\zeta/\partial z$ which makes even the oscillatory solutions proportional to exp ½βz. As a result, if waves occur in an atmosphere in which $\alpha = 0$ and β and u_0 are the same at all heights, the amplitude must become large at great height provided the waves last long enough for a steady solution to be relevant. Certainly long waves produced by the semidiurnal tide do have a very large amplitude at heights around 80 km, but these are outside the present discussion. If α exceeds ½β the difficulty disappears. The meaning of this is that the pressure fluctuations necessary for waves low down in the atmosphere can only occur if the wave amplitude in the layers of very low density at high altitude is very large. However a large increase in velocity with height reduces the amplitude because fast moving fluid is equivalent to dense fluid: this is evident from Equation (5.4.5) where variations in speed can produce variations in $\rho^{1/2}u_0$ as large as those due to variations in density.

The following section shows that the term in β is not a significant difficulty in comparison with the effects of compressibility in the atmosphere.

5.6 THE WAVE EQUATION FOR A COMPRESSIBLE ATMOSPHERE

If we use the full Equation of Continuity (5.1.4), in the place of (5.4.1 and 2) we have, for the compressible case

$$\eta = \frac{\rho_0}{\rho}\left[\frac{du_0}{dz_0}(\nabla z_0)^2 + u_0\nabla^2 z_0\right] + u_0\left[\frac{\partial z_0}{\partial z}\frac{\partial}{\partial z} + \frac{\partial z_0}{\partial x}\frac{\partial}{\partial x}\right]\frac{\rho_0}{\rho} \tag{5.6.1}$$

$$q^2 = \frac{{\rho_0}^2}{\rho^2}{u_0}^2(\nabla z_0)^2 \tag{5.6.2}$$

from which it follows from (5.2.6) that

$$\left(\frac{\rho_0}{\rho}\right)^2\left[\frac{du_0}{dz_0}(\nabla z_0)^2 + u_0\nabla^2 z_0\right] + \frac{\rho_0}{\rho}u_0\left[\frac{\partial z_0}{\partial z}\frac{\partial}{\partial z} + \frac{\partial z_0}{\partial x}\frac{\partial}{\partial x}\right]\frac{\rho_0}{\rho}$$
$$- \frac{\beta_0}{u_0}\left[\tfrac{1}{2}\frac{{\rho_0}^2}{\rho^2}{u_0}^2(\nabla z_0)^2 + gz\right] = \text{constant along a streamline} \tag{5.6.3}$$

We now use Equations (5.4.6 and 7) and assume adiabatic displacements so that we may use (1.8.7) in the form

$$\frac{\rho_0}{\rho} = 1 + \frac{g}{c^2}\zeta, \tag{5.6.4}$$

which is valid for displacements small compared with 30 km in the atmosphere, and obtain

$$-\left(1+2\frac{g}{c^2}\zeta\right)\nabla^2\zeta+(\alpha-\tfrac{1}{2}\beta)\left(1+2\frac{g}{c^2}\zeta\right)\left[(\nabla\zeta)^2-2\frac{\partial\zeta}{\partial z}\right]$$
$$+\frac{g}{c^2}\left(1+\frac{g}{c^2}\zeta\right)\left[\left(1-\frac{\partial\zeta}{\partial z}\right)\frac{\partial\zeta}{\partial z}-\left(\frac{\partial\zeta}{\partial x}\right)^2\right]-\frac{g\beta}{u_0{}^2}\zeta=0. \qquad (5.6.5)$$

The left hand side of this Equation is equal to a constant along each streamline, and this constant has been put equal to zero on the assumption that somewhere in the stream the flow is undisturbed and horizontal so that ζ and its first two derivatives are zero there. We have also neglected terms in the square of $g\zeta/c^2$.

In the atmosphere β and g/c^2 are of the order of 10^{-7} and 10^{-6} cm^{-1} respectively, so that even when $\alpha = 0$ the term in β without g is negligible. This means, physically, that the effects of compressibility on the density produce greater variations in the inertia of a unit volume of air than the stratification does. In practical cases α may be important and must therefore be retained in general, but for the same reasons as we neglect β without g, we can usually neglect most of the terms in g/c^2.

If we put $(1+2\frac{g}{c^2}\zeta)^{-1}=1-2\frac{g}{c^2}\zeta$ and neglect terms in g^2/c^4 compared with $g\beta/u_0{}^2$ we obtain from (5.6.5)

$$\nabla^2\zeta-\left(\alpha-\tfrac{1}{2}\beta-\frac{g}{c^2}\right)(\nabla\zeta)^2+\left(2\alpha-\beta-\frac{g}{c^2}\right)\frac{\partial\zeta}{\partial z}+\frac{g\beta}{u_0{}^2}\zeta\left(1-2\frac{g}{c^2}\zeta\right)=0$$
$$(5.6.6)$$

Since the middle terms are usually negligible anyway except when they contain α, we see that the main effect of compressibility is in the last term. The term in ζ^2 is comparable with that in ζ when ζ is of the order of 5 km in the atmosphere. Thus for application to the atmosphere we write the equation in the form

$$\nabla^2\zeta+\alpha\left[2\frac{\partial\zeta}{\partial z}-(\nabla\zeta)^2\right]+\frac{g\beta}{u_0{}^2}\zeta\left(1-2\frac{g}{c^2}\zeta\right)=0. \qquad (5.6.7)$$

5.7 THE EQUATION OF CONTINUITY IN A COMPRESSIBLE ATMOSPHERE

The following discussion is equally valid if the earth's rotation is taken into account. This is represented by the vector **f**, and the equation of motion is (see Section 4.3)

$$\frac{D\mathbf{v}}{Dt} = -\frac{1}{\rho}\,\mathbf{grad}p + \mathbf{g} + \mathbf{v} \times \mathbf{f} \tag{5.7.1}$$

so that

$$\mathbf{v.grad}p = -\rho\mathbf{v}.\frac{D\mathbf{v}}{Dt} + \rho\mathbf{v.g}$$

$$= -\rho\frac{D}{Dt}\,½q^2 - g\rho w \tag{5.7.2}$$

w being the vertical velocity. For adiabatic motion

$$\frac{Dp}{Dt} = c^2\,\frac{D\rho}{Dt} \tag{5.7.3}$$

and for continuity

$$\frac{D\rho}{Dt} = -\rho\,\mathrm{div}\mathbf{v}, \tag{5.7.4}$$

and from these three equations we obtain

$$\mathrm{div}\mathbf{v} = -\frac{1}{\rho c^2}\frac{\partial p}{\partial t} + \frac{1}{c^2}\left[\frac{D}{Dt}\,½q^2 + gw\right]. \tag{5.7.5}$$

The term in $\partial p/\partial t$ is negligible compared with that in g, because if the local rate of change of pressure is 1 mb min^{-1}, which is only likely to be exceeded at the passage of a sharp front or tornado, the two terms are comparable for a vertical velocity w of only 15 cm sec^{-1}. This is very small compared with what would actually occur if the pressure changes were that large. Thus the vertical displacement of the air in the hydrostatic pressure field produces much greater divergence of velocity in practice than local pressure changes. The term in the material rate of change of $½q^2$ represents the pressure changes which must occur when the air is accelerated. These again are of the order of one or two millibars within any flow field under consideration, and are negligible because that corresponds to vertical displacements of 10 to 20 metres, which are greatly exceeded in cases of interesting dynamics. Consequently the equation of continuity takes the form

$$\mathrm{div}\mathbf{v} = \frac{g}{c^2}\,w \tag{5.7.6}$$

which states that the only important cause of expansion of air parcels is their vertical displacement in the gravitationally produced pressure field.

5.8 SMALL AMPLITUDE WAVES AND THE EFFECTS OF THE EARTH'S ROTATION

We now use a notation appropriate to a small perturbation (u,v,w), τ, ϖ from an undisturbed stream $(U,V,0)$, τ_0, and ϖ_0 in which the undisturbed quantities are all functions of z. We neglect products of small quantities. Because waves of large amplitude are dominated by other effects the earth's rotation is then negligible, and so the perturbation technique is appropriate when the rotation is included. The problem of large amplitude waves with rotation has so far eluded any comprehensible analysis.

The adiabatic equation is

$$\frac{\mathrm{D}\tau}{\mathrm{D}t} = 0 \tag{5.8.1}$$

and the equation of motion is

$$\frac{\mathrm{D}\mathbf{v}}{\mathrm{D}t} = -\frac{1}{\tau}\,\mathbf{grad}\,\varpi + \mathbf{g} + \mathbf{v} \times \mathbf{f}. \tag{5.8.2}$$

A wave motion proportional to e^{ikx} is assumed to be taking place so that when applied to the disturbance

$$\frac{\partial}{\partial x} = ik \tag{5.8.3}$$

but is otherwise zero. The motion is assumed to be two dimensional: this is a limitation in the sense that solutions proportional to $e^{\mu y}$ can be obtained, and represent Kelvin waves travelling along a wall which exerts a pressure on the fluid to counteract the deviating force. These waves can travel in the sea along a coastline or depth discontinuity, but are such a special phenomenon that we shall not discuss them. Such waves could probably occur in the atmosphere along a mountain range, but propagated stable waves which are not stationary relative to the ground are not a very significant phenomenon in the atmosphere. Some cases unaffected by rotation are mentioned later in this chapter, but they have never been reported to be of the Kelvin type in the atmosphere. Of course, in order to have the undisturbed stream in geostrophic balance ϖ_0 does vary in the y direction. Thus the equations for the undisturbed motion are

$$0 = -\frac{1}{\tau_0}\frac{\partial \varpi_0}{\partial x} + Vf, \tag{5.8.4}$$

$$0 = -\frac{1}{\tau_0}\frac{\partial \varpi_0}{\partial y} - Uf, \tag{5.8.5}$$

$$0 = -\frac{1}{\tau_0}\frac{\partial \varpi_0}{\partial z} - g\,. \tag{5.8.6}$$

We have neglected the horizontal components of the earth's rotation because they produce no observable effect: f is the vertical component of **f**, which we assume not to vary horizontally by a significant amount, so that our conclusions are not valid for an area extending over a wide range of latitude.

The full vertical component of (5.8.2) is

$$(\tau_0 + \tau)\left[(U+u)\frac{\partial}{\partial x} + (V+v)\frac{\partial}{\partial y} + w\frac{\partial}{\partial z}\right]w = -\frac{\partial}{\partial z}(\varpi_0 + \varpi) - g(\tau_0 + \tau), \tag{5.8.7}$$

and subtracting out (5.8.6) while neglecting second degree terms in the perturbation, and writing a prime for $\partial/\partial z$, we get

$$\tau_0 ikUw + \varpi' + g\tau = 0. \tag{5.8.8}$$

The x- and y-components obtained in a similar way are

$$\tau_0 ikUu + \tau_0 U'w + ik\varpi - \tau_0 fv = 0, \tag{5.8.9}$$

and

$$ikUv + V'w + fu = 0. \tag{5.8.10}$$

The equations for adiabatic changes (5.8.1), and continuity (5.7.6), are

$$ikU\tau + \tau_0{}'w = 0, \tag{5.8.11}$$

$$iku + w' - \frac{g}{c^2}w = 0. \tag{5.8.12}$$

From the last three equations v, τ, and u may be expressed in terms of w and so (5.8.9) gives ϖ in terms of w thus:

$$ik\varpi = \tau_0 U\left[\frac{g}{c^2}w - w'\right] + \tau_0 U'w + \tau_0 f\left[\left(\frac{V'}{ikU} - \frac{fg}{k^2Uc^2}\right)w + \frac{f}{k^2U}w'\right]. \tag{5.8.13}$$

This may be differentiated with respect to z and compared with (5.8.8) to eliminate ϖ. The result is expressed below as (5.8.16) with the following notation:

$$\alpha = \frac{U'}{U}, \quad \beta = -\frac{\tau_0'}{\tau_0}, \quad \alpha_1 = \frac{V'}{U}, \quad \beta_1 = \frac{T'}{T}, \quad F^2 = \frac{f^2}{k^2 U^2} \tag{5.8.14}$$

$$c^2 = \gamma RT, \text{ which gives } \frac{\partial}{\partial z}\frac{g}{c^2} = -\beta_1 \frac{g}{c^2}. \tag{5.8.15}$$

These quantities represent the basic features of the undisturbed physical situation. α and α_1 may be of the same order of magnitude, and lie between zero and 10^{-5} cm^{-1} in the atmosphere, although they might rise to as large as 10^{-2} cm^{-1} in shallow layers, or even, in billow-producing situations, to 10^{-1} for short periods. β_1 is of the same order of magnitude as β, which in large depths averages around 10^{-7} cm^{-1} or up to 4 times this value in the stratosphere or in shallow layers. g/c^2 is about 10^{-6} cm^{-1}, and in mid latitudes f is about 4×10^{-5} sec^{-1}.

The elimination of all variables but w gives the equation

$$(1 - F^2)w'' - \left[\frac{g}{c^2} + \beta - F^2\left(\alpha + \beta + \frac{g}{c^2}\right) - iF\alpha\right]w'$$

$$+ \left[\frac{g\beta}{U^2} - \frac{U''}{U} - k^2 + \alpha\left(\beta - \frac{g}{c^2}\right) + \frac{g}{c^2}(\beta + \beta_1) - iF\left(\alpha_1(\beta + \alpha) - \frac{V''}{U}\right)\right.$$

$$\left. - F^2 \frac{g}{c^2}(\alpha + \beta - \beta_1)\right] w = 0 \tag{5.8.16}$$

The terms in F are negligible unless F is of the order of unity, which means that the time taken to travel one wavelength is of the order of a day. When that time is a small fraction of a day the effect of the earth's rotation is negligible. This is an important result and applies to a wide range of phenomena besides stable waves, and to convection phenomena in particular. We have noted elsewhere that geostrophic forces can intensify temperature gradients (see Sections 4.7(3), 4.8, and 6.4) but billows, which are on a smaller time scale, are set off by smaller time scale aspects of the medium scale motion. Cloud streets [Section 9.8(1)] are influenced by the geostrophic wind, but the unstable aspects of their motion are not generated by the deviating force, as some people suppose.

When F, which is a kind of Rossby number, is of the order of unity the important terms in (5.8.16) are

$$(1-F^2)w'' + (F^2\alpha + iF\alpha_1)w'$$
$$+\left[\frac{g\beta}{U^2} - \frac{U''}{U} - k^2 - iF\left(\alpha\alpha_1 - \frac{V''}{U}\right)\right]w = 0, \quad (5.8.17)$$

while when $F \ll 1$ they are

$$w'' + \left(\frac{g\beta}{U^2} - \frac{U''}{U} - k^2\right)w = 0. \quad (5.8.18)$$

When the earth's rotation is important, the scale of the motion is likely to be such that the terms in α, α_1, U'', and V'' are not important. Then the equation is

$$(k^2U^2 - f^2)w'' + g\beta k^2 w = 0, \quad (5.8.19)$$

and we see that $\partial/\partial z$ is of the order of $\left(\dfrac{g\beta k^2}{k^2U^2 - f^2}\right)^{1/2}$, which has magnitude of the order of $(g\beta/U^2)^{1/2}$, i.e. more or less the same as when the earth's rotation is not important. The waves may aptly be called **Queney waves** because they were first extensively studied by Queney in 1947. They have a vertical wavelength (within a multiple of 2 or 3) of 2π km and a stream wise wavelength of the order of $2\pi\dfrac{U}{f}$, $= 2\pi \times 250$ km when U is of the order of 10 m sec^{-1} in mid latitudes.

In this case u and v are of the same order of magnitude according to (5.8.10), while u/w is about the ratio of the horizontal to vertical wavelength, according to (5.8.12). Thus the lateral and longitudinal displacements are about 250 times the vertical and the slope of the streamlines is small.

In Queney waves the dominant forces are those due to static stability and the earth's rotation, and the displacements are along surfaces inclined to the horizontal with a gradient of the order of 1:250.

To express (5.8.15), which is the short wavelength case in which f is neglibible, in terms of the vertical displacement ζ, we note that

$$w = \frac{D\zeta}{Dt}, = ikU\zeta \text{ when linearised,} \quad (5.8.20)$$

so that

$$w' = ik(U'\zeta + U\zeta'), \quad (5.8.21)$$

and $$w'' = ik(U''\zeta + 2U'\zeta' + U\zeta''). \tag{5.8.22}$$

From this we obtain the equation for the displacement in the form

$$\zeta'' + 2\alpha\zeta' + \left(\frac{g\beta}{U^2} - k^2\right)\zeta = 0, \tag{5.8.23}$$

which is the linearised version of (5.6.7) assuming ζ to be proportional to $\cos kx$.

In the case where $\alpha = 0$ and $c^2 = \infty$ (no shear, incompressible, or ζ is small enough for $g\zeta/c^2 << 1$) Equation (5.6.7) is linear and can be written

$$\zeta'' + \left(\frac{g\beta}{U^2} - k^2\right)\zeta = 0, \tag{5.8.24}$$

which is identical with the form taken by (5.8.23). The assumptions of perturbation theory are therefore not necessary to obtain this equation. In particular it is not necessary that $u << U$ et cetera, and (5.8.23) is a good approximation to the full equation for fairly large displacements and is exact for arbitrarily large displacements when $\alpha = 0$ and $g\beta/U^2$ is independent of z. The classical derivation of the wave equation by the usual small perturbation methods is therefore both misleading about the conditions in which it is valid, and unduly restrictive as to its application.

The Equation (5.6.7) is obviously non-linear in ζ because of the terms explicitly containing ζ^2. But if α, β, and u_0 are functions of z_0 they become functions of ζ when there is a displacement, so that the most important restrictions that have to be imposed in practice to obtain a linear equation are on the stream profile rather than on the amplitude of the wave.

5.9 LINEAR WAVES OF LARGE AMPLITUDE IN AN INCOMPRESSIBLE FLUID

Equation (5.4.3) can be written

$$\nabla^2 z_0 + S(\nabla z_0)^2 + \ell^2 z = \text{constant along a streamline} = S + \ell^2 z_0, \tag{5.9.1}$$

if $z = z_0$ far upstream, where

$$S = \alpha - \tfrac{1}{2}\beta = \frac{1}{\rho^{1/2}u_0}\frac{\mathrm{d}}{\mathrm{d}z_0}(\rho^{1/2}u_0), \quad \ell^2 = g\beta/u_0{}^2. \tag{5.9.2}$$

We now investigate the possibility that there may be a variable $\psi = \psi(z_0)$ in terms of which this equation is linear. Thus

$$\nabla z_0 = \frac{dz_0}{d\psi} \nabla\psi$$

$$\nabla^2 z_0 = \left(\nabla\psi \frac{d}{d\psi}\right)^2 z_0 = \frac{dz_0}{d\psi} \nabla^2 \psi + \frac{d^2 z_0}{d\psi^2} (\nabla\psi)^2 \tag{5.9.3}$$

$$(\nabla z_0)^2 = \left(\frac{dz_0}{d\psi}\right)^2 (\nabla\psi)^2 ,$$

and the Wave Equation (5.9.1) becomes

$$\nabla^2 \psi + (\nabla\psi)^2 \frac{d\psi}{dz_0}\left[\frac{d^2 z_0}{d\psi^2} + \left(\frac{dz_0}{d\psi}\right)^2 S\right] = \frac{d\psi}{dz_0} [S + \ell^2 (z - z_0)] . \tag{5.9.4}$$

The term in $(\nabla\psi)^2$ disappears if

$$\frac{d^2 z_0}{d\psi^2} + \left(\frac{dz_0}{d\psi}\right)^2 \frac{d}{dz_0} \ln(\rho^{1/2} u_0) = \frac{1}{\rho^{1/2} u_0} \frac{\partial}{\partial\psi} \left(\rho^{1/2} u_0 \frac{\partial z_0}{\partial\psi}\right) = 0 . \tag{5.9.5}$$

This is the case if $\rho^{1/2} u_0 \dfrac{dz_0}{d\psi} = \text{constant} = 1$, say, (5.9.6)

which may be written
$$\frac{d\psi}{dz_0} = \rho^{1/2} u_0, \quad \text{or} \quad \psi = \int \rho^{1/2} u_0 \, dz_0 . \tag{5.9.7}$$

This result is due to Long (1953), and the equation becomes

$$\nabla^2 \psi - S\rho^{1/2} u_0 - \ell^2 (z - z_0)\rho^{1/2} u_0 = 0, \tag{5.9.8}$$

which has the form
$$\nabla^2 \psi - F(\psi) - G(\psi) z = 0. \tag{5.9.9}$$

If this equation is linear in ψ, F and G must be linear, and so

$$\begin{aligned} F(\psi) &= (S - \ell^2 z_0)\rho^{1/2} u_0 = A\psi + B \\ G(\psi) &= \ell^2 \rho^{1/2} u_0 = C\psi + D \end{aligned} \tag{5.9.10}$$

where A, B, C, and D are constants, independent of z_0, and the Equations

(5.9.10) restrict the profiles of ρ and u_0 to certain functional forms of z_0 in the undisturbed stream. The wave equation then becomes, with (5.8.3),

$$\frac{\partial^2 \psi}{\partial z^2} = (A + k^2)\psi + B + (C\psi + D)z ; \tag{5.9.11}$$

and eliminating $\ell^2 \rho^{1/2} u_0$ and S from the equations (5.9.10, 7 and 2)

$$\frac{\mathrm{d}^2 \psi(z_0)}{\mathrm{d}z_0} = A\psi(z_0) + B + (C\psi(z_0) + D)z_0 , \tag{5.9.12}$$

which has solutions of the same functional form as (5.9.11). If A, B, C, and D are chosen (5.9.12) could be solved for $\psi(z_0)$ so that if (5.9.11) is then solved the streamlines ψ = const., i.e. z_0 = constant, can be drawn. The stream profiles of ρ and u_0 can be determined from (5.9.10) but clearly profiles which are of interest can only be found by intelligent integration of (5.9.10) with trial values of the constants A, B, C, and D.

It is fairly easy to find particular cases which are of interest, the most obvious being $S = 0$, $B = C = 0$, $\psi \propto z_0$. This requires that both ℓ^2 and $\rho^{1/2} u_0$ be constant and independent of z_0, which is the case discussed in the following pages. Several cases with more variable values of u_0 and ρ as functions of z_0 can be contrived, but they all require the medium to be bounded by rigid walls above and below because no solution which does not become exponentially large for large z has yet been found. The case in which the disturbance remains bounded for large z, mentioned at the end of 5.5, is that in which S is a constant and u_0 is approximately proportional to z_0; this is not of practical significance.

The solutions of (5.9.12) must be monotonic functions of z_0 if they are to be meaningful because ψ is a stream function defined by (5.9.7). Moreover u_0 must not become zero, for if it did ℓ^2 would be infinite and the solution would have an infinity of zeros in the neighbourhood. What happens at such a level is discussed in 5.17. All the solutions of (5.9.11 and 12) are like the Airy functions, which are either oscillatory or exponential and cannot be algebraic for large z.

The alternative approach of representing a real atmosphere by several superposed layers in each of which the solution is simple is useful only when the displacement is small at the interfaces, because the boundary conditions are non-linear. Normally they are applied at the undisturbed position of the interface but they become intractably dependent on ζ if applied at the displaced position.

Yih has employed linear solutions of the kind implied by this discussion in cases where a liquid is bounded on top by a free surface which, because of the very large density decrease, behaves as a rigid surface. The chief merit of this kind of solution as far as the atmosphere is concerned is in elucidating some of the mechanisms that are important in the atmosphere, as Long has demonstrated. It is to this aspect of the subject that we now turn.

5.10 STANDING WAVES AND ROTORS

The phenomena will be best exemplified by the simplest cases: the possibility of more complicated cases is obvious but no essentially new mechanisms are revealed by them. We use the equation for the vertical displacement ζ because it is linear in the cases treated. The equation for w, the vertical velocity, is linear only for small amplitudes, in the form (5.8.18), because the relationship between ζ and w is non-linear, namely

$$w = \frac{D\zeta}{Dt} = u\frac{\partial \zeta}{\partial x} + w\frac{\partial \zeta}{\partial z} \qquad (5.10.1)$$

in the two-dimensional case. We start with equation (5.6.7) and treat particular cases.

(1) A uniform stratified incompressible stream between horizontal rigid boundaries

The wave equation in this case is, with $\zeta \propto \cos kx$,

$$\frac{\partial^2 \zeta}{\partial z^2} + (\ell^2 - k^2)\zeta = 0 \qquad (5.10.2)$$

and if we write
$$\frac{g\beta}{u_0{}^2} = \ell^2, \quad \nu^2 = \ell^2 - k^2 \qquad (5.10.3)$$

the solution satisfying the boundary conditions $\zeta = 0$ at $z = 0, h$ is

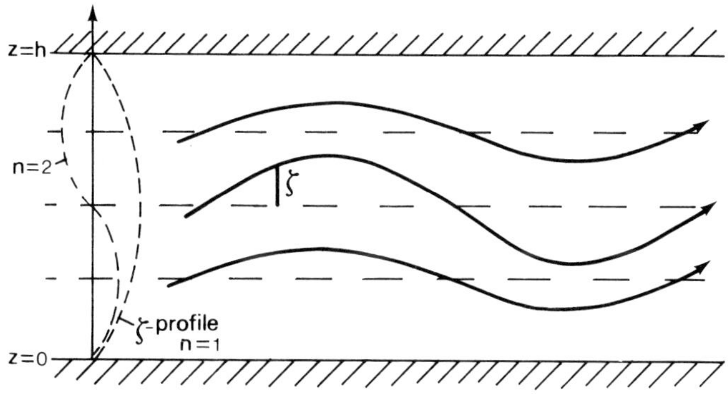

Fig. 5.10.i Coordinates and boundaries for a stratified stream between rigid boundaries. The streamlines relate to the profile of ζ for n = 1.

$$\zeta = A \sin \nu z \cos kx, \qquad (5.10.4)$$

with
$$\sin \nu h = 0, \quad \text{or} \quad \nu h = n\pi. \qquad (5.10.5)$$

This last equation determines the wavelength $2\pi/k$ for each value of n. In Fig. 5.10.i streamlines for $n = 1$ are shown. The profile of ζ has one node within the fluid when $n = 2$, and the flow in each half of the fluid is then like that shown but with opposite phase in the two halves.

There is only one possible value of k if $\pi < \ell h < 2\pi$, none if $\ell h < \pi$, two if $2\pi < \ell h < 3\pi$, and so on. Although there could be a large number of possible values of k, there are usually only one or two in practical cases, and when there are more not all the modes are excited to a significant amplitude.

There can be regions of reversed flow, usually with closed circulations, if the horizontal velocity is somewhere negative, which means that

$$1 - \frac{\partial \zeta}{\partial z} = \frac{\partial z_0}{\partial z} < 0, \quad \text{by (4.1.4)},$$

or

$$\frac{\partial \zeta}{\partial z} > 1 . \tag{5.10.6}$$

In this case the condition is

$$A\nu \cos \nu z \cos kx > 1 \quad \text{somewhere}, \tag{5.10.7}$$

$$\text{which means that} \quad A\nu > 1 \quad \text{somewhere}. \tag{5.10.8}$$

In the case of the first mode $\nu h = \pi$ and so the condition becomes

$$A > h/\pi, \tag{5.10.9}$$

and the reversed flow occurs in the neighbourhood of where $\cos \nu z = \cos kx = 1$ or $\cos \nu z = \cos kx = -1$. In this case there is a closed circulation around these points, each of which is on the boundary, and the circulations are called rotors.

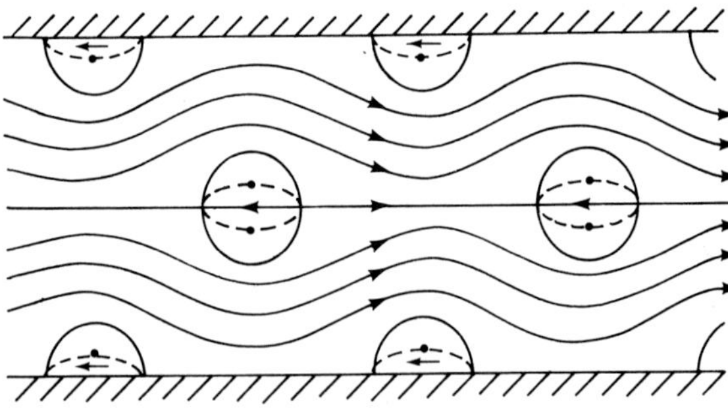

Fig. 5.10.ii The pattern of streamlines for the case n = 2 when the amplitude is large enough for the waves to have rotors. The dotted lines are the boundaries of the regions of reversed flow. The stagnation points at the centres of the rotors are marked and the flow circulates around them: the rotors in the middle of the stream are therefore double-yolked.

The stagnation points in the middle of the circulations are given by

$$\cos kx = \pm 1, \quad A\nu \cos \nu z = \pm 1 \tag{5.10.10}$$

and those on the nodal surfaces by

$$\cos \nu z = \pm 1, \quad A\nu \cos kx = \pm 1. \tag{5.10.11}$$

The line on which the velocity is vertical, which is the boundary of the region in which u is negative and there is backward flow, is given by

$$A\nu \cos \nu z \cos kx = 1. \tag{5.10.12}$$

Figure 5.10.ii shows the flow pattern when $n = 2$. The rotors on the nodal surface $z = h/2$ are double-yolked, and the boundary of the reversed flow is shown.

Where the flow is reversed the fluid is statically unstable if it is stable elsewhere, so that it is to be expected that buoyant convection would occur within the rotor until the density had been made more or less uniform inside. This in turn would produce a small density discontinuity on the boundary of the rotor, which might produce billows by the mechanism described in Chapter 6.

The fluid inside the rotors is described by the mathematics as having a density equal to some of the fluid on the opposite side of the nodal surface. Thus the fluid in the rotors on the bottom is described as having a density, and a value of z_0, from below the bottom. Since this is impossible it must contain fluid which is not statically unstable, and the theory is strictly incorrect in this

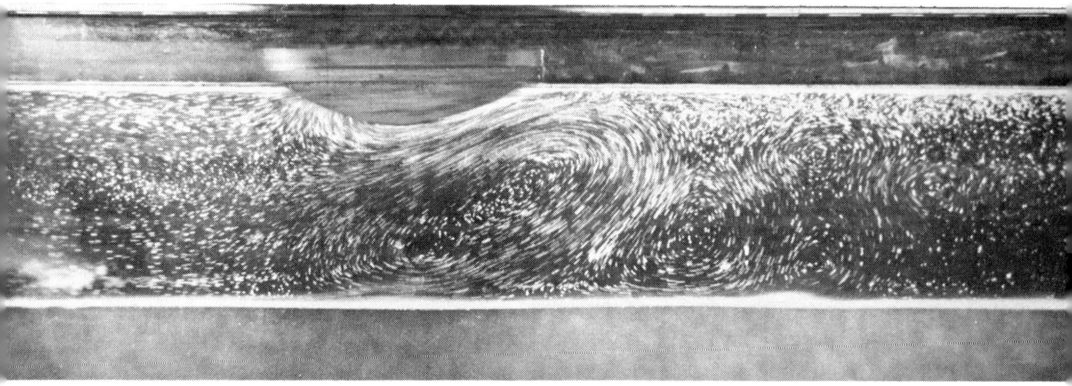

Fig. 5.10.iii A time exposure reveals the vector velocity of particles floating in a water tank. Stratified by the addition of salt. The 'mountain' is moving along to the left, so that the velocity field for a stream 'over' a fixed obstacle is obtained by subtracting the velocity of the mountain from the whole pattern. The disturbance is small on the right because it is near the end of the tank from which the obstacle was started from rest.

detail. In the double-yolked rotors on the central nodal surface the fluid is described as having the same density as fluid in some part of the system, but it is unlikely that if the waves were set up with a large enough amplitude to have rotors the density would be thus distributed.

An important feature of rotor flow is that the flow is very swift in between the stagnation points on the nodal surfaces, and that the crests and troughs of the streamlines are not similarly shaped, the lines close to the nodal surfaces being flattened, while those over or below the rotors having greater curvature.

Experimentally the motion is most readily demonstrated by moving an obstacle along the bottom of a tank of static stratified liquid. Figure 5.10.iii shows the disturbance of a moving obstacle by means of a time exposure.

(2) An infinite atmosphere with a single discontinuity of static stability

It is assumed that the displacement is small enough for the compressibility of the air to be neglected, in which case we can use equation (5.10.2) throughout. We require the amplitude to tend to zero as z tends to infinity and to be zero at $z = -h$. In the upper layer, therefore, the solution is of the form (omitting $\cos kx$)

$$\zeta = Ae^{-\mu z} \tag{5.10.13}$$

where

$$\mu = +(k^2 - \ell_s^{\,2})^{1/2}, \tag{5.10.4}$$

the alternative solution being proportional to $e^{\mu z}$ and unacceptable for large z. The value of ℓ^2 in the upper (superior) layer is $\ell_s^{\,2}$.

In the lower layer we need to have the same form of solution as in Section 5.10(1) in order to have a zero at $z = -h$ and to fulfil the conditions at the boundary between the layers at $z = 0$. These conditions are that the displacement ζ be continuous and that the pressure be continuous. The pressure is given by

$$\frac{\varpi}{\tau} + \tfrac{1}{2}q^2 + gz = \text{constant}. \tag{5.10.15}$$

We write

$$\begin{aligned} q^2 &= u^2 + w^2 \\ &= \left(\frac{\rho_0}{\rho}u_0\right)^2 \left[\left(\frac{\partial z_0}{\partial z}\right)^2 + \left(\frac{\partial z_0}{\partial x}\right)^2\right] \\ &= \left(\frac{\rho_0}{\rho}u_0\right)^2 \left[\left(1 - \frac{\partial \zeta}{\partial z}\right)^2 - k^2\zeta^2\right] \end{aligned}$$

and consider the simplest kind of discontinuity possible at $z = 0$, namely one in which ρ_0 and u_0 are continuous so that ρ is also continuous. Because ζ is continuous, the continuity of ϖ and τ across the interface requires, in view of the equations (5.10.15 and 16), that $\partial\zeta/\partial z$ also be continuous. This result is strictly correct regardless of the amplitude of the displacement of the interface which, being $z_0 = 0$ in this case, lies at $z = \zeta$.

In the lower layer, in which $\ell^2 = \ell_i^2$,

$$\zeta = B \sin \nu(z + h), \quad \text{which} = 0 \text{ at } z = -h, \tag{5.10.17}$$

where

$$\nu = +(\ell_i^2 - k^2)^{1/2} \tag{5.10.18}$$

and so, putting ζ and $\partial\zeta/\partial z$ equal to the same value according to equations (5.10.13 and 17), we obtain at $z = 0$,

$$A = B \sin \nu h, \quad -A\mu = B\,\nu \cos \nu h, \tag{5.10.19}$$

whence

$$\nu \cot \nu h = -\mu\,. \tag{5.10.20}$$

The details of the solution are illustrated in Fig. 5.10.iv.

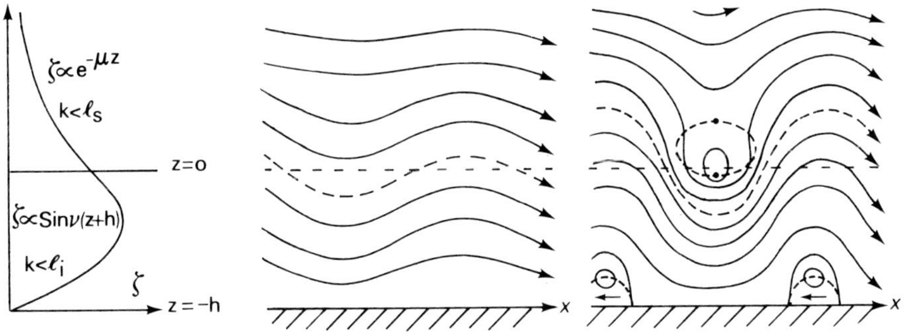

Fig. 5.10.iv On the left is the profile of ζ for the two layer case, the upper layer extending to infinity. In the middle the shape of the streamlines is shown for small amplitude, and on the right where there are rotors. The dashed streamline shows the displaced position of the interface. The boundary of the region of reversed flow passes through the two stagnation points.

There will be a rotor on the bottom boundary ($z = -h$) if

$$\frac{\partial\zeta}{\partial z} = B\nu \cos kx > 1 \text{ somewhere,}$$

or

$$B\nu > 1 \tag{5.10.21}$$

and there will be one at the interface ($z = 0$), where the value of $\partial\xi/\partial z$ has a maximum, if

$$B\nu \cos \nu h > 1\ . \tag{5.10.22}$$

In Fig. 5.10.iv the case of no rotors is illustrated together with the case of rotors at both surfaces. The rotor on the interface is smaller, and if (5.10.21) is satisfied but not (5.10.22) there is a rotor on the bottom boundary only.

The Equation (5.10.20) determines k from the given parameters of the problem, namely ℓ_i, ℓ_s, and h. In Fig. 5.10.v the continuous curve represents the

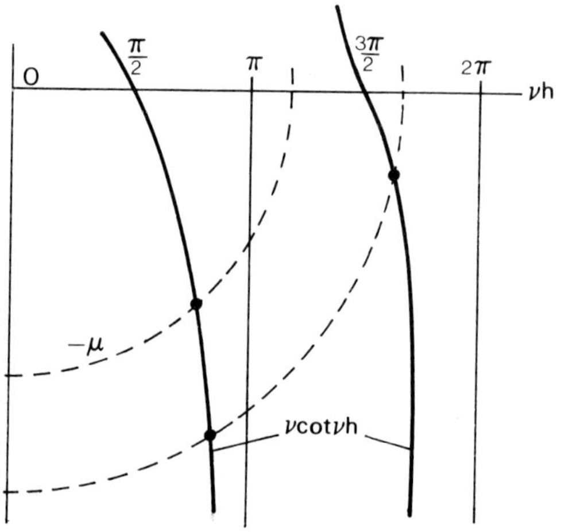

Fig. 5.10.v Solutions of (5.10.20) obtained by plotting the left (continuous curve) and right (dashed) sides as functions of νh.

function $\nu \cot \nu h$ as a function of νh, while the dashed lines represent the function μ as a function of νh, these being circles because

$$\mu^2 \;=\; \ell_i{}^2 - \ell_s{}^2 - \nu^2 . \tag{5.10.23}$$

If the radius of the circle lies between $\frac{1}{2}\pi$ and $\frac{3}{2}\pi$, there is one value of νh at which $\nu \cot \nu h = -\mu$. If it lies between $\frac{3}{2}\pi$ and $\frac{5}{2}\pi$ there are two values. In general there are n values if

$$\frac{(2n-1)\pi^2}{2h} < \ell_i{}^2 - \ell_s{}^2 < \frac{(2n+1)\pi^{\ 2}}{2h} \tag{5.10.24}$$

If $\ell_s^2 > \ell_i^2$ no solutions of this kind, with a zero on the bottom and an exponentially decreasing value for large z, can be obtained; and in all cases where there is a solution

$$\ell_i > k > \ell_s\,. \tag{5.10.25}$$

(3) The general requirement for trapped waves

The previous case is a particular example of waves trapped in the part of the airstream close to the ground. The requirement for at least one wave to be possible is (5.10.24) with $n = 1$, namely

$$\ell_i^2 - \ell_s^2 > \pi^2/4h^2\,. \tag{5.10.26}$$

In the more general case the possibility of the existence of waves can be investigated by considering the case of small amplitude. This is because it is unlikely that a case would be complicated in a way which would make it different from the simple cases where if a periodic wave is possible for a large amplitude it is possble for small amplitude, although the converse is not always true. For small amplitude we may reverse the transformation from w to ζ set out in equations (5.8.20–23) and obtain the equation for w, subject only to the approximation in (5.8.20), namely u, $w \ll U$. Then, from (5.8.18)

$$\frac{\partial^2 w}{\partial z^2} + (\ell^2 - k^2)w = 0 \tag{5.10.27}$$

where now

$$\ell^2 = \frac{g\beta}{u_0^{\,2}} - \frac{1}{u_0}\frac{\mathrm{d}^2 u_0}{\mathrm{d}z_0^{\,2}}\,. \tag{5.10.28}$$

The profile of w, like that of ζ, must asymptote to zero as z tends to infinity for a wave to be trapped, so that $\partial^2 w/\partial z^2$ must be a positive multiple of w, and so $k^2 > \ell^2$. On the other hand, in order to curve back to zero at the ground like the profile of ζ in Fig. 5.10.iii, the curvature of the profile must be reversed in the lower layers of the stream. So in some lower layers we must have either $\ell^2 > k^2$ in a fairly deep layer, or else ℓ^2 must exceed k^2 by a large amount in a less deep layer. Thus in all cases the value of k is equal to some value of ℓ intermediate between the extreme values of ℓ in the profile of ℓ, and there must be enough difference between these extremes for a condition like (5.10.26) to be satisfied.

(4) Trapping by a density discontinuity

We shall see in Chapter 6 that a discontinuity of velocity cannot be maintained, but that discontinuities of density may exist. Methods by which they may be created in the atmosphere are described in Sections 6.4, 11.1(2), and 11.5(2).

In steady motion in two dimensions on each streamline we have

$$\frac{p}{\rho} + \tfrac{1}{2}q^2 + gz = \text{constant} \tag{5.10.29}$$

and in the case of adiabatic motion we may replace p/ρ by ϖ/τ and obtain an analogous result. At an interface the pressure p is continuous and since ϖ is a function of p only for a gas of uniform composition [see (1.6.4)], ϖ must also be continuous at a density discontinuity. ρ and τ are discontinuous. The variable part of p obtained from (5.10.29) is the same on the two sides of the discontinuity. With q and z varying along the streamline, and with z put equal to ζ by choosing the zero of z to be at the undisturbed position of the discontinuity, we thus obtain, with i and s denoting the inferior and superior layers,

$$-\rho_s u_0{}^2 \frac{\partial \zeta_s}{\partial z} + g\rho_s\zeta = \text{constant} - p_s$$

$$= \text{constant} - p_i = -\rho_i u_0{}^2 \frac{\partial \zeta_i}{\partial z} - g\rho_i\zeta. \tag{5.10.30}$$

Here we have put $\zeta_i = \zeta_s = \zeta$, and in each layer by (5.10.16) the variable part of q^2, to first order in ζ, is $-2u_0{}^2\,\partial\zeta/\partial z$.

Thus

$$\rho_i u_0{}^2 \frac{\partial \zeta_i}{\partial z} = g\zeta(\rho_i - \rho_s) + \rho_s u_0{}^2 \frac{\partial \zeta_s}{\partial z} \tag{5.10.31}$$

expresses the continuity of pressure across the discontinuity of density. There are two cases of special interest:

(i) when the upper layer has a very small density and is a gas of approximately constant pressure above a liquid, we may put $\rho_s = 0$ or $p = \text{constant}$ on the surface. In this case

$$\frac{\partial \zeta_i}{\partial z} = \frac{g\zeta_i}{u_0{}^2} \quad \text{at the surface.} \tag{5.10.32}$$

This result implies that we are discussing external waves, that is waves on the free surface, and for them the internal stratification is of negligible importance. If we are to discuss internal waves we have to impose the boundary condition that the free surface is displaced negligibly compared with the internal displacements and the boundary condition becomes $\zeta = 0$ there. The pressure variations along the surface are maintained by a displacement of it which is small and is calculated from (5.10.32) after $\partial\zeta/\partial z$ has been obtained on the assumption that ζ is zero there.

(ii) when the density discontinuity is a small fraction of the density we may put

$$\rho_s = (1 - \epsilon)\rho_i \tag{5.10.33}$$

and neglect squares of ϵ. Then (5.10.31) may be written

$$\frac{\partial\zeta_i}{\partial z} = \frac{\epsilon g\zeta}{u_0^2} + \frac{\partial\zeta_s}{\partial z} \tag{5.10.34}$$

A discontinuity of density behaves like a very thin layer in which β is very large, and so the slope of ζ may be reversed on passing through it. We may therefore have, for example, two layers with static stability small in the upper layer and zero in the lower layer and a stable density discontinuity at the interface. This is a common situation in the atmosphere when sunshine warms the ground and produces thermal convection in the lower layer. In that case,

$$\zeta_s \propto e^{-\mu z}, \quad \zeta_i \propto \sinh k\,(z + h) \tag{5.10.35}$$

so that the waves vanish at great height and are of zero amplitude at the ground $z = -h$. The interface condition (5.10.34) is then, at $z = 0$

$$k \coth kh = \frac{\epsilon g}{u_0^2} - \mu \tag{5.10.36}$$

if u_0 is constant throughout, and μ is defined by (5.10.14). There is a value of h satisfying this condition if $\coth kh > 1$, i.e. if

$$\epsilon g > u_0^2(k + \mu) \tag{5.10.37}$$

Theoretical treatments tend to regard the determination of k as the objective of the analysis, but in practice the length of waves can be measured if they are observed and u_0 and μ can then be obtained from standard soundings of the air, approximately. The theory thus determines the value which ϵ must have exceeded, the actual value being obtainable from (5.10.36) if h is known roughly.

5.11 UNTRAPPED WAVES

Waves may be set up in a stream by flow over obstacles but they are not necessarily trapped. In Section 5.12 we shall discuss how the trapped waves are excited to a greater degree, but all the wavelengths which may be though of as corresponding to the Fourier components of the obstacle shape have to be considered. These have a finite magnitude at the ground, but the same boundary conditions apply at an interface as for trapped waves. We shall illustrate the properities of untrapped waves by means of the two layer model of Section 5.10(2): the theory for more complicated or simpler cases follows the same line.

In the upper layer, denoted by suffix s, there is no exponentially decreasing solution because

$$\ell_s^{\,2} > k^2. \tag{5.11.1}$$

We therefore have to consider the physical properties of the two waves

$$\zeta_s = Ce^{\pm i\nu_s z + ikx} \quad \text{(real part)} \tag{5.11.2}$$

where $$\nu_s^{\,2} = \ell_s^{\,2} - k^2 .$$

If the upper layer extends up to infinity there is no reflection of the waves at an upper boundary, and so it is usually argued that the 'radiation condition' must apply, namely that energy must be propagated upwards by the waves because the disturbance is originated at the bottom. This means that the fluid below a given Bernoulli surface exerts a force in the upstream direction on the fluid above that surface so that work is done on the fluid above as it streams over the stationary wave pattern.

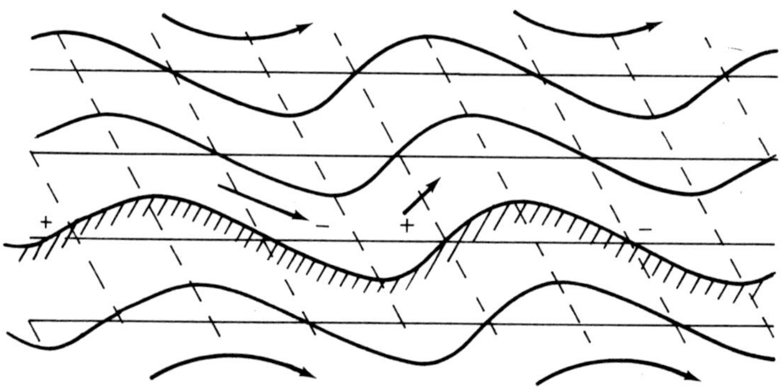

Fig. 5.11.i Waves with the crests in surfaces inclined upstream have a negative pressure anomaly on the downslope where the velocity is larger. The fluid below any given streamline therefore exerts a force in the upstream direction on the fluid above it.

Waves corresponding to the positive sign in (5.11.2) have the surfaces containing their crests and troughs sloping in the upstream direction (Fig. 5.11.i). As a result the velocity on the downward slope of a stream surface is greater than on the upslope areas, and so the pressure is greater on the upstream side of a hump than on the downstream side. There is a force on the fluid above in the upstream direction, which means that the hump exerts a drag force on the fluid stream above it, and the fluid exerts a force on the hump in the direction of the stream. In such waves momentum is transferred downwards. The trapped waves considered in the previous section have the same phase at all heights and there is no drag on the stream, which must be the case because the bottom is horizontal and can exert no horizontal force on the fluid. Waves corresponding to the negative sign in (5.11.2) have crests sloping backwards and these cannot be set up by the action of an obstacle in a stream except by reflection at an interface in the fluid. In the case of trapped waves the reflection is total and the downward travelling wave is equal to the upward travelling one.

The waves in the two layer model of Section 5.10(2), above, take the form

$$\zeta_i = A\cos(kx + \nu_i z) + B\cos(kx - \nu_i z) \tag{5.11.3}$$

which is not zero for all x for any value of z, but has the same amplitude at all heights in the lower layer. If ζ and its vertical derivative are continuous at the interface, at $z = 0$, we must have

$$\left.\begin{aligned} C &= A + B \\ \nu_s C &= \nu_i A - \nu_i B \end{aligned}\right\} \tag{5.11.4}$$

so that

$$A = \frac{\nu_i + \nu_s}{2\nu_i}C, \quad B = \frac{\nu_i - \nu_s}{2\nu_i}C. \tag{5.11.5}$$

This shows that the downward proceeding wave, whose amplitude is B, is smaller than the upward proceeding one. In the case where they are equal $\nu_s = 0$, or $k = \ell_s$, and the amplitude in the upper layer is independent of z. That case can be a standing wave over level ground if $\nu_i h = \frac{1}{2}\pi$, for then ζ_i vanishes at $x = -h$. But this is an unlikely special case, for it would be required that

$$\frac{\pi^2}{4h^2} = \nu_i^2 = \ell_i^2 - k^2 = \ell_i^2 - \ell_s^2 \tag{5.11.6}$$

which is a condition upon the stream profile. It corresponds to the limiting case of (5.10.26).

The implications for the stratopshere of untrapped waves have been discussed by Berkshire and Warren (1970).

5.12 DECAYING WAVES

The question arises whether decaying waves may be investigated by assuming that $\zeta \propto e^{(-m+ik)x}$, where m and k are assumed to be positive. In that case, taking the wave equation in the form (5.4.9) neglecting $(\nabla\zeta)^2$ and the term in $\frac{1}{2}\beta$,

$$\nabla^2\zeta + \alpha\frac{\partial\zeta}{\partial z} + \ell^2\zeta = 0\,, \tag{5.12.1}$$

and so if there is a solution of the form $\zeta \propto e^{(\mu+i\nu)z}$ we must have

$$(-m+ik)^2 + (\mu+i\nu)^2 + \alpha(\mu+i\nu) + \ell^2 = 0. \tag{5.12.2}$$

This requires that

$$m^2 - k^2 + \mu^2 - \nu^2 + \alpha\mu + \ell^2 = 0 \tag{5.12.3}$$

and

$$\nu(\mu + \tfrac{1}{2}\alpha) = mk \tag{5.12.4}$$

which means that since m and k are positive, ν and $\mu + \frac{1}{2}\alpha$ must have the same sign. In the case where $\alpha = 0$, this means that if the wave amplitude decreases with height ($\mu < 0$) it is a downward progressing one. Consequently the energy is coming down to the ground, and the ground is exerting a force on the stream in the direction of the stream. If there is a force in the opposite direction, a wave decaying exponentially downstream must increase exponentially upwards, and the crests must be displaced upstream with increasing altitude.

Decaying waves of this kind might occur behind an obstacle whose presence set up the waves, but they are not usually of great importance compared with waves which do not decay downstream. Furthermore, because they increase exponentially upwards they cannot be properly discussed in terms of steady waves, and a time-varying solution to the equations is required for full interpretation of them. That is outside the scope of this book.

5.13 LEE WAVES BEHIND AN OBSTACLE

The method using Fourier integrals is illustrated by means of waves in an enclosed stream of the kind considered in Section 5.10(1). The roof is assumed to be flat and we shall consider the effect of a hump on the bottom boundary, which we call the ground. The solution, see (5.10.4), is

$$\zeta = -\zeta_{-h} \sin \nu z / \sin \nu h, \tag{5.13.1}$$

where the coordinate system places the ground at $z = -h$. This means that if

$$\zeta_{-h} = \int_0^\infty F(k) \cos kx \, dk \qquad (5.13.2)$$

is the function of x representing the value of ζ at $z = -h$, thereby defining the shape of the lower boundary, the displacement for other values of z is

$$\zeta = -\int_0^\infty F(k) \cos kx \frac{\sin \nu z}{\sin \nu h} \, dk. \qquad (5.13.3)$$

This conclusion is only valid if the equation of which (5.13.1) is the solution is linear in ζ, for only in that case can the components of the integral be added together to give a solution for the complete boundary shape. The solutions of this section are not valid when the equation is not linear. In (5.13.3) ν is a function of k which takes imaginary values when $k > \ell$, and may be replaced by μ in that case to give

$$\begin{aligned} \mu^2 &= k^2 - \ell^2 \\ \zeta &\propto \sinh \mu z \end{aligned} \qquad (5.13.4)$$

in the place of (5.13.1).

We now take the best known and most widely used obstacle shape. This is given by the following Fourier integral which is readily evaluated, namely

$$\begin{aligned} \zeta_{-h} &= ab \int_0^\infty e^{-kb+ikx} \, dk \quad \text{real part} \\ &= a \frac{b^2}{b^2 + x^2} . \end{aligned} \qquad (5.13.5)$$

The shape is shown in Fig. 5.13.i.

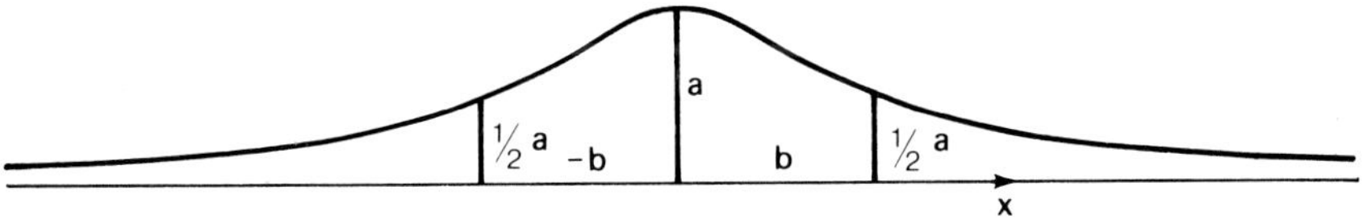

Fig. 5.13.i The function ζ defined by (5.13.5).

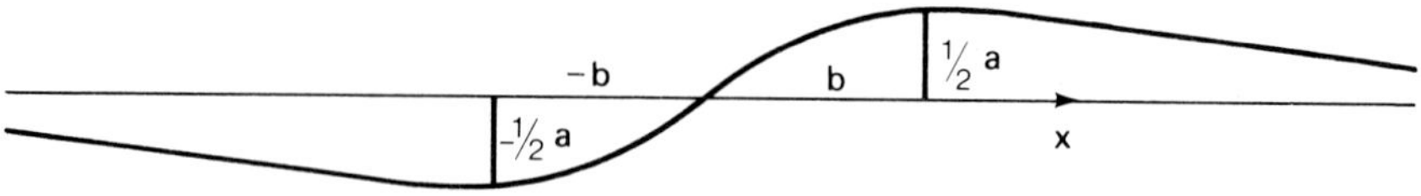

Fig. 5.13.ii The function $-\zeta$ defined by (5.13.6).

The parameter a is the maximum value of ζ which occurs at $x = 0$, while b is called the half-width, which is the distance from the maximum at which the height is $\frac{1}{2}a$. The imaginary part of the same integral, namely

$$\zeta_{-h} = iab \int_0^\infty e^{-kb+ikx}\, dk \text{ real part}$$

$$= -a \frac{bx}{b^2 + x^2} \tag{5.13.6}$$

represents the profile of a ridge next to a trough, with dimensions shown in Fig. 5.13.ii. This is of less practical use because it decreases like x^{-1} for large x, which is rather slowly. According to (5.13.3), therefore, the profile (5.13.5) gives rise to a displacement at height z equal to

$$\zeta = -ab \int_0^\infty e^{-kb+ikx} \frac{\sin \nu z}{\sin \nu h}\, dk \tag{5.13.7}$$

$$= -ab \int_0^\ell e^{-kb+ikx} \frac{\sin \nu z}{\sin \nu h}\, dk$$

$$-ab \int_\ell^\infty e^{-kb+ikx} \frac{\sinh \mu z}{\sinh \mu h}\, dk. \tag{5.13.8}$$

Since the ground is at $z = -h$, z is always negative and when $k = \ell$ $\sin \nu z/\sin \nu h = \sinh \mu z/\sinh \mu h = z/h$ because $\mu = \nu = 0$. The integrand has a pole where $\nu h = n\pi$, which is where $\zeta_{-h} = 0$ (see 5.10.4). The residue at this pole is equal to

$$R = e^{-k^*b+ik^*x} \sin \nu^* z \Big/ \left[\frac{\partial}{\partial k} (\sin \nu(k) h) \right]_{k=k^*} \tag{5.13.9}$$

where the star denotes that $k = k^*$, which is such that

$$\nu^* h = n\pi. \tag{5.13.10}$$

By deforming the contour of the integration in the k-plane as indicated in Fig. 5.13.iii, evidently iR will appear as part of the integral (5.13.7) when it is evaluated. Since the real part of iR is proportional to $\sin k^*x$, evidently a train of waves extending to infinity exists, which we call lee waves.

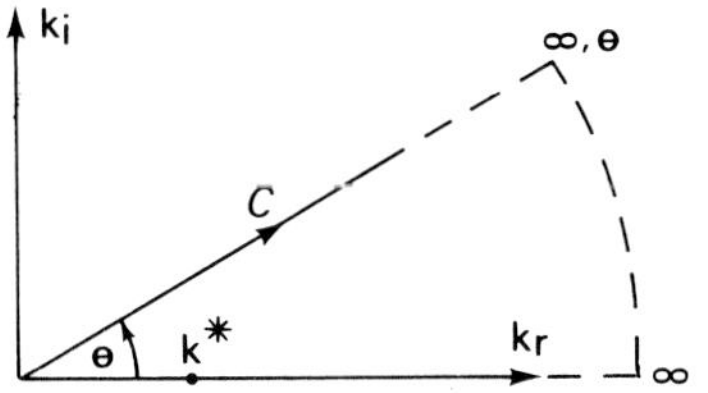

Fig. 5.13.iii The integral along the real axis to ∞ is deformed to the integral along the radius marked C. This invokes half the residue at the pole k* of the integrand.

Since R is a solution of the wave equation and is zero on the boundary $z = -h$ because of (5.13.10), any multiple of it may be added to a solution of our problem without invalidating it. In consequence some other condition must be invoked to determine the magnitude of R in the final solution. The integral we seek to evaluate is an integral from 0 to ∞ along the real k-axis. The pole is on the contour and so, since it subtends an angle π at the pole, the integral (5.13.7) is equal to

$$-\int_0^{\infty,\theta} - \int_{\infty,\theta}^{0} abe^{-kb+ikx} \frac{\sin \nu z}{\sin \nu h}\, dk + \pi i R. \tag{5.13.10}$$

in which θ is positive if x is positive, for otherwise the integrand would be exponentially large for a large imaginary part of k. If x is negative, θ must be negative and the negative sign must be taken for the residue because the contour subtends an angle $-\pi$ at the pole. Thus the difference between the values of the integral for large positive and negative x is $2\pi i R$. A practical solution is one in which there is no disturbance at large negative x, that is far upstream. In that case the solution upstream of the obstacle is simply the integral without the residue in the formula above, and downstream the term is $2\pi i R$.

This result is equivalent to saying that the pole is really above the real axis in the k plane and contains a positive imaginary part k_i, even if of only infinitesimal magnitude. In that case the residue contains the factor $e^{-k_i{}^*x}$, and the lee waves decay slowly in the downstream direction. If they have such a property the flow approaching the obstacle would have no waves in it, and would therefore have no waves far upstream, as we have supposed. Lamb (§§242, 245) described such a result for lee waves on the surface of water by introducing a representation of friction by damping force proportional to the velocity to give an exponential decay. However such a justification of the result is artificial in that it relates only to the particular formula for friction which, as Lamb pointed out, is known to be incorrect, and is credible only because it gives the result which we wish to have for other reasons.

The integral along the arc at ∞ is zero because of the factor $e^{-k_r b - k_i x}$ which makes it exponentially small. There remains, therefore, in (5.13.10) the integral along the radius C at angle θ. To evaluate it approximately we note (i) that it contains the negative exponential to which we have just referred, (ii) that it is oscillatory because of the factor $e^{i(k_r x - k_i b)}$, which means that most of the contribution to the integral is made when $|k|$ is small, and (iii) that the factor $\sin \nu z/\sin \nu h$, being a function of k^2, has zero derivative with respect to k at the origin. Therefore we may insert the value of this factor at the origin ($k = 0$) and get the approximate formula

$$-ab \frac{\sin \ell z}{\sin \ell h} \int_0^{\infty,\theta} e^{-kb+-ikx}\, dk = -a \frac{b^2}{b^2+x^2} \frac{\sin \ell z}{\sin \ell h}, \tag{5.13.11}$$

and the complete solution, with (5.13.9), noting that $\partial\nu/\partial k = -k/\nu$ and $\cos \nu^* h = (-1)^n$, is

$$\zeta = -a \frac{b^2}{b^2+x^2} \frac{\sin \ell z}{\sin \ell h} + (-1)^{n+1}\, 2\pi \frac{ab\nu^* e^{-k^*b}}{k^*h} \sin \nu^* z \, \sin k^* x \tag{5.13.12}$$

for $x > 0$, the second term being absent for $x < 0$. $n = 1$ for the case of no internal nodal surfaces, and when $n = 2$ there will be an additional second term of the form shown.

This approximate formula gives the correct value of ζ on the boundary $z = -h$, and also at the upper boundary at $z = 0$, but it is clearly incorrect near $x = 0$ at intermediate values of z. This is because the lee wave term has a discontinuous slope at $x = 0$, being absent for $x < 0$. Since

$$-bk + ikx = [-(b\cos\theta + x\sin\theta) - i(b\sin\theta - x\cos\theta)]\ |k| \tag{5.13.13}$$

the best approximation would appear to be obtained by maximising the negative exponent which is proportional to $b \cos \theta + x \sin \theta$, giving $\tan \theta = x/b$. But this also renders the oscillatory component equal to zero, so that the approximation is not good if the integration path passes near to the pole, as it does when x, and therefore θ, is small. Near the pole the assumption that $\sin \nu h$ may be replaced by the constant $\sin \ell h$ is in serious error. Therefore the formula (5.13.11) is not good for small x away from the boundaries $z = -h, 0$.

More strictly the formula is not good for small x when b is small because we have assumed that the integrand has a large negative exponent at distance k^* from the origin. Thus bk^* must be large (of the order of 2 or more, say), otherwise the discontinuity in the lee wave slope at $x = 0$ is not small.

These inaccuracies can be remedied, except when b is rather small compared with the lee wavelength, by drawing smooth streamlines near $x = 0$ and ignoring the discontinuity in slope there. However when b is small there are further waves in the streamlines not taken into account near $x = 0$.

To calculate the waves for a model stream like that in Section 5.10(2) we note that the displacement for each k is given by the following formulas, the interface being at $z = 0$ and the ground at $z = -h$, similar to (5.11.3),

$$\zeta_i = (Ae^{i\nu z} + Be^{-i\nu z})e^{ikx} \quad \text{real part} \tag{5.13.14}$$

$$\begin{aligned} \zeta_s &= Ce^{-\mu_s z + ikx}, \qquad k > \ell_s \\ &= Ce^{i(\nu_s z + kx)} \qquad k < \ell_s \end{aligned} \tag{5.13.15}$$

At $z = 0$, ζ and its derivative are continuous so that when $k < \ell_s$

$$2A\nu = (\nu + \nu_s)C, \quad 2B\nu = (\nu - \nu_s)C, \tag{5.13.16}$$

and so

$$\zeta_i \propto (\nu \cos \nu z + i\,\nu_s \sin \nu z)e^{ikx} \quad \text{(real part)}. \tag{5.13.17}$$

When the bottom boundary condition at $z = -h$ is given by (5.13.5) the total disturbance in the lower layer is given by

$$\zeta_i = ab \int_0^\infty e^{-kb + ikx} \frac{\nu \cos \nu z + i\,\nu_s \sin \nu z}{\nu \cos \nu h - i\,\nu_s \sin \nu h}\, dk \quad \text{(real part)} \tag{5.13.18}$$

together with the lee wave term which we shall consider in a moment. If we evaluate this by the same approximation as before, we write $\nu = \ell_i, \nu_s = \ell_s$ in the integrand and obtain

$$\zeta_i = abp\int_0^\infty e^{-kb}\,[\cos kx(\ell_i^2\cos\ell_i h\cos\ell_i z - \ell_s^2\sin\ell_i h\sin\ell_i z)$$

$$- \ell_i\ell_s\sin kx(\cos\ell_i h\sin\ell_i z + \sin\ell_i h\cos\ell_i z)]\;dk$$

$$= \frac{ab^2p}{b^2+x^2}(\ell_i^2\cos\ell_i h\cos\ell_i z - \ell_s^2\sin\ell_i h\sin\ell_i z)$$

$$+ \frac{abxp}{b^2+x^2}\ell_i\ell_s\sin\ell_i(z+h) \tag{5.13.19}$$

where

$$p^{-1} = \ell_i^2\cos^2\ell_i h + \ell_s^2\sin^2\ell_i h\,. \tag{5.13.20}$$

The lee wave term is obtained from the integral expressed in terms of μ_s instead of ν_s, because for a wave to be trapped $k^* > \ell_s$. We therefore replace $i\nu_s$ in (5.13.17) by $-\mu_s$ and obtain for the total disturbance

$$\zeta_i = ab\int_0^\infty e^{-kb+ikx}\frac{\nu\cos\nu z - \mu_s\sin\nu z}{\nu\cos\nu h + \mu_s\sin\nu h}\,dk. \tag{5.13.21}$$

The equation for k^* is that the denominator of the integrand be zero, namely

$$\nu^*\cot\nu^* h = -\mu_s^* \tag{5.13.22}$$

which is, of course, the same as (5.10.20). The lee wave term may be simplified by means of (5.13.22) and is found from the residue of the integrand in (5.13.21) to be

$$2\pi\frac{ab\nu^{*3}\mu_s^{*}e^{-k^*b}\sin\nu^*(z+h)\sin k^*x}{k^*\sin^2\nu^*h\,[\nu^*(\mu^*h-1) - \mu^{*2}(1+k^*h)]}\,. \tag{5.13.23}$$

In the upper layer the lee wave is proportional to $e^{-\mu_s^* z}$ and is equal to the value in the lower layer at $z = 0$. It is thus equal to

$$2\pi\frac{ab\nu^{*3}\mu_s^{*}e^{-k^*b-\mu_s^*z}\sin k^*x}{k^*\sin^2\nu^*h\,[\nu^*(\mu^*h-1) - \mu^{*2}(1+k^*h)]}\,. \tag{5.13.24}$$

The integral in the upper layer is equal to (5.13.18) at $z = 0$ and is proportional to $e^{\,i(\nu_s z + kx)}$, and is therefore

$$\zeta_i = abp\int_0^\infty e^{-kb}\,[\cos kx(\ell_i^2 \cos \ell_i h \cos \ell_i z - \ell_s^2 \sin \ell_i h \sin \ell_i z)$$

$$- \ell_i \ell_s \sin kx(\cos \ell_i h \sin \ell_i z + \sin \ell_i h \cos \ell_i z)]\ dk$$

$$= \frac{ab^2 p}{b^2+x^2}(\ell_i^2 \cos \ell_i h \cos \ell_i z - \ell_s^2 \sin \ell_i h \sin \ell_i z)$$

$$- \frac{abxp}{b^2+x^2}\ell_i \ell_s \sin \ell_i(z+h) \tag{5.13.19}$$

where
$$p^{-1} = \ell_i^2 \cos^2 \ell_i h + \ell_s^2 \sin^2 \ell_i h\,. \tag{5.13.20}$$

The lee wave term is obtained from the integral expressed in terms of μ_s instead of ν_s, because for a wave to be trapped $k^* > \ell_s$. We therefore replace $i\nu_s$ in (5.13.17) by $-\mu_s$ and obtain for the total disturbance

$$\zeta_i = ab\int_0^\infty e^{-kb+ikx}\,\frac{\nu \cos \nu z - \mu_s \sin \nu z}{\nu \cos \nu h + \mu_s \sin \nu h}\,dk. \tag{5.13.21}$$

The equation for k^* is that the denominator of the integrand be zero, namely

$$\nu^* \cot \nu^* h = -\mu_s^* \tag{5.13.22}$$

which is, of course, the same as (5.10.20). The lee wave term may be simplified by means of (5.13.22) and is found from the residue of the integrand in (5.13.21) to be

$$-2\pi\,\frac{ab\nu^{*2}e^{-k^*b}}{k^*(1+\mu_s^*h)}\cot \nu^*h \sin \nu^*(h+z)\sin k^*x. \tag{5.13.23}$$

In the upper layer the lee wave is proportional to $e^{-\mu_s^* z}$ and is equal to the value in the lower layer at $z = 0$. It is thus equal to

$$-2\pi\,\frac{ab\nu^{*2}e^{-k^*b}}{k^*(1+\mu_s^*h)}\cos \nu^*h e^{-\mu_s^* z}\sin k^*x. \tag{5.13.24}$$

The integral in the upper layer is equal to (5.13.18) at $z = 0$ and is proportional to $e^{i(\nu_s z + kx)}$, and is therefore

$$\zeta_s = ab \int_0^\infty e^{-kb+ikx+i\nu_s z} \frac{\nu}{\nu \cos \nu h - i\nu_s \sin \nu h} \, dk \text{ (real part).} \qquad (5.13.25)$$

By the same approximation this is found to be equal to

$$\frac{ab^2 p\ell_i}{b^2 + x^2} (\ell_i \cos \ell_i h \cos \ell_s z - \ell_s \sin \ell_i h \sin \ell_s z)$$

$$+ \frac{abxp\ell_i}{b^2 + x^2} (\ell_i \cos \ell_i h \sin \ell_s z + \ell_s \sin \ell_i h \cos \ell_s z) \qquad (5.13.26)$$

which is equal to (5.13.19) at $z = 0$, and has the lee wave term (5.13.23) added to make up the complete solution for $x > 0$.

Clearly a more complicated model than this, such as a three layer model, introduces considerable algebraic complexity, but is not in principle any different. Also a variety of obstacle shapes can be employed by using the appropriate Fourier integrals. Wallington (1955) gives an example of this. Scorer and Klieforth (1958) employed a three layer model of a special kind in order to obtain a better approximation to the boundary condition at the interface: they chose a case in which not only were ζ and $\partial\zeta/\partial z$ continuous at an interface where the amplitude was large, but in which $\partial^2\zeta/\partial z^2$ was also continuous. This device made the boundary conditions at the displaced position of the interface more nearly satisfied.

All the formulas in this example contain the factor be^{-k^*b} in the lee wave. This has a maximum for given k^* when $bk^* = 1$, which shows that the hump which excites the lee wave to the greatest amplitude is that which is matched to it in this way: the half width is about one sixth of a wavelength (Fig. 5.13.iv).

Reference is made to papers by Scorer (1949 and 1953b) for worked examples. A wave pattern from the former reference is shown in Fig. 5.13.v. Sawyer (1960) made calculations in which a more accurate estimation of the

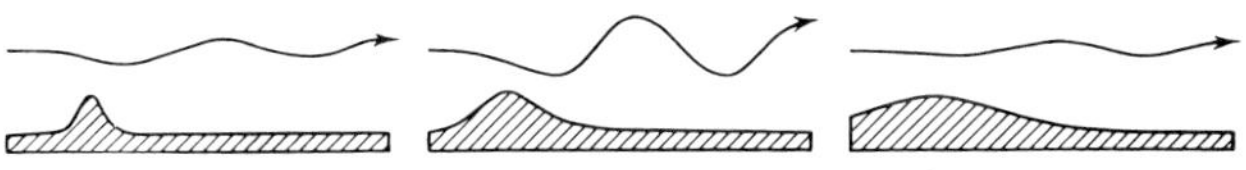

Fig. 5.13.iv The amplitude of the lee waves generated by an obstacle of given height is a maximum when the half width is about π^{-1} the wavelength, for the shape given in Fig. 5.13.i.

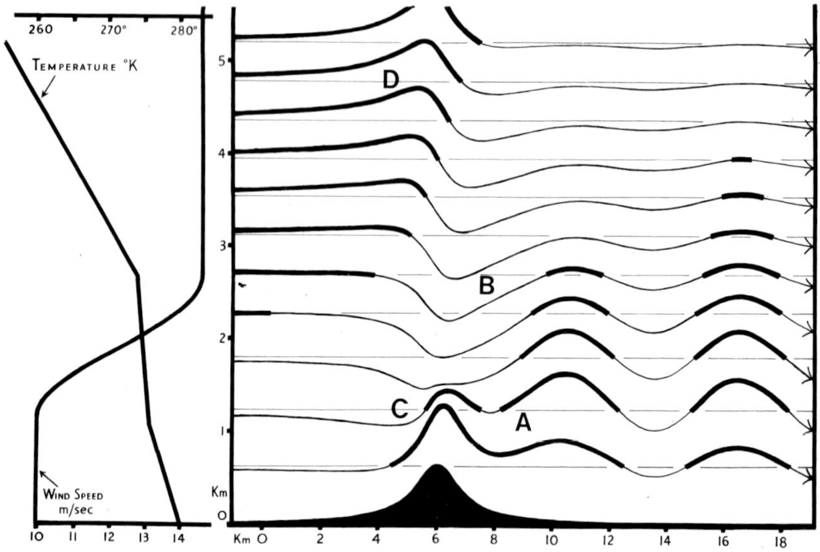

Fig. 5.13.v A calculation of the lee waves due to a two-dimensional mountain ridge in a case in which there exist trapped (lee) waves. The wind and temperature profiles are shown on the left. In this case the effect of a three layer model was obtained by a two layer model. In the upper half of the lower layer the quantity ℓ^2 defined in (5.10.28) was kept constant although U_0 varied. This avoided having a discontinuity in U_0 at the interface.

integral was obtained by direct numerical integration. All his diagrams show the displacement ζ calculated for constant values of z, and use is not made of the fact that the theory is valid for fairly large ζ. They have therefore not shown rotors where they are implied by the numbers; in some of Sawyer's diagrams streamlines are shown intersecting and this must not be interpreted as an error, but simply as a graphical representation of the calculated values of ζ.

5.14 EFFECT OF WIND PROFILE ON LEE WAVES

We consider the wave equation in the form

$$\frac{\partial^2 \zeta}{\partial z^2} + 2\alpha \frac{\partial \zeta}{\partial z} + (\ell^2 - k^2)\zeta = 0. \tag{5.14.1}$$

If there is a gradient of wind then unless β varies in such a way as to keep ℓ^2 $(= g\beta/u_0{}^2)$ constant, ℓ^2 is a function of z_0, which is a function of ζ, and so the equation is nonlinear. However, the effect on ℓ^2 is just the same as if β were varying and we may gain some insight into the effect of a wind gradient by examining the effect of the term α while assuming that ℓ^2 is constant, or varying negligibly. If both ℓ^2 and α are constants then the solution is

$$\zeta \propto \exp(-\alpha \pm \sqrt{\alpha^2 - \ell^2 + k^2})z \tag{5.14.2}$$

which shows that if α is positive, so that the wind increases with height like $e^{\alpha z}$, the effect is twofold. The waves decrease more rapidly with height through the term $e^{-\alpha z}$ and at the same time the value of k at which various properties are present decreases when k^2 is replaced by $\alpha^2 + k^2$. Thus a smaller range of the longer wavelengths has a sinusoidal variation with height; wavelengths for which $\ell^2 > k^2 > \ell^2 - \alpha^2$ have an exponential rather than a sinusoidal variation with z.

If α is negative this last property is the same, but we now have ζ proportional to $e^{\alpha z}$. This is the consequence of the property that faster moving fluid behaves as if it had greater inertia than slower moving fluid, and if the upper layers are slower moving the wave amplitude has to be larger to produce the pressure field necessary to maintain the waves below. Thus an increase in windspeed with height means that the wave amplitude decreases, and at the same time the effect of increasing ${u_0}^2$ in ℓ^2 is the same as if β were decreased, and so again the range of wavelengths for which the vertical variation is sinusoidal is reduced. We have already noted that to trap lee waves it is required that the value of ℓ^2 in the upper layers be small: we now see that an increase in wind with height in the upper layers makes trapping in the lower layers more effective if ℓ^2 is larger down below.

If we restrict the discussion to small amplitude waves we may write the equation as one for the vertical velocity by means of (5.8.20) namely

$$\zeta = w/iku_0 \, . \tag{5.14.3}$$

In the case in which $\beta = 0$ this gives the equation

$$\frac{\partial^2 w}{\partial z^2} - \left(\frac{1}{u_0}\frac{\partial^2 u_0}{\partial {z_0}^2} + k^2\right) w = 0 \tag{5.14.4}$$

to obtain which we have to assume equality between $\partial/\partial z$ and $\partial/\partial z_0$, which is only permissible if ζ is small. From (5.14.4) it appears that if $\partial^2 u_0/\partial z_0^2$ were negative an effect would be produced similar to that of a positive value for β (see 5.10.28). If $\partial^2 u_0/\partial z^2$ is constant and equal to m^2 then the solution of (5.14.4) is

$$w \propto \sin n(z - z_1) \tag{5.14.5}$$

where

$$n^2 = k^2 - m^2 . \tag{5.14.6}$$

But at the same time
$$u_0 \propto \sin m(z - z_2) \qquad (5.14.7)$$

and since $m > n$ the vertical distance between zeros of u_0 is less than between the zeros of w, or, to put it another way, the curvature of the profile of u_0 is greater than that of w. The consequence is that if we seek a situation in which standing waves are made possible by the curvature of the velocity profile instead of by a positive β in the lower layers, we shall obtain a zero of w at the ground at a lower altitude than a zero of u_0. But a zero of u_0 in the stream creates an impossible situation because of (5.14.3). Therefore the curvature of the velocity profile alone cannot be contrived to trap standing waves in the bottom layers of air.

It has been suggested by some authors that a positive value of $\partial^2 u_0/\partial z_0{}^2$ has the same effect as a negative value of β, and is therefore destabilising. This is an incorrect conclusion. The effect is simply to alter the range of values of k for which various properties of stable waves are exhibited. Provided u_0 does not have a zero, however, the presence of layers in which $\partial^2 u_0/\partial z_0{}^2$ is comparable with $g\beta/u_0$ does sometimes make possible the trapping of waves of a particular wave number k which would not be trapped if the wind profile were not curved (see Scorer 1949, for example).

There is a well known theorem (see, e.g. Drazin and Howard 1966) that a stream of constant density may be unstable for some wavelengths if there is a point of inflexion in the velocity profile, and is not unstable otherwise for two dimensional waves. This means that for instability there must be a layer in which the wind gradient is a maximum, i.e. a layer of greater vorticity than above and below, and the phenomenon which results is discussed in detail in Chapter 6.

On the other hand it is no obvious matter to make general statements about the circumstances in which complicated velocity profiles may make standing waves possible. A recent paper by Banks, Drazin, and Zaturska (1976) illustrates the complexity of the problem when a maximum and a minimum of velocity occur in the stream away from the boundaries.

5.15 LEE WAVE PATTERNS IN THREE DIMENSIONS

Because the calculations become very complicated in every case, we illustrate the principle behind them by the simplest case occuring in nature, the pattern of waves behind a disturbance, such as a ship, moving across the surface of deep water. The pattern is standing still relative to the ship and may be thought of as lee waves. The velocity, U, of waves on deep water is given by

$$U^2 = g/k \qquad (5.15.1)$$

and if a ridge-like disturbance is inclined at angle ϕ to the normal to the stream, the component $U \cos\phi$ across it produces waves. The component $U \sin\phi$ parallel to it may be added to it after the wave form has been discovered. If the lee waves are proportional to $-\sin kx$, x being distance downstream in a direction perpendicular to the ridge (see Fig. 5.15.i), then the following two equations are derived, for a point P whose polar coordinates are (r, θ), from the definition of X and (5.15.1):

$$X = r\cos(\phi + \theta); \quad k = g/U^2 \cos^2 \phi \tag{5.15.2}$$

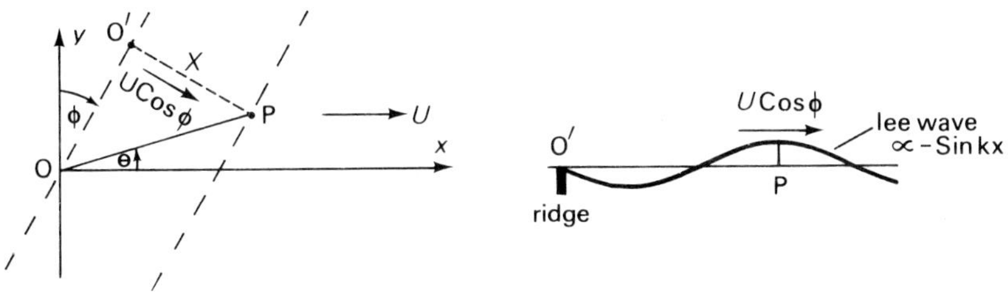

Fig. 5.15.i The left diagram shows the coordinates for the lee wave at P parallel to the ridge OO′ inclined at angle ϕ to the normal to the stream. A set of ridges at different angles ϕ together make an isolated obstacle at O: the lee waves are calculated by summing the individual lee waves.

In the right diagram is the elemental lee wave in vertical section through O′P.

We now envisage the disturbance centered on O to be made up of a system of ridges at angle ϕ, $-\pi/2 < \phi < \pi/2$, each of which produces its own lee waves. The total wave pattern behind the disturbance is the sum of all the lee waves. If P is on the crest of a wave in the total wave pattern, the locus of P is the envelope of the individual lee wave crest lines, on which

$$kX = N \tag{5.15.3}$$

where N is the phase, in this case a crest. Equally the nodes and troughs of the total pattern are the envelopes of the individual node and trough lines, ϕ being the parameter determining the individual lee waves. The envelope is discovered as the locus of the intersection of the lines (5.15.3) with the lines for the neighbouring value of ϕ, and this is obtained by solving (5.15.3) together with its derivative

$$\frac{\partial}{\partial \phi}(kX) = 0, \tag{5.15.4}$$

These two equations are readily found, from (5.15.2) to be

$$r \sec^2 \phi \cos(\phi + \theta) = a = NU^2/g \tag{5.15.5}$$

$$2 \tan \phi = \tan(\phi + \theta) \tag{5.15.6}$$

If we now put $x = r \cos \theta$, $y = r \sin \theta$, the locus of the phase lines in the total lee wave pattern is found to be

$$\begin{aligned} x &= a \cos \phi \,(2 - \cos^2 \phi) \\ y &= a \sin \phi \,(1 - \sin^2 \phi) \end{aligned} \tag{5.15.7}$$

the parameter ϕ ranging from $-\pi/2$ to $\pi/2$.

The most interesting feature of this wave pattern, which is shcwn in the diagram below, (5.15.ii), is that it is confined between two values of θ. Since from (5.15.6)

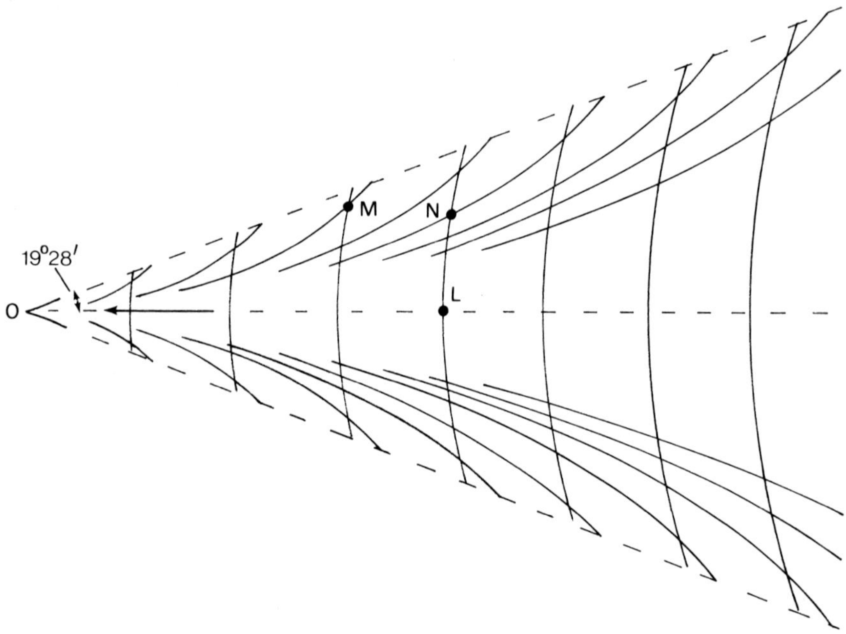

Fig. 5.15.ii The crest lines of the total lee wave pattern of a disturbance centered at O travelling over the surface of deep water. The lines of constant phase for other phases are similar. The points where the diverging phase lines cross the transverse ones, such as M, are where the maximum amplitude of the wave occurs. A second maximum is sometimes found at N, but this must not be confused in the case of a ship with the wave crests of type M from the stern of the ship.

The transverse waves at points like L are large, but in the case of a well-designed ship travelling at its design speed the total disturbance primarily due to the bow and the stern, is minimised there to reduce the wave resistance.

$$\frac{y}{x} = \tan\theta = \frac{\tan\phi}{1 + 2\tan^2\phi} \tag{5.15.8}$$

it is evident that θ is zero when $\phi = 0$ or $\pm\pi/2$, and its maximum occurs when

$$\tan^2\phi = \tfrac{1}{2} \tag{5.15.9}$$

and

$$\tan\theta_{max} = 2^{-3/2}$$

or

$$|\theta_{max}| = 19°28' \tag{5.15.10}$$

All the phase lines in the wave pattern have the same shape and are obtained from successive values of N, which is increased by 2π in proceeding from one crest to the next.

These formulas are simple because of the simple nature of equation (5.15.1) which expresses the relationship between wave length and speed on deep water. In a stratified fluid the simplest relationship is (5.10.5) for the case of flow between horizontal rigid boundaries. (5.15.2) is then replaced by

$$X = r\cos(\phi + \theta); \quad k^2 = \frac{g\beta}{U^2\cos^2\phi} - \frac{n^2\pi^2}{h^2} \tag{5.15.11}$$

and (5.15.3, 4) become correspondingly much more complicated, while as ϕ approaches $\pm\pi/2$ we may put n successively equal to 2, 3, ... and obtain higher modes of lee waves contributing to the pattern. The contribution of these higher modes is usually negligible because they are restricted to a small range of ϕ and their amplitude is usually smaller in the two dimensional case from which they are derived. They will therefore be neglected in further discussion and we shall write $n = 1$ in (5.15.11) and corresponding equations for other cases.

Some details of the wave pattern are revealed by a different approach as follows. An arbitrarily shaped disturbance at O may be represented by

$$\zeta(r,\theta) = \int_{-\pi/2}^{\pi/2} \Phi(\phi) f(X) \mathrm{d}\phi \tag{5.15.12}$$

where $f(X)$ is the profile of the individual component ridge and $\Phi(\phi)$ is its amplitude, varying with ϕ. Thus if $\Phi = 1$, so that all the ridges are the same, and have the cross section of a thin 'wall' of unit cross section area, at distance r from the origin the height of the obstacle given by (5.15.12) is proportional to r^{-1}. If the 'wall' has width b then the obstacle is flat topped for $r < b$ and for $r > b$ its height decreases and becomes proportional to r^{-1} at large r.

If the lee waves of the ridges are proportional to $-\sin kX$ then the total wave pattern is

$$\int_{-\pi/2}^{\pi/2} \Phi(\phi) i e^{ikX} \zeta(z,\phi)\, d\phi \tag{5.15.13}$$

where $\zeta(z,\phi)$ represents the variation of the component waves with height, which may vary with ϕ.

To evaluate this integral we make use of the principle of stationary phase. We write

$$kX = k^*X^* + \tfrac{1}{2}(\phi-\phi^*)^2 (k^*X^*)'' + 0(\phi-\phi^*)^3 \tag{5.15.14}$$

where the star denotes the value of the functions obtained by putting $\phi = \phi^*$, and ϕ^* is the value of ϕ for which

$$(kX)' = 0, \quad \text{i.e. } (k^*X^*)' = 0 \tag{5.15.15}$$

where the prime denotes $\partial/\partial\phi$. Since e^{ikX} is an oscillatory function of ϕ, the major contribution to the integral will be made in the neighbourhood of the value of ϕ satisfying (5.15.15) because, since the phase kX is stationary only at that point in the range of integration, e^{ikX} is very slowly changing. A good approximation to the integral is therefore obtained by replacing the variables $\Phi(\phi)$ and $\zeta(z,\phi)$ by the values they have at $\phi = \phi^*$. We also make use of the formula

$$\int_{-\infty}^{\infty} e^{im^2\xi^2}\, d\xi = \frac{\sqrt{\pi}}{m}\frac{1+i}{\sqrt{2}} = \frac{\sqrt{\pi}}{m} e^{i\frac{\pi}{4}} \tag{5.15.16}$$

where i may be replaced by $-i$ throughout. Thus is m^2 if negative the right hand side is replaced by $\dfrac{\sqrt{\pi}}{|m|} e^{i\frac{\pi}{4}}$. In our case, therefore, the approximate value of (5.15.13) (real part) is

$$\Phi(\phi^*)\zeta(z,\phi^*) \left(\frac{2\pi}{|(k^*X^*)''|}\right)^{1/2} \cos\left(k^*X^* \pm \frac{\pi}{4}\right) \tag{5.15.17}$$

the upper or lower sign being taken according to whether $(k^*X^*)''$ is positive or negative. ϕ^* is now our parameter, and we see that for the locus of constant phase $k^*X^* = \text{constant} = N$, which is (5.15.3). Also (5.15.15) is equivalent to (5.15.4). The maximum value of θ in (5.15.8), which is (5.15.4) written for the particular case of the wave pattern on deep water, was obtained by differentiation with respect to ϕ. This means that at the outer boundary of the pattern $(k^*X^*)''$ is zero and changes sign as ϕ passes through the value giving the maximum of θ. Consequently the sign in (5.15.17) is changed there and $\pi/2$ is added to the phase N. For this reason the waves in Fig. 5.15.ii are drawn so that the transverse and diverging waves are out of phase there by $\pi/2$.

In arriving at (5.15.17) we have replaced the limits of integration in (5.15.13) by $\pm\infty$, on the grounds that the oscillations of the integrand well away from the point of stationary phase contribute negligibly.

The fact that the amplitude is large near the boundary is made plain in (5.15.17), although the approximation fails at the boundary where $(k^*X^*)''$ is zero. Therefore we would expect to see the waves most pronounced in this neighbourhood. A more accurate calculation of the value of the integral in that region using the term in $(k^*X^*)'''$ and $(\phi - \phi^*)^3$ by a method described by Scorer (1950) has been given by Warren (1961). The formula (5.15.17) also indicates how the amplitude of the waves is related to the function Φ. The transverse waves, which are generated for $0 < |\phi| < \phi^*$, are greatest when the obstacle has a larger extent across the stream, whereas an obstacle elongated in the downstream direction has larger values of Φ when $|\phi| > \phi^*$. Thus a long narrow boat, such as a rowing eight, produces only the diverging waves, while a squat flat-sterned boat generates only the transverse waves. These features are illustrated in Fig.s 5.15.iii and iv.

The same kind of treatment may be applied to the lee waves behind a hill in a stratified stream capable of supporting standing waves. An example was given by Scorer and Wilkinson (1956). The maximum value of θ beyond which there is no disturbance varies according to the circumstances and in the example cited it was shown that the wave pattern could spread as much as $\pm 45°$ or only to $\pm 1°$ if the parameters were suitably chosen. A picture of a very narrowly spreading pattern is given in Fig. 5.15.v.

The pattern is not symmetrical when $\Phi(\phi)$ is not symmetrical about $\phi = 0$. The form of disturbance due to an oval obstacle inclined at an angle to the stream is given by Scorer (1956). It is clear that there exist innumerable possible different cases, the main features of which may be elucidated by these methods. It should also be remembered that methods of this kind are not worth pursuing in much greater detail because it is impossible in any practical case to specify the boundary conditions and the airstream profile with any precision. Berkshire (1975) has applied similar methods to the study of three dimensional wave patterns in the stratosphere, in order to obtain qualitative results, but there can be little prospect of calculations of this kind being of any use except in the examination of causes in a well documented case long after the event.

Fig. 5.15.iii The wave pattern due to a fast moving motor boat seen from the Eiffel Tower. This shows the diverging waves only.

Fig. 5.15.iv The wave pattern produced by a tugboat, composed mainly of the transverse waves.

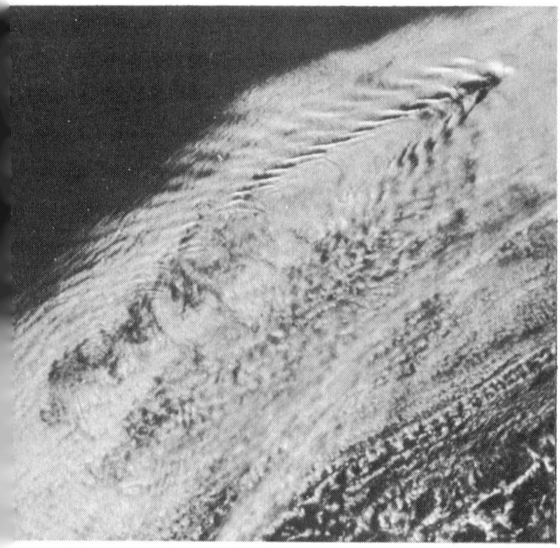

Fig. 5.15.v A satellite (NOAA5) photograph of waves in the lee of Jan Mayen in a northerly airstream. The additional waves on the west side are due to the smaller peak (see Fig. 2.11.iii).

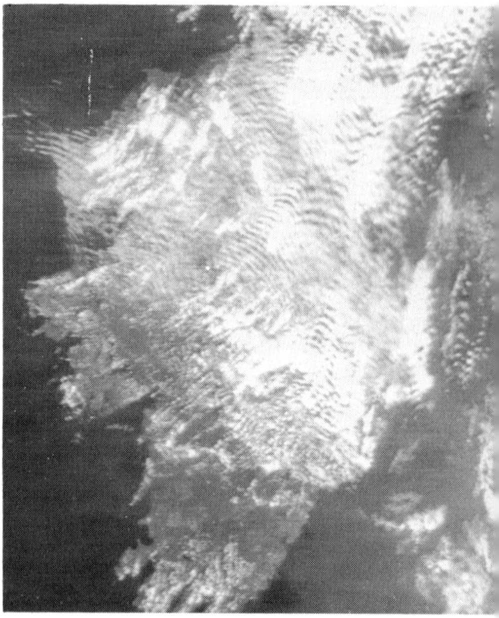

Fig. 5.15.vi A satellite (NOAA 4) photograph of clouds over Ireland with a prominent ship-wave pattern with transverse waves downwind of the Mourne mountains. (852 m). [This and the previous photograph reproduced courtesy of Department of Electrical Engineering, Dundee University).

5.16 SOME IMPORTANT PROPERTIES OF LEE WAVES

The waves over and near to the obstacle depend to a large extent on the shape of the obstacle. The real difficulty about applying the kind of theory described here lies in the nature of the assumptions. Thus we have assumed that the profile of the stream far upstream of the obstacle can be specified and can be assumed to be undisturbed. If an obstacle is placed suddenly in a stream it is observed in all cases to send a wave, or bore, travelling upstream which alters the height of the Bernoulli surfaces in the stream arriving at the obstacle. It cannot be assumed that a given stream and a given obstacle are compatible and that no disturbance of that kind will occur before a steady state is reached.

It has also been assumed that a steady state can be reached, yet no argument has ever been given to justify this. This is particularly important in the case of separation of the flow from the surface, because periodic flow may then occur (see Section 5.18), rather like the shedding of vortices behind a cylinder.

Nevertheless waves over mountains and lee waves do occur, and waves upstream have never been recorded without doubt. The formulas obtained for the lee waves, for example (5.13.11), show that the lee wave amplitude can be as large as, or even larger than, the obstacle height. In the case of waves trapped in a two layer model, the formulas, such as (5.13.19) and (5.13.23), show that when the stream profile is such that waves are trapped, the disturbance over the mountain may rise to a larger maximum than in a stream in which no waves are trapped.

In exceptional cases the lee waves may be considerably larger than the height of the obstacle, and the parameters in the formulas can be juggled to achieve this result in a theoretical example. In practice the largest lee waves occur when the airstream flows down from a plateau without separation from the surface. This is because the waves represent energy added to the stream. In our solutions this energy appears to be added without any source. While this can be accounted for in the case of flow between two rigid boundaries by a pressure drop in the downstream direction, in the case of a stream of infinite depth in which no disturbance is imposed at the top this explanation is not acceptable. In the atmosphere there could be an acceleration of the flow by a deviation towards low pressure across the isobars, but such an acceleration cannot be effective in a distance which is very small compared with that travelled by the fluid in a day. On the other hand the theory is a small amplitude one for the case of no upper boundary, so that energy considerations are not applicable because energy is represented by second degree terms.

The most effective cause of large lee waves other than descent from a plateau is to have several parallel ridges spaced at a whole number of lee wave lengths apart. Thus the wave amplitude may be doubled by a second obstacle one wavelength downstream of the first.

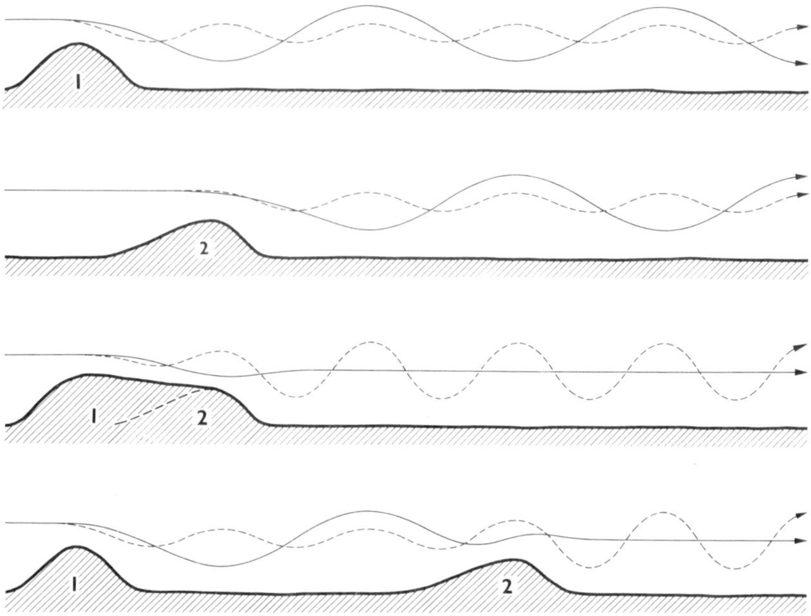

Fig. 5.16.i The mountain 1 produces the lee waves shown for two airstreams one having twice the wavelength of the other. When mountain 2 is added the lee waves may be increased or decreased. In the case shown they are doubled for the shorter and cancelled out for the stream having the longer wavelength.

Equally a second obstacle placed at an odd number of half wavelengths downstream could exactly cancel the lee waves of the first obstacle, and leave none. In general a second obstacle merely alters the phase of lee waves. In Fig. 5.16.i two mountains 1 and 2 have two airstreams depicted, the one with dashed lines having half the lee wavelength of the other. When mountain 2 is 1 or 3 wavelengths downstream the lee waves are added, but at ½ or 1½ wavelengths downstream the lee waves cancel out.

Since the component of velocity along a two dimensional ridge is unaffected by the ridge and retains its undisturbed value in each Bernoulli surface, a given airstream profile may produce lee waves when it crosses two parallel ridges at one angle but not when it crosses at another angle. A rare combination of circumstances may cause very large lee waves to occur, and then a very strong wind is produced in the wave trough with almost no wind, or even reversed flow in a rotor, at the surface under the wave crests. Such a lee wave removed the roofs from houses in a suburb of Sheffield although some had stood undisturbed for over 50 years. The gale which achieved this was considered the more remarkable because of the areas of calm on each side of it not far away. The rare circumstance in this case was actually the coincidence of the lee wavelength in the airstream with the spacing of three successive ridges.

When insects are carried by a wind over a mountain range they may settle more readily in the calm areas under a lee wave crest than in the strong wind in the trough, and their distribution on arrival may look suspiciously as if it had been contrived by an enemy releasing them in rows. There is also a local tradition in the Isle of Man that if the roof is blown off a barn or house this is because the fairies are displeased at its location in their territory. The theory is 'proved' when the roof is blown off again, after repair, by a very localised wind which does not occur elsewhere, even on other occasions.

One very common feature of lee waves is that they are more pronounced at night. The sky is typically filled with wave clouds at dawn, and they disappear and are replaced by convection clouds during the day. As the sun sets the wave clouds often return. Billows associated with wave clouds show the same diurnal variation. The most pronounced gales in lee waves occur at night, and it is at night that fairies remove the roofs from barns.

One of the most striking examples of the onset of waves in the evening is experienced at the gliding site at Great Hucklow near Sheffield (see Roper 1952), close to the area where the roofs were blown off. The site is situated on the top of the second ridge and experiences lee waves in the evening. These used to be called evening thermals because they were thought to be associated with warm air rising up the west facing slope which was the last piece of ground to be exposed to the rays of the setting sun. Actually the site often finds itself in the almost calm air under the wave crest of the first lee wave from the ridge upwind (see Fig. 5.16.ii). While it is difficult to get a glider aloft by the simple device of

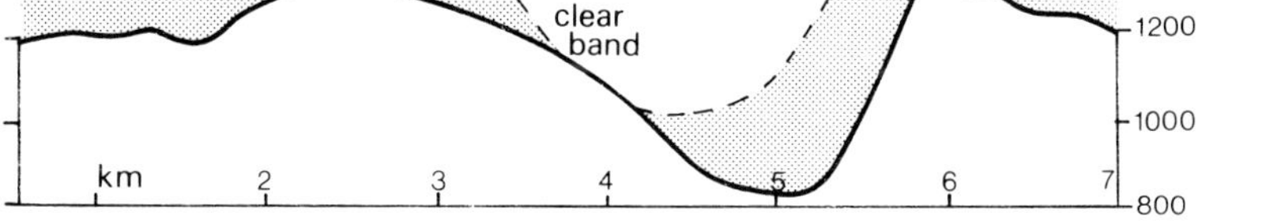

Fig. 5.16.ii The waves in an airstream are often visible at a haze top. This example, from Roper (1952), shows the observed behaviour on an evening when waves were restablished and the wave occurred so as to produce a calm over the second ridge top. Because lift was found by glider pilots the phenomenon was given the name 'evening thermal' at a time when waves were not well understood. This 'Clear Band' could have been partly an optical illusion, the haze not being visible in the shadow on the east side of the hill in the evening.

catapulting it from the top of the ridge into an upslope wind, it is often very easy for those already airborne at a few hundred metres above the site to find the upslope side of the wave and soar to greater heights with it.

The early explanation of this phenomenon was based on the assumption that destruction of the stable stratification in the lowest layers by sunshine on the ground during the morning was reversed, thus decreasing the value of ℓ^2 near the ground and making trapping impossible. It now seems more likely that it is the result of the ending of separation of the flow from the lee side of the obstacle. This is discussed in Section 5.18. The reason why the first explanation in terms of waves is unlikely to be correct is that the stable stratification produced by cooling the ground affects only a very shallow layer of air close to the ground during the first few hours. On the other hand, the warming of the ground in the morning can alter the stratification over a greater depth in a short time.

In the Eden Valley in the north of England, where lee waves have long been known, the cloud in the first lee wave moves away from the mountain ridge during the morning and finally disappears, local lore asserting that it never crosses the river (see Manley 1945). Thus the lee wave length increases (k^* decreases) as would be expected, because the depth of the stable layer at the bottom of the airstream is reduced from below by daytime convection, and the time required for an oscillation is increased. Alternatively we may argue, according to the ideas set out in Section 5.10(3), that since the layer of large β is reduced in thickness, a smaller value of k is required in order to achieve a zero at the ground. This is because that increases the curvature of the ζ-profile in the layer in which $\ell > k$ and reduces the curvature in the opposite direction in the layers in which $k > \ell$. As the layer is thinned, it becomes impossible to achieve a zero of ζ at the ground, and the lee waves disappear.

A common experience of glider pilots, noticed very early in the days of exploiting lee waves for soaring, is that the lee wave amplitude has a maximum at about the height of the top of a layer of cloud if one is present. Very often the cloud top is the upper limit of a layer of air which has been stirred by convection from the ground, and the density (temperature) discontinuity which exists at its top is the cause of the presence of waves (see Section 5.10(4)). As is explained in Chapter 11, a cloud top is usually a level from which heat is lost by radiation into space and so the layer below it is continuously cooled, and a stable temperature discontinuity is established. Since this is, in effect, a shallow layer of very great static stability, it is one in which the curvature of the ζ-profile is very large and negative, and so it is almost inevitably a level at which the profile has a maximum.

Since a layer of cloud may be continuous, that is to say it is thick enough not to be evaporated in the wave troughs, the fact that waves are present is sometimes not visible from the ground. If the layer of air in which the cloud is formed is well mixed, all stream surfaces have the same condensation level and the cloud base is horizontal. On rare occasions a moist layer lies above a drier layer, and then waves are visible on the cloud base.

A common observation in mountainous areas is that when the air streams down off a plateau, sometimes a layer of cloud existing over the plateau disappears because the air at the level of the cloud also descends. But equally often a layer of cloud is seen over the low ground and not over the plateau, or even occasionally there is a hole in a layer of cloud over a mountain ridge and not elsewhere. This is called the Főhn window (Lücke) (Fig. 5.16.iii), because the blue sky above can be seen only through two or more cloud layers in the neighbourhood of the downslope region. The formula (5.13.25) for the displacement of the streamlines near to the ridge shows that the phase of the displacement is reversed with a period of $2\pi/\ell_s$ in z, and it is a detail of the case whether the displacement is upwards or downwards over the edge of a plateau at the level at which the cloud layer happens to be.

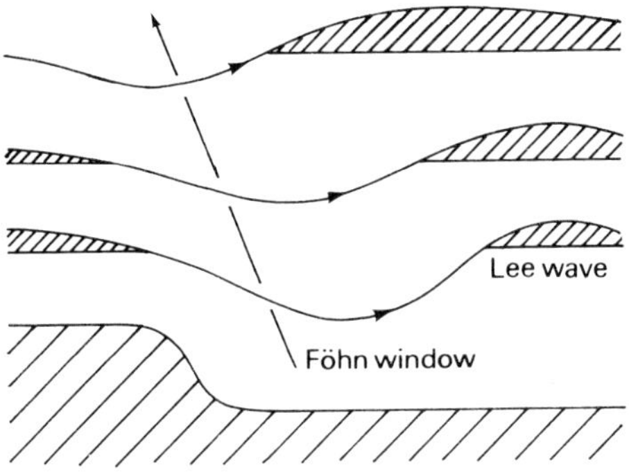

Fig. 5.16.iii Whether there is a lee wave visible in cloud at a low level or not there is sometimes a higher layer at which the air ascends above its condensation level when the air at ground level descends. A 'window' to the sky then appears over the lee slope, and is often referred to in the literature as a 'fohn lucke'.

5.17 A SECOND KIND OF TRAPPED WAVE

If at some altitude the wind is zero relative to the ground and below the air is streaming over an obstacle, we cannot find an acceptable solution of the wave equation in such a case because ℓ^2 becomes infinite when u_0 is zero. Also, unless $\partial u_0/\partial z_0$ is zero at that altitude, α also becomes infinite.

The mathematical solutions of the equation (5.6.7) have nodal surfaces packed ever closer together as the altitude of zero u_0 is approached from below (Scorer 1951), and the magnitude of ζ exceeds the distance between nodal surfaces, close to the zero of u_0, which means that rotors are implied. The only alternative is to have a cat's eyes pattern with a row of vortices at the level of zero u_0 with waves above and below (see Fig. 6.2.i bottom diagram). See also Queney (1955) and Gerbier and Berenger (1961).

Various suggestions have been made as to what probably happens. Several authors, e.g. Booker and Bretherton (1967), have given the level of zero u_0 the rather unhelpful and unspecific name 'critical level'. They guess that any wave energy propagated from below is absorbed at this level, which implies that momentum is communicated to the air there and that the waves are not reflected. The theory used to back up this idea is very speculative, is valid only for very small wavelengths, and cannot be applied to typical lee wave lengths in the atmosphere. Since momentum would be communicated to the air in which the kinetic energy is absorbed, the presence of layers of zero u_0 is thought to imply that the air at that level is decelerated by this absorption and that much of the drag of a mountain is applied to the air at that level. This conclusion is based on incomplete mathematics not valid for the wavelengths in question.

Perhaps the most important possibility is that if the wind gradient is large below the zero of u_0 it might be confined to a sufficiently shallow layer for it to behave like a discontinuity of velocity with zero velocity above, in the case of waves whose length is large compared with the depth of the shear layer. Although this would mean that it would be likely to break down into billow motion (see Chapter 6) it could still, for longer wavelengths, behave like a statically stable discontinuity of density and velocity with stagnant air above. In that case it could support standing waves, and the air could be in smaller scale unsteady motion near the top of the layer. This possibility was envisaged in some early calculations (Scorer 1949(1)), and waves having such characteristics have been observed from time to time by glider pilots (e.g. at the Long Mynd, Scorer 1953 a).

Although the theory is incapable of a proper explanation, several authors have described the phenomena in diagrams which summarise their cumulative experience. The diagrams in Fig. 5.17.i are from a paper by Gerbier and Berenger (1961), and Forchtgott (1949) gave diagrams reproduced by Scorer (1952) based on observations of moving rotors.

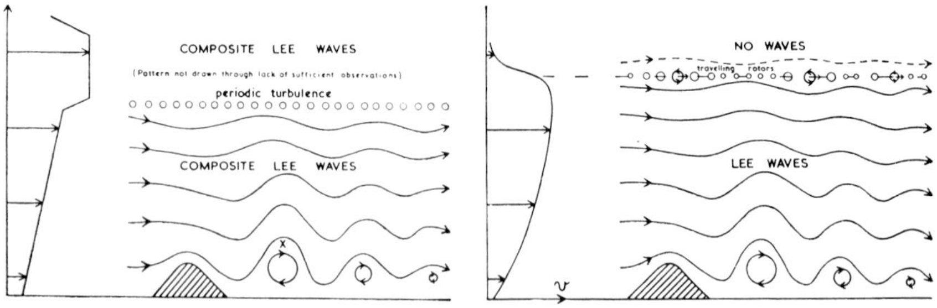

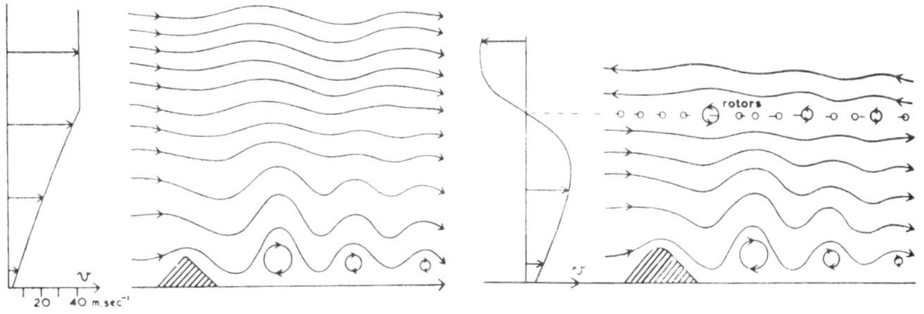

Fig. 5.17.i Four schematic diagrams indicating the flow pattern observed with four different velocity profiles in studies by Gerbier and Berenger in the French Alps. It is a matter of argument whether the rotors observed at the level of zero velocity are billows or rotors of the kind envisaged by Queney (1955) as 'cat's eye' rotors. Rotors decrease in size downwind in lee waves, and in the first diagram there is a longer wavelength evident at higher altitude.

5.18 SEPARATION OF THE FLOW FROM THE SURFACE

The theory of waves has so far made no useful progress with unsteady wave phenomena, and none is in prospect because of separation (see Scorer 1955). The phenomena are so complicated and varied that they are best described by simple diagrams.

(1) Salient edges

In this case the topography has a sharp edge which induces separation. This can sometimes be due to a wall. It may take place on the lower or upper side of a cliff top or in a street with tall buildings on both sides (Fig. 5.18.iv).

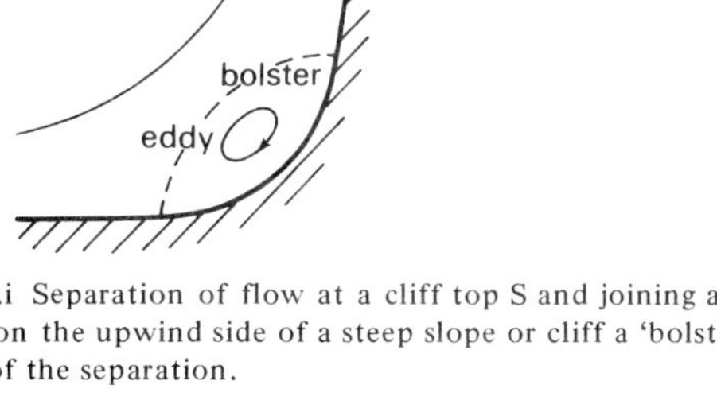

Fig. 5.18.i Separation of flow at a cliff top S and joining at J. Where the flow diverges on the upwind side of a steep slope or cliff a 'bolster' eddy may occur because of the separation.

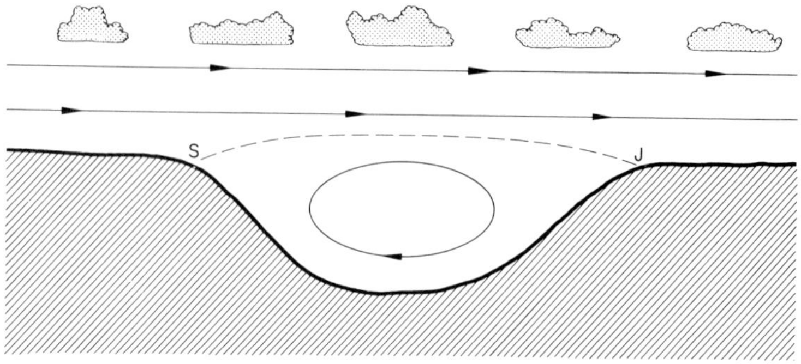

Fig. 5.18.ii Flow separating on a lee slope, the eddy filling a valley. There is no effect on the flow above.

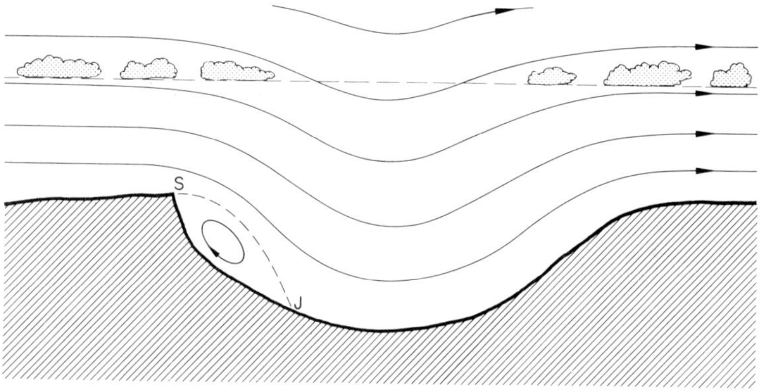

Fig. 5.18.iii A small eddy may be produced by a salient edge but the large valley may nevertheless produce down motion which evaporates the cloud over it.

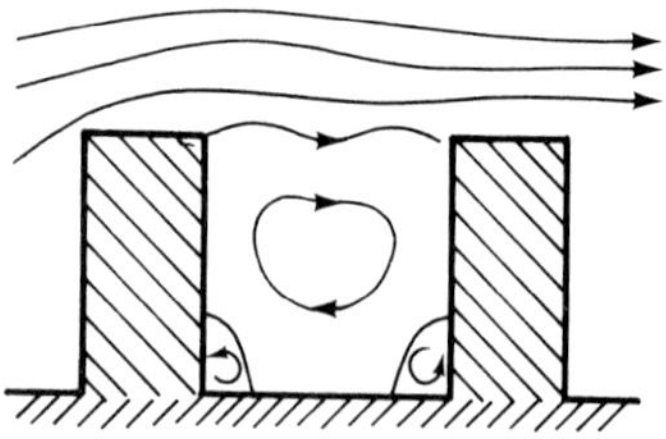

Fig. 5.18.iv In a street standing eddies may occur, including corner eddies driven by the main one. But the exterior flow is not usually steady, and contains eddies shed by other obstacles upwind; furthermore it contains a wind component along the street which influences the pattern of separation. Since the architecture is also variable, this diagram is only a first indication of the kind of motion to be looked for.

(2) Corner eddies

These are eddies formed in a boundary layer at the foot of a steep hill or a building (Figs. 5.18.i and iv). They are sometimes referred to as 'bolster' eddies. It can be demonstrated that they are due to the velocity profile of a boundary layer, and that the separation is like that produced in a boundary layer by placing a solid surface upstream of a wedge. Bolsters are then induced at the apex of the wedge: the effect is illustrated by Fig. 5.18.v and vi.

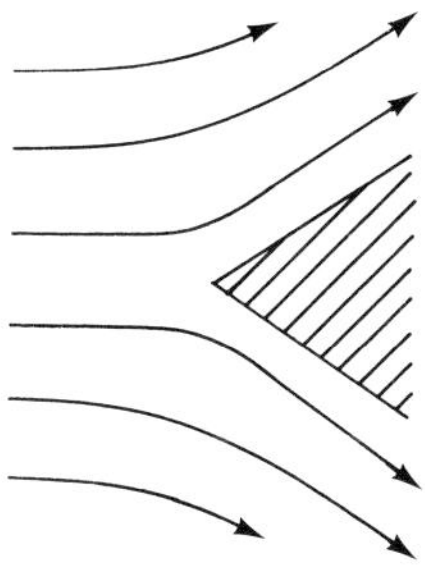

Fig. 5.18.v This shows the flow produced by a sharp corner facing the wind.

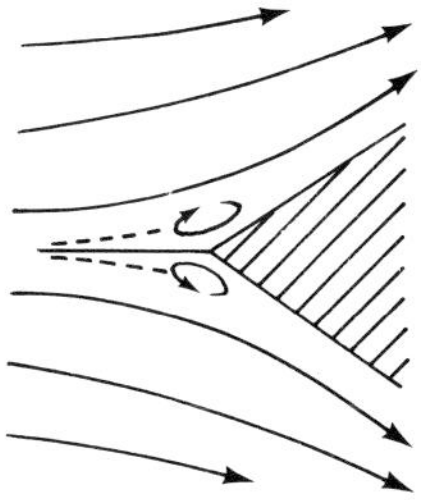

Fig. 5.18.vi If a solid surface is placed upstream of the apex the drag of its surface may lead to the formation of bolster eddies.

(3) Lee eddies

When the flow separates on the lee side of an obstacle because the main stream is widening out and moving towards higher pressure, the separating streamline is usually unsteady because it is in a region of maximum vorticity. The eddies are sometimes steady, particularly if there is a good salient edge to

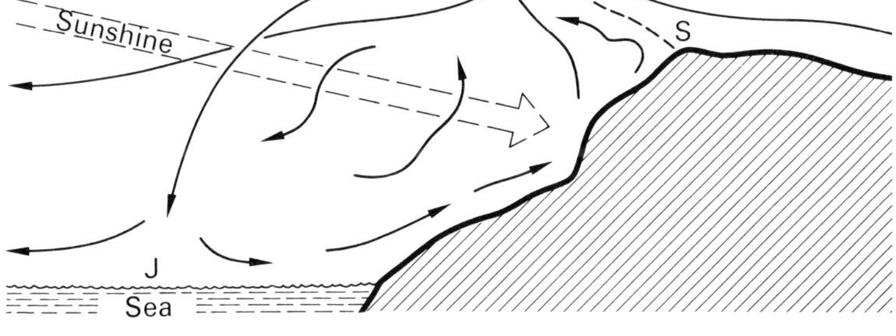

(a)

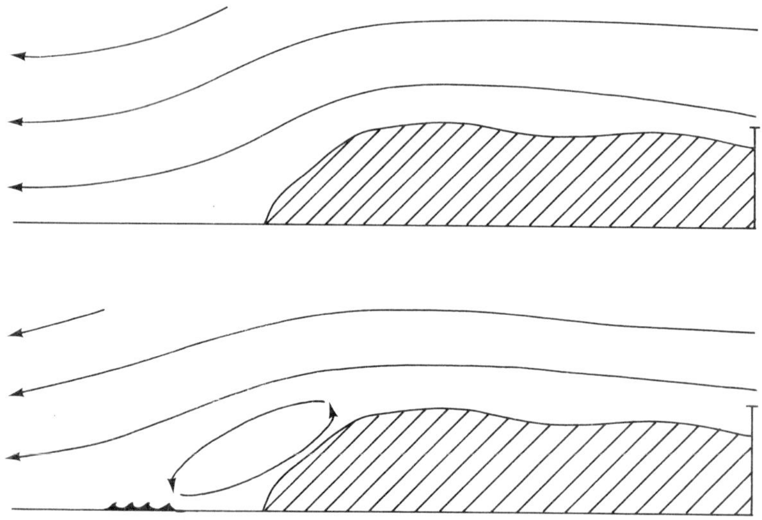

(b)

(c)

Fig. 5.18.vii Diagram (a) depicts the complicated situation where a sun-facing slope may have separation at the top caused by a salient edge or by an anabatic wind, or both. The anabatic wind produces thermals which are intermittent, and sometimes produce cumulus clouds. The eddy is very unsteady, and at the point of joining the wind may be gusty and creates cat's paws on the sea surface, or on the opposite side of a valley in bracken, grass, or a standing crop.

(b) on the other hand shows the simplest situation of fairly smooth flow out to sea. (c) shows the simpler case of an eddy, and the flow often oscillates between (b) and (c) in a way that can be detected by putting wind socks or smoke generators on the slope.

fix the point of separation, but with a fluctuating outer boundary. Others are shed, often periodically. Some parts of a coastline have a sharp cliff top (Fig. 5.18.vii(a)) where an eddy is fixed, while on other well rounded parts (Fig. 5.18.vii(b), (c)) sometimes there are eddies and sometimes not. The presence of eddies can have a significant effect on lee waves, for the position at which rejoining takes place may be determined by minimising the wave energy.

On the other hand wave clouds have often been observed to come and go periodically, and periodic upsurges of wind followed by slackening are experienced on the lee side of hills. This could be due to the periodic shedding of eddies from a hill, and in this case, the first lee wave cloud is observed to drift slowly down wind and disappear, to be replaced at a point closer to the hill by a new cloud. This sequence is depicted in Fig. 5.18.viii which is taken from Scorer (1955).

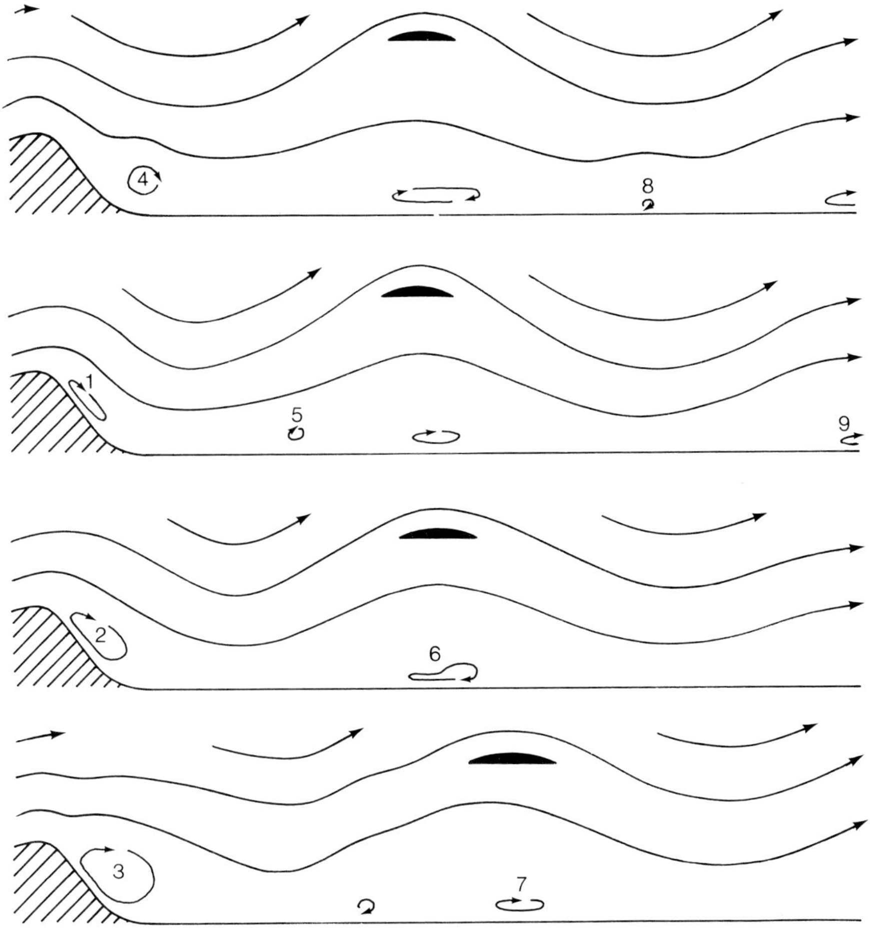

Fig. 5.18.viii The release of eddies may be periodic, with a consequent change in the posititition of the lee waves. The numbers show the successive positions of the eddy that is shed. A sequence of this kind is repeated so that the lee wave cloud leaps forward and then drifts backward as the lee eddy grows.

(4) Curl-over

This title refers to the behaviour on top of a cliff up which the wind is blowing. Such a cliff top is a favourite site for glider launching and so pilots wish to land downstream of the cliff edge but close to it. Whether there is simply a widening of the stream tubes (Fig. 5.18.ix) and a consequent reduction in wind strength, or whether there is an actual reversal of wind because of an eddy, does not make much difference to the pilot who in either case experiences a decrease in airspeed as he lands. The phenomenon is therefore called a 'clutching hand', because it causes a rapid increase in rate of descent (see Scorer 1951(b)).

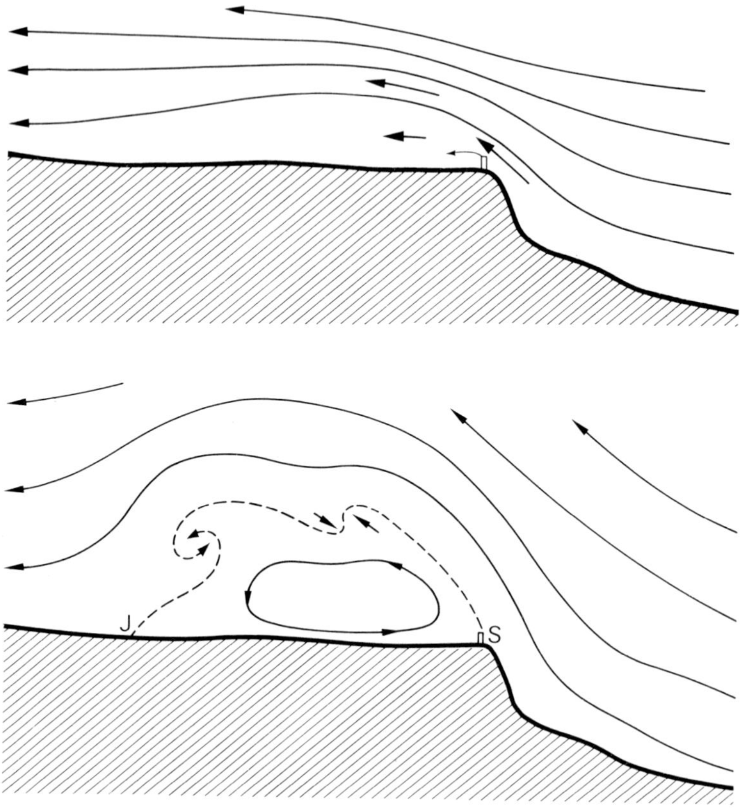

Fig. 5.18.ix (a) Flow at a hill top widening out but not separating. The motion aloft in which there may be waves influences this behaviour and (b) depicts an eddy with billow-like vortices travelling down the separation streamline.

The eddy formation is more common in stable air, for example when the ground is cold, and less common when it is sunny and the air is turbulent near the surface. But in any case it is a rather unsteady phenomenon, and even if the eddy is more or less permanent, its boundary is likely to be subject to billow instability.

(5) Cat's paws

The point at which the flow rejoins the surface is often seen on water where it looks like a giant paw quietly sliding forward. The point moves downstream behind an eddy that has separated and creates a sudden increase in wind strength which ruffles the water surface. Cat's paws indicate that the release of eddies is not simply a downstream drift of the flow pattern due to stationary eddies, but contains fronts (see Section 9.9 and Fig. 4.8.ii for comparison) behind which air with greater velocity descends to the surface (see Fig. 5.18.x).

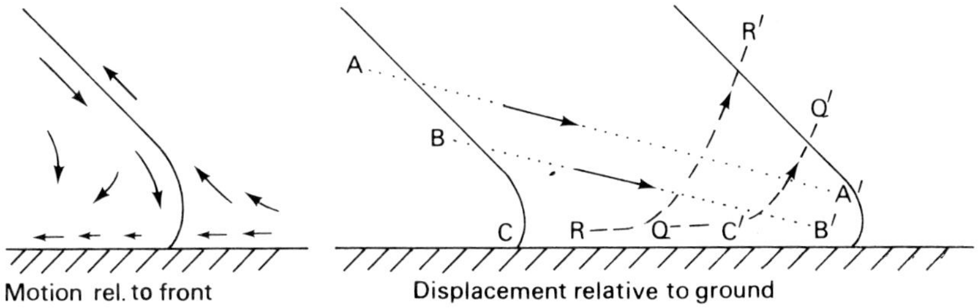

Fig. 5.18.x A cat's paw is a front at which air from higher up descends to the surface causing a sudden increase in wind. They are best seen on water because of the sudden ruffling of the surface by short waves, but they are also easily seen on fields of standing grain. They occur in the lee of obstacles and in storm downdraughts. The air ahead of the front (Q, R) is lifted from the surface while the air behind is being laid down as a carpet of swifter moving air (A, B, C). The primes denote the positions of air particles when the front has moved to the right hand position.

(6) Diurnal effects

Anabatic and katabatic winds have a very marked effect on separation phenomena. A katabatic wind, which begins when the ground cools as the sun goes down, is very shallow, but is nevertheless sufficient to prevent separation from the lee slope of a hill. The most interesting aspect of this is that as a result lee waves may begin to occur, and the large wind speed which occurs in the first lee wave trough at the foot of the hill has often been thought to be the katabatic wind itself. A well developed example was described by Lee and Neumark (1952) at Dunstable where a glider soared to 4,000 ft in the lee wave behind a 200 ft downslope.

Anabatic winds very readily induce separation, particularly if the lee slope is facing the sun (see Fig. 5.18.vii(a)). This is a major cause of the disappearance of wave clouds by day.

(7) Blocking

According to the theory outlined earlier in this chapter, the disturbance in the neighbourhood of a ridge has a phase variation in the z-direction with a period $2\pi/\ell$. This means that there must be a nodal surface at some height not greater than $2\pi/\ell$ above the undisturbed bottom of the stream, and if a mountain ridge extends above this height there is no flow of the stream which can be described by the theory, even in its arbitrary amplitude form. Long suggests that probably blocking occurs in that case, for there is no possible systematic displacement of the air above the ridge which can produce a low enough pressure for the surface air to reach there (the Froude number is much larger than unity).

Blocking is the main cause of Főhn winds, in which the air descends from an altitude not far from that of the ridge top on the upwind side to the surface on the lee side (Fig. 5.18.xi). The warmth of Főhn winds is ascribed in many text-

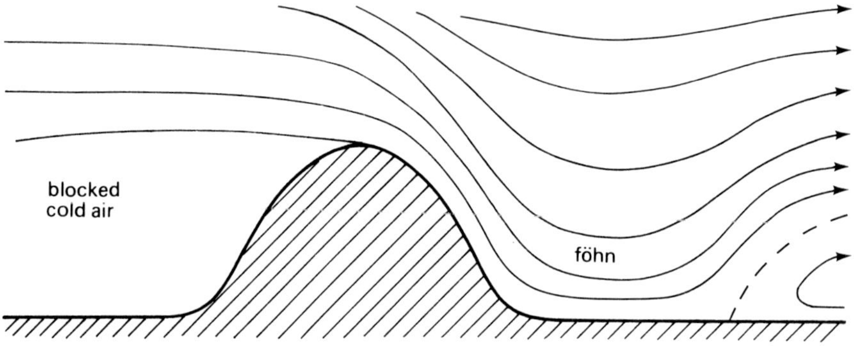

Fig. 5.18.xi The warm air associated with the föhn and Chinook winds usually originates from an altitude near to the mountain top when a cold air mass is blocked by a range of mountains. Separation often occurs under the first lee wave, and this creates a stark contrast between the gale at the foot of the slope and the calm or reversed flow in the rotor.

books to the release of the latent heat of rain falling on the ridge. It is shown in Section 11.6(2) that this explanantion is wrong in most cases even when there is rain, but it is obviously wrong in the many cases where there is not rain and more so when there is not even any cloud above the mountain ridge. Typical Főhn and Chinook winds occur when a cold air mass is blocked by a mountain range.

A cold air mass on the lee side may cause separation there, the cold air remaining stagnant in a valley on the lee side of the mountain. The top surface in such a case is usually a cloud top, radiation from which produces a stable density discontinuity (see Section 11.1(2)), and it is tilted by lee wave motions in the airstream above it. Sometimes, even if it is not significantly tilted but in the strong winds of a wave trough, billows may be produced at the stable layer. The presence or otherwise of anabatic winds on the part of the lee slope above the stable layer may cause a ruffling of the top of the layer, particularly if the cloud there is very thin and the anabatic wind is being generated by the sunshine on the slope below the layer (Fig. 5.18.xiii). Sometimes shallow pools of very cold air may remain in hollows or under the light surface winds in lee wave crests, with the boundary between them and the warm főhn air moving around and causing very large changes in temperature, both up and down. This is most common in the first warm föhn of spring when cold air is first carried away but the ground is still very cold.

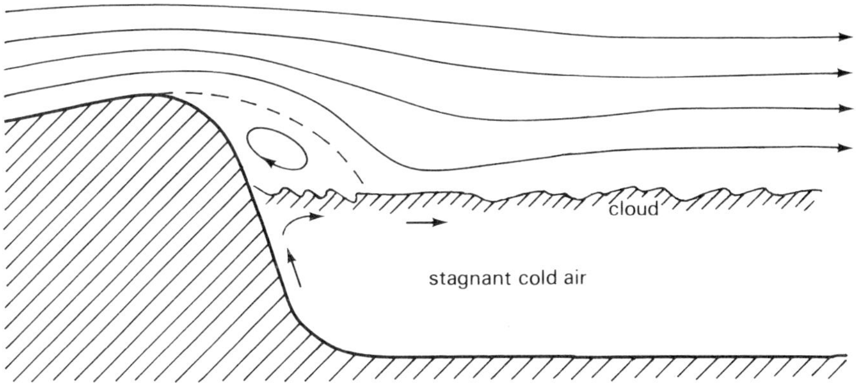

Fig. 5.18.xii Often a cold air mass occupies the valley or plain behind a mountain range. It is often capped by a cloud layer, and the position of billows or other disturbances on the cloud top is determined by the behaviour of eddies behind the mountain crest.

(8) Three dimensional effects

There is almost no systematic way of discussing the flow behind a three dimensional obstacle of arbitrary shape when separation occurs. Each case has to be observed to be discussed. On buildings with courtyards or towers, on ships with a variety of salient edges, on the ends of mountain ridges, the flow depends not only on the shape but also on the level of turbulence in the air and its stratification. The best advice is to learn to use any tracer that is going to elucidate the flow: this is not evasive advice because the effectiveness of a tracer depends very much on the user, and there is lots of fun to be had with smoke, soap bubbles, balloons, snow, flags, etc. A flow pattern suggested by Forchtgott (1950) for a circular hill is shown in Fig. 5.18.xiii, based on observations.

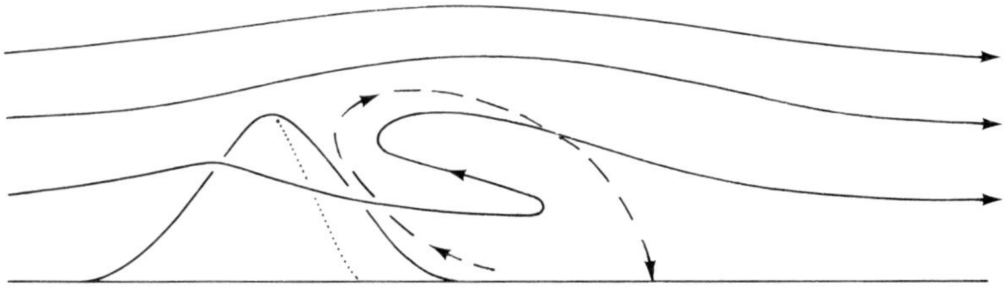

Fig. 5.18.xiii The flow up the lee slope of an isolated hill may not be in the form of a lee eddy but a circulation through the lee slope region in which some particles are recirculated perhaps a few times before escape.

(9) Separation Vortices

A line of separation may be the base of a vortex sheet, if the airstreams approaching from the two sides have different components of velocity along the line. The upward acceleration from the line, which is a line of stagnation points, stretches the vortex lines which are vertical and end on the line. The intensity of the sheet is thus increased and individual vortices may become quite violent. Dust whirls and whirls of blown snow often occur on the lee slopes, and sometimes bits of loose turf or tufts of vegetation are carried into the air by a vortex. These vortices are usually shortlived. The motion is depicted in Fig. 5.18.xiv.

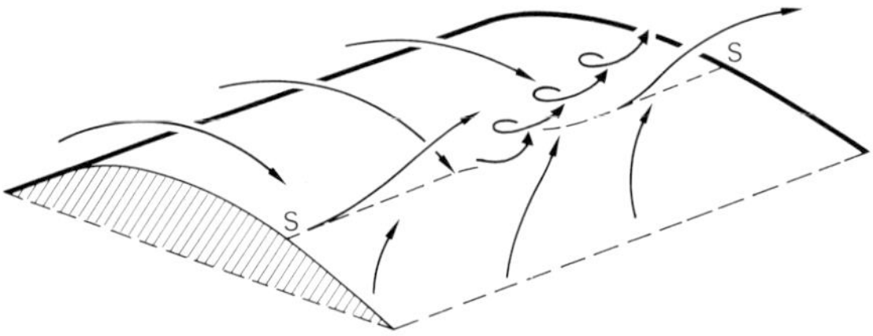

Fig. 5.18.xiv When separation occurs on the lee slope vortices are sometimes intensified by the stretching and dust or snow devils may be formed. Similar vortices have been observed on the ascending side of a rotor.

5.19 SOME THEORETICAL DIFFICULTIES

In attempting to arrive at a definitive boundary condition to be applied at great height in the atmosphere, many theoreticians have thought that it is enough to define a condition that gives a unique answer. Its validity depends on whether the conditions imposed are those actually prevailing. The popular one is the 'radiation condition', according to which the solution is chosen which corresponds to the application of a force on the obstacle in the direction of the stream (as discussed in Section 5.11 and used in Section 5.13). This is the solution arrived at by allowing a time dependent solution to settle down, because all the downward proceeding waves which may be present die away if there is no source of them aloft in the stream (i.e. no downward radiation).

There are, however, two important steady situations in which this criterion is inapplicable and which cast doubt on its universal applicability. No theory has ever been presented in which the structure of the airstream changes with time from one capable of trapping waves to one which is not. Nor has any ever depicted an airstream structure in which the direction of the wind in the lower layers veers with time from along a ridge to across it when the upper layers blow across it all the time. There is also the case of blocking and the possibility of an obstacle causing a modification of the upstream profile which is not encompassed by theory at present.

The first case is the flow down from, or up on to, a plateau. The solution may be obtained easily from the integral already used, namely (5.13.5), by integrating with respect to x. The profile (in Fig. 5.13.i) then represents the magnitude of the slope of the new function which is

$$\begin{aligned} \zeta &= ab \int_0^\infty e^{-kb} \int e^{ikx}\,\mathrm{d}x\,\mathrm{d}k \quad \text{(real part)} \\ &= ab \int_0^\infty \frac{1}{k} e^{-kb} \sin kx\,\mathrm{d}k \end{aligned} \tag{5.19.1}$$

$$\begin{aligned} &= \int \frac{ab^2\,\mathrm{d}x}{b^2 + x^2} \\ &= ab \tan^{-1} \frac{x}{b} \end{aligned} \tag{5.19.2}$$

The arbitrary constant of integration has been chosen as zero so that ζ shall be zero at $x = 0$. This formula represents flow up on to a plateau whose height for large positive x is πab greater than for large negative x. Flow down off a plateau is obtained by using a negative value for a. b is the distance at which the height is half the distant value, and the integral represents an abrupt step up when $b = 0$. A brick-shaped obstacle can be obtained by writing $x + c$ for x giving a step up at $x = -c$, and then adding the same function with the sign of a changed and $x - c$ written for x, to give a step down at $x = c$.

The theoretical difficulty arises because, according to the 'radiation condition', the solution to be taken is that obtained by integrating (5.13.18) and (5.13.24) with respect to x. Since the second term is then proportional to log $(b^2 + x^2)$, it grows indefinitely large as x becomes large. The implication of this is that indefinitely intense rotors are produced which are not closed as x increases. This might be interpreted as blocking. The only alternative is to apply the boundary condition at large $|x|$ instead of at large z and to reject this logarithmic part of the solution, which is zero at the bottom boundary and was only added to give the radiation condition at large z. In that case an acceptable solution is obtained in which there is no change of phase of the component waves with height.

When this is done we obtain a solution which is like blocking when the amplitide a is made large enough. As shown in Fig. 5.19.i the flow consists of a system of 'jets' which is produced when an infinitely along obstacle is introduced into a stratified stream. The Froude number is not so large that only blocking occurs, for when it is very large the flow is slow and virtually horizontal outside the layer of the obstacle.

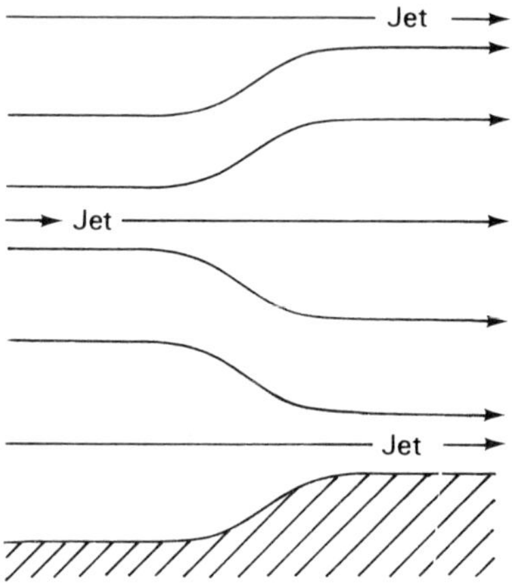

Fig. 5.19.i The formal solution for flow over the step represented by (5.19.2) is approximately as shown, but it does not satisfy the radiation condition at large height. The full solution with satisfactory conditions at large $|x|$ has additional waves in the streamlines which are usually small (see Queney 1948, and discussion of 5.13.12).

The second case in which the radiation condition cannot be applied is when there is a layer of zero stability anywhere in the stream, and this of course could easily occur over warm ground or in a cloud layer. In that case $\beta = 0$ and unless α is large the only variation with height for waves is in amplitude: the phase remains constant, because $k > \ell$ for all k. Another way of looking at it is to note that because the phase cannot vary with height in the neutral layer it is impossible for any horizontal drag to be communicated through the layer and so upward proceeding waves cannot be maintained. Perhaps this is why waves are not common when thermal convection occurs. However, it must be emphasised that this difficulty does not apply to trapped lee waves, and there are plenty of airstreams possible in which there is an adiabatic layer and in which trapping is possible (see Scorer 1953b), although the adiabatic layer necessarily decreases the amplitude of the lee waves. The other part of the solution is still open to argument, but any steady solution is unreliable in view of the likelihood of separation in a stream with a neutral layer in it.

5.20 OTHER KINDS OF WAVE

(1) Rossby Waves

If the flow on a rotating earth is strictly horizontal and the same at all heights with no thermal wind, a flow which is termed **barotropic**, there is no extension by the flow of the vertical vortex lines. Therefore the total or absolute vorticity is constant following the motion. This vorticity is the sum of the vorticity relative to the earth, denoted by the vertical component ζ, and of the Coriolis parameter f, which is the vertical component of the earth's vorticity.

Thus $$\zeta + f = \text{const along a particle path.} \tag{5.20.1}$$

If we now return to equation (5.2.6) and note that for practical purposes the term in $q^2\beta$ has been ignored by comparison with $g\beta z$, which is a form of the Boussinesq approximation, then for the case of an incompressible fluid $\rho_0 = \rho$, and the equation is

$$\eta - \frac{g\beta}{u_0} z = \text{const.} \tag{5.20.2}$$

Rossby pointed out that waves of small amplitude would be possible in the case leading to (5.20.1) if f were set equal to βy, where y is measured northwards. Since this is approximately the case over a small range of latitudes, although by (4.1.11) strictly $f = 2\,\Omega \cos\theta$, it is a useful simplification and has become known as the β-plane approximation. It converts (5.20.1) into

$$\zeta + \beta y = \text{const} \tag{5.20.3}$$

which is exactly analogous with (5.20.2). The difference in sign of the terms is simply due to the coordinate systems used, the roles of y and z having been interchanged. Therefore waves having properties somewhat determined by the north and south boundary conditions can be expected to be found in zonal winds (i.e. winds blowing along lines of latitude). Clearly the theory does not depend on the amplitude being small, and rotors can be confidently looked for. These are, in fact, found and are termed **blocking anti-cyclones** because they have the appearance of making the main flow pass around them. They are stationary relative to the large topographical and thermal features of the world, such as the Rocky Mountains and the warm Gulf Stream. The linear dimensions of the waves are determined by the magnitude of the coefficient β and the wind speed u_0, and are of the order of $(u_0/\beta)^{1/2}$, just as the linear dimensions of the gravity waves are of the order of $(u_0{}^2/g\beta)^{1/2}$ for (5.20.2). Since Rossby's β is of the order of 7×10^{-8} km^{-1} sec^{-1} and in strong westerly winds in the upper air u_0 is of the order of 3×10^{-2} km sec^{-1}, the wave dimension is around 1000 km, and a wavelength typically may be $2\pi \times 10^3$ km.

(2) Sonic Waves: big bangs

An explosion which creates a pressure disturbance of the order of 1 mb at a distance comparable with the radius of the earth is called a **big bang**, this being detected by a barograph used for ordinary meteorological purposes. A little bang produces a pulse which is detected only on special microbarographs. The classicial big bang is the eruption of Krakatoa in 1883 and the classical little bang is the great Siberian meteorite of 30 June 1908. Astapowitsh (1934) is good reading on this topic. Recently H- and A-bombs have been in each of these categories respectively. A little bang is of the order of 100 to 1000 km^3 expansion, a big bang up to 10^4 km^3. Such a large explosion as 100 km^3 produces an outward displacement of only 160 m at a distance of 10 km, and so at large distances small perturbation theory is adequate. 10^{24} ergs is required to introduce 10^3 km^3 at atmospheric pressure.

The explosion generates a shock wave which produces a hemispherical hump in the atmosphere, and when this has reached up into the stratosphere a gravity wave begins to travel outwards like the rings of waves expanding when something is thrown into a pond. This wave travels roughly with the speed of sound, c, which is proportional to $T^{1/2}$ (see 1.8.8) and is about 316 m sec^{-1}. This value is intermediate between the speed in the bottom of the troposphere and that in the stratosphere, which may be up to 70°C colder. Waves travelling at different speeds are theoretically possible, but do not correspond to a source located at the ground and are not therefore of observable amplitude.

According to conventional perturbation theory the perturbation is separable into a function of height and a function of the horizontal coordinates for any given frequency σ (i.e. proportional to $e^{i\sigma t}$). Scorer (1950(b)) used horizontal distance r from the source and found that the modified pressure, defined in (1.6.4), satisfied

$$\varpi = \varpi(z)\bar{\varpi}(r), \tag{5.20.4}$$

$$\left(\cot\frac{r}{a} + \frac{\partial}{\partial r}\right)\frac{\partial\bar{\varpi}}{\partial r} = k^2\bar{\varpi} \tag{5.20.5}$$

$$\varpi'' + \left(\frac{g\tau_0'' + \sigma^2\tau_0'}{g\tau_0' + \sigma^2\tau_0} + \frac{g}{c^2}\right)\varpi' + \left(g\frac{\tau_0'}{\tau_0} + \sigma^2\right)\left(\frac{1}{c^2} - \frac{k^2}{\sigma^2}\right)\varpi = 0 \tag{5.20.6}$$

the motion being assumed adiabatic, a being the earth's radius, and $\tau_0(z)$ being the undisturbed value of τ, and a prime denoting $\partial/\partial z$. The vertical velocity is given by

$$w = \frac{i\sigma}{\sigma^2 \tau_0 + g\tau_0{}'} \varpi' \qquad (5.20.7)$$

and this is zero at the ground for waves travelling away from the bang. Equation (5.20.6) has solutions which are oscillatory at the bottom of troposphere but die away exponentially upwards in an isothermal stratosphere, and have speeds σ/k, k being a radial wavelength, equal to c so that the coefficient of ϖ in (5.20.6) changes sign at some intermediate level.

The higher frequencies travel more slowly, and so a pulse generated at a point, which may be represented by a double Fourier integral over all values of σ and k, starts out as a single wave but is gradually dispersed into a large number of oscillations of increasing frequency. An atmosphere having a troposphere with a constant lapse rate and a isothermal stratosphere extending to infinite height cannot propagate uniform trains of waves in this way when the period is less than about 2 minutes. However, shorter oscillations appear at the ground in a pulse composed of superposed waves of all frequencies down to a period of about 13 sec. Fig. 5.20.i shows an enlarged version of the Siberian meteorite waves as observed in England, prepared by Whipple (1930).

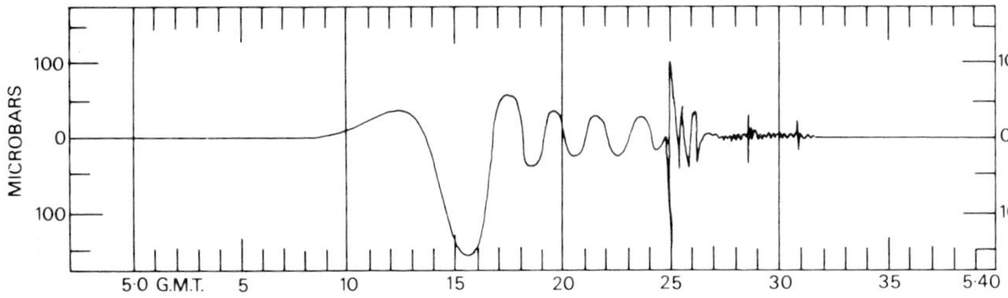

Fig. 5.20.i The pulse measured on microbarographs in England which travelled outwards from the point of impact of the Great Siberian Meteorite. (A composite diagram by Whipple from 5 recordings). It contains the main characteristics of many pulses observed since after nuclear explosions.

For periods down to about 2 min an atmosphere like this model acts as a wave guide. Waves of higher frequency are observed, travelling rather more slowly, and these are due to reflection in the warmer layers of the stratosphere where the temperature increases between about 30 and 50 km. Audible frequencies have been observed to be reflected at these levels (Whipple 1935), the classical case being the gunfire in Flanders heard in England in the Great War. Yamamoto (1957) studied these in more detail after observing the pulses from several nuclear explosions.

The duration of the whole pulse at one earth radius distance from the source of an instantaneous pulse is of the order of 20 minutes, and most of the energy passes in half that time. Therefore the earth's rotation plays a negligible part in the mechanics of their propagation. This is not true of tidal waves (see Section 11.11), which have a period of several hours.

The importance of sonic waves is that they occur whenever a part of the atmosphere expands. Thus a wave travels outwards from an island heated in the sunshine over which the air is warmed. This causes the density of the heated air to decrease, and the phenomena of thermal convection follow. Hence the motion due to the expansion of the heated air is a sonic wave which travels away from the site of the heating with the speed of sound, and is quite separate from, and does not interfere with, the buoyancy induced motion which follows. For practical meteorological purposes, therefore, any non-adiabatic temperature change may be assumed to produce immediately a corresponding density change at constant pressure (Scorer 1952(b)). For example, the outflow causing the low pressure over heated land which leads to a sea breeze is an expansion wave travelling away from the land with the speed of sound, and is not observable as a wind anywhere.

(3) Storm pulses

Most air masses are capable of transmitting waves, like standing waves, which travel through the air with a speed of the same order as typical winds. The speed varies greatly with wavelength; that is why standing waves have such a narrowly determined wavelength. Consequently there is very considerable dispersion of a pulse.

A pulse may be created by a sudden downdraught from a large thunderstorm if the upper air from which it descends is moving with a different speed from a lower layer into which it falls, causing a blockage. Fig. 5.20.ii shows the consequences as observed at Abingdon, near Oxford, as a result of thunderstorms in the English Channel at night. The range of wavelengths trapped when travelling in a particular direction may be small, and in the case shown it appears to be wavelengths in the neighbourhood of 3 km. It is significant that the storms were at night over the sea, for in this case it means that the cumulus had no roots at the surface, and were generated above the condensation level in air with a different velocity from that of the surface layers.

If there is no difference in velocity between the two layers, nor a marked stable layer, waves are not trapped and the only consequence of the downdraught will be a cold outflow, similar mechanically to a sea breeze front (see Section 9.9).

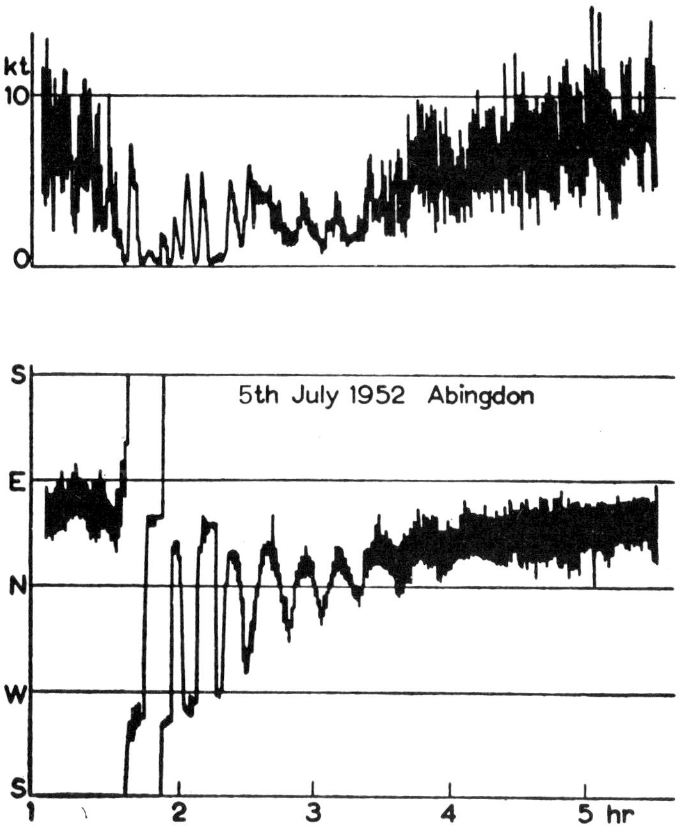

Fig. 5.20.ii The wind speed and direction recorded at Abingdon at the passage of gravity waves emanating from nocturnal thunderstorms over the English Channel. [Reproduced by courtesy of I. J. W. Pothecary, Meteorological Office]

Chapter 6

BILLOW MECHANICS

6.1 INSTABILITY OF A VORTEX LAYER

The classical discussion of the instability of a horizontal surface at which the density and velocity are discontinuous is as follows:

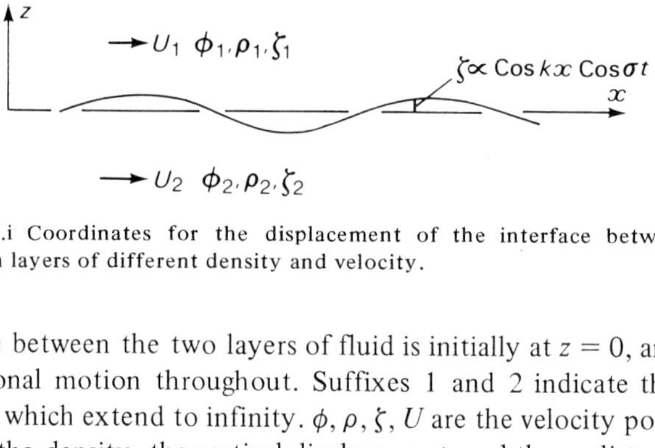

Fig. 6.1.i Coordinates for the displacement of the interface between two uniform layers of different density and velocity.

The interface between the two layers of fluid is initially at $z = 0$, and each layer is in irrotational motion throughout. Suffixes 1 and 2 indicate the upper and lower layers, which extend to infinity. ϕ, ρ, ζ, U are the velocity potential of the disturbance, the density, the vertical displacement, and the undisturbed uniform horizontal velocity respectively, and the displacement of the interface is assumed to be proportional to $\cos kx$ and to $\cos \sigma t$, represented by $e^{i(kx-\sigma t)}$, the real part of the equation being taken.

The fluids are supposed incompressible so that the equation of continuity is

$$\text{div}\,\mathbf{grad}\phi \;=\; \nabla^2\phi \;=\; 0. \qquad (6.1.1)$$

Since the disturbance is assumed to be small, so that all powers of the disturbance higher than the first can be ignored, the Lagrangian operator is, for two-dimensional motion, with velocity $(U + u, w)$

$$\frac{\mathrm{D}}{\mathrm{D}t} = \frac{\partial}{\partial t} + (U+u)\frac{\partial}{\partial x} + w\frac{\partial}{\partial z} = i(-\sigma + kU) \qquad (6.1.2)$$

when applied to the disturbance, which is taken to be proportional to $e^{i(kx-\sigma t)}$. To satisfy (6.1.1) the disturbance must be proportional to $e^{\pm kz}$, The positive sign holding in the lower layer and the negative sign in the upper layer; otherwise the disturbance would be infinitely large for large positive or negative z. From this it follows that

$$w = \frac{\partial\phi}{\partial z} = \frac{D\zeta}{Dt} = -i(\sigma - kU)\zeta \tag{6.1.3}$$

$$\begin{aligned} &= -k\phi \qquad \text{for } z > 0 \\ &= \quad k\phi \qquad \text{for } z < 0. \end{aligned} \tag{6.1.4}$$

and so

$$\phi_1 = i\frac{\sigma - kU_1}{k}\zeta_1; \quad \phi_2 = -i\frac{\sigma - kU_2}{k}\zeta_2. \tag{6.1.5}$$

By Bernoulli's equation (1.7.9) with ρ = constant and

$$q^2 = \left(U + \frac{\partial\phi}{\partial x}\right)^2 + \left(\frac{\partial\phi}{\partial z}\right)^2$$

we have $$\frac{p_1}{\rho_1} + U_1\frac{\partial\phi_1}{\partial x} + \frac{\partial\phi_1}{\partial t} + g\,\zeta_1 = \text{constant in the upper fluid} \tag{6.1.6}$$

with a similar equation with suffix 2 for the lower layer. The constants $U_1{}^2$ and gz for each streamline (at height $z + \zeta_1$) are absorbed into the right hand side, and the second order terms in ϕ neglected for small disturbances. The variable part of the pressure is given by

$$-p_1 = \rho_1[i(kU_1 - \sigma)\phi_1 + g\zeta_1] = \rho_1\left(\frac{(\sigma - kU_1)^2}{k} + g\right)\zeta_1 \tag{6.1.7}$$

and $$-p_2 = \rho_2[i(kU_2 - \sigma)\phi_2 + g\zeta_2] = \rho_2\left(-\frac{(\sigma - kU_2)^2}{k} + g\right)\zeta_2. \tag{6.1.8}$$

We have already imposed sufficient boundary conditions at large postive and negative z, which would be slightly more complicated mathematically if they had been replaced by rigid boundaries at finite distance. In order to obtain the dispersion equation, which relates σ to k, or the wave length to the wave velocity, we apply the physical constraints at the interface, namely

$$p_1 = p_2, \quad \zeta_1 = \zeta_2 \qquad \text{at } z = 0.$$

Consequently
$$\rho_1\left(\frac{(\sigma - kU_1)^2}{k} + g\right) = \rho_2\left(-\frac{(\sigma - kU_2)^2}{k} + g\right) \tag{6.1.9}$$

which is a quadratic equation for σ of which the solutions are

$$\sigma = \frac{k}{\rho_1 + \rho_2}\left(\rho_1 U_1 + \rho_2 U_2 \left[\pm\ (\rho_1 U_1 + \rho_2 U_2)^2 - (\rho_1 + \rho_2)[\rho_1 U_1^2 + \rho_2 U_2^2 - \frac{g}{k}(\rho_2 - \rho_1)]\right]^{1/2}\right). \tag{6.1.10}$$

If σ turns out to be complex then the disturbance, which is proportional to $e^{i\sigma t}$, will contain a real positive exponential which indicates instability because the disturbance will grow without limit. When a finite size is achieved the higher order terms become important and the present mathematics is no longer applicable, and we shall study subsequent development by other means. Instability occurs when the quantity inside the larger square bracket in (6.1.10) is negative, and this is the case if

$$(\rho_1 U_1 + \rho_2 U_2)^2 < (\rho_1 + \rho_2)[\rho_1 U_1{}^2 + \rho_2 U_2{}^2 - \frac{g}{k}(\rho_2 - \rho_1)],$$

i.e.
$$\frac{(\rho_2 - \rho_1)(\rho_2 + \rho_1)}{\rho_1 \rho_2}\frac{g}{k} < (U_1 - U_2)^2. \tag{6.1.11}$$

This condition shows that there is always a range of wave number, k, giving unstable waves if $U_1 \neq U_2$. The condition (6.1.11) is fulfilled for all wave numbers if $\rho_2 < \rho_1$, which is obvious because it is the case of static instability. Instability in the statically stable case is called Helmholtz instability (often, but unnecessarily, coupled with the name of Kelvin).

When the discontinuity of density is of small magnitude we may write

$$\rho_2 - \rho_1 = \Delta\rho, \quad \rho_2 + \rho_1 = 2\rho, \quad \rho_1\rho_2 = \rho^2 \tag{6.1.12}$$

and so if $U_1 - U_2 = \Delta U$, the condition (6.1.11) is that instability occurs when

$$k > k_c = 2g\frac{\Delta\rho}{\rho}/(\Delta U)^2. \tag{6.1.13}$$

Waves of length shorter than $2\pi/k_c$ are unstable.

This treatment has many shortcomings in practice. If the wavelength is very small the gradients of velocity are increased and viscous effects become important. Furthermore, a discontinuity of velocity becomes smudged into a thin layer with a large gradient of velocity, or vortex layer. The treatment is not valid when the wavelength is small and comparable in dimension with the thickness of the layer, if only because we have assumed each layer to be in irrotational motion, of uniform speed, in the undisturbed state. The theory gives no indication of the nature of the development when the waves become large; and in the case of the surface of water ruffled by the wind the effects of surface tension can be overriding.

More sophisticated treatments by Drazin (1958, 1960, 1973), Miles (1958, 1959), Howard (1961), Warren (1968, 1974) and others have considered the case in which there is a transition zone from one layer to the other, and have arrived at criteria corresponding to (6.1.13) but with a lower limit for the wavelengths that are unstable. They thus find the maximum value for a quantity similar to that on the right of (6.1.13) for which some instability is possible. They are also concerned with criteria for the onset of instability. Our objective is different. We are concerned not with the exact criterion for instability but with its mathematical form so that we may study what has to happen in nature to bring the instability about. We are also concerned to know what the subsequent motion is like. The following treatment is from Scorer (1971).

We represent the vortex layer by a layer of thickness Δz in which there is a velocity gradient η, which is the vorticity component normal to the planes of the motion. There is also a density gradient represented by β, so that

$$\Delta\rho = \beta\rho\Delta z, \qquad \Delta U = \eta\Delta z. \qquad (6.1.14)$$

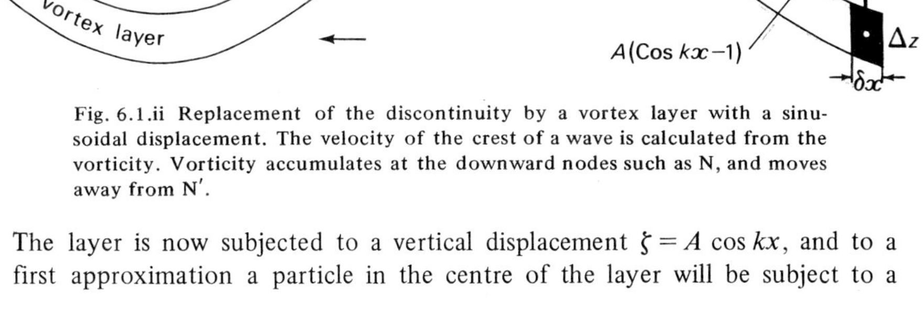

Fig. 6.1.ii Replacement of the discontinuity by a vortex layer with a sinusoidal displacement. The velocity of the crest of a wave is calculated from the vorticity. Vorticity accumulates at the downward nodes such as N, and moves away from N′.

The layer is now subjected to a vertical displacement $\zeta = A \cos kx$, and to a first approximation a particle in the centre of the layer will be subject to a horizontal velocity $u_0 \cos kx$, having been initially at rest with the fluid moving in opposite directions above and below. This motion is due entirely to the vorticity contained in the layer, and the material of the layer is subject to its own velocity field.

The coordinates of an element of the layer, of length δx, relative to the crest of the wave which is at displaced height A, are $[x, A(\cos kx - 1)]$. It possesses vorticity $\eta\Delta z\delta x$, and imposes on the particle at the crest of the wave a horizontal component of velocity equal to

$$\frac{\eta\Delta z\delta x}{2\pi x}\frac{A(1-\cos kx)}{x}. \qquad (6.1.15)$$

To obtain this we have assumed that $A(1 - \cos kx)$ is small compared with x. The velocity of the particle at the wave crest, denoted by u_0, is therefore

$$u_0 = \int_{-\infty}^{\infty} \frac{\eta\Delta zA(1-\cos kx)}{2\pi x^2}\,\mathrm{d}x$$

$$= \frac{\eta\Delta zA}{\pi}\int_{-\infty}^{\infty}\frac{\sin ½kx}{x^2}\,\mathrm{d}x$$

$$= -\frac{\eta\Delta zA}{\pi}\left[\frac{1}{x}\sin^2 ½kx\right]_{-\infty}^{\infty} + \frac{\eta\Delta zA}{2\pi}\int_{-\infty}^{\infty} k\frac{\sin kx}{x}\,\mathrm{d}x. \qquad (6.1.16)$$

The first term in this last expression vanishes because it represents the integral of an odd function of x which vanishes at infinity. The integral in the second term is a well known form and is equal to $k\pi$ (see for example 5.19.2). The horizontal component of the velocity of the fluid in the vortex layer is therefore given by

$$u = u_0 \cos kx = ½\eta\Delta zAk \cos kx\,. \qquad (6.1.17)$$

The vertical velocity at the crest is zero because the vorticity producing the motion is mirrored in the z axis. This is an interesting disagreement with the classical treatment described above, which appears to allege that the amplitude of the waves grows but the shape (sinusoidal) remains unaltered. It may be noted that the wave velocity is zero if the two layers have undisturbed velocity $\pm U$, for if ρ_1 is roughly equal to ρ_2 the real part of σ is zero in (6.1.10). Also w is seen in (6.1.3) to be out of phase with ζ so that the vertical velocity of the crest is zero although it is growing exponentially. The growth mechanism is therefore not treated in the classical theory.

Since u in (6.1.17) varies in the horizontal direction, the vorticity carried by this velocity tends to accumulate at the downward sloping nodes such as N in Fig. 6.1.ii, and is drawn away from the upward sloping nodes N'. The rate of accumulation of vorticity per unit length of the layer is equal to $-\eta\Delta z \dfrac{\partial u}{\partial x}$, which is equal to

$$\tfrac{1}{2}(\eta\Delta z)^2 A k^2 \sin kx \tag{6.1.18}$$

At the same time the vorticity in the layer is being altered by the action of gravity on the sloping density gradient, at a rate represented by the term $\mathbf{R} \times \mathbf{g}$ in equation (1.4.1). This rate of creation of new vorticity is therefore $\beta g \dfrac{\partial \zeta}{\partial x}$ per unit area of the cross section of the layer, which is a rate per unit length of layer equal to

$$g\beta\Delta z \frac{\partial}{\partial x} A \cos kx \quad = \quad -g\beta\Delta z A k \sin kx \tag{6.1.19}$$

If the sum of (6.1.18 and 19) is postive when sin kx is positive there will be an accumulation of vorticity at the downward sloping nodes, towards which it is being advected. There will be a corresponding flight of vorticity from the upslope nodes. Since such a redistribution of vorticity will tend to make the crests move upwards and the troughs downwards, the disturbance will grow. Equally, if the sum of (6.1.18 and 19) is negative when sin kx is positive there will be an increase of vorticity at the upslope nodes and the crests will be moved downwards. Hence the disturbance will not grow, but will travel as a stable wave or oscillate about the equilibrium undisturbed position. The condition for instability is therefore

$$\tfrac{1}{2}(\eta\Delta z)^2 k > g\beta\Delta z$$

or

$$k > k_c \quad = \quad 2g\beta/\eta^2\Delta z. \tag{6.1.20}$$

It is necessary to add that a thick vortex layer is not unstable for very short wavelengths. If we imagine that a mechanism is operating which gradually increases η, the vorticity of the layer, the wavelength which will become unstable first will be at the long wavelength end of the unstable range represented in (6.1.20). This is, of course, the same condition as (6.1.13) because of

(6.1.14). If we now call the thickness of the vortex layer h, (6.1.20) implies that

$$\tfrac{1}{2}k_c h = \frac{g\beta}{\eta^2} = \mathrm{Ri} \qquad (6.1.21)$$

where Ri is the Richardson number of the layer (see Section 7.5).

The classicial studies of Miles, Howard and Warren have shown that instability cannot set in unless the Richardson number is less than ¼ somewhere, and so if it is roughly equal to this value we have that, roughly,

$$\frac{2\pi}{k_c} = 4\pi h. \qquad (6.1.22)$$

This gives a crude estimate of the wavelength most likely to appear in a vortex layer of thickness h. The mechanism whereby the Richardson number is decreased so as to cause instability is discussed later in this chapter.

Any calculation leading to a conclusion like (6.1.22) must of necessity be very rough because actual layers do not have a uniform Richardson number throughout but one that varies through the depth of the layer. Criteria about instability must depend not only on the value of the Richardson number but on its distribution. This is because there need be no instability in a stream in which $\mathrm{Ri} < \frac{1}{4}$ everywhere, the obvious example being the case of uniform density and uniform shear in which $\mathrm{Ri} = 0$ everywhere. The instability arises, according to this analysis, because there a concentration of vorticity into a layer. The vorticity is collected into the nodes by the instability and the layer is rolled up into 'billows', which is the name given to this phenomenon to distinguish it from gravity waves caused by an outside disturbance such as a hump in the boundary, or violent convection. Billows always occur several at a time, and as we shall see, are the principal cause of mixing in a stably stratified medium.

The observed phenomenon, very common in cloud forms, is striking because of the sudden appearance of a disturbance of a clearly defined single wavelength. There is not an existing unstable situation subject to a fortuitous perturbation leading to the sudden appearance of the instability. Rather a gradual destabilisation takes place, and the disturbance grows as soon as a very small range of wavelengths becomes unstable. Viscosity plays no significant role in the determination of the smallest wavelength that is unstable in the atmosphere or ocean where the phenomenon is very common (see Woods and Wiley 1972). Even if the destabilisation continues after the first unstable waves have appeared, the growth rate will simply be increased, other wavelengths will not appear because the initial unstable wave has a start on them and will be completely dominant as the growth is exponential.

6.2 THE GROWTH OF THE CAT'S EYES PATTERN

We shall now make a rough study of the development of the unstable waves with a view to getting an estimate of the final outcome. According to the classical theory the waves grow in amplitude only. We have seen that the growth is caused by, and itself causes, the accumulation of vorticity at the nodes in which the upper fluid is moving down the slope. A series of experiments by Thorpe (1968) showed many of the important features of billows which are not easy to observe in nature. Woods (1962) demonstrated their existence under water, and Scorer (1972) has illustrated the various cloud forms that are created by billow motion.

As a result of the stretching of the vortex layer at the upslope nodes, the rotation is intensified at the downslope nodes towards which the vorticity is carried. The subsequent evolution is rapid, and is indicated in Fig. 6.2.i.

At the third stage shown the fluid at A is denser than that at B and so the fluid in the rolled up vortices becomes statically unstable and becomes mixed up. In the final stage the cat's eyes pattern is achieved. The mixed fluid is gathered into discrete rolls with a wave motion above and below, in which the phase of the waves has shifted a quarter of a wavelength downstream on each side so that the wave is of opposite phase above and below.

In the third stage it is possible to observe a considerable winding up of the vortex into a sort of 'swiss roll', especially when the density difference is small and the shear large. When the density difference is large, mixing begins soon after stage 2. In the theory the formulas would be modified by the introduction of a factor tanh kd, d being the depth of one of the fluid layers if it were of finite depth. When $kd > 1.5$, tanh $kd = 1$ with an error of less than 10 per cent, and so if the depth of the layer exceeds about a quarter of a wavelength the presence of a rigid boundary has little effect on the wavelength. In Thorpe's experiments this effect was present in shallow layers, but in deeper layers the wavelength was determined by the thickness of the vortex layer, as would be expected from the theory.

In the final stage the vorticity is obviously distributed throughout the mixed fluid in a manner that cannot be determined. The flow outside is not much affected because the amount of the vorticity is the same, whatever its distribution. Therefore it is not a bad approximation to assume that it is all collected at the centres of the cells, and on that basis the classical formula for the motion of a row of vortices can be used. The complex potential is given by Lamb (1932, Art. 156) to be

$$\phi + i\psi = -2iA \log \sin \tfrac{1}{2}k(x + iz) \tag{6.2.1}$$

and the horizontal and vertical velocities are

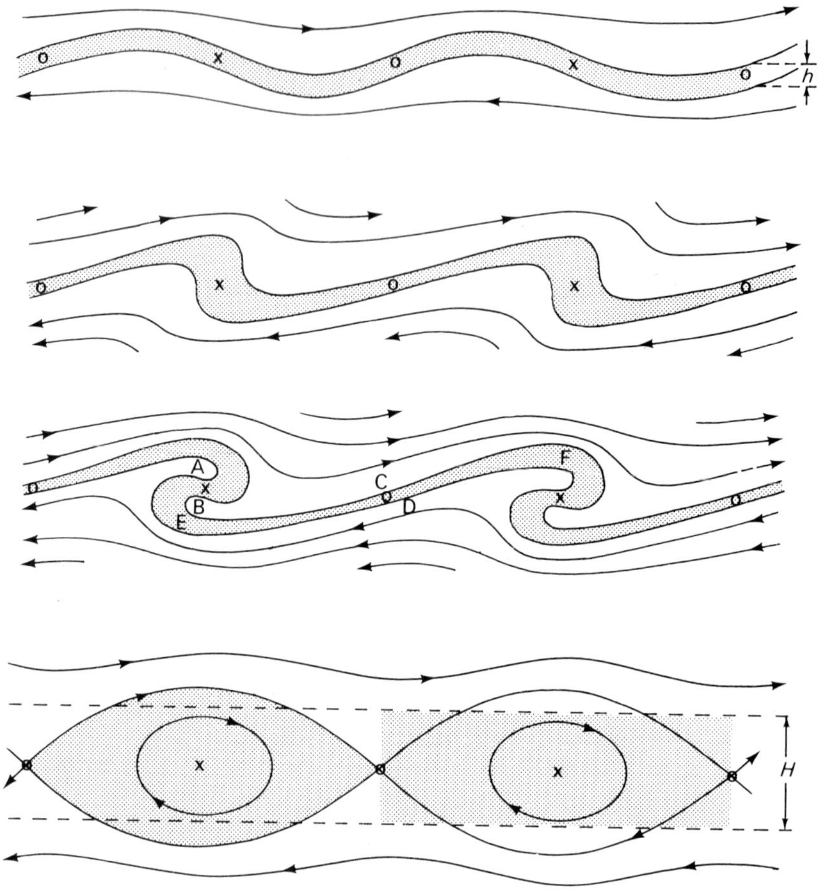

Fig. 6.2.i The evolution of billows from a sinusoidal perturbation of the vortex layer, which rolls up at the downslope nodes. The fluid in the rolls is gravitationally unstable and soon begins to mix up to form discrete lumps of mixed fluid in the cat's eyes pattern shown in the last diagram: eventually this fluid flattens out under gravity into a layer of depth H indicated on the right of that diagram.

$$U + u = Ak \frac{\sinh kz}{\cosh kz - \cos kx}$$

$$\rightarrow Ak \operatorname{sgn} z \, [1 + \exp(-k|z|) \cos kx] \qquad (6.2.2)$$

for large z, and

$$w = Ak \frac{\sin kx}{\cosh kz - \cos kx} \rightarrow Ak \exp(-k|z|) \sin kx \qquad (6.2.3)$$

for large z, U being the velocity which transports the fluid horizontally and u the part dependent upon x. The value of U at a distance of a few wavelengths above and below the pattern does not affect the motion as $\tanh kz \to 1$. Above and below the pattern the motion is sinusoidal except very close to the cell boundary. Lamb discussed the stability of the pattern and showed that any disturbance which takes one or more of the vortices out of line will grow. This is because they would be subject to a velocity transporting them further out of line in exactly the same way as we saw that the distortions of the vortex layer grow. When there is no stabilising density gradient, as in Lamb's case, disturbances of all wavelengths are unstable; but in our case, where there is a stable density gradient, and in the atmosphere and ocean where the layers are usually stably stratified throughout, the unstable wavelengths are excluded. The point of noting Lamb's analysis is to obtain a rough estimate of the volume within the closed circulations. It turns out (Scorer 1951) that the streamlines in the final stage have equation

$$z = \frac{1}{k}\cosh^{-1}\left[\exp(\psi/A) + \cos kx\right] \tag{6.2.4}$$

and those marking the boundary of the circulations are given by

$$\psi = A \log 2 \tag{6.2.5}$$

The centres of the circulations have $\psi = -\infty$, but of course in practice they rotate solidly near the centre with the same total vorticity. If the fluid becomes mixed within the circulations and then subsequently flattens out as a horizontal layer of thickness H, equating the cross section areas before and after

$$\pi H = \int_0^{\pi/k} 2\cosh^{-1}\left[2 + \cos kx\right]\,dx \tag{6.2.6}$$

from which

$$\frac{2\pi}{k} \simeq 2.7\,H, \tag{6.2.7}$$

and comparing this with (6.1.22) we have

$$H = 4.6\,h \tag{6.2.8}$$

The effect of the whole evolution, therefore, is to multiply the thickness of the vortex layer by the factor 4.6, and to mix it up into a uniform fluid, with a fairly large density step at the top and the bottom of the new layer.

The largest density gradient produced in the sequence of events is first around the upslope node, the fluid between C and D in stage 3 being stretched and thinned. Later a large density discontinuity is produced all the way from E to F on each wave, and it is that shape that is clearly seen by radar when billows occur in the atmosphere (see Fig. 6.2.ii). Still later the vortices become visible when the mixing first begins, and large density gradients are produced in the early stages of mixing by the juxtaposition of masses of different density. Finally the fluid within the cat's eye circulations becomes more uniform, and the radar sees the outer boundary of the cells where the density increment becomes concentrated.

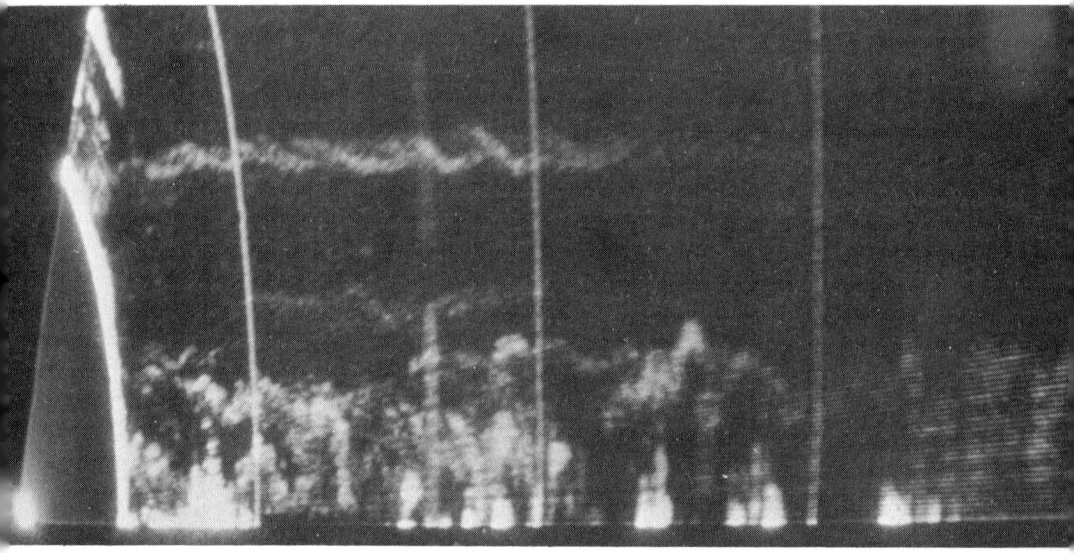

Fig. 6.2.ii A radar picture of billows at 3.5 km with convection below in clear air. The range lines are at 5 n mi intervals, and the near area is blanked off to avoid glare. [Photograph by courtesy of J. J. Hicks, John Hopkins University]

6.3 THE GENERATION OF BILLOW INSTABILITY

The generation of vorticity in the atmosphere is dominated by the effect of gravity. In equation (1.4.1) the effects of viscosity and divergence of velocity are quite negligible, and in two-dimensional motion there is no gradient of $\mathbf{v}$ in the direction of ω, so the vortex lines are not stretched but remain perpendicular to the planes of the motion. The acceleration of the fluid is small compared with gravity because it is produced by buoyancy forces (Boussinesq approximation) and the vorticity equation becomes

$$\frac{D\boldsymbol{\omega}}{Dt} = \mathbf{R} \times \mathbf{g} \tag{6.3.1}$$

or

$$\frac{D\eta}{Dt} = q\frac{\partial \eta}{\partial s} = g\beta \sin \psi \tag{6.3.2}$$

where η is the magnitude of the vorticity, s is measured along a particle trajectory, q is the fluid speed, β measures the magnitude of the density (or potential temperature) gradient and ψ is the inclination of the isentropic surfaces (the stream surfaces in adiabatic motion) to the horizontal (Fig. 6.3.i).

If the motion is steady

$$\frac{\partial \eta}{\partial s} = \frac{g\beta}{q} \sin \psi \tag{6.3.3}$$

and for a small element of arc $\delta s = \delta z/\sin \psi$:

therefore

$$\delta\eta = \frac{g\beta}{q}\,\delta z\,. \tag{6.3.4}$$

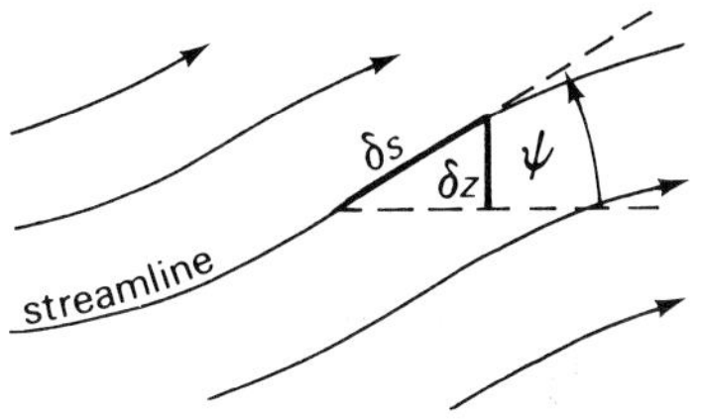

Fig. 6.3.i Coordinates for the derivation of the equation (6.3.3) for the growth of vorticity in inclined flow.

In steady motion in two dimensions the speed and the density (potential temperature) gradient are both proportional to the closeness of the streamlines and so we can write

$$\frac{q}{\beta} = \text{const} = \frac{q_0}{\beta_0} \tag{6.3.5}$$

the suffix 0 referring to some position far back along the streamline.

The situation is of special interest when there is a thin layer in which β_0 is large and in which there is, to begin with, very little shear, so that it is stable. If it is tilted, in a wave motion produced by flow over a mountain for example, the additional vorticity generated is directly related by (6.3.4) to the vertical displacement of the streamline from its original level, and the Richardson number is given by

$$\mathrm{Ri} = \frac{g\beta}{(\eta_0 + \delta\eta)^2} = g\beta\bigg/\left(\eta_0 + \frac{g\beta_0}{q_0}\,\delta z\right)^2 . \tag{6.3.6}$$

When η_0 is small so as to be of little importance (and Ri is initially very large and the flow stable), we have, in the displaced state

$$\mathrm{Ri} = \frac{q^2}{g\beta(\delta z)^2} = \frac{qq_0}{g\beta_0(\delta z)^2} . \tag{6.3.7}$$

In motion through a given wave pattern, with given δz, the smallest values of Ri will be found where β takes its largest value, and also, of course, where δz is largest and q smallest. But δz and q are circumstances determined by factors on a much larger scale than the depth of the layer and β in it.

It was known from the very early days (around 1950) in which clear air turbulence was studied by flight of aircraft through it, that it seemed to occur most often in layers of air where static stability was greatest. This was a source of puzzlement until it was appreciated (Scorer 1961) that, although β was a stabilising influence, the vorticity η which it produced appeared as η^2 in the Richardson number. Thus as soon as the stable layers are tilted, the large static stability becomes a destabilising influence.

In unsteady motion the equation corresponding to (6.3.4) is

$$\delta\eta = g\beta\delta t \sin\psi \tag{6.3.8}$$

where δt is the time for which the tilt ψ is maintained. This explains why the factor q^{-1} appears in (6.3.4), for a slow motion through a given wave pattern will maintain the tilt of a parcel of air longer.

It has often been thought that a horizontal acceleration of the flow, which decreases the spacing of the streamlines, has a significant effect on the Richardson number. If the vorticity is perpendicular to the flow it remains unaltered in the accelerated flow because the vortex lines remain the same length. Meanwhile the value of β is increased as the isentropic surfaces are brought closer together. Consequently the Richardson number is increased and the layer is stabilised. The motion is therefore destabilised by a deceleration.

On the other hand, if there is only a streamwise component of vorticity, ξ, in a flow not varying in the transverse (y-) direction, it is increased by an acceleration, and stretching, in the direction of the stream (x). Thus using (1.4.6) for ω we have

$$q\frac{\partial \xi}{\partial s} = \xi\frac{\partial q}{\partial s}, \quad \text{or} \quad \frac{\delta \xi}{\xi} = \frac{\delta q}{q} \tag{6.3.9}$$

and

$$\mathrm{Ri} = \frac{g\beta}{\xi^2} = \frac{g\beta_0 q/q_0}{\xi_0{}^2(1+\delta\xi/\xi_0)^2} = \frac{g\beta_0}{\xi_0{}^2}\frac{q}{q_0}\left(1+\frac{\delta q}{q_0}\right)^{-2} \tag{6.3.10}$$

$$= \mathrm{Ri}_0\left(1-\frac{\delta q}{q}\right) \tag{6.3.11}$$

roughly. In this case of transverse shear, an increase in speed therefore decreases the Richardson number even though the static stability is increased. Since this is a contrary effect to that on the component η, it is not likely that acceleration can be taken with confidence to be the cause of a change in Ri: in any case the effect is likely to be small compared with any effect of tilting when β is large.

In the most general kind of motion likely in the free atmosphere or ocean, accelerations are likely to be small. However, tilting can produce large effects, especially in long wavelengths or in the special case in which eddies are shed from the lee of mountains and the waves travel downstream, and are therefore slow-moving relative to the air.

In the equation for lee waves the most significant terms are (e.g. 5.8.24)

$$\nabla^2\zeta + \frac{g\beta}{u_0{}^2}\zeta = 0 \tag{6.3.12}$$

where ζ is the vertical displacement in two-dimensional motion in the xz-plane. At layers where β is much larger than at other levels, the curvature of the profile of ζ as a function of z has a maximum curvature, unless ζ happens fortuitously to be small at that level. These levels are therefore likely to be those at which the gradient of the ζ-profile is reversed and ζ has a maximum magnitude. Thus in addition to being the level at which Ri is most increased when tilting occurs, the layers of large β are usually those at which the wave amplitude, and therefore δz, is a maximum. This again has a maximum effect on Ri. Mountain waves are therefore a very likely site for the occurrence of billows. Their evolution is depicted in Fig. 6.3.ii.

Indeed glider pilots have known for a long time that the top of a layer of stratocumulus is usually the level at which lee waves have their maximum amplitude. This is because the radiation from the cloud top cools all the air below (by downward convection) and creates a layer of large static stability at its top, where billows are often seen; in fact cloud layer tops are their most common site in waves. (See Scorer *Clouds of the World*, Chapter 6).

Fig. 6.3.ii The diagram shows the typical stage of evolution of billows in the crest of a wave, and below in the trough of a wave where they are not usually visible in cloud.

6.4 THE DYNAMICAL GENERATION OF LAYERS OF LARGE STATIC STABILITY

In this section we illustrate a very economical type of argument which can be used in many stability problems where there is no viscosity or boundary present to introduce particular dimensions into the dispersion equation for waves in a medium. The results about to be obtained can, of course, be obtained by classical perturbation theory.

We consider the case of an airstream moving horizontally in geostrophic balance, but with static stability β and vertical and horizontal gradients of wind $\partial u/\partial z$ and $\partial u/\partial y$, the flow being in the x-direction. We then consider the work done against the various forces which arise when a displacement is made. If this work turns out to be negative the displacement will support itself and represents a form of instability. As we shall see, some of these types of displacment produce large gradients of potential temperature which, as we have just seen, are likely causes of billows and therefore of clear air turbulence, which is billows in almost all cases.

A parcel displaced a vertical distance c would experience a buoyancy force $g\beta c$, and so in being displaced this force would absorb an amount of work equal to

$$\int_0^c g\beta c \, \mathrm{d}c = \tfrac{1}{2} g\beta c^2 . \tag{6.4.1}$$

As a result of this displacement it would have a velocity anomaly $c\dfrac{\partial u}{\partial z}$, and would therefore experience a deviating force due to the earth's rotation which differs from that on its surroundings by amount $fc\dfrac{\partial u}{\partial z}$, where f is the vertical component of the earth's rotation. From this force it would receive work if it travelled a distance b in the direction of the force equal to

$$fbc\frac{\partial u}{\partial z}\,. \tag{6.4.2}$$

But in being displaced a distance b in the y-direction it would acquire a velocity anomaly $b\left(\dfrac{\partial u}{\partial y}-f\right)$ in the x-direction (bx absolute vorticity), and if f is assumed to be constant over the distance of the displacement, the work done in the displacement b in the y-direction by the corresponding anomaly of the deviating force is

$$\int_0^b fb\left(\frac{\partial u}{\partial y}-f\right)\mathrm{d}b \;=\; \tfrac{1}{2}f\left(\frac{\partial u}{\partial y}-f\right)b^2\,. \tag{6.4.3}$$

If the total work done in the displacement (b,c) in the y- and z-directions is positive, such displacements are unstable, and circulations around the streamlines can develop because a positive amount of work would also be done in the displacements $(-b, -c)$ of other parcels, necessary to complete the circulations. Adding the three components (6.4.1, 6.4.2 and 6.4.3) we obtain the condition for instability, namely

$$\tfrac{1}{2}f\left(\frac{\partial u}{\partial y}-f\right)b^2 \;-\; fbc\frac{\partial u}{\partial z} - \tfrac{1}{2}g\beta c^2 \;>\; 0\,. \tag{6.4.4}$$

The flow is unstable when $c = 0$ if $\dfrac{\partial u}{\partial y} > f$. This is simply a statement that in space the flow, which is rotating with the earth, has circulation decreasing along the outward normal to the curved flow (see Section 3.12). Putting this in local terms, the direction of rotation of the tangent moving with the particle is in the opposite direction, in space, to the vorticity of the fluid, the component normal to the earth's surface only being considered. We have also ignored the curvature of the streamlines relative to the earth (and therefore the cyclostrophic force). More typically $\partial u/\partial y$ is of the order of $\tfrac{1}{2}f$.

But c/b can be negative, and so also can $\partial u/\partial z$. A real value of c/b exists to satisfy (6.4.4) if

$$\left(\frac{\partial u}{\partial z}\right)^2 \geq g\beta\left(1 - \frac{\partial u}{\partial y}/f\right). \tag{6.4.5}$$

The case of just no instability is determined by writing equality in (6.4.4 and 6.4.5), and then

$$\frac{c}{b} = \left(\frac{f\left(f - \frac{\partial u}{\partial y}\right)}{2g\beta}\right)^{1/2}. \tag{6.4.6}$$

This quantity is the aspect ratio of the secondary cells which would develop in this geostrophically moving airstream. To get an idea of the shape we put in typical values, as follows:–

$f = 5 \times 10^{-5}\ \text{sec}^{-1}$ (at latitude 45° roughly)

$g = 10^3\ \text{cm sec}^{-2}$ roughly

$\beta = 10^{-7}\ \text{cm}^{-1}$ (when $\partial T/\partial z$ is about $\tfrac{1}{3}\,\Gamma$)

$\dfrac{\partial u}{\partial y} = \tfrac{1}{2}f = 2.5\ \text{m sec}^{-1}$ per 100 km

and then $$\frac{c}{b} = 2.5 \times 10^{-3} \tag{6.4.7}$$

The range of values of $\partial u/\partial y$ and β which are common make this quantity vary by a factor of about 3. The corresponding value of the vertical gradient of wind speed given by equality in (6.4.5) is $(\tfrac{1}{2}g\beta)^{1/2}$, and so

$$\frac{\partial u}{\partial z} = 0.7 \times 10^{-2}\ \text{sec}^{-1} \tag{6.4.8}$$

which represents a gradient of 7 m sec^{-1} per km. This is quite common, and gradients larger than this would increase the degree of instability. An increase in $\partial u/\partial y$ to more nearly equal to f reduces c according to (6.4.6), and so the cells are flattened, but even when $\partial u/\partial y = 0$ the value in (6.4.7) is not altered significantly for the purposes of this discussion. In the case when $\partial u/\partial y = 0$, the criterion (6.4.5) is sometimes expressed in the form of a Richardson number, thus

$$\mathrm{Ri} = \frac{g\beta}{(\partial u/\partial z)^2} \leq 1\,, \tag{6.4.9}$$

and (6.4.4) can be written

$$\mathrm{Q}^2 + \frac{2}{\mathrm{Ri}^{1/2}}\mathrm{Q} + 1 < 0 \tag{6.4.10}$$

where

$$\mathrm{Q} = \frac{c}{b}\,\frac{(g\beta)^{1/2}}{f}\,, \tag{6.4.11}$$

and if $\mathrm{Ri} < 1$, (6.4.10) is satisfied for the range of values of c/b satisfying

$$1 - (1 - \mathrm{Ri})^{1/2} < \mathrm{Ri}^{1/2}\mathrm{Q} < 1 + (1 - \mathrm{Ri})^{1/2}. \tag{6.4.12}$$

The sign of $\partial u/\partial z$ is immaterial, and the places where a large magnitude is likely to occur are where a jet stream is being decelerated as it catches up slower moving air. In the northern hemisphere, where f is positive, and where $\partial u/\partial y$ is positive on the left of the jet stream, the most likely place for this instability to occur is on the left of a jet stream, particularly at the exit, because there the velocity anomaly is in excess of the value assumed in calculating (6.4.3) where geostrophic flow was assumed. Indeed ageostrophic flow will certainly enhance these effects.

The consequence of this secondary motion in flat cells is to distort the temperature field so as to produce layers of increased and decreased vertical gradient where it was previously uniform. Since layers where β is large are more stable statically they will be the boundaries of cells and so the value of β will be further increased there. According to (6.4.7) the cells are very flat, and if they are 100 km in horizontal extent they will have a depth of the order of $250 \overset{\times}{\div} 3$ metres, the more stable layers being thinner.

Since they are produced by the deviating force of the earth's rotation they require a time of the order of a day or two at least to appear and it is unlikely that they would be directly responsible for the low Richardson numbers which could cause billows. We need to look to mountain waves or fairly vigorous motion near fronts, where the isentropic surfaces become more inclined, for the causes of moderate or intense clear air turbulence.

Nevertheless very light clear air turbulence is often observed far from mountains and fronts, and is not uncommon in the stratosphere, with a wavelength of the order of 100-200 m. Two comments need to be made in this connection. First of all, if the tilting of the stable layers is so slight that the

Richardson number is only very slowly reduced, the degree of instability (growth rate) will be very small because the disturbances will have plenty of time to grow without the instability becoming greater. In the case of mountain waves the tilting can be so large that a very large growth rate is produced before disturbances have had time to grow. When one of these slight disturbances has taken place it leaves a very stable layer at its top and bottom, and if the original tilting is still present slight turbulence could well be produced there. Consequently very slight turbulence could be a feature of a layered air mass for many hours.

Secondly, it must be recognised that billows are usually quickly evolving in time so that an observation of clear air turbulence may show very definite periodicity if the billows are traversed at an early stage. However, if the flight through them is along the length and at a stage when the cat's eyes pattern is degenerating into a horizontal layer of depth H, the turbulence will not appear to be organised. It is probably not necessary to invoke the possibility of any other kind of turbulence in clear, stably stratified, air.

We are thus led to a more thorough study of turbulence in nature and its effects, which is the main theme of part 2 of this treatise.

Part 2

TURBULENT PHENOMENA, CLOUDS, AND DISPERSION

Chapter 7

TURBULENCE

7.1 THE OUTSTANDING DIFFICULTY

In the first (1895) edition of *Hydrodynamics* Lamb wrote (Art. 311), 'It remains to call attention to the chief outstanding difficulty of our subject.' The same sentence heads the completely rewritten treatment of the last (1932) edition (Art 365). Interesting though it is, Lamb's discussion is not particularly illuminating to the student, whereas Brunt, by contrast, has managed an orderly presentation which is a mixture of good physical concepts and simple mathematical treatment (*Physical and Dynamical Meteorology*, Ch XII, 1939). The simplicity is illusory: it is a tribute to Brunt's unsurpassed skill as a teacher that a sense of certainty seems to be imparted and that the analysis conveys the impression of achievement. By making no pretence that all motions which someone has categorised as turbulence are susceptible to a single treatment, and by considering certain selected situations where light can be shed on the mechanics, one can both advance and avoid the error of thinking that **turbulence** must be one thing and that any property it is discovered to possess in one context must be possessed by it in another. Turbulence has almost acquired the characteristics of a 'Loch Ness monster' (see Section 12.2).

Writing in 1962 Sutton also said that '**turbulence** ... constitutes the outstanding difficulty of hydrodynamics' (*The Challenge of the Atmosphere*, p. 129). Since there are many turbulent motions for which a full analysis can never be obtained, we have to employ a new approach. Batchelor's *The Theory of Homogeneous Turbulence* is a good example of a treatment of a selected topic in the vast realm of turbulent motions, and it has often been unfairly criticised because its techniques have very limited usefulness in practical cases where the turbulence is generally not homogeneous. It is however very enlightening on the process of decay, which we are scarcely concerned with in this book.

Prandtl's mixing length theory gives a simple mathematical description of turbulent motion near boundaries and the transfer of momentum to the fluid; this, together with Reynolds's eddy transfer theory, stimulated the solution of problems by K-theory (well exemplified in Brunt's treatment). The feeling of warm familiarity which K-theory engenders in engineers and others who are well versed in the theory of conduction and of viscous fluids made inevitable the extravagant flirtations with that theory in which some authors have engaged (e.g. Hinze 1959).

A completely different approach to turbulence, started by Reynolds's study of the onset of turbulence in pipes, is concerned with hydrodynamic instability. Much of this work is very sophisticated and erudite, and discusses the conditions in which small disturbances will grow and cause a transition from laminar to turbulent flow. We need to know something about this approach because the transition is very common in the atmosphere, but it must be emphasised that, although engineers may seek to establish fairly exact criteria for the transition, we are more concerned with the mechanisms which change the state from stable to unstable than with the mechanics of transition itself. Nature is not concerned with the critical state in which there may or may not be eddies, but engages in a destabilisation or stabilisation over many minutes or hours. When the motion becomes unstable it is so violent that it bears no resemblance to the form of the growing disturbances at the initial moment of breakdown of laminar flow.

In Chapter 8 we treat details of cases in which there are some useful simple certainties. Although we shall make some appropriate remarks about turbulence in the neighbourhood of solid boundaries, there are many texts on this part of the subject which we shall make no attempt to summarise. Turbulence away from or in the absence of boundaries is almost completely neglected in the literature. Indeed many discussions of such cases (for example, of the motions which cause inhomogenieties of refractive index in the atmosphere and give rise to the scattering of radio waves) are bedevilled by concepts developed in connection with boundary layer turbulence or wind tunnel turbulence created by flow through grids. The free use of the word turbulence to categorise such motion is one cause of the mischief. Some people always assume that the theory of turbulence taught and used in one context must be applicable to complicated fluctuating motions in a quite different context, especially when the same instruments (e.g. hot wire anemometers) are used in both cases.

It is regrettably necessary to warn the reader against two errors of logic which are common. The first is that a statement about one particular kind of turbulent motion may be true (i.e. is not necessarily untrue) even though it may be shown to be obviously untrue in a different kind of turbulent motion. Thus it has been suggested that since flying insects have no navigational ability, turbulence must disperse them just as it disperses smoke and aphids in the atmosphere. But some insects are concentrated by motion which in other contexts may validly be called turbulence (see Section 12.6). Equally some complicated wave motions may have many of the appearances of turbulent motion but may yet fail to disperse or diffuse anything carried by the fluid.

It has been shown that decaying homogeneous turbulence has, under certain circumstances, a range of eddy sizes in which their energy, E, is related to their wave number, k, according to

$$E \propto k^{-5/3}. \qquad (7.1.1)$$

If a specimen of turbulence is then measured and found to obey this relationship it is sometimes argued that it is homogeneous and decaying, and that the eddies observed lay in the range for which (7.1.1) holds. Furthermore, these statements are sometimes said to be roughly true when (7.1.1) is found to be roughly true. The absurdity of such a conclusion is clear when we note that a single sharp change in velocity may be analysed into Fourier components. Then

$$E \propto k^{-6/3}, \qquad (7.1.2)$$

which, in practical cases where fluctuations of velocity are subjected to Fourier analysis, is scarcely distinguishable from (7.1.1). The logical incompleteness of the conclusion should be apparent to all scientists; yet the temptation to assert that an explanation which appears to predict the observed fact must be correct is often found irresistible.

7.2 DEFINITIONS OF TURBULENCE

It is not my pupose to arrive at a definition of turbulence; too much breath at conferences has already been expended on that. My aim is simply to clarify some of the ideas so that two things will be plain. First, that there is no single definition which ought to be universally adopted, and secondly that there is enough incompatibility among the definitions to make us wary of applying to other cases conclusions arrived at in the study of some example of motion called turbulent.

The idea of turbulence arises because some motions cannot be precisely described. Furthermore, it would not be useful to have a precise description of them because it would be too complicated. What we desire is to talk about the effects of such motions in practical terms, which means arriving at overall or statistical descriptions or conclusions without discussing the details. To some people this has apparently meant that we may not or cannot discuss the details of the mechanics and must therefore represent the effects of turbulence by parameters. However, there is no guarantee that a representation by parameters has any value, for in some cases we can only evaluate the parameters afterwards, and predictions are not facilitated. Lamb's description of it as the 'outstanding difficulty' carries the implication, or at least the hope, that one day this difficulty will be resolved. In fact the resolution seems to be taking the form of learning to live with impossibilities rather than inventing a tractable theory.

This coming to terms with our limitations is actually much more satisfactory than it may seem because it amounts to making our objectives sensible. It is obvious that the deeper we probe into something the more complex it seems to be, and in the case of indescribably complicated motion the only road to simplification is to decide that there shall be limits to the complexity we are prepared to study. Life is too short to spend it sorting out the infinite details of the flow on a single occasion, and anyone who did analyse them could have no assurance that the next occasion would be the same, nor that anyone else would be interested anyway.

The degree of complexity beyond which we decide not to go in our studies is a matter of choice. Up to that point we think of the flow as being laminar and susceptible to description in all detail, but beyond that point we call it turbulence, possessing complexities whose overall effect only will interest us.

Thus we arive at the idea that there is in existence a mean motion with some fluctuations added. The mean motion is fully described, but of the fluctuations we only have statistical information. Consequently there is an element of choice: we decide how to make the division between the mean motion and the fluctuations. There is no reason why two people should not make different divisions, either because they set up different criteria for doing so, or because they had different purposes for their theory. As a result the turbulence (the fluctuating part) is not a clear cut objective property of the motion but a matter of arbitrary definition for our convenience. We may be driven by the logic of the situation to make the definition in a certain way, and then it may appear to be a recognition of a real feature of the flow. However the logic is not general; it is peculiar to the situation and its relationship to us, to our skills, and our wants.

A quite different definition to the effect that 'turbulence is a complex motion which causes diffusion' is a very practical one. It recognises that when we stir the milk into the tea or the dye into the paint it is the result of a very complex pattern of motion that uniformity of composition is arrived at, and that the mixing is much more effective than molecular diffusion. Diffusion is an irreversible event. Furthermore, when we wave a spoon about in the tea we produce spoon or cup sized motions which quickly degenerate into much more complicated motions with smaller size scales associated with them. The motion which spreads tobacco smoke would be described as turbulence: it is not long before the smell can be detected in every corner of a room. This is also an arbitrary definition in the sense that we have already made up a definition of diffusion, which we think of as a complicated motion which transports fluid particles around and mixes them up irreversibly. Sometimes turbulence is described as 'the random part of the motion', and again there is implied a previous choice as to what is random, for random merely means 'containing details we shall not enquire into'.

The turbulence diffuses the fluid itself and any characteristic attached to it such as colour, chemical constitution, together with airborne pollution such as smoke, smell, water vapour and so on. Fluid parcels also possess heat, density, linear and angular momentum, and kinetic energy of rotation and of linear motion relative to their surroundings. There are features not attached permanently to the parcels: some may be diffused by molecular transfer mechanisms, or by the action of pressure gradients at a much greater rate than would occur through the molecular diffusion of the material. To cite extreme cases, fluctuations of density may be communicated through a fluid at the speed of sound by pressure waves, whereas fluctuations of concentration of a pollutant can only be moved by the movement of the material.

The velocity of the fluid at a point might be fluctuating because of eddies which transport the fluid elements or because of waves passing in all directions and causing all the fluid elements to oscillate about their mean positions and return to them when the fluctuations have finished. In the first case, any property carried by the fluid parcels would be diffused; in the second case, the distribution would be as it was before the waves began to pass through the fluid.

There is a very obvious difference between these two types of motion. Waves can be thought of as being produced by the movement of the boundary of the fluid, and if for the moment we consider the case in which the viscosity produces negligible effects, any surface within the fluid may be thought of as the boundary (of the region of interest, say). If the fluid is at rest initially, or is in motion but possesses no vorticity, and is of uniform density (or potential temperature in the case of adiabatic motion) no vorticity will be produced. The motion will therefore be irrotational throughout and will also be reversible. Thus if there were a boundary of very complicated shape moving past the fluid, or if the boundary were flexible and made to move in a very complicated way, the fluid would be subjected to very complicated fluctuations of velocity and the parcels subjected to very complicated changes of shape. However, the complexity of the motion would correspond with the complexity of the motion of the boundary, and would not be more complicated at any stage. Also, the fluid could be brought back exactly to its initial position by simply reversing the movement of the boundary back to the beginning.

By contrast, a body of fluid with a fixed boundary enclosing it, set into eddying motion by some means such as a spoon moved through it, would diffuse throughout its whole volume any concentrated material which initially polluted only a small part of it. In the absence of viscosity the motion which would produce this diffusion could take place in the reverse direction, but the prescription for setting up that motion which would reverse diffusion that had already taken place becomes more and more complicated as time proceeds. It is easy to distinguish between earlier and later states through which the fluid passes when diffusion occurs, but in the case of the irrotational fluctuating motion the later state is essentially no more complicated than an earlier state. Likewise, if waves were travelling back and forth without any dissipation or loss of energy in a box containing fluid, it would be impossible to tell by inspection whether a cine film of it was being shown forwards or backwards. A casual observer might describe a wavy sea surface or a wobbling jelly as being in turbulent motion because of the random nature of the velocity fluctuations of a particle or at a point, but the motion is essentially different from that in which eddies are causing diffusion. Our discussion will be restricted to cases where vorticity is present and where the direction in time is obvious.

This leads us to what is probably the simplest definition of turbulence, namely **chaotic vorticity**. This still retains the arbitrary human element in the word chaotic, for often the experienced eye sees order in a scene which to the untrained observer is chaos. However, there is a degree of complexity beyond which no one wishes to probe when analysising the details of the motion of a fluid. When there is a fairly complicated pattern of vorticity present, and in some cases when there is a very simple pattern (such as a plane vortex sheet), the motion will become progressively more complicated. The element of arbitrariness in our definition then is considerably reduced because the motion will be turbulence before very long by any definition, unless viscosity stops it becoming complex by destroying it. Thus if we waft a spoon once through a bucket with a blob of ink in it we set up turbulence and the blob becomes a diffusing cloud which obviously qualifies for the title 'turbulent'. If we waft the spoon through a bucket of golden syrup in the same way a blob of black treacle in its midst would at worst be slightly distorted, but would certainly not assume a shape of indescribable complexity. An artist could reproduce its shape in all essentials with ease and a technician could measure it. This is because the viscosity is so large that eddies of the size of the bucket, or smaller, cannot turn through more than a small fraction of one rotation before they are brought to a standstill.

A very simple way of determining whether a distribution of vorticity is chaotic or not is as follows. We determine a grid of points such that we are not interested in investigating motion on a smaller scale than the grid spacing. If the vorticity vector is highly correlated over several neighbouring grid points it means that an orderly pattern of motion is present on a scale larger than the grid spacing. If we make a Fourier series such that it takes the value of the velocity at all the grid points along one coordinate line, the motion must be on a much larger scale than the grid length if only the first few terms in the series are required to give a good fit. If, on the other hand, the final term used corresponding to wavelengths of twice the grid length has an amplitude as large as the others, the motion has not been approximately described by giving the values at the grid points. It means that if the number of grid points were doubled twice as many terms might be required. Therefore much of the motion was being missed by using the original grid size, this being the part of the motion we didn't want to know about except in so far as it affected the motion on a larger scale, for example by diffusing momentum and behaving like viscosity.

The special feature of motion possessed of vorticity is that it contains energy of motion in the fluid which cannot be removed from it by suitable motion at the boundaries, and which equally cannot be set up simply by motion of the boundaries. The motion of the spoon through the cup of tea or bucket of water is special in that it has salient sharp edges or lines along which the flow separates from the boundary, and vorticity can stream into the fluid from the line of separation.

Before studying the mechanics of the turbulent motion composed of chaotic vorticity it is necessary to point out that progress can be made only if the motion can be divided effectively into a mean motion, which is intelligible, and a random motion which is chaotic. We may wish to set the dividing length, or time, scale between the two parts of the motion for our own convenience, but it is possible that there may be limits set by the nature of the motion itself. Thus there may be a very orderly motion on a large scale such as the flow down a straight pipe of circular section. Superimposed on this there may be fluctuations which clearly cause the diffusion of a filament of ink released into the pipe. There are no eddies much larger than the radius a of the pipe, and no fluctuations of period longer than na/U where n is not much larger than unity, and U is the average speed derived from the flux $\pi a^2 U$. The natural division into a mean flow, $u(r)$, and fluctuations is achieved by defining $u(r)$ as the average speed at radius r either at a fixed section over a time much longer than na/U or over a length of pipe much greater than r.

Such a natural division is not necessarily possible because eddies of all sizes may be present. This would mean that if a small number of smaller eddies or a large fraction of one larger eddy was sampled in an attempt to measure the mean by taking an average over a certain length or time, the value obtained would depend very much on the starting and ending points of the sample. If a very large number of eddies is included in the sample another half eddy will make a negligible difference to the mean value obtained: equally, if only a very small fraction of an eddy is sampled a slightly larger fraction will obtain very nearly the same value. Then samples obtained from different parts of this large eddy would merely serve to reveal the shape of the eddy, which would be seen as part of the mean motion. Thus eddies whose scale is very different from the sampling scale sort themselves naturally into random fluctuations, which we do not study in detail, and aspects of the mean motion. The technique fails or is very limited in usefulness if there are eddies present of scale near to the scale of the sampling procedure, which in our previous example was the grid spacing.

We now turn our attention to the mechanisms by which chaotic vorticity produces the effects which we ascribe to turbulence.

7.3 REYNOLDS STRESSES; MIXING LENGTH; DEGENERATION OF EDDIES

The mean value of a quantity q which varies with s, which may be a space or time coordinate, is defined at $s = s_0$ by

$$\overline{q}(s_0) = \frac{1}{2\sigma}\int_{s_0-\sigma}^{s_0+\sigma} q(s)ds\,. \qquad (7.3.1)$$

This is the average value of q over the interval of length 2σ surrounding the point or instant s_0. To be meaningful this definition must make $\overline{q}$ independent of σ over a fairly wide range of values of σ. If σ is too large we shall be unable to detect variations of $\overline{q}$ from one value of s_0 to a nearby value; if σ is too small $\overline{q}$ will be found to fluctuate rapidly as s_0 varies, and the whole purpose of our definition is to define a mean value quite distinct from the fluctuations. The success of the following analysis depends on the existence of a significant range of values of σ which lead to the same value of $\overline{q}$. σ must lie in a gap in the spectrum of eddy size present in the motion.

We may now write

$$q = \overline{q} + q' \qquad (7.3.2)$$

where q' is the fluctuation. By definition

$$\overline{q'} = 0 \qquad \text{and } \overline{\overline{q}} = \overline{q}\,. \qquad (7.3.3)$$

For two components of velocity v_i and v_j we have

$$\begin{aligned} \overline{v_i v_j} &= \overline{(\overline{v_i} + v_i')(\overline{v_j} + v_j')} \\ &= \overline{\overline{v}_i\overline{v}_j} + \overline{\overline{v}_i v_j'} + \overline{\overline{v}_j v_i'} + \overline{v_i' v_j'} \\ &= \overline{v}_i\overline{v}_j + \overline{v_i' v_j'} \end{aligned} \qquad (7.3.4)$$

for by (7.3.3) $\overline{\overline{v}_i v_j'} = \overline{v}_i\overline{v_j'} = 0.$

The equations of motion (1.3.26) and continuity $(\partial v_i/\partial x_i = 0)$ for an incompressible fluid may be combined to give the following

$$\rho\frac{\mathrm{D}v_i}{\mathrm{D}t} = \rho\left(\frac{\partial v_i}{\partial t} + v_j\frac{\partial v_i}{\partial x_j}\right)$$

$$= \rho\left(\frac{\partial v_i}{\partial t} + \frac{\partial}{\partial x_j}(v_i v_j)\right)$$

$$= \rho F_i + \frac{\partial}{\partial x_j} p_{ij}, \tag{7.3.5}$$

and if we now take the mean value of each term, the fluctuations only appear in the product term and we have, for the case in which ρ is a constant,

$$\rho\left(\frac{\partial \bar{v}_i}{\partial t} + \frac{\partial}{\partial x_j}(\bar{v}_i \bar{v}_j)\right) = \rho \bar{F}_i + \frac{\partial}{\partial x_j}\left(\overline{p_{ij}} - \rho\overline{v_i' v_j'}\right). \tag{7.3.6}$$

This equation is of the same form as (7.3.5), but is an equation for the acceleration of the mean motion in terms of the mean body force $\bar{F}_i$, and the gradients of the mean stresses $\overline{p_{ij}}$ with an additional term $-\rho\overline{v_i' v_j'}$ added. This additional term is called the Reynolds stress, and (7.3.6) is Reynolds's equation for the mean motion.

The quantity p_{ij} is the force in the i-direction on unit area normal to the j-direction, and represents the rate of transfer of i-momentum in the j-direction by the viscous stresses. The mean value of the viscous stress is defined so that the fluctuations in the stress have zero mean value. This is the same as the stress that would be produced by viscosity if the velocity were everywhere equal to the mean velocity if there are no fluctuations in viscosity. The quantity $\rho\overline{v_i' v_j'}$ is the average rate at which i-momentum represented by $\rho v_i'$ is transported by the velocity v_j' in the j-direction. It is therefore called the eddy stress, and when there are eddies it is usually very much larger than the mean viscous stress $\overline{p_{ij}}$, except when $i = j$, for then the mean normal pressure is much larger than the eddy normal pressure represented by $\overline{\rho v_i'^2}$.

The mechanism of the eddy shear stress is simple. If there is a gradient of the mean velocity v_i in the j-direction, the fluctuation v_i' is correlated with the fluctuation v_j' by the gradient. Thus if v_x increases in the y-direction (see Fig. 7.3.i), a parcel of fluid moving in the y-direction possesses, on the average,

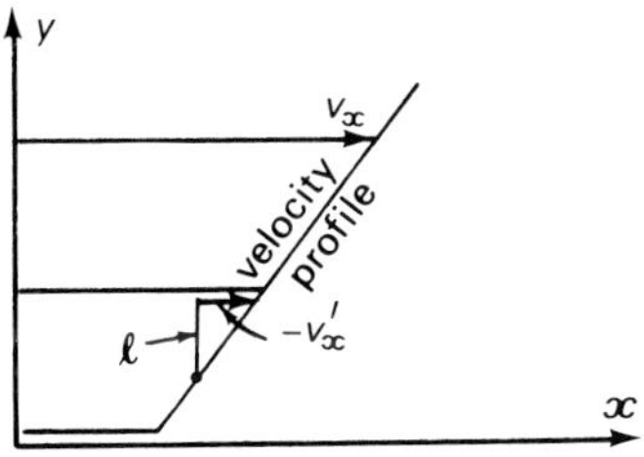

Fig. 7.3.i If the gradient of mean velocity is positive, a parcel displaced a distance ℓ in the positive direction acquires a negative velocity anomaly $v'_x = -\ell\, \partial v_x/\partial y$.

a smaller v_x than the surroundings into which it is moving, and it therefore carries a negative anomaly of x-momentum.

If we imagine the parcel to travel a distance ℓ, called the mixing length, before it mixes with its new surroundings, its velocity anomaly, v_x' is $-\ell\partial v_x/\partial y$. The mixing length is not the same for all parcels but only represents a statistical average; furthermore it could well be that the length is different for the transfer of different properties. Thus the parcel may share its momentum with its surrounding more readily through pressure gradients than it shares its heat, or its content of some pollutant that has to be shared by molecular or small scale mixing.

It has sometimes been argued, as it was originally by Prandtl, that the velocity fluctuations have no preferred direction, and therefore v_y' is the same as v_x'. This is based on observations, backed up by some theory, that the velocity fluctuations quickly tend to this state, but the argument which now follows is valid provided only that the various components are in the same proportion at all distances from the boundary. The shear stress τ, which is the force along the boundary on unit area parallel to the boundary, is equal to the rate of transfer across the boundary of momentum along it. Thus if we accept Prandtl's assumption that the only quantity which determines the eddy size and the mixing length is y, the distance from the boundary

$$\tau = \rho\overline{v_x' v_y'} = \kappa\rho y^2 \left(\frac{\partial v_x}{\partial y}\right)^2, \tag{7.3.7}$$

which is sometimes written as

$$\kappa\rho y^2 \frac{\partial v_x}{\partial y}\left|\frac{\partial v_x}{\partial y}\right| \tag{7.3.8}$$

in order to ensure the correct sign. Physical considerations make it obvious what the sign is. κ is a nondimensional number which relates ℓ to y, and if the turbulence is generated by the work done against the eddy shearing stress and is dissipated into heat via smaller eddies, it may be supposed that equilibrium is reached and that κ always takes the same value. The exercise is simply one of dimensional analysis in that case, and there are many alternative derivations. Thus von Karman proposed that the eddy viscosity was determined by an appropriate combination of the features of the velocity profile which generates the turbulence, and suggested that

$$\tau = \kappa\rho\left(\frac{\partial v_x/\partial y}{\partial^2 v_x/\partial y^2}\right)^2 \left(\frac{\partial v_x}{\partial y}\right)^2. \tag{7.3.9}$$

This formulation, however, amounts to the same thing, for in the case of both (7.3.7) and (7.3.9) we find that if τ is supposed independent of y, so that there is no pressure gradient in the x-direction, the velocity profile is given by

$$v_x = \frac{1}{\kappa}\left(\frac{\tau}{\rho}\right)^{1/2} \log y + \text{const.} \tag{7.3.10}$$

This is the logarithmic profile of velocity near a boundary and is to be expected when no other boundary is near, and at a sufficient distance from the boundary for the flow to be turbulent. That means that the eddies must be large enough for them not to be dominated by viscosity. It also means that the velocity gradient at larger distances from the wall is smaller because the flow is subject to a greater eddy viscosity: this is clearly not the case in many natural flow fields, and there are other causes for the decrease of velocity gradient, such as are described in Chapter 4. κ is called **von Karman's constant** and is equal to about 0.4.

We are thus led to consider what happens to turbulence, and what is the real mechanism of the Reynolds stress, for the fluid is a continuum and the mixing length theory is no more than a picture for the dimensional analysis. The results can be obtained directly from the assumptions as to what factors determine the motion, and in this case it is assumed that the eddies and the motion are on a sufficiently large scale for the effects of viscosity to be negligible. They are valid for a large enough Reynolds number, which in this case is

$$\text{Re} = \frac{(\tau/\rho)^{1/2} y}{\nu}, \tag{7.3.11}$$

there being no other velocity specifiable for the system than $(\tau/\rho)^{1/2}$, which is called the **friction velocity** and is often denoted by $u_\star$. Clearly viscosity becomes important for small y. ν was defined on page 34.

A turbulent field of velocity is composed of eddies, or fluctuations, of a wide range of sizes. The scale of the turbulence can be measured by its largest and its smallest eddies. A large eddy will cause correlations between the velocities of particles at two different points in it, and so the macro-scale of turbulence is the distance at which there is no correlation of the velocity fluctuations. In the atmosphere this scale is larger than the size of the earth because wind systems such as the trade winds and westerlies extend right round it.

Small eddies are quickly dissipated by viscosity; indeed it is possible to create smaller eddies in water or air than in treacle because of this. The greater the energy input the smaller the eddies that are created, and if the turbulence is left to itself the size of the smallest eddy increases as the motion is turned into heat by viscosity. If ϵ is the rate at which energy is being fed into the eddies which are dissipating it as heat, then the quantity

$$(\nu^3/\epsilon)^{1/4} \tag{7.3.12}$$

is a length, which may be regarded as a measure of the smallest eddy size.

If the eddies are regarded as a spectrum of scales of motion which can be represented by a Fourier analysis, it would be expected that the interaction of two eddy sizes would be to create both larger and smaller eddies. Furthermore, at large enough scales for viscosity to be negligible, the flow is theoretically reversible because there is no dissipation to heat, and so the motion which generates small eddies out of large ones can be reversed and generate large ones out of small ones. This cannot happen in practice because the situation in which only the larger ones exist is special, and time cannot be reversed so as to make the flow tend towards it. For example the complicated flow generated in billows, or any other unstable flow pattern, cannot be reversed because the complex pattern which would lead to it cannot be set up. We now discuss how this comes about, for it is part of the mechanics of increasing disorder and degeneration of motion into heat.

A characteristic of vortex motion is that the line of particles joining two given particles is ultimately stretched. Thus, in the case of plane shearing motion in which the velocity is (Fig. 7.3.ii)

$$v_x = \alpha y, \qquad v_y = v_z = 0, \tag{7.3.10}$$

the particles at A and B are moved apart, whereas the particles at A and C are moved nearer together until the upper particle arrives at D, and from then onwards they are moved apart, for ever. If the particle at B were allowed to move away for a century and the flow then put into reverse, after a century of moving towards A it would move away from A for ever.

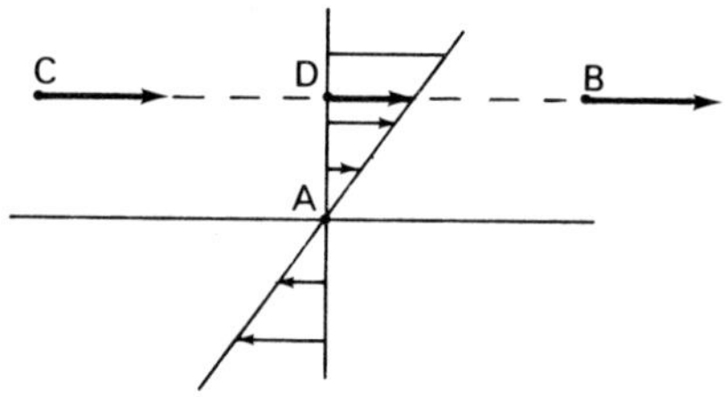

Fig. 7.3.ii Even the line AC, which is compressed by the motion at first, is ultimately subjected to stretching from the position AD.

Turbulence contains vorticity because it is a motion in which kinetic energy has been added *in situ* without increasing the orderly transport of the fluid (Section 1.1 p. 24). It is chaotic vorticity, because if it were orderly it would be possible to describe it in every detail. In calling it chaotic we decide not to describe every detail because it serves no purpose to do so, although theoretically it could be done. Thus, if we were to describe the distribution of vorticity by giving the vorticity vector at every point of a grid, there would be an orderly pattern which we could subtract out, together with a departure from it which we would call the fluctuations. The fluctuations at neighbouring points of the grid would not then be correlated. There would be eddies present which would be on a scale smaller than the grid length and could not be described by specification of velocity or vorticity at the grid points. The intensity and some other properties of the small eddies would, however, be described statistically by the information at the grid points.

An eddy stress is a mechanism whereby energy of the mean motion, or of a large eddy (which is the same thing as part of the mean motion if we choose to probe into it in detail), is transferred to the eddies. The vortex lines are carred with the motion, except on the small scale, where each component of the vorticity is conducted by the viscosity as well. The eddies are motions around the vortex lines. After a finite time, which in practice means at all times except in specially contrived situations which cannot persist indefinitely, the length of every vortex line is being increased by the motion due to all the others and the density of vorticity is continuously increasing. This is made obvious by consideration of the kinetic energy of an individual eddy represented by a single vortex line, for when a cylinder of fluid surrounding a segment of the line is

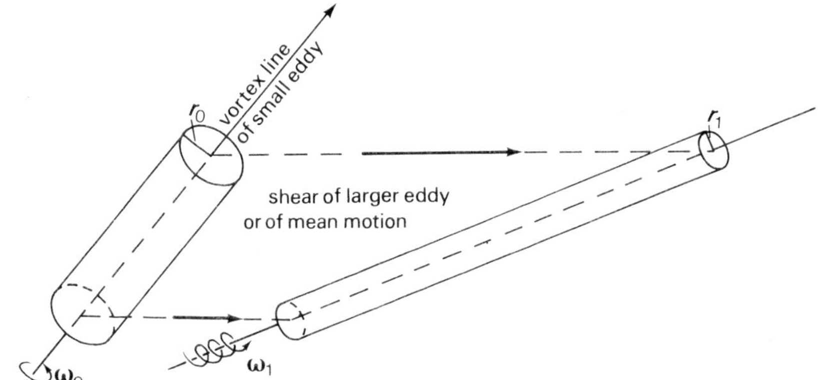

Fig. 7.3.iii Stretching of vortex lines and increase of kinetic energy.

stretched its angular momentum is conserved. Fluid originally at distance r_0 with angular velocity ω_0 around the line (see Fig. 7.3.iii) has angular velocity ω_1 when it is at distance r_1, where

$$\omega_1 = \frac{r_0^2}{r_1^2}\omega_0 . \qquad (7.3.14)$$

The initial and final kinetic energies of unit mass are therefore

$$½(r_0\omega_0)^2 \text{ and } ½(r_1\omega_1)^2, = \frac{r_0^2}{r_1^2}½(r_0\omega_0)^2, \qquad (7.3.15)$$

and so the kinetic energy is increased in the ratio r_0^2/r_1^2.

As a consequence of the increase in density of vorticity the viscous dissipation of the motion is increased. Therefore there is a continuous reduction in the sizes of the eddies and a transfer to them of kinetic energy from the larger eddies or from the mean motion at the same time.

It is now evident that turbulence which produces a Reynolds stress cannot be isotropic: if it were, for every eddy that was being stretched there would be another that was being compressed along its vortex lines. Such a situation cannot persist because after a finite time the eddies could no longer be paired in this way and the majority would be in a state of being stretched. The fluctuations of velocity produced by an eddy consisting of a single vortex line can be considered for the case of plane shear flow, and the argument is the same for any flow containing vorticity. On the near side of the vortex line AB a positive value

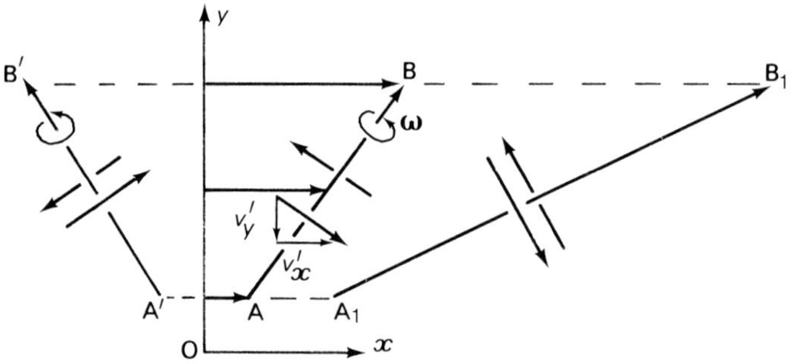

Fig. 7.3.iv Velocity fluctuations due to the vortex lines of a small eddy.

of v_x' is produced and is associated with a negative value of v_y'. On the paired vortex line A′B′ a positive v_x' is associated with a postive v_y' and so the momentum transfer $\rho\overline{v_x'v_y'}$ is of the opposite sign. But after a finite time A′B′ takes up the former position of AB, which has moved by then into the position A_1B_1, and both are giving rise to a shear stress of the same sign. Vortex lines parallel to either of the axes produce no shear stress; a vortex line along Ox produces no v_x' and one along Oy no v_y'. Also a vortex line in any direction in the planes of the motion (y = const) remains unstretched in the case of this simple motion.

Unit length of vortex line is stretched at the maximum rate if it is inclined at 45° to the planes of the motion, which is important in the production of sound in a compressible fluid (see Section 7.11). The shearing motion itself is giving such eddies energy at the maximum rate, and they may therefore be thought of as shear-induced. On the other hand, vortex lines produced by flow around obstacles are created parallel to the lines along which separation occurs. This is shown in Fig. 7.3.v. In general this is most likely to be at right angles to the flow. Therefore obstacles in the flow produce vorticity with a significant component at right angles to the shear vector produced by the boundary on which they stand. Mechanically produced turbulence is thus

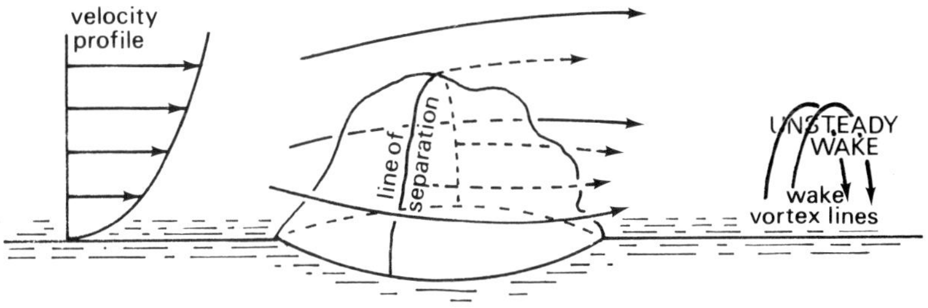

Fig. 7.3.v The vortex lines shed in the wake of an obstacle from the line of separation have a vertical component and are therefore stretched by the shear due to boundary drag in the main stream, and produce an eddy shear stress as soon as the vortex sheet in which they lie becomes billowed.

produced by roughnesses on the boundary, for their wakes extract energy from the mean shear flow produced by the boundary. The mean shear flow itself makes no contribution to the chaotic vorticity unless it is unstable, but it is likely to become unstable to smaller, newly generated, disturbances if it becomes curved (see Section 3.10).

By contrast, in shear flow over the ground with horizontal streamlines, the eddies produced by gravity necessarily have vortex lines which are horizontal, and are therefore like the vortex lines produced by viscosity in creating no Reynolds stress. However, when a thermal (see Chapter 8) penetrates to a higher level through shear, the vortex lines are tilted and do then produce a stress (see Section 9.6).

7.4 PROPERTIES OF TURBULENCE

The most important feature of turbulence is that it causes diffusion of any physical property carried by the fluid, such as heat or pollution. This is due to the stretching of all lines of particles so that any compact mass of fluid is drawn out into a sheet, stretched and distorted in a complicated way. Thereafter, when the thickness of the sheet is reduced to less than the size of the smallest eddy, further diffusion is effected by molecular motion. The turbulence achieves diffusion by a mixing motion, but the concentration gradient of any property originally occupying a limited compact volume of the fluid is always increased by the turbulence. This is because surfaces of different constant concentration are brought closer together as the sheets of polluted fluid are stretched by both large and small scale eddies, and by the mean motion.

Because all concentration gradients are increased, so also are all velocity gradients, and the rate of dissipation of energy is increased in a given eddy by the reduction in its size. The Reynolds number of an eddy remains constant because when its size is reduced in accordance with (7.3.14), its characteristic velocity $r\omega$ is increased, and their product is the circulation. The Reynolds number, $r^2\omega/\nu$, thus remains constant. It may be thought of as a measure of the number of rotations it has to make while its velocity is halved, and so its lifetime is reduced when it is made smaller and its angular velocity increased. The dimension along the vortex line is increased, but that is irrelevant, being large compared with the diameter anyway.

If a section of a turbulent fluid were taken and contours of constant vorticity component normal to the section drawn, the viscosity would be seen to diffuse and cancel out areas of negative and positive vorticity. As time passes the scale of the pattern is reduced and it becomes more complicated, the areas of positive and negative vorticity becoming smaller and more intermingled, and therefore more rapidly reduced by viscosity. The rate of dissipation by viscosity is determined by the rate at which small intense eddies are produced and intensified by stretching. The size of the smallest eddies is determined by the rate at which energy is fed down the size scale, which determines how small an eddy becomes before it performs those rotations which will completely exhaust its kinetic energy. The rate of cascading of energy down the size scale is determined not at all by how it is dissipated, but by the rate at which it produces smaller eddies from big ones.

In between the shear of the mean motion (or of the large eddies into which energy is fed by some sort of instability), and the smallest eddies (whose size is proportional to $(\nu^3/\epsilon)^{1/4}$), there may be a large range of eddy sizes having an energy spectrum equal to $\overline{u'^2}\,E(k)$, where $\overline{u'^2}$ is the mean square of the velocity fluctuations and k is the wave number. If this is a function of k and ϵ (which has the dimensions of veloc^2 $\times$ time^{-1}), which seems reasonable because ϵ is the manifestation of its effect, we find that the only combination of k and ϵ having the dimensions of acceleration (which is required because $\int\overline{u'^2}\,E(k)\mathrm{d}k$ has dimensions of veloc^2 being the kinetic energy, the density being taken as unity in this discussion) is

$$\overline{u'^2}\,E(k) \;\propto\; \epsilon^{2/3}k^{-5/3} \tag{7.4.1}$$

This is Kolmogorov's $-5/3$ law. It has been applied with incorrect logic in a multitude of situations. It is applicable only to decaying turbulence in the inertial range, usually called subrange, of eddy sizes remote from both the largest and smallest eddies. It is valid only for homogeneous turbulence, which means that it is not usually valid for shear flows, particularly when near to a boundary.

Since there is a considerable body of theory valid for homogeneous isotropic turbulence, there is a temptation to apply it in situations where turbulence seems to be present. Many people have sought to demonstrate that the turbulence is homogeneous and isotropic and in the inertial range by measuring the spectrum and finding it to be roughtly proportional to $k^{-5/3}$. But their logic is faulty here because many quite different fluctuating systems have very similar spectra, if only because their scale is roughly the same. The long wavelength end of any spectrum in nature is determined by the length of the sample used to determine it, and the form of the fluctuations is determined by the degree of phase correlation of the different wavelengths. Thus a single discontinuity of velocity has a spectrum proportional to $k^{-6/3}$, which is scarcely distinguishable in any practical case from $k^{-5/3}$. The process of averaging and of expressing all the features of a turbulent flow by statistics only hides many important characteristics. In nature there are usually boundaries nearby, or else the medium is stratified, or it has a significant mean velocity gradient, or there are no gaps in the spectrum which make the taking of averages and the definition of a mean easy. All these factors make it very unlikely that the turbulence is either homogeneous or isotropic. It is usually either undergoing rapid development from a state of no turbulence to one of intense turbulence, or destruction by viscosity or by radiation of waves.

Any medium capable of transmitting waves can carry away the energy of turbulence, and thereby destroy it. A compressible fluid radiates sound waves from the pressure fluctuations caused by the stretching of the vortex lines: the pressure fluctuations also produce displacements which radiate gravity waves in a medium with a density stratification. In the former case, the sound waves represent an almost completely negligible fraction of the energy of the flow; in the latter case, quite a large fraction of the energy can be radiated as gravity waves.

7.5 RICHARDSON'S CRITERION FOR THE NON-PERSISTENCE OF TURBULENCE

In a stratified fluid, turbulent motion displaces parcels of fluid so as to create potential energy. Thus if a layer of air which is stratified with warm air on top is stirred up by turbulence, the potential temperature becomes more nearly uniform and so the upper layers are cooled and the lower layers warmed. The gravitational potential energy of the whole mass is increased by the lifting of its centre of gravity. At the same time, if there exists a horizontal shear flow, this mean flow will do work against the eddy stresses, and energy will be fed into the turbulence. Richardson argued that if the energy supplied in this way were not as much as the work being done against gravity, the turbulence would die out. Actually it would die out with a lower degree of stratification because a considerable fraction of the turbulent energy is used up in the cascade to smaller eddies and its subsequent dissipation by viscosity.

If a parcel of unit mass is displaced a vertical distance ζ in a stratified fluid of static stability $g\beta$ the work done against gravity is

$$\int_0^{\zeta} g\beta z \, dz = \tfrac{1}{2} g\beta\zeta^2 \tag{7.5.1}$$

because the restoring force due to a displacement z is $g\beta z$. The kinetic energy of the relative motion of the parcel displaced a distance ζ, in a shear of magnitude $dU/dz = \eta$, is

$$\tfrac{1}{2}(\eta\zeta)^2 . \tag{7.5.2}$$

If the expression (7.5.1) exceeds (7.5.2), the turbulence cannot be maintained, and so there will be no turbulence if the Richardson number which is the ratio of these quantities, denoted by Ri, exceeds unity, i.e. if

$$\text{Ri} = \frac{g\beta}{\eta^2} > 1. \tag{7.5.3}$$

The converse is obviously not true, and it is incorrect to say that laminar flow in which $\text{Ri} < 1$ will be unstable and become turbulent, for when $\beta = 0$ and η is the same at all heights the motion is stable and has a uniform vorticity, η, which cannot be redistributed to set up eddies. On the other hand, we saw in Chapter 6 that if the vorticity is concentrated into layers, these layers may be rolled up into vortices which are themselves gravitationally unstable. We also saw that such instability may be induced in layers in which Ri is initially infinite if they are tilted. In nature the interesting criterion is not whether a flow is unstable, but whether there are mechanisms operating which will change it so as to make it so.

Many authors have shown that the flow must have a Richardson number less than ¼ somewhere if it is to be unstable, an idea conjectured by G. I. Taylor; again, it must be remarked that many flows with smaller Richardson numbers are stable. The most concise demonstration of this result is given by Howard (1961; the short paragraph 2 of the paper).

The Richardson number is a local number which may vary across a flow field, and it has no global significance, such as the Reynolds number has, for in the case of viscous steady flow the motion is coupled throughout. In stratified fluids the flow of each layer is less dependent on the others.

Attempts have been made from time to time to find a similar criterion for the stability of cylindrical flow in which centrifugal forces play a role rather similar to gravity. These have been unsuccessful because the instability is essentially more complex. In horizontal shear flow the most unstable disturbance is a two-dimensional one in vertical planes containing the shear vector. In cylindrical flow unaffected by gravity the acceleration **f** changes direction along the path of the fluid. Consequently axisymmetric (toroidal) disturbances take on a different nature from two-dimensional ones in which the disturbance is in planes perpendicular to the axis of the undisturbed motion. Thus the criterion developed in Section 3.10 cannot be expressed in terms of a single Richardson number.

7.6 TURBULENCE GENERATED BY STANDING WAVES

Momentum can be transferred across the mean flow by steady gravity waves provided that the surfaces containing the crests are inclined to the verical (see Section 5.11). In the theory it is supposed that the momentum flux is upwards, away from the obstacle on the ground which experiences a force in the direction of the flow, called wave drag. This is the radiation condition, which is appropriate in a medium which extends to infinity. In the atmosphere two mathematical difficulties arise. The first is that the decrease of density upwards means that the waves slowly tend towards very large amplitude making the problem very non-linear. Thus it is usually supposed that the waves overturn or are absorbed by the very large viscosity at heights in the atmosphere above 150 km. The second difficulty arises when the velocity of the waves relative to the air is zero at some altitude. This is very common simply because it means that at that height the wind is zero relative to the obstacle setting up the waves. It is not as great an issue as the two-dimensional theory suggests because in three dimensions a rotation of the wind with height produces a zero wind relative only to an infinitesimal two-dimensional component of the obstacle.

Nevertheless, in the two-dimensional case energy is being transmitted upwards, and cannot pass through the level of zero velocity (see Section 5.17). This energy must therefore be added to the air at this level of zero velocity and must manifest itself as energy resident in the motion of that air, namely vorticity, which is turbulent. There is no acceleration of the flow between this level and the ground, but the air here must be subject to an upstream acceleration due to the momentum it absorbed, and the mean velocity profile modified as a result.

7.7 APPLICATION OF TURBULENCE THEORY TO THE ATMOSPHERE

Reynold's equation is only valid if there is a range of eddy sizes in the atmosphere which contain almost no energy, otherwise without such a gap in the spectrum the definition of the mean value is not applicable. This difficulty appears to have been circumvented by the general assumption first that whatever the turbulence is like, its effects can be represented by an eddy transfer coefficient (K-theory), and secondly that the motion obeys the same equations as for a molecular transfer coefficient, but with a coefficient 10^3 or 10^4 times as great. However, the problem of defining the mean is not circumvented by not mentioning it, and the coefficient has no meaning without that definition. Nor does the representation by the coefficient give any information about how the coefficient varies in space, with height in particular, nor about the rate of transfer in different directions. The rate of transfer in the atmosphere often varies by several orders of magnitude within a very few hours, and is very different for different transferred quantities. Thus momentum can be transferred by wave motions in a stratified fluid while pollution is not transferred at all. On the other hand, gravitationally produced turbulence may transfer material much more effectively than momentum (see Section 9.6). The application of eddy transfer coefficients to the atmosphere is a convenient way of describing mechanisms that are not understood, but it has little usefulness. Any situation in which it appears to be useful is likely to be one in which a prediction is made that a future occasion will be rather like a past one when the effects were measured.

An exception to this general condemnation of K-theory as being of little use is where new formal theoretical results are obtained. An example is Saffman's (1962) treatment of the effect of transverse shear aloft on the distribution of pollution downwind of a source. He calculated the path taken by most or all of the pollution reaching the ground at a point displaced sideways from the line downwind of the source. This could be via the upper layers of air with turbulent transfer upwards and transverse displacement with the mean upper wind, followed by downward turbulent transfer to the ground, rather than by horizontal transfer by eddies across the mean wind direction near the ground. The formula has no practical use in the field.

The variety of forms of turbulence is legion, and their effects so different, that there can be no single theory for 'turbulence'. We study various mechanisms which are common in the atmosphere, and are often called turbulence, in the remaining chapters of this book.

7.8 TURBULENCE IN ROTATING FLUIDS

The question arises whether turbulence can produce effects in rotating fluids which are fundamentally different from effects of viscosity. The outstanding problem of how hurricanes are started is the most important practical question. A hurricane is a centre of low pressure caused by rotation, and the air moving into the vortex near the sea surface is cooled by the adiabatic expansion which causes a flux of heat and water vapour from the sea beneath which is warmer than this cooled air. The low pressure centre is therefore a source of heat, most of which appears in the release of latent heat in the production of copious rain. The heat source contributes to the maintenance of the low pressure, which is produced by the sucking down of the tropopause and the consequent presence of a core of warm stratospheric air (see Section 11.10).

Once the hurricane is started it is therefore easy to find mechanisms which will keep it going, but these mechanisms depend on the existence of a low pressure centre. Therefore we look for mechanisms which will cause the concentration of vorticity and thereby produce small vortices with low pressure centres.

The predominant feature of areas in which hurricanes begin is the presence of large cumulonimbus clouds. In addition to their vertical air currents which produce rain, the air from which the rain is released spreads out horizontally at the tropopause with speeds of the order of 5-10 m sec^{-1}. At the same time downdrafts are produced by rain and these spread out at the sea surface. Large parcels of air are therefore transferred horizontally by these anvils (the spreading out cloud tops), and these may be studied in a crude way by mixing length theory.

A mass of fluid rotating initially as a solid annulus between radii a and b with angular velocity Ω, subject to the radial displacement of parcels of fluid which are subsequently mixed with their new surroundings, would achieve a final distribution of velocity in which all parcels had the same moment of momentum about the axis. This happens in a tea cup, where the radial displacements are caused by the secondary flow in the boundary layer at the bottom (see Section 3.8), but in that case there is a decrease in the total kinetic energy and angular momentum of the fluid.

In the case we are envisaging there is an energy source in the buoyancy forces produced by the release of latent heat in the clouds. Work is done by gravity, and when the cloud anvils spread out at the tropopause (or at some lower altitude) they possess kinetic energy of horizontal motion which is used up in moving across the isobars against the stable stratification due to the rotation. As the radially displaced parcels become mixed into their new surroundings the final state of zero vorticity is achieved, in which the circumferential velocity v is equal to k/r. Since all the forces due to buoyancy are vertical and due to mixing are internal, the total moment of momentum remains constant and consquently

$$\tfrac{1}{2}\pi\Omega(b^4 - a^4) = \pi k(b^2 - a^2) \tag{7.8.1}$$

or

$$k = \tfrac{1}{2}\Omega(b^2 + a^2)\,. \tag{7.8.2}$$

Initially $v = r\Omega$, finally $v = k/r$, and $\partial p/\partial r = v^2/r$; from which we find that, if Δp_1 and Δp_0 are the final and initial pressure differences between radii a and b, with unit density

$$\Delta p_1 = k^2 \frac{b^2 - a^2}{2b^2 a^2}, \qquad \Delta p_0 = \tfrac{1}{2}\Omega^2 (b^2 - a^2)\,; \tag{7.8.3}$$

or

$$\Delta p_1 = \tfrac{1}{4} \frac{(b^2 + a^2)^2}{b^2 a^2} \Delta p_0 \triangleq \tfrac{1}{4} n^2 \Delta p_0 \tag{7.8.4}$$

if $b = na >> a$.

The vorticity was originally distributed uniformly with magnitude $\tfrac{1}{2}\Omega$; finally it takes the form of two vortex sheets at $r = a,b$, each sheet containing half the original vorticity, assuming that the fluid inside $r = a$ and outside $r = b$ has not been altered.

There thus exists the possibility that convection clouds confined to a finite region might generate a centre of very low pressure in the middle by concentrating half the vorticity there. An interesting experiment to investigate this matter was devised by McEwan (1976).

The difficulty of testing this possibility arises mainly because there is no information to indicate whether the displacements in anvils are indeed primarily in a radial direction in the region of convection. In the next section we show that without some preferential orientation there is no effect.

7.9 MIXING LENGTH THEORY APPLIED TO A ROTATING FLUID

We assume that the flow is circular (Fig. 7.9.i) with circumferential velocity $v(r)$. A particle is now given additional initial radial and tangential velocities u_0', v_0', and moves in the environmental pressure field for time τ, after which it mixes with its new environment. The velocity anomaly which it possesses after time τ determines the effect of the mixing on the mean motion. This is a momentum mixing theory, and follows Prandtl's model. The stress is represented by $\overline{u'v'}$ and it is assumed that the velocity anomalies are small. The pressure gradient $\partial p/\partial r$ is equal to v^2/r.

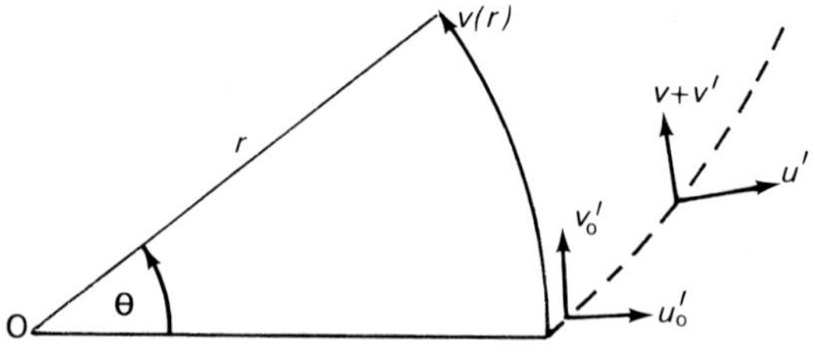

Fig. 7.9.i The dashed line shows the track of a particle displaced by being given an initial perturbation of velocity (u'_0, v'_0).

The equations governing the motion of the particle are Bernoulli's equation and the conservation of angular momentum, namely

$$\frac{\partial}{\partial r}\tfrac{1}{2}(\mathbf{v}+\mathbf{v}')^2 + \frac{v^2}{r} = 0 \tag{7.9.1}$$

and

$$v + v' = k'/r \tag{7.9.2}$$

where k' is a constant for the particle in its free trajectory in the pressure field of the mean flow. We differentiate (7.9.2) and then eliminate k'. Since

$$\mathbf{v}+\mathbf{v}' = (u', v+v') \tag{7.9.3}$$

we deduce that

$$(v+v')\frac{\partial}{\partial r}(v+v') + u'\frac{\partial u'}{\partial r} + \frac{v^2}{r} = 0$$

and

$$\frac{\partial}{\partial r}(v+v') = -\frac{k'}{r^2} = -\frac{v+v'}{r},$$

so that

$$-\frac{(v+v')^2}{r} + \frac{v^2}{r} + u'\frac{\partial u'}{\partial r} = 0.$$

Therefore

$$u'\frac{\partial u'}{\partial r} = v'\left(\frac{2v}{r} + \frac{v'}{r}\right) \tag{7.9.4}$$

and

$$\frac{\partial v'}{\partial r} = -\left(\frac{v+v'}{r} + \frac{\partial v}{\partial r}\right) = -\left(\zeta + \frac{v'}{r}\right), \tag{7.9.5}$$

where

$$\zeta = \frac{\partial v}{\partial r} + \frac{v}{r} = \text{vorticity of the mean flow.} \tag{7.9.6}$$

The stress may be written as the time mean value of

$$u'v' = u_0'v_0' + \Delta r\left[\frac{\partial}{\partial r}(u'v')\right]_0 + \tfrac{1}{2}(\Delta r)^2\left[\frac{\partial^2}{\partial r^2}(u'v')\right]_0 + 0(\Delta r)^3 \tag{7.9.7}$$

and

$$\Delta r = u_0'\tau + \tfrac{1}{2}\left[u'\frac{\partial u'}{\partial r}\right]_0 \tau^2 + 0(\tau^3) \text{ after time } \tau \tag{7.9.8}$$

because $\partial u'/\partial t = u'\partial u'/\partial r$. Thus we obtain

$$u'v' = u_0'v_0' + \tau\left[u'\left(u'\frac{\partial v'}{\partial r} + v'\frac{\partial u'}{\partial r}\right)\right]_0$$

$$+ \tfrac{1}{2}\bar{\tau}^2\left[v'\left(\frac{2v}{r} + \frac{v'}{r}\right)\left(u'\frac{\partial v'}{\partial r} + v'\frac{\partial u'}{\partial r}\right) + u'^2\left(u'\frac{\partial^2 v'}{\partial r^2} + 2\frac{\partial u'}{\partial r}\frac{\partial v'}{\partial r}\right.\right. \qquad (7.9.9)$$

$$\left.\left. + v'\frac{\partial^2 u'}{\partial r^2}\right)\right]_0 + 0(\tau^3).$$

We assume that the eddies arc symmetrical about $r = \text{const}$ and $\theta = \text{const}$, which means that

$$\overline{u_0'v_0'} = 0\,; \qquad (7.9.10)$$

furthermore, on taking the mean at $t = 0$, $\overline{u_0'} = \overline{v_0'} = 0$,

and so v'^2u', v'^3, $\dfrac{v'^3}{u'}$, $\dfrac{v'^4}{u'}$, $\dfrac{v'^5}{u'}$ all vanish in the mean at $t = 0$.

Therefore $$\overline{u'u'\frac{\partial v'}{\partial r}} = \overline{-u'^2\left(\zeta + \frac{v'}{r}\right)} = \overline{-u'^2}\zeta, \qquad (7.9.11)$$

$$\overline{u'v'\frac{\partial u'}{\partial r}} = \overline{v'^2\left(\frac{2v}{r} + \frac{v'}{r}\right)} = +\overline{v'^2}\,\frac{2v}{r}, \qquad (7.9.12)$$

and $$v'\left(\frac{2v}{r} + \frac{v'}{r}\right)\left(u'\frac{\partial v'}{\partial r} + v'\frac{\partial u'}{\partial r}\right)$$

$$= \left(\frac{2v}{r}v' + \frac{v'^2}{r}\right)\left[-u'\left(\zeta + \frac{v'}{r}\right) + \frac{v'^2}{u^2}\left(\frac{2v}{r} + \frac{v'}{r}\right)\right] \qquad (7.9.13)$$

$$= 0 \text{ in the mean at } t = 0.$$

Also we find that

$$u'^2 u' \frac{\partial^2 v'}{\partial r^2} = -u'^3\left(\frac{\partial \zeta}{\partial r} - \frac{\zeta}{r} - \frac{2v'}{r^2}\right),$$

$$u'^2 \frac{\partial u'}{\partial r}\frac{\partial v'}{\partial r} = -2u'v'\left(\frac{2v}{r} + \frac{v'}{r}\right)\left(\zeta + \frac{v'}{r}\right), \tag{7.9.14}$$

and

$$u'^2 v' \frac{\partial^2 u'}{\partial r^2} = u'v'\left[-\frac{v'^2}{u'^2}\left(\frac{2v}{r} + \frac{v'}{r}\right)^2 - \left(\zeta + \frac{v}{r}\right)\left(\frac{2v}{r} + \frac{v'}{r}\right) + v'\left(\frac{\partial}{\partial r}\frac{2v}{r} - \frac{\zeta}{r} - \frac{2v'}{r^3}\right)\right],$$

all of which vanish in the mean at $t = 0$ because they are composed of odd powers of u'.

Consequently $$\overline{u'v'} = \left[-\overline{u_0'^2}\zeta + \overline{v_0'^2}\frac{2v}{r}\right]\tau + 0(\tau^3). \tag{7.9.15}$$

Calculation also shows that the lowest powers of u' and v' in the term in τ^3 are

$$\frac{4v\zeta}{3r}\left[\overline{u_0'^2}\zeta - \overline{v_0'^2}\frac{2v}{r}\right]\tau^3, \tag{7.9.16}$$

the higher terms being of the order of v'/v in comparison with those retained. Since this has the same factor as the term in τ, no new phenomena are introduced by retaining it because it will vanish too if the term in τ vanishes. We may now consider various cases.

(1) Isotropic eddies: $\overline{u_0'^2} = \overline{v_0'^2} = q'^2$.

$$\overline{u'v'} = -q'^2 \tau e_{r\theta} \tag{7.9.17}$$

where $e_{r\theta}$ = fluid deformation rate $= \dfrac{\partial v}{\partial r} - \dfrac{v}{r} = \zeta - 2\dfrac{v}{r}$, and this is zero only in solid rotation. Thus the eddies will produce a stress as long as the rotation is not solid, and so they have the same effect as viscosity.

(2) Eddies composed of radial displacements only: $\overline{v_0'^2} = 0$.

$$\overline{u'v'} = -q'^2\tau\zeta, \tag{7.9.18}$$

and the stress vanishes only when $\zeta = 0$, which is when $v = k/r$. In order to achieve the result envisaged in Section 7.8, therefore, the eddies must produce predominantly radial rather than tangential displacements of fluid parcels.

(3) Eddies composed of tangential displacements only: $\overline{u_0'^2} = 0$.

$$\overline{u'v'} = \frac{2v}{r} q'^2\tau = (\zeta - e_{r\theta})q'^2, \tag{7.9.19}$$

and the stress only vanishes if r is very large. This may be taken to imply that eddies of this kind, in which parcels are pushed in both directions along the tangents to the flow, tend to straighten out the flow, presumably because all particles are displaced radially outwards as a result of displacements along the tangents. Certainly no effect on the profile $v(r)$ is indicated. ζ can only be equal to $e_{r\theta}$ when the flow is straight.

(4) Rectilinear flow: $r = \infty$.

In this case ζ is equal to the shear and we obtain the classical result of mixing length theory, namely

$$u'v' = -q'^2\tau\frac{\partial v}{\partial r}, \tag{7.9.20}$$

the mixing length being $q'\tau$.

7.10 GENERATION OF VORTICES BY EDDIES

We see from the cases described in Section 7.9 that with isotropic eddies, namely anvils which spread out equally in all directions, the region in which they are generated tends to be made to rotate like a solid. If its total angular momentum is not changed and it is not initially rotating like a solid, the consequence will be to alter the velocity at the boundary of the region and generate a vortex sheet there. Since air in the trade winds does not originate recently all from the same latitude and the intensity of convection is not uniform, it is to be expected that vortex sheets, and individual vortices as a consequence, will be generated on the boundaries of regions of intense convection.

The eddies might be made non-istropic by the alignment of the convection clouds into streets so that they are more nearly two dimensional, the main horizontal displacements being perpendicular to the lines of cloud. Any deformation of the air tends to draw out a compact region of air into a long line so that it is not surprising that clouds in the tropics are usually observed to be arranged in lines. Streets also tend to form along the wind shear, usually not far from the wind direction. Even a block of air in rectilinear flow which is uniform relative to the earth possesses the earth's rotation in addition, and so the state of zero vorticity may be approached. This means that in the northern hemisphere a uniform stream with lines of cumulonimbus along it could generate zero absolute vorticity in itself by accelerating the fluid on the left and decelerating that on the right. This could generate vortex sheets, and subsequently individual vortices along the boundaries of the region of convection.

Attempts have been made in the laboratory to produce results of this kind, but it should be appreciated that the outcome must either depend very much on the amount of anisotropy in the horizontal velocities generated in the eddies, or upon large spatial variations in the intensity of the eddies (either isotropic or anisotropic).

A criticism of this crude and simple analysis is that it does not appear to conserve the vorticity of the parcels displaced by the eddies. As in the case of classical mixing length theory however, the ultimate state which will not be changed is the same whether the parcels conserve either momentum (or moment of momentum in this case) or their vorticity, because the final state has uniform or zero vorticity.

7.11 ELASTIC PROPERTIES OF VORTEX TUBES IN TURBULENCE

A vortex is capable of transmitting waves along its length. To study this property we take the simplest possible case. With cylindrical polar coordinates (r,θ,z) we suppose a fluid velocity $(0,r\Omega,0)$ in the undisturbed state where $\Omega = \Omega(r)$, and add a small perturbation velocity (u,v,w). The equations of motion in the three directions for uniform density are:

$$\frac{\mathrm{D}u}{\mathrm{D}t} - \frac{(v + r\Omega)^2}{r} = -\frac{1}{\rho}\frac{\partial p'}{\partial r} = \frac{1}{\rho}\frac{\partial p_0}{\partial r} - \frac{\partial p}{\partial r}, \tag{7.11.1}$$

$$\frac{\mathrm{D}}{\mathrm{D}t}(v + r\Omega) + \frac{u(v + r\Omega)}{r} = 0, \tag{7.11.2}$$

$$\frac{\mathrm{D}w}{\mathrm{D}t} = -\frac{1}{\rho}\frac{\partial p'}{\partial z} = -\frac{\partial p}{\partial z}, \tag{7.11.3}$$

where for convenience we have represented the perturbation pressure by ρp, thus the pressure is

$$p' = p_0 + \rho p. \tag{7.11.4}$$

If there is a perturbation proportional to exp i (kz-σt), corresponding to waves travelling along the axis of rotation with velocity σ/k, then with axial symmetry,

$$\frac{\mathrm{D}}{\mathrm{D}t} = \frac{\partial}{\partial t} + u\frac{\partial}{\partial r} + w\frac{\partial}{\partial z}$$

$$= -i\sigma \text{ for the perturbation linearised,} \tag{7.11.5}$$

$$= u\frac{\partial}{\partial r} \text{ for the undisturbed motion.} \tag{7.11.6}$$

The linearized equations are

$$i\sigma u + 2\Omega v = \frac{\partial p}{\partial r} \tag{7.11.7}$$

$$-i\sigma v + u\frac{\partial}{\partial r}(r\Omega) + u\Omega = 0 \tag{7.11.8}$$

$$i\sigma w = \frac{\partial p}{\partial z}, \tag{7.11.9}$$

and the equation of continuity is

$$\frac{\partial}{\partial r}(ru) + \frac{\partial}{\partial z}(rw) = 0. \tag{7.11.10}$$

Eliminating p, v, and w, we find that u satisfies

$$\frac{\partial^2 u}{\partial r^2} + \frac{1}{r}\frac{\partial u}{\partial r}\left[\frac{1}{r^2} + k^2\left(1 - \frac{4\Omega^2 + 2r\Omega\Omega'}{\sigma^2}\right)\right]u = 0 \tag{7.11.11}$$

where

$$\Omega' = \partial\Omega/\partial r. \tag{7.11.12}$$

If waves can be trapped in the tube and propagated along it, u must vanish on the axis, must also have an oscillatory form for small values of r so that it may rise to a maximum, and must have an exponential form for large r so that the perturbation may die away at infinity. The analysis is simplified if we write

$$\phi = r^{1/2}u, \qquad x = kr, \qquad S = 2\Omega/\sigma \tag{7.11.13}$$

and then in a region in which Ω is constant (7.11.11) takes the form

$$\frac{\partial^2 \phi}{\partial x^2} = \left(\frac{3}{4x^2} + 1 - S^2\right)\phi . \tag{7.11.14}$$

In a region of zero vorticity in which

$$\Omega = \frac{k}{r^2}, \qquad \Omega' = -\frac{2\Omega}{r} \tag{7.11.15}$$

we have

$$\frac{\partial^2 \phi}{\partial x} = \left(\frac{3}{4x^2} + 1\right)\phi . \tag{7.11.16}$$

The simplest case in which waves are trapped is one in which a region of solid rotation (constant Ω) is surrounded by irrotational motion extending to infinity. If the coefficient of ϕ in (7.11.14) for some values of r is negative and in (7.11.16) for large r is positive, there will exist solutions of the kind desired. This will happen if the region in which Ω is constant extends to large enough r. The larger the value of S (and Ω) the smaller the radius in which waves are trapped. It is required at the very least that $S^2 > 1$,

i.e.

$$2\Omega > \sigma \tag{7.11.17}$$

and so tubes of this kind cannot trap oscillations of frequency greater than the vorticity of their core (2Ω). Much smaller frequencies are trapped at a small radius which satisfies

$$S^2 > 1 + \frac{3}{4x^2} . \tag{7.11.18}$$

If we now imagine a turbulent shear flow, such as that shown in Fig. 7.11.i which is a shadowgraph of a jet emerging into still air, the shear represents a very small frequency disturbance, and so distortions of the ends of tubes which are visible as a tangled spaghetti will be propagated along them. If the end of a tube is pulled away at one end, a stretching wave will be propagated along it as if it were elastic. The pressure on its axis is reduced when the stretching occurs

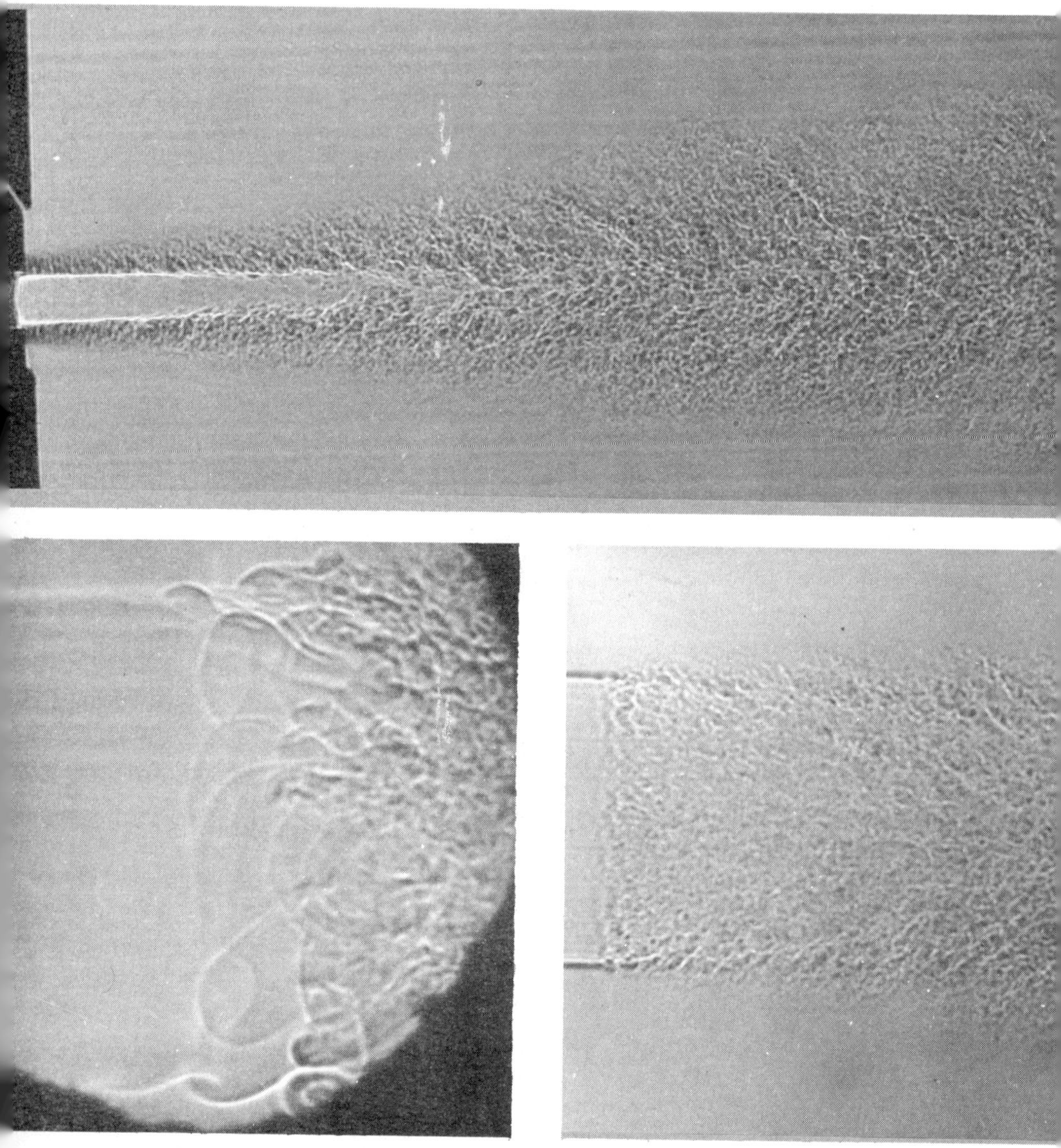

Fig. 7.11.i Shadowgraph of turbulence on the boundary of a high speed subsonic jet of hot air emerging from a circular hole. When the vortex sheet rolls up into a tangle of vortex tubes the most intense are those which are stretched most rapidly, and these are the ones which lie at 45° to the direction of shear. The lower photographs show the beginning of the conversion of the vortex sheet into billows and the vortices thereby generated into loops and tangles. [Photographs by courtesy of J. Mollo-Christensen, M.I.T.]

because of the increased rotation, and as the tension of the tube, represented by the force on a cross section of it, is increased. If such a stretching were relaxed the recently stretched tubes would provide a tension equivalent to a system of forces tending to restore the distortion just imposed on the spaghetti. Hence we would expect this turbulence to have viscoelastic properties since the tension would die away in being propagated along the tubes in an uncoordinated way because of the tanlged structure of the eddies.

In the picture 7.11.i the jet emerges surrounded by a vortex sheet which rapidly breaks down into a tangled mass of eddies. The most intense eddies are those which are most rapidly stretched in the shear zone, and these are inclined at 45° to the axis. The rapid stretching increases the rotation which decreases the pressure on the axis. There is a corresponding increase in volume of the eddy which is almost explosive in character, and which emits a sound pulse. In this way jet noise is generated. The theory of jet noise simply calculates the pressure fluctuations which are generated as if the fluid were incompressible, and then applies them to a compressible fluid. With the greatest fluctuations being in eddies aligned at 45 ° to the axis, the sound wave fronts will tend to be oriented in the same direction and propagated in the direction at right angles, and will travel predominantly out along the 45° cone from the orifice. This explanation assumes that there will be larger eddies of intermediate size which stretch the small eddies in bundles, and this can be seen to be happening in the picture.

The shadowgraph picture shows there to be a smallest scale of turbulence in which the large temperature gradient of the original vortex sheet is rolled up, initially in billows, and the core made uniform by molecular conduction. The most intense vortices have the lowest axial density because of the low pressure. This turbulence is clearly anisotropic because of the large shear.

Chapter 8

PARTLY TURBULENT FLOW

8.1 A RISING BUBBLE OF AIR IN WATER

The flow is partly turbulent when fluid is injected with momentum, or with buoyancy which creates momentum, into an otherwise undisturbed fluid environment.

Our first example is that of a void introduced into a uniform medium otherwise at rest. The word 'void' comes from the study of fluidised beds consisting of heavy particles suspended by their aerodynamic drag in an upward stream of fluid. As a whole the medium behaves very like a fluid of larger density. Thus a medium of sand particles suspended in an upward stream of air can have a free surface above which there are no sand particles and upon which larger bodies of smaller density than the sand and air combined can float as if on a liquid. In order to maintain the fluidised bed, air has to be pumped through small holes or through a porous surface at the bottom, and it is easy, particularly if the minimum flux of air needed to fluidise the bed is much exceeded, to cause bubbles of air to rise through the bed like a bubble of air in water. Such a bubble is called a void because it contains no particles of the bed. We now discuss a large air bubble in water.

The air in the bubble has a density about one thousandth of the density of water, and so the motion of the air inside the bubble produces negligible variations of the pressure inside. This pressure may therefore be regarded as a constant. If a steady bubble shape is maintained, the flow of the water around it must be such as to produce a constant pressure on the surface. We have seen (Section 2.11) that on the front of a body moving through a fluid the flow is approximately irrotational. We have also seen that the pressure on a sphere moving vertically through water of density ρ with velocity w is (e.g. (2.3.1) but including the gravity term)

$$p = p_0 - g\rho z - \frac{9}{8}\rho w^2 \sin^2\phi, \qquad (8.1.1)$$

where p_0 is the pressure far from the sphere, at the level where $z = 0$, and is a constant (see Fig. 8.1.i(a)).

Large bubbles are observed to be approximately spherical on top and to have a more or less flat bottom on which there is an irregular motion. They are lens-shaped, and the rim is in fluctuating motion although the upper surface is smooth (see Fig. 8.1.iii).

The effect of surface tension is not a significant factor in determining their overall shape when the bubbles are a few centimetres in diameter, but the rim of

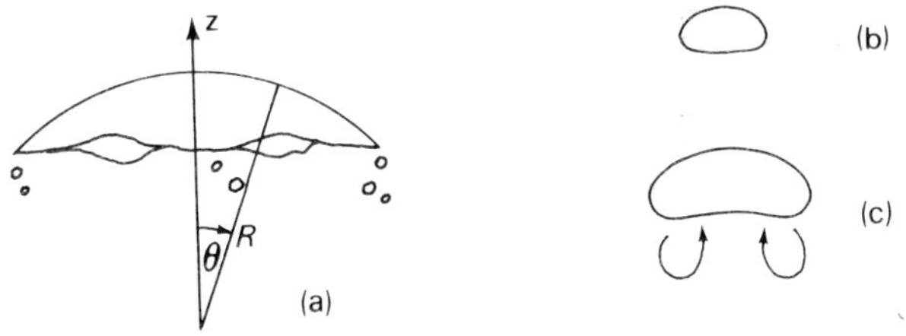

Fig. 8.1.i (a) The coordinates used for a large (10 cm say) spherical cap bubble. The base of the bubble is irregular and unsteady with small bubbles circulating in the wake.

(b) The shape of a small bubble (< 2 cm) for which surface tension is important and makes the periphery rounded.

(c) When the fluid viscosity is increased (Reynolds number decreased) a moderately large bubble has a rounded periphery and may have a concave lower surface.

those less than about 2 cm in diameter is noticeably rounded (Fig. 8.1.i(b)). If a more viscous liquid than water is used, the turbulence in the wake is reduced and the wake is made more nearly steady. It then appears to contain a ring distribution of vorticity beneath the bubble, and the base of the bubble is humped up in the middle (Fig. 8.1.i(c)).

The zero of z is put at the top of the bubble. In order for p to be equal to a constant, since

$$z = -R(1 - \cos\theta) \tag{8.1.2}$$

and
$$\sin^2\theta = (1 - \cos\theta)(1 + \cos\theta) \triangleq 2(1 - \cos\theta)\,, \tag{8.1.3}$$

roughly for θ less than about 30°, and very accurately for much smaller values of θ, it is seen from (8.1.1) that the pressure is a constant provided that

$$w = {}^2/_3(gR)^{1/2}, \tag{8.1.4}$$

where R is the radius of curvature of the upper surface of the bubble.

This formula for the velocity of rise of the bubble, first given by R. M. Davies and G. I. Taylor (1950), gives it in terms of the size of the bubble provided the angular aperture of the bubble (the value of θ at the rim) is known. Bubbles can be produced with apertures ranging from around 25° up to over 40°, and so the volume is not determined by knowledge of R alone.

Large bubbles with diameters up to around 20 cm can be made fairly easily by filling a dome-shaped saucer with air under water and allowing it to upturn so as to pour the air upwards.

These bubbles have many fascinating properties. They can be studied by causing them to rise through patches of dye in a water tank so that the motion in the wake is made visible. The wake is roughly parabolic in shape as illustrated in Fig. 8.1.ii, with a strong upcurrent up the middle where some particles move upwards at about twice the speed of the bubble. If a small bubble is released

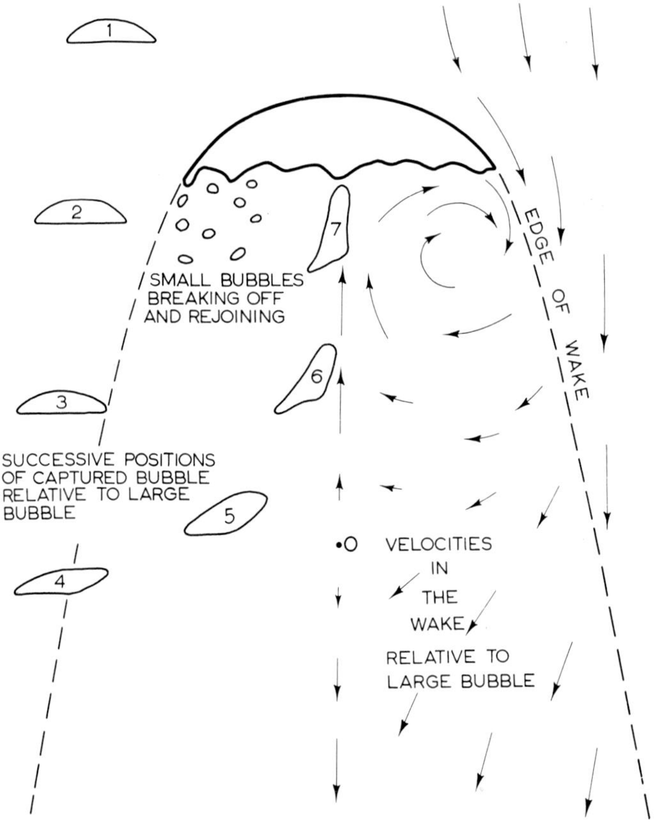

Fig. 8.1.ii On the left are shown 7 successive positions of a smaller bubble overtaken by an ascending larger one and a group of smaller bubbles which circulate close to the edge, sometimes breaking away from and sometimes rejoining the main bubble. On the right is shown the mean velocity around which there are fluctuations associated with the unsteady rim and base of the main bubble. The velocities are relative to the bubble and 0 is a stagnation point.

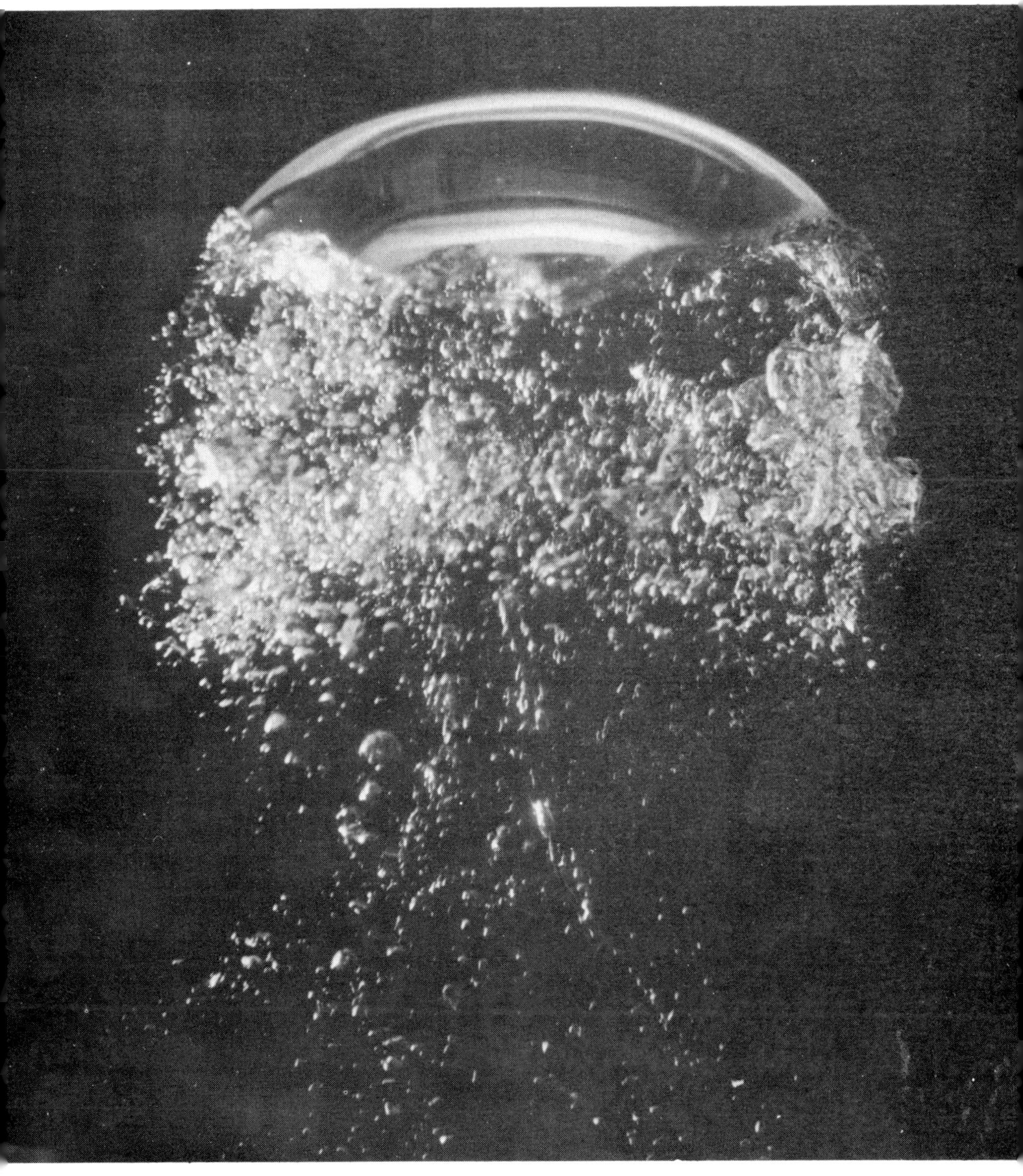

Fig. 8.1.iii A bubble of air of about 15 cm radius rising through water. The position of the small bubbles in the wake indicates that, although it is very turbulent, it is dominated by a vortex ring circulation.

ahead of a large one it can be seen first to be pushed aside as the large one overtakes it, but later its edge will impinge on the edge of the wake. Soon after that it is carried rapidly up the centre of the wake and is distorted in shape by the deformation of the water around it. When the velocity of the small bubble becomes large the importance of virtual mass is apparent, for a small bubble rising up the wake with a sufficient speed relative to the water can be observed to pass through the large bubble an emerge from its upper surface! This is in contrast to the behaviour of a bubble with small velocity relative to the water, which behaves as if it possessed negative mass (See Section 2.9).

If we wished to extend this analysis to a bubble of liquid of different density ρ' which is immiscible with water, we might expect a formula for the velocity such as

$$w = \tfrac{2}{3}\left(g\,\frac{\rho-\rho'}{\rho}\right)^{1/2} \tag{8.1.5}$$

where we have replaced g by the buoyancy force due to the density difference. This result may be invalidated by the inertia forces in the interior liquid, which produce a variable pressure on the bubble surface, by skin effects, or by mixing across the surface. In this book we are only concerned with fluids which are miscible, such as hot air in cool air or salt water in fresh water, and with cases in which the density difference is small.

8.2 AXISYMMETRIC THERMALS

A thermal is a body of buoyant fluid in a fluid with which it can mix. The name was given by glider pilots to those masses of warm air rising from hot ground warmed by sunshine in which they can soar. In the atmosphere buoyant convection produces a very complicated motion (Chapter 9), but here we are concerned with the behaviour of one isolated thermal in an otherwise undisturbed environment. Fig. 8.2.i shows successive pictures of a laboratory thermal.

A very simple theoretical treatment is sufficiently useful where more complicated theories have led to no more useful results, largely because thermals differ from one another quite considerably in a way that cannot be predicted.

The assumptions of the theory are justified by its success. We observe that a thermal in water made visible by a white precipitate or dye is apparently in turbulent motion and resembles a cumulus cloud which has a cauliflower-like appearance on its upper surface. We therefore assume that the molecular viscosity is of no importance. We also observe that the thermal grows in size as it moves vertically, but retains roughly the same shape. Because of the turbulence we suppose that the initial distribution of density in the thermal when it was small has a negligible influence on the distribution later on when it is much

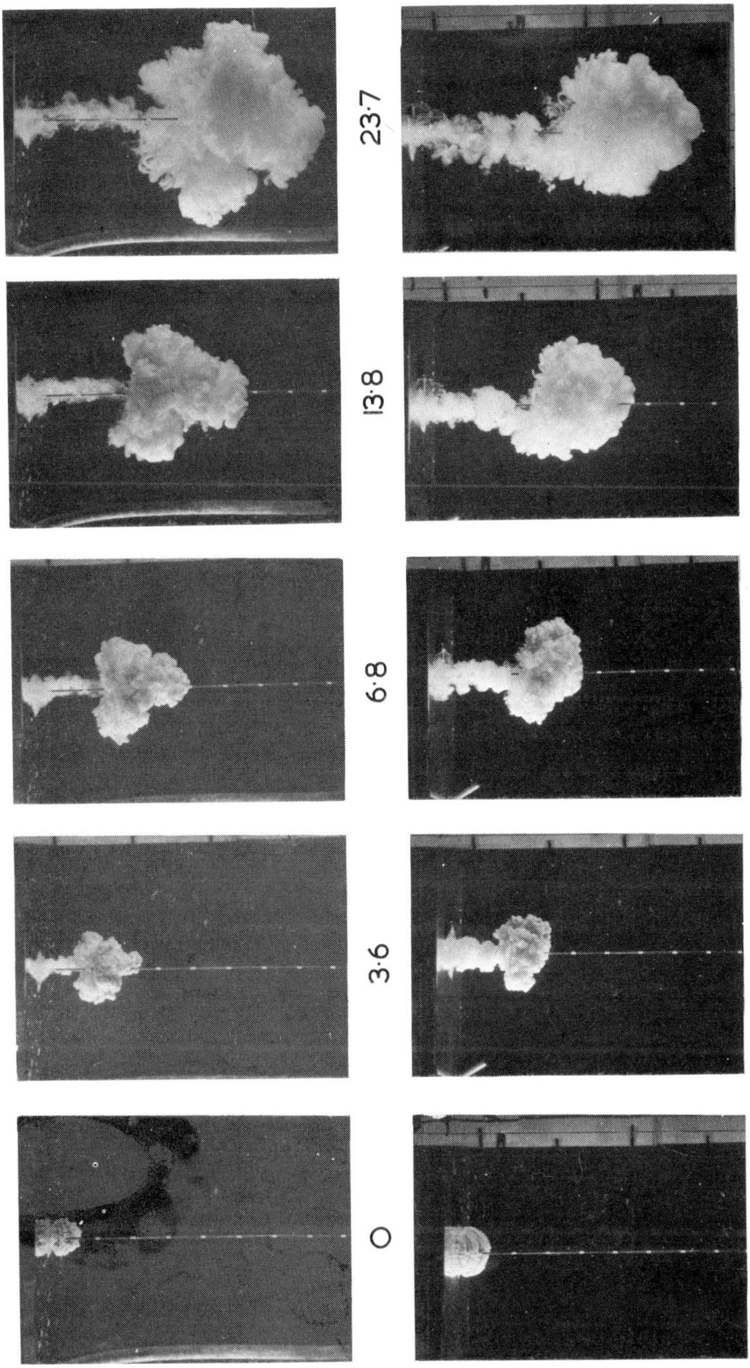

Fig. 8.2.i Two thermals of sodium sulphate solution descending through water, made visible by the addition of barium chloride which gives a white precipitate of barium sulphate. The column of material left behind is due to the manner of release from a spherical cup which is rotated, leaving some more dilute material near the surface. The verical rod is marked at intervals of 10 cm. Reflections can be seen in the top surface of the water. The numbers indicate time in seconds from the moment of release.

larger, and when some of the individual eddies can be seen to be larger than the thermal was at its creation or moment of release. It is reasonable to suppose, therefore, that the original configuration is forgotten (provided that it was compact) and that all thermals tend to take up the same shape, with the same mean velocity and density distributions. All the velocities, including typical eddy velocities, will then be determined by the total buoyancy and overall size of the thermal: thus to have the right dimensions the velocity must be of the form

$$w = C(gBR)^{1/2}, \tag{8.2.1}$$

where B is the average buoyancy such that if the volume is mR^3, the total buoyancy (or weight deficiency) is $gBmR^3$. C is a constant which depends on how w and R are defined. Thus if R is defined as the radius of the largest horizontal cross section and w is w_c, the vertical velocity of the cap or uppermost point of the thermal, we find from experiments and C is about 1.2.

Because the rate of growth of the radius, $\mathrm{d}R/\mathrm{d}t$, is a velocity, it must have similar mathematical form to w. Since the total buoyancy is constant,

$$gB = gB_0R_0^{\,3}/R^3, \tag{8.2.2}$$

where suffix 0 refers to the values at a particular moment and so we have

$$R\frac{\mathrm{d}R}{\mathrm{d}t} = C'(gB_0R_0^{\,3})^{1/2}, \tag{8.2.3}$$

or

$$R^2 \propto t, \tag{8.2.4}$$

the constant of proportionality being $2C'(gB_0R_0^{\,3})^{1/2}$. C′ is a different constant, namely that appropriate to the rate of increase of radius.

If z is the height of the uppermost point (the cap) of the thermal, by the same argument we have that

$$z^2 \propto t \tag{8.2.5}$$

and so we can write

$$z = nR. \tag{8.2.6}$$

This conclusion is a direct consequence of (8.2.1), which was obtained by a dimensional argument. If other factors such as viscosity, ν, influenced w, we could not arrive at such a simple result. This is because wR/ν, which is the Reynolds number, is dimensionless and so C, instead of being a universal constant, could be a function of the Reynolds number.

In (8.2.4) and (8.2.5) we have assumed that the constant of integration is chosen so that R and z are zero at $t = 0$. Consequently the thermal is behaving as if it once possessed an infinitesimal size and was growing up from a virtual point origin. This is clearly impossible, principally because the density cannot be negative, and so B defined by

$$B = \frac{\rho - \rho'}{\rho} \tag{8.2.7}$$

cannot exceed unity. Therefore the motion cannot possibly be extrapolated backwards beyond the size at which the total buoyancy is achieved by zero interior density.

On the other hand, we can imagine a negative thermal, namely a body of heavier fluid, to the density of which there is no theoretical upper limit. But in that case we would be no longer justified in making the Boussinesq approximation and introducing B only through g, because the motions of the fluid inside the thermal produce pressure forces proportional to the density. Since B is itself a non-dimensional number, in that case C could be a function of it, just as it might be of the Reynolds number. This analysis is therefore valid only if the Boussinesq approximation is valid and the accelerations of the fluid are small compared with g (see Section 2.12). Since the fluid accelerations are produced by the buoyancy forces in the first place all we need to ensure is that

$$B \ll 1 \tag{8.2.8}$$

In practice the temperature excesses in an atmospheric thermal are of the order of 1°C and so B is around 0.003. In a model experiment using saturated salt water to produce a negative buoyancy in fresh water, B is around 0.15 at first; but it soon becomes much less when mixing occurs. If alcohol, or some other fluid lighter than water and miscible with it, is used to produce positive buoyancy, again B is comparable with unity only at the very early stages. Thus it does not matter whether experimental measurements are made on thermals of positive or negative buoyancy. In the early stages the thermals are gradually assuming the final form which we have supposed they will take up, and measurements at that stage will be strictly relevent only to the particular case and the kind of apparatus used to produce it.

To conduct an experiment we have to release the thermal without disturbing the surrounding fluid. This is readily done with a hemispherical cup made of thin metal and partly filled with buoyant fluid, and which is pivoted about a horizontal diameter. The cup is then inverted rapidly so as to release a spherical-cap-shaped body of buoyant fluid. If it is released from the surface, the level of fluid inside and outside the cup must be made equal, otherwise a vertical impulse is given when the buoyant fluid is 'dropped' in. The cup must be overturned rapidly for otherwise a sideways impulse is given to the thermal as the fluid is poured from the cup.

Successive positions of the thermal can be recorded by cine photography and then traced successively on to graph paper, as in Fig. 8.2.ii. The value of

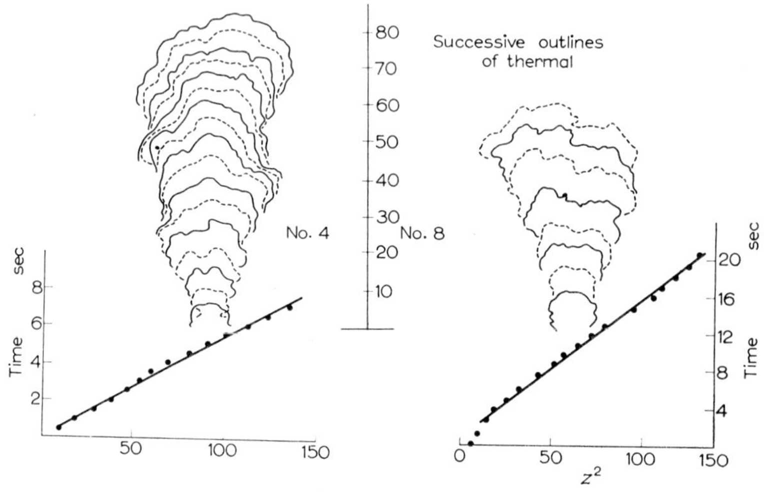

Fig. 8.2.ii Outlines of successive positions of the two thermals drawn from cine film. Below are plotted graphs of z^2 against time.

z at which $R = 0$ can then be determined. If we plot z^2 against t we find that the points lie on or very close to a straight line. z can be defined as the most forward visible point of the thermal, and it is observed that as successive knobbles on the surface of the thermal emerge from its surface the extreme point advances faster than the average. At the same time it is seen that the rate of advance decreases as the foremost knobble is displaced by the circulation of the motion towards the side. Thus the fluctuations from the straight line represent the turbulence in the thermal itself, though otherwise they are found to follow a straight line well. Using (8.2.6), equation (8.2.5) can be written

$$z^2 = \frac{2nC}{m^{1/2}}(gB_0mR_0{}^3)^{1/2}t = t/k \tag{8.2.9}$$

in which $gB_0mR_0{}^3$ is the total buoyancy, or weight deficiency. To check the validity of this theory experimentally, thermals of several different total buoyancies were measured and from each a value of k was determined. This was plotted against the total buoyancy. In the first set of experiments done (Scorer 1957) the total buoyancy varied from 1 to 45 gm wt so that k varied by a factor of about 7, and the quantity $2nC/m^{1/2}$ varied by up to 16% on either side of the mean value (Fig. 8.2.iii).

In order to see whether C is a universal constant we need to examine the variations of n and m. According to the rather simple idea that all thermals will tend to a universal shape, it is to be supposed that n and m are universal constants giving the distance from the virtual origin to the cap and the volume of the thermal as a multiple of a power of the horizontal radius. The shape factor m is difficult to measure. It is roughly equal to 3 because a thermal is flattened by comparison with a sphere for which the value is $4\pi/3$, and any variations which might be used will have a small effect on the calculations since it appears as $m^{1/2}$. On the other hand n is easy to measure and shows considerable variation from thermal to thermal, even though it usually appears to remain constant for the measurable life of a given thermal. In the early experiments it

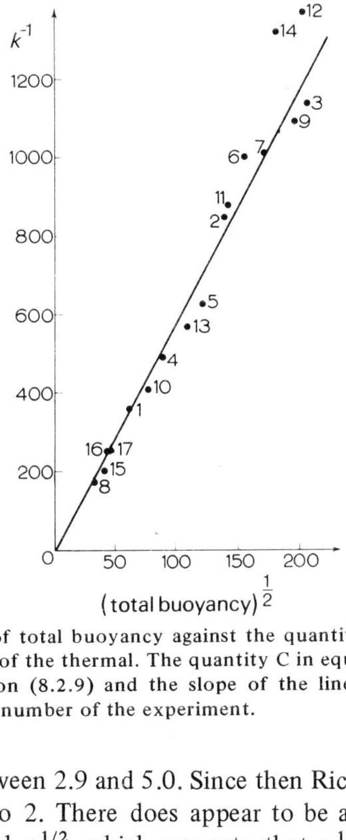

Fig. 8.2.iii A plot of total buoyancy against the quantity $1/k$ measured from the rate of advance of the thermal. The quantity C in equation (8.2.1) is determined from equation (8.2.9) and the slope of the line in this diagram. The number is the serial number of the experiment.

was found to vary between 2.9 and 5.0. Since then Richards has obtained values up to 7.3 and down to 2. There does appear to be a correlation between the value of $2nC/m^{1/2}$ and $n^{1/2}$ which suggests that $n^{1/2}C$ is roughly constant (Fig. 8.2.iv). A theoretical rationalisation of this is that instead of putting the volume equal to mR^3 as in (8.2.2) we should put it equal to mR^2z/n, in which case equation (8.2.1) becomes

$$w = C(gB_0R_0{}^2z_0n)^{1/2}/z = \frac{Cn^{1/2}}{m^{1/2}}(\text{total buoyancy})^{1/2}t, \qquad (8.2.10)$$

or
$$z^2 = 2\frac{Cn^{1/2}}{m^{1/2}}(\text{total buoyancy})^{1/2}t. \tag{8.2.11}$$

Figure 8.2.iii shows the manner in which k^{-1} was found to vary with total buoyancy $(gB_0mR_0{}^3)^{1/2}$. Figure 8.2.iv shows how the ratio of these quantities for each experiment was related to n, and the curve represents a function proportional to $n^{1/2}$. Some of the variation is thereby accounted for, but no smooth line representing a simple function can be drawn to pass closer to all the points.

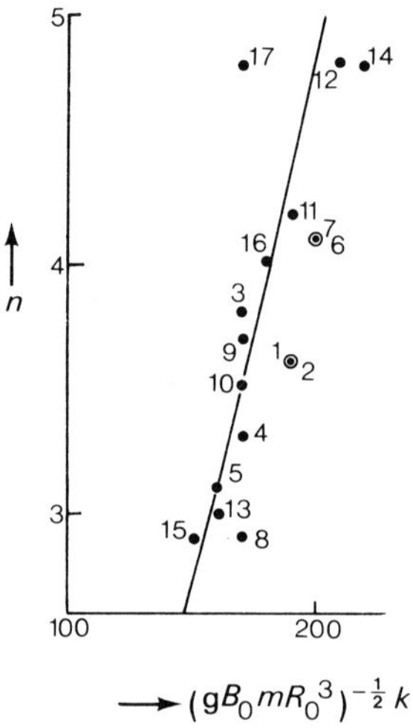

Fig. 8.2.iv The variation of the slope of the line from the origin to points in Fig. 8.2.iii plotted as a function of n. The curve represents a function proportional to $n^{1/2}$.

On the other hand, Richards (1961) added many more experiments to those included in Fig. 8.2.iv and found that the quantity $k^{-1}(B_0mR_0{}^3)^{-1/2}$ varied like $n^{3/2}$ over a greater range of values without increased scatter. His results imply that the constant C (in equation 8.2.1 and 8.2.10, for example) is not strictly constant and equal to about 1.2, but is more accurately given by

$$C = 0.36\,m^{1/2}n^{1/2} = 0.63\,n^{1/2}, \tag{8.2.12}$$

if m is taken to be about 3.

It is not very useful to take the variation in n into account anyway, because we do not at present have any very certain idea of what determines n. The points in Fig. 8.2.iii lie close to the line which gives

$$2Cn/m^{1/2} = 180g^{-1/2} \quad \text{CGS} = 5.7\,, \tag{8.2.13}$$

and with $n = 4$, $m = 3$, we thus obtain, approximately

$$C = 1.2, \tag{8.2.14}$$

or

$$Cn^{1/2} = 2.4 . \tag{8.2.15}$$

One significant conclusion from the results shown in Fig. 8.2.iii is that there is no tendency to deviate from a straight line through the origin in any consistent way. This may be cited as evidence that no viscous effects were important in the experiments, and since such measurements as have been made on clouds lead to the conclusion that the value of C is about 1 although their total buoyancy is about 10^{10} times as great and the medium was air, it appears that in all cases the turbulent effects dominated the viscous effects (see, for example, Malkus and Scorer 1955).

A Reynolds number for a thermal is wR/ν, which is a constant according to (8.2.10), since $z \propto R$. Thus the ratio of the inertia forces to the viscous forces is constant for a given thermal, and if the viscous forces are negligible at one stage in its life they are negligible throughout its life. It could be argued that they may be important in determining the internal velocity pattern without significantly affecting the parameters such as C, which refer to its overall behaviour. If, for example, a small drop of ink in a glass of water turned out to have the same value of C, this would not prove that it was turbulent. The belief that the laboratory thermal with total buoyancies of a few grams weight in water is similar to atmospheric thermals is justified because we can see the turbulent eddies in action.

8.3 THE PROPERTIES OF THERMALS

The value of n sometimes changes rather abruptly in the life of a thermal. A rather larger than usual protuberance may appear on one side and become detached by not being completely entrained into the circulation as it moves round to the rear. The value of n then decreases. The value of n sometimes increases when the thermal becomes widened by one or two large protuberances which do become completely entrained in the rear. There therefore appears to be no certainty that a thermal will retain its characteristic value of n even though it seems to have preserved it up to a certain moment.

The circulation of a thermal may be defined as the circulation around a circuit as shown in Fig. 8.3.i. This is composed of a section of its axis of symmetry through it together with a loop completing the circuit in the irrotational

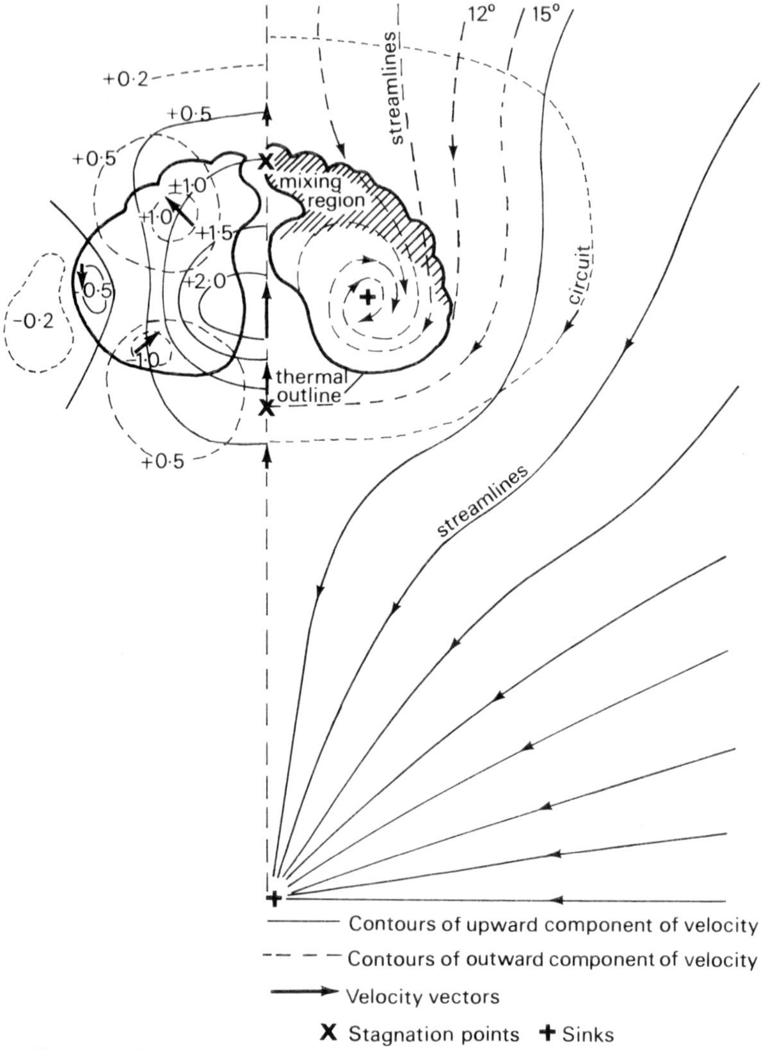

Fig. 8.3.i The right half of the diagram shows the flow of the fluid relative to the outline of the thermal which is kept fixed in size, but with turbulent protuberances moving around it and the flow up its axis wobbling from side to side as it decelerates towards the cap to form protuberances.

Fluid inside a cone of roughly 12° angle is entrained into the front, and between that and a cone of roughly 15° it is entrained into the rear. Fluid outside this is represented as moving into a sink at the origin. The fluid inside goes into a sink at the point where the fluid is instantaneously at rest relative to the thermal outline.

The circuit marked is completed up the axis and is used for defining the circulation of the thermal, which is constant.

On the left contours of upward and outward velocity components are shown expressed as multiples of the rate of the rise of the cap, and vector velocities are indicated at significant points.

flow outside where its position does not affect the magnitude of the circulation. The magnitude of this circulation which is given by the integral,

$$\oint_{OAPO} \mathbf{v.ds} \propto wR \tag{8.3.1}$$

is constant throughout the life of the thermal because $w \propto R^{-1}$ for any characteristic velocity. The circulation is only changed by the action of forces which produce accelerations around the circuit, as indicated by the first term on the right hand side of equation (1.1.15), and the only forces which can operate in this case are the buoyancy forces. Consequently this result implies that there are no buoyancy forces on the fluid up the axis of symmetry.

At first sight this seems to contradict the obvious fact that generally some buoyant fluid can be found on the axis somewhere. It can only be explained by the assumption that the eddies produce Reynolds stresses which act like viscosity, reducing the circulation so as to offset exactly the increase caused by buoyancy on the axis. An alternative way of expressing the same fact is to say that in a thermal a path must exist leading up through the middle of the thermal on which there is no buoyancy (i.e. no variation from the external density). The circulation measured using such a path, wherever it may be at the moment of measurement, is then a constant for the thermal.

Although a compact body of buoyant fluid is released, after a short time a hole through the middle appears, which can easily be seen when the buoyant material is made cloudy (Fig. 8.3.ii). A simple way to do the experiment is to use either milk or a fine white precipitate, easily obtained by adding a little barium chloride to a thermal prepared from sodium sulphate (used instead of the usual sodium chloride to make the water denser) so as to form a dense precipitate of barium sulphate. The adding of silver nitrate to a sodium chloride solution to form a precipitate of silver chloride is not recommended. Silver salts are expensive and silver chloride tends to flocculate and settle in the spherical cup if there is any delay in releasing the thermal while internal motions are allowed to die away.

It used to be thought that the turbulent protuberances on a thermal were produced primarily by the unstable density gradient there as seen in the first picture of Fig. 8.2.i, but this is in fact a very minor cause. In the first place, the most vigorous protuberances appear near the centre of the cap of the thermal where the unstable density gradient is least; and secondly, identical protuberances appear on the front of a puff, where there are no density gradients (see 8.6).

Initially a thermal is a compact body of warm air or other buoyant fluid. When it is released in a model experiment in water it quickly developes the cauliflower appearance on its surface, and this must be due to static instability.

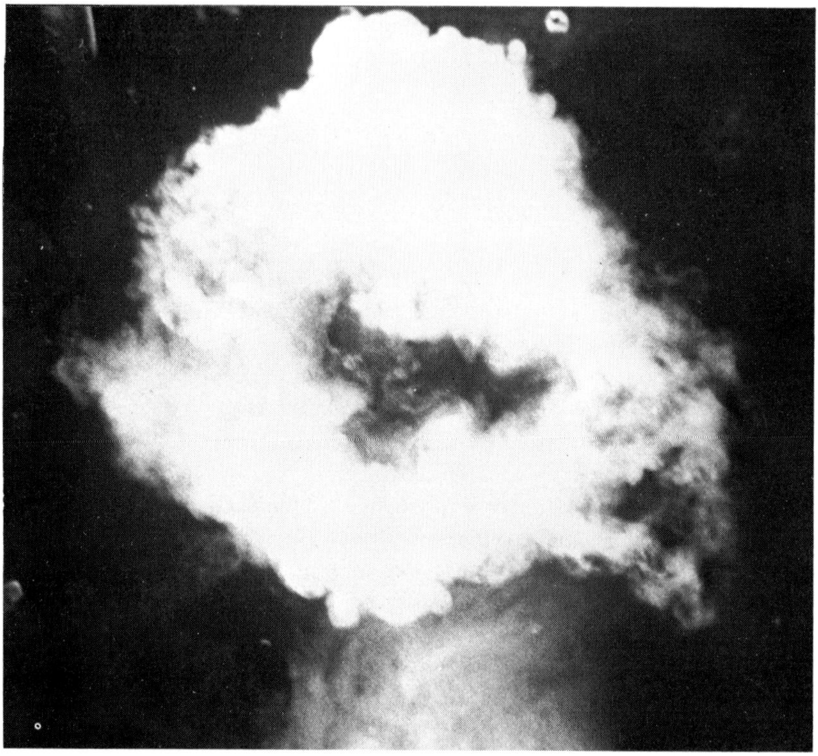

Fig. 8.3.ii A view of a thermal from above showing the doughnut shape with a clear view through the middle.

But the vorticity generated in the form of rings around the buoyant fluid, which cause it to penetrate through the surroundings, also cause fluid from the surroundings to penetrate from the rear through to the front so that the buoyant fluid forms a torus-like mass (see Fig. 8.3.ii). It differs from both a laminar vortex ring and a spherical vortex, as well as from anything in between. This is because the fluid penetrating through the middle becomes turbulent in the region of deceleration and diffluent curved flow, and mixes with the buoyant fluid. The fast moving stream through the middle meanders about in the diffluent region and draws buoyant fluid into the eddies. This stream therefore distributes itself all around the front of the thermal with protuberances of the mixture passing round the circulation first on one meridian, then on another.

Thus there is a turbulent region all over the front of the thermal and marked particles of fluid in its path can be seen to be mixed into the front. There is a cone of roughly 12° semi angle from the virtual origin of the thermal such that all particles inside it are mixed into the front of the thermal as it advances. Another cone of angle roughly 15° is such that all the particles between it and the smaller cone flow into the rear of the thermal as it advances. All particles outside this larger cone are not engulfed by the thermal. The angles of these cones must not be regarded as having unique values because they vary with the angle of spread of the thermal itself. This may pass up a cone of angle as small as 8° or as large as 26°, but is most commonly around 16°. About 60 per cent of the fluid entrained enters the front; but this is not to be taken as an exact figure, and individual cases could easily range from 50 to 70 per cent.

All thermals are thought of as being geometrically identical, the only differences being in size, R, and in the magnitude of their buoyancy, B, (which is expressed as a fraction of the weight of the environmental fluid it displaces). A cine film of one thermal could be projected to a size and with a speed to represent any thermal, and we may think of the thermal as being fixed in size with the world around it shrinking into the virtual origin or being entrained into the thermal. This representation is given in the right half of Fig. 8.3.i and in Fig. 8.4.vi.

8.4 FURTHER THERMAL EXPERIMENTS

A thermal turns itself inside out as it rises about 1½ diameters, which means that every parcel of its passes through the mixing region and becomes diluted when it travels that distance. Inside, the core shaped like a torus is continually displaced backward through the thermal because fluid is removed by mixing at the front and transferred to the rear.

If a particle inside a thermal has a very small fall speed relative to the surrounding fluid, it will remain inside because some exterior air below the thermal is always entrained into it. With a larger fall speed its position in the thermal would determine whether or not it would fall out the next time round. Woodward (1959) studied this question and her conclusions are summarized in Fig. 8.3.i and Fig. 8.4.vi.

A question of considerable interest in connection with the upward transport of pollution when the gases containing it are buoyant is how much will be carried through an 'inversion', as a density discontinuity is usually called in this context (see Fig. 8.4.iv). Richards (1961 and 1963 a and b) found the following results experimentally in water tanks. Y, called the yield, is the fraction of the material of the thermal which continues to ascend after passing through the inversion; S is the amount of stratification. It is expressed as the ratio of the density difference across the inversion to the average density difference which the thermal would have had from the surroundings at the level of the inversion if the inversion had not existed but instead the layer in which the thermal started had continued. For axisymmetric thermals it was found that (see Fig. 8.4.ii)

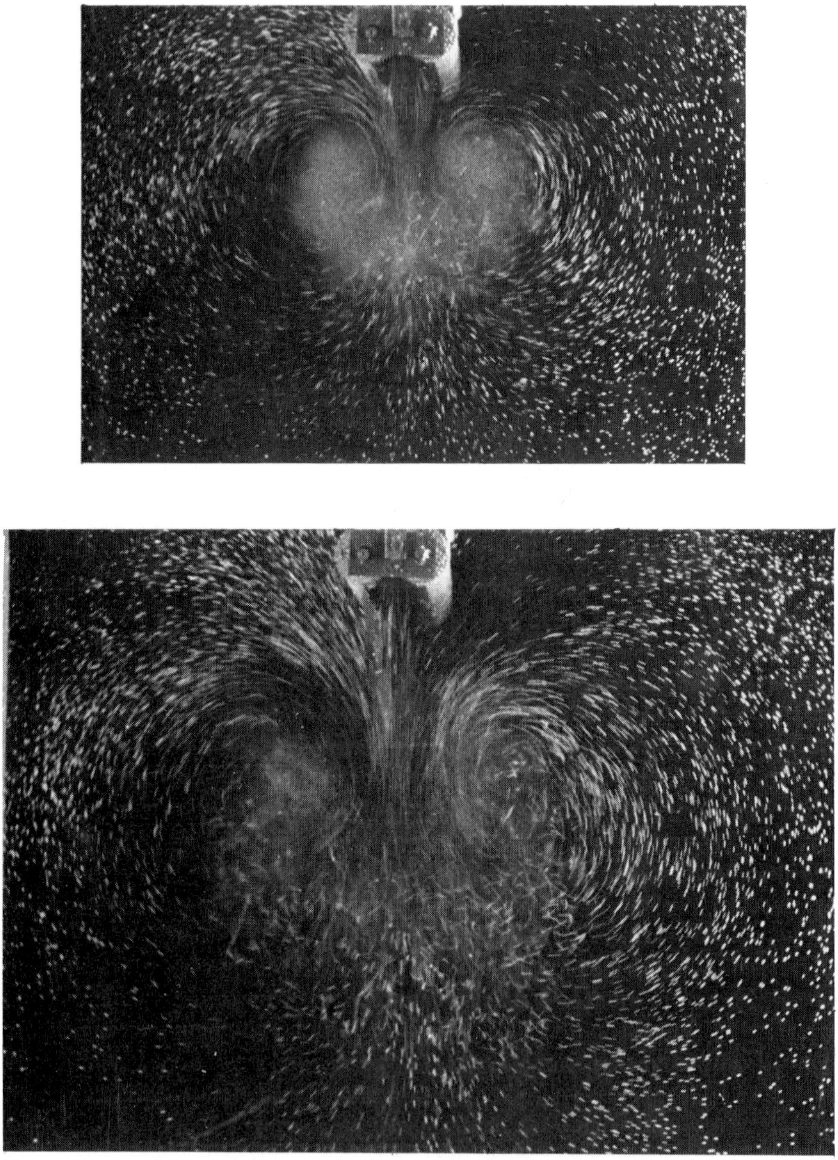

Fig. 8.4.i The motion due to a two-dimensional thermal revealed by a time exposure of particles illuminated in a section. The region of turbulence is marked by more erratic vectors, the region of buoyancy is marked by slight cloudiness, which is more evident in the upper photograph closer to the release trough.

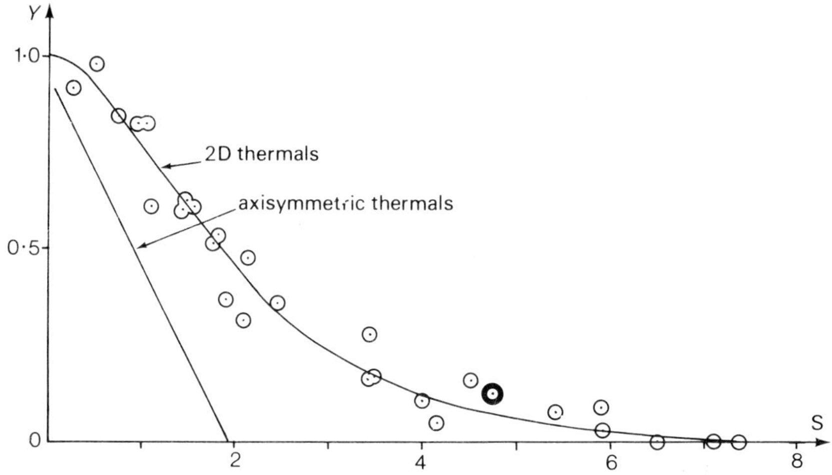

Fig. 8.4.ii Graphical representation of the experimental results of Richards (1963a) represented in equations (8.4.1 and 2).

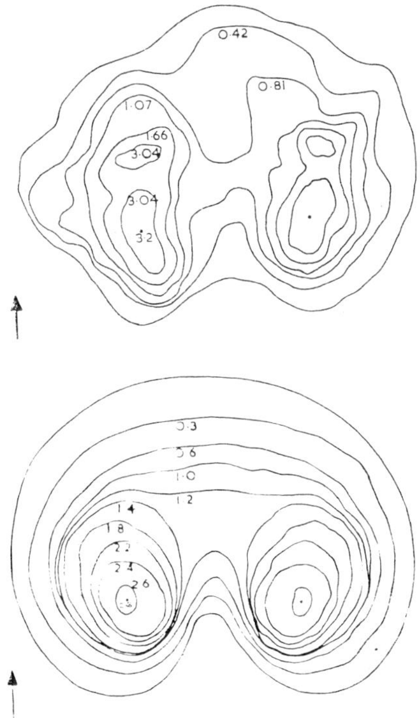

Fig. 8.4.iii Two typical mean cross sections of the density of a two-dimensional thermal expressed as multiples of the mean buoyancy, measured by Richards (1963a).

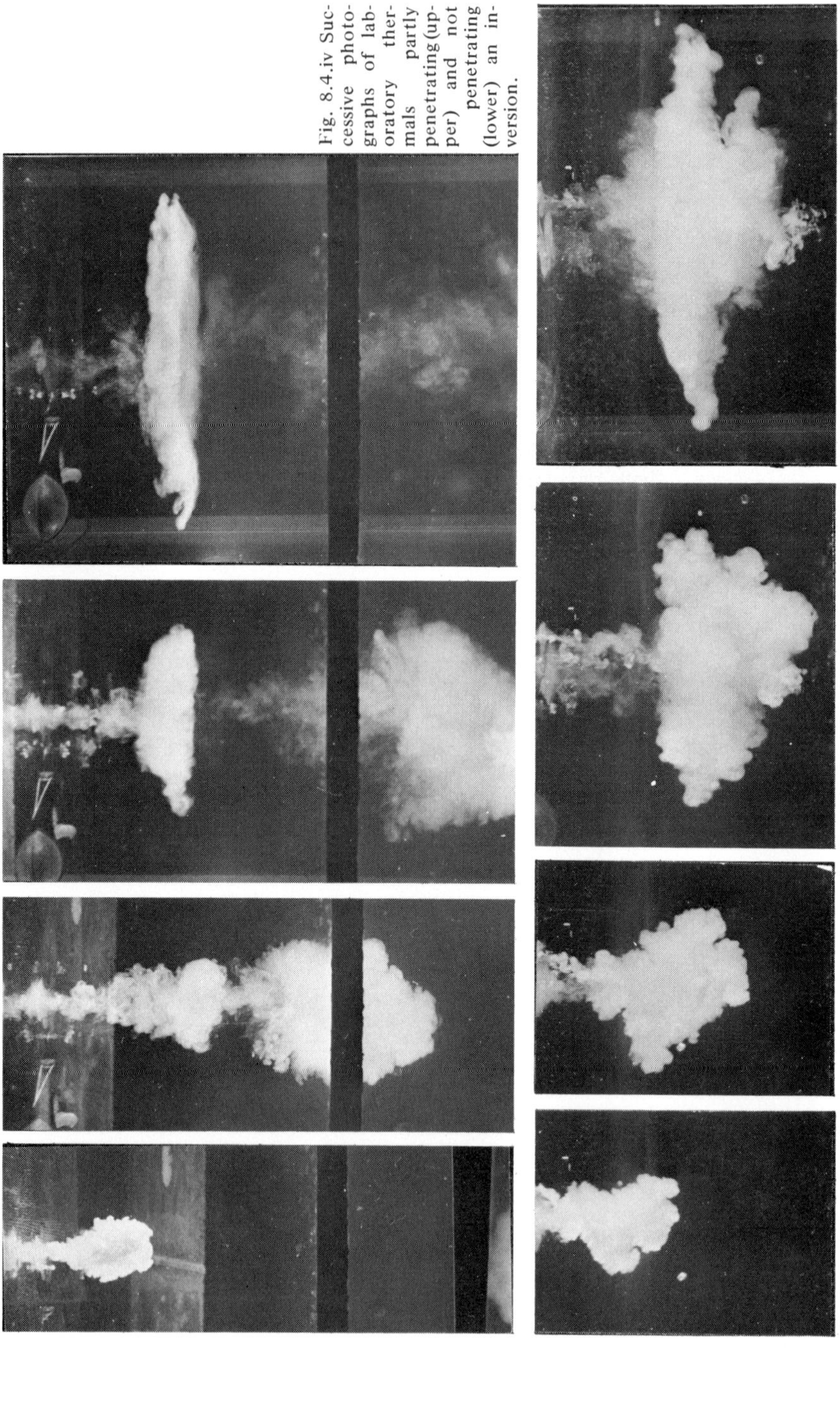

Fig. 8.4.iv Successive photographs of laboratory thermals partly penetrating (upper) and not penetrating (lower) an inversion.

$$Y = 0.95 - \tfrac{1}{2}S \text{ when } 0.1 < S < 1.9$$

$$Y = 0 \text{ when } S > 1.9 \qquad (8.4.1)$$

$$Y = 1 \text{ when } S = 0$$

and for 2D (i.e. cylindrical) thermals (see Section 8.8 for the 2D cases)

$$Y = (0.93\,S + 1)\exp(-0.93S), \qquad (8.4.2)$$

Y being zero for practical purposes when S exceeds about 7.5.

The distribution of density in a thermal is not uniform, and clearly a thermal will not penetrate a sufficiently strong inversion, but it will completely or almost completely pass through one sufficiently weak. These equations express what happens for inversions of intermediate strength. Axisymmetric thermals are less effective at penetrating, particularly when S exceeds unity. This is probably because in the 2D case the fast moving fluid in the centre of the thermal has about the average density of the thermal (see Fig. 8.4.iii) whereas in the axisymmetric case it has about the same density as the surroundings (see Fig. 8.3.ii). The forward momentum of fluid parcels is more closely correlated with their buoyancy in the 2D case. Furthermore, after penetrating, many of the 2D thermals became a row of separate fragments which would, of course, again amalgamate into a 2D thermal as they grew larger by entrainment. Thus the variations in density along the length of the cylindrical thermal act in favour of the most buoyant parts penetrating although the material would be stopped if it were uniformly distributed. This explains how some material can penetrate even when S is much larger than 2.

The flow pattern can in any case only be represented rather roughly because thermals differ from one another and have large fluctuations in their own motion. The chief cause of fluctuations is where the exterior fluid advancing up the middle from the rear meanders first one way and then another producing large knobbles, or protuberances, on the front cap. These are then displaced round the sides as their successors follow them. Each large knobble seems to have smaller knobbles on it with yet smaller knobbles on them, and so on, each knobble behaving rather like the larger one on which it sits. This appears to be a characteristic of the region of declerating motion because turbulent puffs (see Section 8.6) behave in a similar way without buoyancy.

Surrounding the thermal, which can be defined easily by making the fluid visible, is a region of irrotational motion produced as if by a body moving through it. There is no wake such as is found behind a rigid body because there is no drag on its surface. The velocity field outside can be crudely represented as due to a doublet of strength μ with potential $\mu \cos\theta/r^2$. This is the same as the irrotational flow (see Section 2.2) due to a sphere of radius a moving with velocity U, where $2\mu = Ua^3$. In the context of a thermal, the doublet strength is proportional to wR^3. Since wR is constant, the flow is becoming stronger outside at a given distance r from the centre of the thermal in proportion to R^2 which increases like t. This is because momentum is being added to the motion at a constant rate equal to the total buoyancy. Thus although the thermal itself is declerating ($w \propto t^{-1/2}$) it is getting larger and the exterior velocity at a fixed distance from its centre is increasing.

The velocity field as measured experimentally by Woodward (1959) is incorporated in Fig. 8.3.i. It can now be appreciated that because the motion, or any other feature of a thermal, can be displayed in relation to the outline of the buoyant fluid, this outline may as well remain fixed in a diagram. Fig. 8.3.i (right half) is drawn to show the motion of particles relative to this outline. Particles are either entering the thermal and becoming incorporated into it or are passing round it and entering a sink at the virtual origin of the thermal. It is not possible to define precisely in such a diagram a streamline which would separate particles entering the front from those entering the rear because the knobbliness of the thermal makes this boundary diffuse.

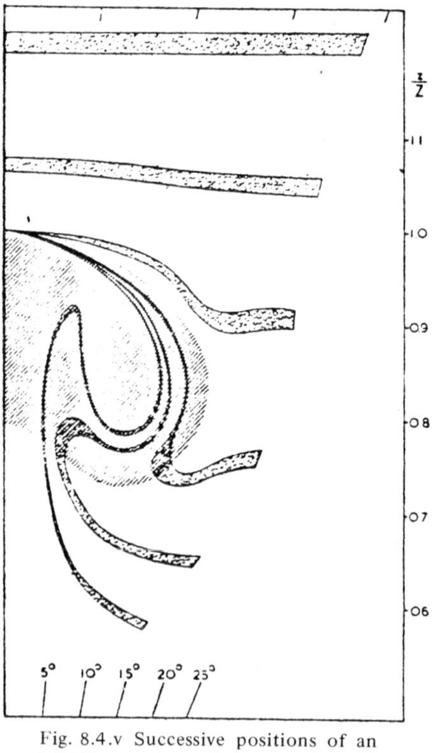

Fig. 8.4.v Successive positions of an initially horizontal layer of fluid above a thermal shown relative to the growing thermal (Woodward 1959).

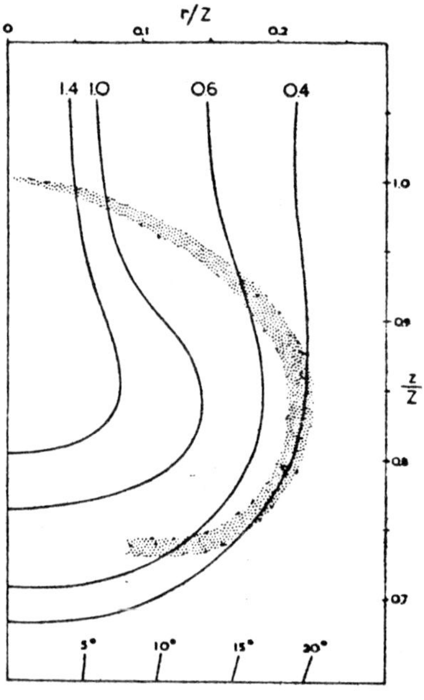

Fig. 8.4.vi The boundaries of the region outside which a particle with a given fall speed will fall out of the bottom of the thermal. The thermal outline is indicated by shading. The fall speeds are expressed as multiples of the velocity of rise of the cap of the thermal (Woodward 1959).

In using these diagrams it must be appreciated not only that the angles at which the thermals expand differ but so also does the shape. Thus a very flat thermal expanding up a cone of the same angle as a less oblate one would have a smaller value of n, because that relates the distance of the cap from the origin to the maximum horizontal radius.

In the upper part of the thermal near to the top boundary the motion is very unsteady. At the same time a horizontal surface of fluid through which the thermal penetrates is stretched around it, by the steady component of the motion, as shown in Fig. 8.4.v. Thus the mixing is achieved by the simultaneous mechanisms of penetrative eddy motion and stretching out of compact bodies of fluid. Duing the ascent of 2 or 3 diameters a single blob of fluid may be mixed into the whole body of the thermal.

Particles with a fall speed relative to the fluid in which they are immersed may be carried around the core of the thermal several times before they fall out of it. Woodward used her measurements to calculate Fig. 8.4.vi, which shows where a particle would have to be so as not to fall out of the bottom of the thermal. An interesting consequence of the inflow into the base of a thermal concerns those particles which are carried up by a thermal and will eventually fall out as the rate of its rise decreases. In descending through the inflow they are carried towards the axis. Thus they all fall back to the starting level close to the virtual origin, even though they may have been spread out over a much larger horizontal distance while they circulated within the thermal.

8.5 FLYING IN THERMALS

In the development of soaring, glider pilots have had a variety of pictures in their minds as they have exploited thermals in the atmosphere to gain height. At first they were thought of as more or less spherical masses of air called bubbles, which rose through the environment. If a glider pilot was lucky enough to enter near the top he would then be able to sink down through it until he emerged from the bottom. Thereafter he would no longer be able to gain height with it.

Since solid bodies passing through air have turbulent wakes, it was thought that thermals also had wakes, and as a pilot emerging from the bottom would feel turbulent air this accorded with the pattern of experience. It fitted with the idea that a thermal was rather like an air bubble in water as well. This analogy saw all the buoyancy as being at the top with a ring vortex behind this lens-shaped buoyant mass in which smooth updraughts could sometimes be found. Below was a turbulent wake with the upward velocity gradually decreasing at greater distance below the thermal itself. The analogy with the air bubble was modified by saying that, since the warm air can be mixed with the environment, mixing (clearly seen to take place on the top of a cumulus cloud), would occur and would erode the bubble. This would cause it to decrease in size and eventually waste away. This theme is developed in the chapter on clouds (Chapter 11). Such an eroded thermal would have a vortex ring-like wake with turbulence below it.

It was a common experience that on first entering a thermal finding its centre by circling and manoeuvring to obtain the maximum updraught appeared to be easy. Turbulence was experienced when it became difficult to centre on the thermal and find the best updraught. Some pilots believed thermal upcurrents to be elongated like a sausage in the vertical direction, but that particular idea has not had much support. This is because no one has ever produced a buoyant mass of fluid which could retain such a shape: the lower part always caught up with the upper part and made a more spherical shape. The question whether thermals can be plumes of warm air (see Section 8.10) is discussed in Chapter 9.

When laboratory experiments on thermals were performed it became obvious that the turbulence was on the top (the front) and that there was no wake below (in the rear) at all. At about this time the conventional picture, with its strong upcurrents near the top and a turbulent wake with weak upcurrents below, was being questioned because the less experienced pilots at gliding championships had acquired the habit of watching the champions and, when they circled, of circuling under them. Instead of having the advantage over novices the champions soon found themselves surrounded by them, unable to get away. This was remarkable because when the champions came into the turbulent region it appeared that the others had a stronger upcurrent lower down, although they should have been beyond the thermal altogether.

The argument was then advanced that thermals tended to travel in groups, in series up the same path for some reason, and that the novices below who seemed to be doing so well were in the next thermal of the series. At the same time, since thermals are due to rising hot air, it was supposed that temperature measurements would reveal vertical velocity to be well correlated with temperature excess. Some pilots fitted thermometers on their wingtips and tried to locate updraughts by turning towards higher temperatures indicated by a sensitive dial showing the temperature difference between the wingtips. They met with no success that could not be ascribed to a greater awareness of vertical velocity differences between the wingtips, and there were many frustrating failures: frustrating because pilots continued to believe that the method must be successful since excess temperature and strong updraught should go together.

It fact the most rapidly rising air is the coolest of all, because the environment of thermals is slightly stably stratified (see Section 9.4). Hence that air which has been displaced upwards the greatest distance either ahead of the thermal or by entrainment up its centre is coldest.

All the experiences of pilots except those gained near to the ground were resolved and put in place by the thermal model which emerged from the laboratory experiments. It is rather funny that pilots used to speak of having reached the top of a thermal when it became turbulent, meaning that they had gone as high as they could in that one and imagining that they were sinking out of the bottom of it. They had in fact reached its top.

Captain N. Goodhart (1958) pointed out the very important fact that in the region of confluence, where the horizontal indraught in a thermal is as strong as the updraught, a pilot circling with his wings tilted at 45° to the horizontal will be able to use the indraught component as an updraught. Equally, in the diffluent region, which is the turbulent part of the thermal, he will regard the outflow as if it were a downdraught when circling. He will also find it very easy to centre on the lower part of the updraught because of the inflow. However, above the middle of the thermal it will be more difficult to do so, even though the vertical component of the velocity of the air is the same as below.

In using a thermal the pilot is most likely to discover it by entering near the bottom. When he banks to circle in the upcurrent he will find that it appears to increase on account of the confluence. By contrast it will tend to disappear if he circles around it at the top, in the diffluence. He exploits the thermal by rising in the upcurrent, whose maximum strength is equal to about 2.2 × the rate of rise of the thermal. When the air becomes turbulent it is time to look for another thermal because he cannot continue to ascend at more than about half the former rate: at the same time those circling below in the same thermal will be able to use it to great advantage.

In making use of the diagram showing the velocity distribution in a thermal it must be appreciated that in the upper part, where the velocity is fluctuating, it may be possible to soar momentarily near the cap in an upcurrent twice as strong as the rate of rise of the cap, and that this upcurrent may quickly fade as it emerges from the top. The magnitudes given in Fig. 8.3.i represent averages for the meandering upcurrent.

If all thermals were like the model they would be very intense and very small near to the ground, and would have a velocity which decreases like z^{-1} upwards. In fact there must be modifications both near the ground where the thermals originate and higher up where they must overlap as they increase in size. These are discussed in Section 9.3.

The exploitation of thermals by birds and insects is discussed in Sections 12.2, 3, 4, and 6.

8.6 PUFFS

A puff is a turbulent body of fluid ejected into an otherwise undisturbed environment. A vortex ring is not a puff because it is not turbulent. Assuming that there is a unique configuration and that the total momentum remains constant, we must have the following relationships between its forward velocity w, its size R, and the distance travelled z:

$$wmR^3 = \text{constant} = w_0 m R_0^3 \tag{8.6.1}$$

$$\frac{dR}{dt} \propto w = \frac{dz}{dt}. \tag{8.6.2}$$

From this it follows that

$$z = nR \tag{8.6.3}$$

$$\frac{w}{w_0} = \frac{R_0^3}{R^3} \tag{8.6.4}$$

and

$$z^4 = 4w_0 R_0^3 n^3 t \tag{8.6.5}$$

The suffix 0 denotes values at a particular time, mR^3 is the volume, and n is a constant (at least for each puff) defining the angle of spread. The above argument can be presented in this abbreviated form because we have knowledge of the consequences of the relevant suppositions gained from our study of thermals. There is a dimensional argument hidden away in (8.6.1), which includes the supposition that viscosity is of no importance.

The circulation is proportional to wR, which is proportional to z^{-2}. Consequently the circulation, and therefore the Reynolds number, is decreasing like $t^{1/2}$. This means that before long the viscous forces will take over and the flow will become predominantly laminar.

There are two other important features of puffs which must be mentioned. Richards (1965) measured the velocity distribution in and around puffs made in water in the laboratory and found that it was the same as in a thermal, to within the accuracy of measurement possible and meaningful in such a situation. The overall appearance is also very similar, there being a mixing region at the front as in a thermal. Typically $m = 3$ and $n = 4$, but n varied between 1.7 and 6.3, which is about as much as in the case of thermals. The pattern of motion, therefore is not significantly dependent on how it is produced, for the thermal looks the same even though it is decelerating less because the buoyancy is feeding in new momentum. The rapid internal circulation and mixing ensure the continual redistribution of the momentum.

The second feature is that they are difficult to produce in the laboratory because it is easier to produce a vortex ring from a circular orifice. At the edge of an orifice there is separation of the flow and a vortex sheet streams away from it. If a short collar-shaped piece of vortex sheet is emitted it quickly rolls up into a vortex ring, especially if the length emitted is small compared with the radius of the ring. Richards emitted a slug of fluid longer than the width of the orifice from a rectangular orifice, and in this way ensured that the flow was turbulent. Other configurations are possible, but two things must be remembered. Some distance is needed to form the model puff, and the motion will soon become laminar because of the rapid decrease of Reynolds number. Good puffs are made by the firing of guns.

8.7 BUOYANT PUFFS

If we assume that the configurations of a thermal and a puff are precisely the same, it may be expected that a puff possessing buoyancy will in due course become a thermal if the original momentum is vertical because no change in form is required. In that case the momentum is of the form

$$kwmR^3 = kw_0mR_0^{\ 3} + gB_0mR_0^{\ 3}t \qquad (8.7.1)$$

where k is a number which gives the momentum as a multiple of the velocity multiplied by the volume. Using (8.6.2) we therefore obtain

$$z^4 = 4n^3R_0^{\ 3}\left(w_0t + \frac{1}{2k}gB_0t^2\right) , \qquad (8.7.2)$$

which for small t is like (8.6.5) but for large t is like (8.2.11). The circulation and the Reynolds number are proportional to wR, which in terms of time is given by

$$wR = \frac{n^{1/2}R_0^{\ 3/2}}{2}\frac{w_0 + \dfrac{gB_0}{k}t}{\left(w_0t + \frac{1}{2}\dfrac{gB_0}{k}t^2\right)^{1/2}} \qquad (8.7.3)$$

This decreases when t is small like $t^{-1/2}$, but tends to the limit $n^{1/2}R_0^{\ 3/2}g^{1/2}B_0^{\ 1/2}/(8k)^{1/2}$ for large t.

For the study of puffs emitted with buoyancy at an angle to the vertical the reader is referred to the work of Richards (1968, 1970).

8.8 TWO DIMENSIONAL THERMALS AND PUFFS

An exactly analogous treatment may be used to study two-dimensional cases. A two-dimensional thermal can be created in a laboratory by having a cylindrical trough mounted on a horizontal axis which is quickly inverted after being loaded with buoyant (heavy) liquid. It is found that it is better to have either a divided trough or two parallel troughs which tip their fluid towards each other in order to avoid imparting a net sideways momentum to the fluid.

As for the axisymmetric thermal, considering unit length, we now have

$$z = nR \tag{8.8.1}$$

and

$$w = C(gBR)^{1/2}, \tag{8.8.2}$$

with

$$BR^2 = B_0 R_0^{\,2} \tag{8.8.3}$$

for the conservation of total buoyancy. This leads to

$$\frac{dz}{dt} = C(ngB_0 R_0^{\,2}/z)^{1/2}$$

or

$$z^{3/2} = \tfrac{3}{2}C(ngB_0 R_0^{\,2})^{1/2} t\,; \tag{8.8.4}$$

and since

$$w \propto z^{-1/2} \tag{8.8.5}$$

the circulation and the Reynolds number are proportional to

$$wz \propto z^{1/2} \propto t^{1/3}, \tag{8.8.6}$$

which increases with distance and time. Two-dimensional thermals therefore remain turbulent, and if not initially fully turbulent will become so.

The general configuration of 2D thermals is very similar in cross section to axisymmetric thermals and puffs, except that the angle of spread is greater, a typical value of n being 2. At the same time there is no complete penetration to the front of exterior fluid entrained at the rear; otherwise there could be no buoyancy forces operating along the axis of symmetry to increase the circulation according to (8.8.6). Richards has found that for 2D thermals n may range between 1.3 and 3.2 while for non-buoyant 2D puffs it ranges between 2.6 and 5.7, with typical values of about 2 and 4 respectively.

Figure 8.4.i shows a 2D thermal with its motion pattern made apparent by taking a time exposure of bright particles in an illuminated cross section.

The 2D puff satisfies the following:

$$z = nR \tag{8.8.7}$$

and

$$wR^2 = w_0 R_0^{\,2}, \tag{8.8.8}$$

according to which

$$z^3 = 3n^2 w_0 R_0^{\,2} t \tag{8.8.9}$$

and
$$wR = nw_0R_0{}^2z^{-1} \propto t^{-1/3}, \tag{8.8.10}$$

showing that the Reynolds number decreases, although more slowly than in the case of the axisymmetric puff.

For a 2D buoyant puff the equation analogous to (8.7.1) is

$$kwmR^2 = kw_0mR_0{}^2 + gB_0mR_0{}^2t \tag{8.8.11}$$

from which
$$z^3 = 3nR_0{}^2\ wt + \tfrac{1}{2}\frac{gB_0}{k}t^2 \tag{8.8.12}$$

and a calculation for the Reynolds number shows a decrease followed by an indefinite increase.

Without a specific experiment we have no assurance that a thermal which is very slow because it originated as a puff which had been slowed down and which became laminar at low Reynolds number will necessarily become turbulent again. However, the implications of this argument are that it will do so when it becomes large enough, and the Reynolds number of the eddies becomes large on this account.

8.9 JETS

The style of analysis is now established and we may proceed quickly with the cases of different kinds of continuous source of momentum or buoyancy.

The axisymmetric jet of fluid entering an otherwise motionless fluid of the same density is easily observed to be conical. This is inevitable if the only factors determining the motion are the momentum flux and the distance from the orifice. We may be at distances so far from the orifice that all trace of its shape and of the velocity profile across it have been lost because of the eddies, and the eddies may be intense enough for the viscous effects to be negligible. If these two provisions are met we must find that the width R is proportional to distance from the virtual origin, and that the eddy size characteristic of the motion at any point must also be proportional to it. Therefore

$$z = nR. \tag{8.9.1}$$

Likewise all velocities at a given distance bear the same relationship to one another as at all other distances. If fluid emerging from the orifice is marked (see Fig. 8.9.i), a cone, whose radius we take to define R, will be seen clearly setting the boundary beyond which marked fluid does not penetrate. Outside it the motion is irrotational, even if it can be observed to be unsteady near to the cone. Inside the cone the average motion and pattern of fluctuations is similar at all distances: this is a consequence of the simplicity of the situation which excludes

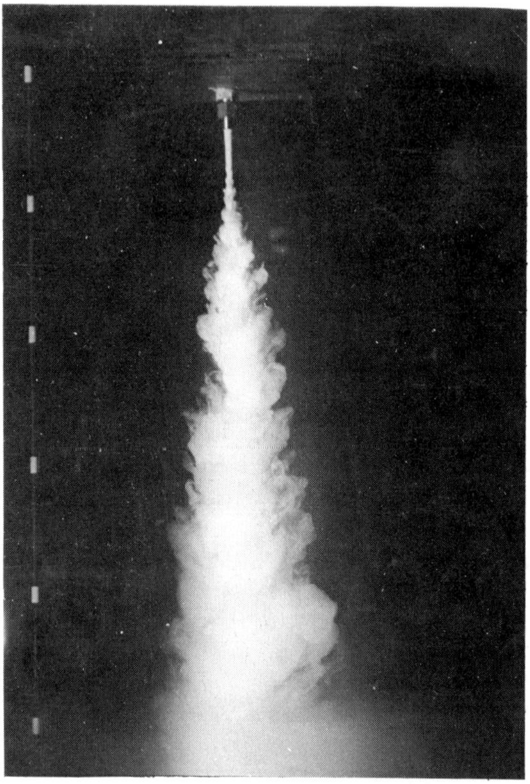

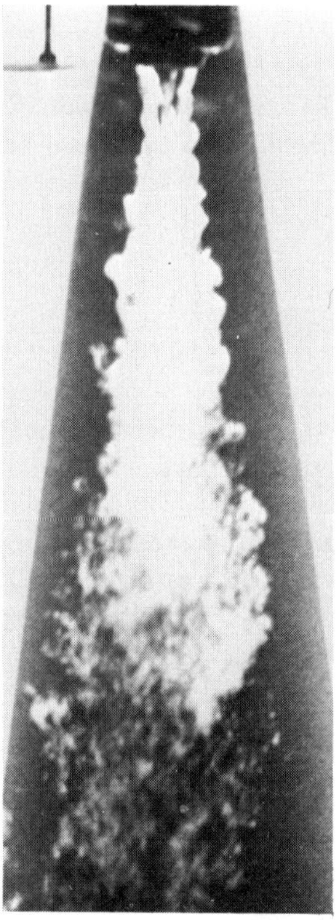

Fig. 8.9.i Buoyant fluid emerging from a nozzle into an otherwise undisturbed tank of water. Notable features which are referred to in 8.9–13 are

(1) the slight convergence of the fluid as it moves away from the nozzle according to (8.13.7);

(2) the beginning of instability when the velocity reaches the value given by (8.10.1), the constant being given in the table at the end of the chapter;

(3) the growth along a cone with eddy size increasing in proportion to the cone width;

(4) the shape of the eddies indicating a more rapid flow up the axis of symmetry.

A jet possessing no buoyancy but emitted with large momentum from the orifice has a very similar appearance except that the turbulence begins close to the nozzle.

In this case the fluid was salt water containing $BaSO_4$ preciptitate, having a larger density than the fresh water in the tank, but the appearance would be the same with upward motion and buoyancy. The scale of this experiment is indicated by the 10 cm intervals on the left of the picture.

The photo on the right is of a plume of colourless salt water. Its background is a dark wedge, but the illumination from each side is refracted so as to reveal the plume. This shadowgraph shows the converging stage of a wider plume emerging with a velocity lower than that appropriate to its buoyancy.

any influence of compressibility. Thus the analysis is not valid for supersonic jets.

Since the momentum flux is proportional to the square of the velocity and the cross section area

$$w^2 z^2 = w_0^{\,2} z_0^{\,2} = \text{constant}, \tag{8.9.2}$$

or

$$w \propto z^{-1}. \tag{8.9.3}$$

The Reynolds number is equal to

$$\frac{wR}{\nu} = \text{constant}, \tag{8.9.4}$$

but the circulation (measured up the axis, out along a radius and back to the origin through the irrotational fluid) is not a useful concept because it is logarithmically large at $z = 0$. The constancy of the Reynolds number means that if the jet is turbulent near the orifice it will remain turbulent up to infinity (unlike a puff).

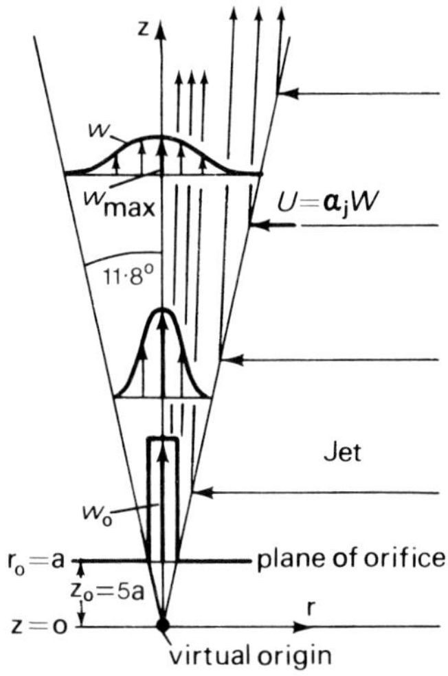

Fig. 8.9.ii Coordinates used to describe a jet. At the orifice the velocity has the 'top-hat' profile, but this quickly becomes bell-shaped and is fairly well represented by a gaussian or cosine curve. On the right are shown the mean streamlines. The virtual origin, at the vertex of the enveloping cone is about 5 orifice radii from the orifice.

The inflow velocity, denoted by U at the boundary at distance R, may be assumed to be horizontal. For dimensional reasons it is proportional to w and therefore to z^{-1} and R^{-1}. This accords with the continuity of momentum flux across a section normal to the axis, for the increment is zero for a displacement dz in the direction of the axis. It also accords with the equation of continuity, for the increment in upward flux of volume must equal the horizontal flux inwards through depth dz. Thus (see Fig. 8.9.ii)

$$\mathrm{d}(wR^2) \propto 2\pi RU\mathrm{d}z. \tag{8.9.5}$$

We use the proportionality sign because we have not defined precisely what velocity is represented by w.

The flow is fluctuating from time to time and from place to place so that we may write

$$w = \overline{w}(r,z) + w'(r,z,t) \tag{8.9.6}$$

where $\overline{w}$ represents the time-average value at a point. w has a characteristic profile at each value of z defined by the function $w_1(r)$, so that

$$\overline{w} = w_1(r)z^{-1} \tag{8.9.7}$$

and

$$w' = w_1{}'(r)z^{-1}f(t). \tag{8.9.8}$$

observation shows that the vertical velocity falls to around zero at any point from time to time, therefore w' is of the same order of magnitude as $\overline{w}$. The consequence of this is that the time-averaged momentum flux, which is

$$\overline{\int_0^R \int_{\theta=0}^{2\pi} w^2 r\,\mathrm{d}\theta\,\mathrm{d}r} = 2\pi \overline{\int_0^r (\overline{w}^2 + w'^2 + 2\overline{w}w')\mathrm{d}r},$$

$$= 2\pi \int_0^R (\overline{w}^2 + \overline{w'^2})\mathrm{d}r, \tag{8.9.9}$$

$$= \dot{M}_m + \dot{M}_f$$

where θ is the azimuth in the plane $z = \text{constant}$. $\dot{M}_m$ and $\dot{M}_f$ are the momentum fluxes attributable to the mean and fluctuating motions respectively. It is assumed that the time averaging is done over a long enough period for all averages to be independent of θ, so that $\overline{w'} = 0$. Since w' is of the same order of magnitude as $\overline{w}$ the expression (8.9.9) is very dependent on the eddies.

If it be supposed that the jet contains some sort of passive pollution of concentration σ whose flux is constant up the jet, then

$$w\sigma R^2 = \text{constant}, \tag{8.9.10}$$

and so

$$\sigma \propto w \propto R^{-1} \propto z^{-1}. \tag{8.9.11}$$

It appears therefore that pollution is diluted in exactly the same manner as momentum, but without making measurements there is really no justification for any assumption that the profiles of w and σ are the same. On the contrary, it is clear that in the case of a puff the correlation is very poor because the fastest moving fluid on the axis is unpolluted. The assumption that momentum is diluted in the same way as any quantity such as pollution attached to the fluid is known as Reynolds's analogy. It is only justified as far as dependence upon z of the quantities is concerned, and it implies nothing about correlation between σ' and w'. If we write similar equations to (8.9.7 and 8.9.8) for σ, the pollution flux is seen to be

$$2\pi \int_0^R (\overline{w}\,\overline{\sigma} + \overline{w'\sigma'})\mathrm{d}r. \tag{8.9.12}$$

In this expression the second term contributes a smaller fraction of the whole than the second term in (8.9.9) because w' and σ' may be not well correlated. On the other hand $\overline{\sigma}$ and $\overline{w}$ are observed to have very similar profiles (i.e. are proportional to roughly the same function of r).

The motion outside the jet may be assumed to be horizontal if the orifice is in a plane wall so that there is a constraint present not contributing any dimensional factors. In that case the radial inward velocity takes the value

$$u = UR/r \tag{8.9.13}$$

at all values of z for $r > R$. This ensures that the pressure field can exist to produce this motion, the isobaric surfaces being cylinders. Inside the jet the time average streamlines are approximately as shown in Fig. 8.9.ii, and have been measured by several workers since Schmidt (1941) first embarked on this kind of analysis.

The form the streamlines take where the orifice is not in a rigid plane boundary capable of supporting this pressure distribution has not been analysed. This analysis is valid for the case of an orifice at the end of a long tube of negligible diameter. Any other configuration such as an orifice at the tip of a cone, or at the end of a pipe at a finite distance from a plane boundary, would introduce additional non-dimensional numbers which would destroy the simplicity of the analysis. However it cannot be correct in practice for a nozzle on the end of a long tube because the pressure field behind the nozzle is not the same as in front due to entrainment, and so the outside flow cannot be plane.

The value of n in (8.9.1) is, for practical purposes equal to 5. It cannot vary from case to case, as it does for isolated thermals and puffs, because it is a phenomenon averaged over a sufficiently long time to obtain a unique value. In theory, therefore, all jets are similar and a cine film of one can be used to represent them all by varying the speed and magnifying it as required. The exterior motion in the case where there is a plane wall is the same as for a uniform distribution of sinks along the axis, i.e. a line sink. Hence it is reasonable to suppose that the motion outside the jet when there is no wall is roughly that due to a semi infinite line sink. At points close to the jet boundary this is not very different from the case of an infinite line sink because of the small angle of spread.

A two dimensional jet is produced by a line source of momentum. The constancy of momentum flux requires that

$$w^2 z = \text{constant}, \tag{8.9.14}$$

and so

$$w \propto z^{-1/2} \propto R^{-1/2}. \tag{8.9.15}$$

The Reynolds number is proportional to wz which increases up the jet like $z^{1/2}$, and so it is always turbulent beyond a certain point, although it may not be nearer the source. We are, however, in some difficulty because the inflow velocity decreases upwards like $z^{-1/2}$, which means that the same horizontal pressure gradient is not required to set up the motion at all heights. Since the motion is two dimensional there need be no horizontal acceleration towards the jet if the exterior motion is strictly horizontal. Thus horizontal exterior motion is a theoretical possibility although it is rotational and therefore could not be set up by the jet in otherwise motionless surroundings. In practice there is likey to be some descending motion outside the jet because its finite existence prevents it setting up the theoretical state which has no pressure gradients.

A new jet sometimes called a starting jet (Fig. 8.9.iii) is susceptible to simple analysis. It consists of an upper portion which looks like the front of a puff and a lower conical portion which looks like a jet. It retains its similarity so that its total volume is proportional to z^3. The total upward momentum is proportional to the time since it began, and since all velocities are proportional to one another the upward velocity of its cap can be taken as representative.

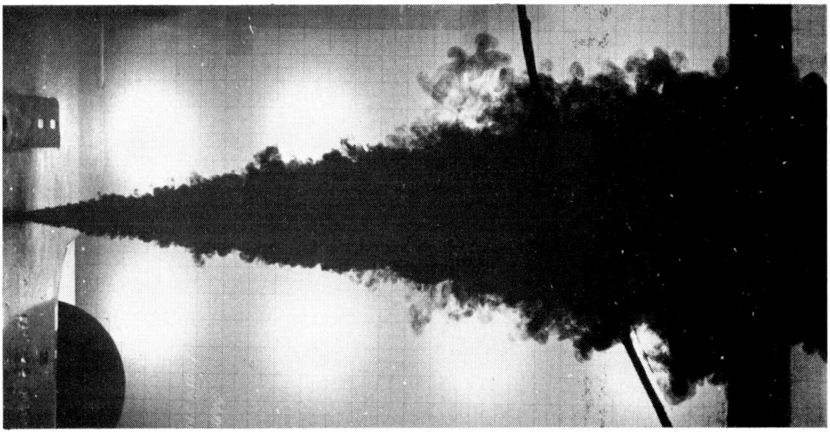

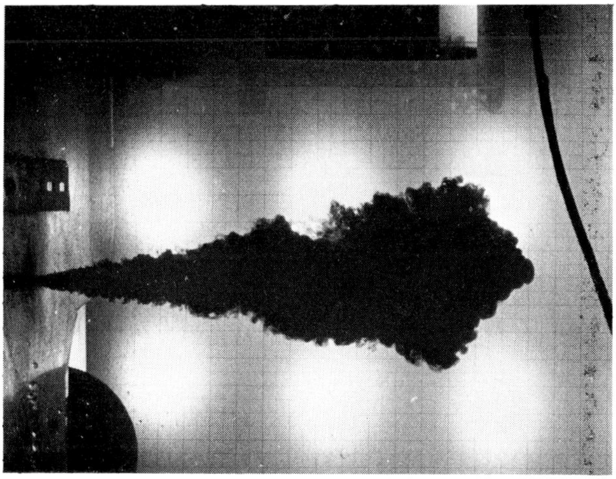

Fig. 8.9.iii The line diagram shows successive positions of the outline of a new jet (from Spackman 1964). A middle and a late position of the jet are shown. Occasional protuberances extend in the early stages beyond the cone indicated by the straight lines, but when the cap is far away they are entrained inside the cone.

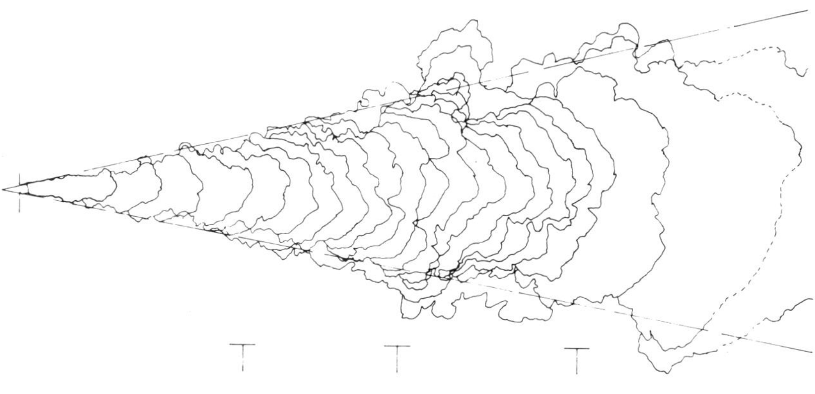

Therefore $$wz^3 \propto t \tag{8.9.16}$$

and $$w = \frac{dz}{dt}, \tag{8.9.17}$$

so that $$z^2 \propto t. \tag{8.9.18}$$

This relationship and (8.10.10) below was accurately confirmed by Spackman (1964). The Reynolds number, and the circulation measured around a circuit such as MNPQM in Fig. 8.9.iv, is a constant because

$$w \propto z^{-1} \tag{8.9.19}$$

as in the jet. The behaviour of the top dome-shaped part is more like a thermal than a puff because momentum is fed into its base to retain (8.9.18 and 8.9.19); otherwise it would be retarded according to (8.6.5).

If the source continues, the behaviour at fixed distances from it becomes like that of a jet as the top moves further away; if the source stops, the fluid already set in motion becomes a puff.

Spackman found the following features:

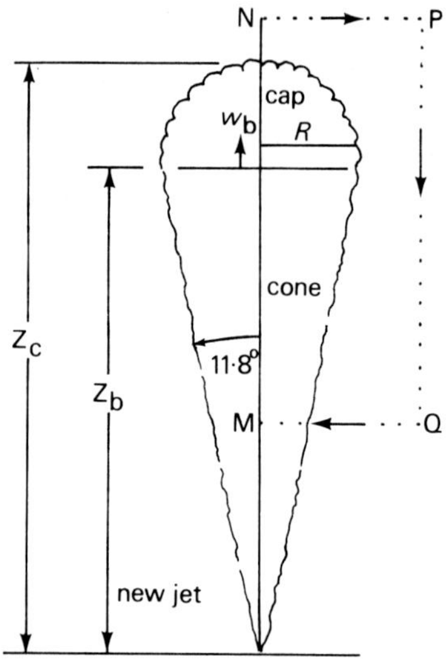

Fig. 8.9.iv Definitions and dimensions for a new jet. The circuit MNPQN may be used to define the circulation of the system, M being the mid-point between the virtual origin and the base of the cap, or some similarly defined point.

(a)

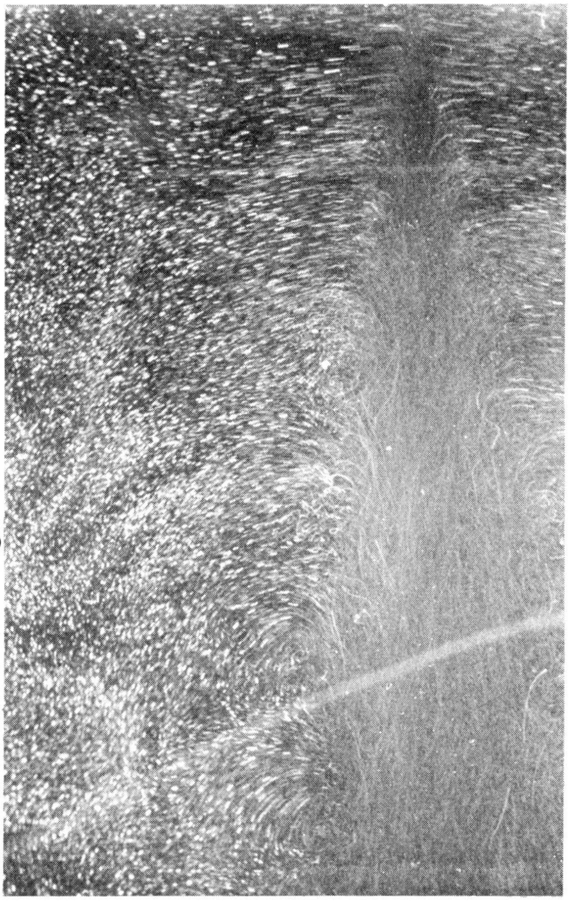

(b)

Fig. 8.9.v (a) A time exposure of a new jet, the displacement of particles floating in the water indicating the velocity vector. Note the variations around the large protuberances and the outflow ahead of the cap.

(b) A time exposure of a plume of unmarked fluid showing the displacements of particles illuminated in a plane of the motion. Note particularly the greatly increased velocity inside the plume and the disturbance of the exterior fluid due to the larger eddies. (From Spackman 1964).

For a jet $\theta = 11.8^\circ \pm 0.3^\circ$, giving a mean value of n equal to 4.76. Using measurements of Ricou and Spalding (1961) he found that $\dot{M}_f/\dot{M}_m = 0.74$, which is the ratio of the momentum fluxes due to the fluctuations and the mean flow. For a new jet he obtained a value for the constant of proportionality in (8.9.18), namely

$$z_c{}^4 = 180 \frac{\dot{M}}{\rho} t^2 \qquad (8.9.20)$$

where $\dot{M}$ is the momentum flux through the orifice. Shape factors were as follows

$$\begin{aligned} z_c/z_b &= 1.3 & \mathrm{vol}_{\mathrm{cap}} &= 0.012 z_c{}^3, \\ \mathrm{d}z_c/\mathrm{d}t &= 2W_b & \mathrm{vol}_{\mathrm{cone}} &= 0.021 z_c{}^3 \end{aligned} \qquad (8.9.21)$$

He also made estimates of the entrainment velocity on the cap, and found it to vary from a maximum on the axis to the value 0.11 W_b at the junction with the cone, with a mean value on the cap of 0.6 W_b. W_b is the mean vertical velocity at $z = z_b$, the mean being defined as for (8.11.3), below. Rather more than a quarter of the entrained fluid enters over the cone. The fluctuations of velocity in the irrotational flow outside are illustrated in the time exposure picture in Fig. 8.9.v.

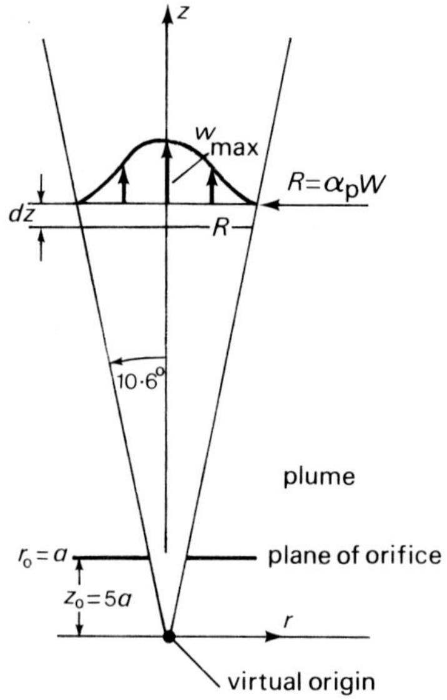

Fig. 8.10.i Coordinates and definitions for a plume.

8.10 PLUMES

For convenience the motion produced by a steady source of buoyancy is called a **plume**, the word jet being used for the case of no buoyancy. An axisymmetric plume has a constant flux of buoyancy (or pollution), denoted by B, in the direction of its axis, and the increase in momentum flux over a height dz is equal to the buoyancy force on the slab of thickness dz. Thus (see Fig. 8.10.i).

$$wBR^2 = \text{constant}, \tag{8.10.1}$$

and

$$d(w^2R^2) \propto BR^2 dz. \tag{8.10.2}$$

Since, for the same reasons as before, $R \propto z$, these equations require that

$$w \propto z^{-1/3} \tag{8.10.3}$$

$$B \propto z^{-5/3}. \tag{8.10.4}$$

Thus it is seen by comparison with (8.9.11) that a plume slows down less rapidly than a jet, because of the continuous application of the buoyancy force, but it is diluted more rapidly because of the additional mixing induced by density gradients. Since the velocity is not proportional to R^{-1} the inflow is different at different heights at a given radius outside the plume, and it is therefore not strictly horizontal.

The most extensive measurements on gaseous plumes, which confirmed the relationships (8.10.3 and 8.10.4) were made by Rouse, Yih and Humphreys (1952). It is impossible to set up a plume with laminar flow because it is essentially unstable, the Reynolds number being proportional to wz, i.e. to $z^{2/3}$, which increases rapidly up the plume. That line of argument is not complete, however, for it only implies that there could be a laminar region lower down in a plume which is turbulent higher up, because the Reynolds number decreases downwards. This question is discussed further in Section 8.12. Certainly experience confirms the conclusion that plumes cannot be prevented from becoming turbulent.

A **two dimensional plume** is produced by a horizontal line source of buoyancy. The equations corresponding to (8.10.1 and 8.10.2) are

$$wBR = \text{constant} \tag{8.10.5}$$

and

$$\mathrm{d}(w^2R) \propto BR\,\mathrm{d}z, \tag{8.10.6}$$

which, with $R \propto z$, imply

$$w = \text{constant} \tag{8.10.7}$$

and

$$B \propto z^{-1}. \tag{8.10.8}$$

Equation (8.10.7) means that the maximum, mean, and typical eddy velocities are the same at all heights although the profile, which is of the same mathematical form, is spread over a width proportional to height. The inflow velocity is the same at all heights; but since this does not correspond to the pressure field produced by the buoyant fluid above the flow cannot be strictly horizontal, unless the vertical accelerations in the plume are exactly matched to the field there.

If the inflow velocity is u, and the plume is set up in a cross wind of strength u_0, it may be argued that the plume is simply translated sideways as it rises, that its centre rises up a plane inclined at angle $\tan^{-1} u_0/w$ to the vertical and that the inflow is $u + u_0$ and $u - u_0$ on the two sides. This idea is naive in the sense that there is no single upward velocity, w, which will determine the angle at which it will be inclined. Moreover the plume is inclined to gravity so that the pressure field at each height is altered. When u_0 exceeds u no inflow can occur on the downwind side, and so the plume is attached to the ground there. A very practical application of this idea was made in FIDO, the system for clearing fog from an airfield runway in no wind. A line source of heat in the form of paraffin burners was installed on each side of the runway, and each produced a wind towards the other equal to u. In effect they entrained all the air in between, quickly coalescing to form a single source located half way between them at approximately the same height (see Fig. 8.10.ii). There was, as a consequence, a region of heated air in which the fog was evaporated over the runway and in which an aircraft could be landed visually. The flow pattern is similar in all such cases; all that was required was enough heat to evaporate the fog in the entrained air.

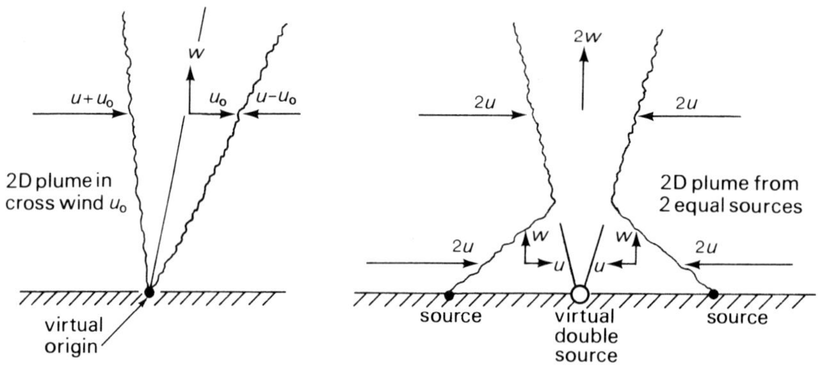

Fig. 8.10.ii A two-dimensional plume in a cross wind u_0 shown adding this velocity to the inflow velocity u, which is the same at all heights. Thus two parallel line sources of heat (buoyancy) form a single plume by entraining each other: they cover the ground between them with a region of heated air, above which the plume rises as if from a double strength source at the middle.

A two dimensional plume is very turbulent because the Reynolds number is proportional to z and increases rapidly upwards.

A **new axisymmetric plume** is similar to a new jet in appearance but has a total upward momentum increasing at a rate proportional to the total buoyancy already emitted. Thus with characteristic buoyancy B and a constant emission rate beginning at $t = 0$

$$\frac{\mathrm{d}}{\mathrm{d}t}(wR^3) \propto gBR^3 \propto t \tag{8.10.9}$$

or

$$gBmR^3 = Ft$$

where mR^3 is the volume and F is the buoyancy flux at the orifice.

Since $w = \dfrac{dz}{dt}$ and $R \propto z_c$ we must have

$$z_c \propto t^{3/4}, \tag{8.10.10}$$

$$w \propto z_c^{-1/3} \propto t^{-1/4}, \tag{8.10.11}$$

and

$$B \propto z_c^{-5/3} \tag{8.10.12}$$

The laws of dilution of buoyancy (and pollution) and momentum appear to be the same as for an axisymmetric steady plume (see 8.10.3 and 8.10.4), but z_c is the overall height, and w and B are characteristic values of the new plume. The thermal-like upper part is fed with buoyancy and momentum from below and is slowed down less rapidly than a thermal in which $w \propto z^{-1} \propto t^{-1/2}$, $B \propto z^{-3}$. The Reynolds number increases like $z^{2/3}$ and so the phenomenon must become turbulent.

A two-dimensional new plume follows similar relationships which now take the form

$$\frac{d}{dt}(wR^2) \propto gBR^2 \propto t, \tag{8.10.13}$$

which leads to

$$z_c \propto t \tag{8.10.14}$$

$$w = \text{constant} \tag{8.10.15}$$

and

$$B \propto z^{-1}. \tag{8.10.16}$$

These resemble (8.10.7 and 8.10.8) as expected.

It may be noted that for any of the buoyancy driven systems we may write

$$w \propto (gBR)^{1/2} \tag{8.10.17}$$

as an alternative to one of the equations used above. This is based on simple dimensional analysis.

Spackman found that for a new axisymmetric plume, in the notation defined at the end of 8.9,

$$\theta = 10.4^\circ \pm 0.6^\circ;$$

$$z_c^{\,4} = 39Ft^3, \quad \text{where } F = \text{buoyancy flux at the orifice;}$$

$$\left.\begin{aligned} \text{vol}_{\text{cap}} &= 0.0084\, z_c^{\,3}, \\ \text{vol}_{\text{cone}} &= 0.017\, z_c^{\,3} \end{aligned}\right\} \begin{aligned} &m' = 0.0254 \\ &\text{where } m'z_c^{\,3} \text{ is the volume} \end{aligned}$$

$$z_c/z_b = 1.3$$

$$\mathrm{d}z_c/\mathrm{d}t = 1.3\, W_b.$$

He also estimated the entrainment and found that the contribution over the cone is more important than over the cap. This is because it has a mean value of 0.26 W_b on the cap, which is about the value at the top of the cone. About 60 per cent of the fluid entrained enters over the cone.

8.11 BUOYANT JETS

In an axisymmetric plume there is a momentum flux appropriate to the buoyancy flux which causes it. If the momentum flux exceeds the appropriate value the system will behave like a jet in that the momentum will decrease more rapidly and the dilution proceed less rapidly than in a plume. When the appropriate value is reached it will behave like a plume.

We desire an anlysis for the region in which the transition takes place. Because the dimensional analysis produces single power relationshsips between the quantities, it is often assumed that this type of relationship is a natural one, and some researchers assume a power relationship and use experiments to measure what the power is. It should be emphasised that when the situation is as simple as the cases we have been discussing, that is simple enough to produce single power proportionalities, the analysis will always reveal what the law is. We must perform experiments to find out all the other relationships, in particular those relating to the geometrical shape of the phenomena, the profiles of velocity and buoyancy across horizontal sections, and the magnitudes of, and correlations between, fluctuations of velocity and buoyancy (or pollution).

If w and b are the instantaneous values of upward velocity and buoyancy, the time averaged fluxes of momentum and buoyancy are

$$mR^2W^2 = \int_0^\infty 2\pi r\overline{w^2}\,\mathrm{d}r = \int_0^\infty 2\pi r(\overline{w}^2 + \overline{w'^2})\mathrm{d}r \qquad (8.11.1)$$

and

$$\frac{1}{g}\Phi = kR^2WB = \int_0^\infty 2\pi r\,\overline{wb}\,\mathrm{d}r = \int_0^\infty 2\pi r(\overline{w}\,\overline{b} + \overline{w'b'})\mathrm{d}r \qquad (8.11.2)$$

where the bar denotes time averaging and the prime the departure from the time average. Φ is defined by (8.11.2); m and k are constants of proportionality which have to be measured. R is defined, as at the beginning of Section 8.9, to be the radius beyond which the material of the buoyant jet does not penetrate at a given height. W and B are a mean velocity and buoyancy defined so that the flux of velocity (momentum at unit density) is $\pi R^2 W$ and the total buoyancy in a section of thickness dz is $gBR^2\,\mathrm{d}z$. The mean horizontal inflow velocity at height z and radius R is defined as U, and the entrainment ratio α is defined by

$$U = \alpha W. \tag{8.11.3}$$

The equation of continuity may be expressed by stating that the change in upward volume flux is equal to the inward volume flux; thus

$$\frac{\mathrm{d}}{\mathrm{d}z}(\pi R^2 W) = 2\pi R U = 2\pi\alpha R W. \tag{8.11.4}$$

If we put in the formulas for the vertical variation of R, W, and U with z, which for a jet are (8.9.1 and 8.9.2) we find that, using (8.11.3), for an axisymmetric jet

$$\alpha = \frac{1}{2n}. \tag{8.11.5}$$

On the other hand if we use (8.10.3 and 8.10.4) we find that for an axisymmetric plume

$$\alpha = \frac{5}{6n}. \tag{8.11.6}$$

The numerical quantities α, n, m, k are constants for each separate case but there is no means of telling what their values are and whether they differ from one case to another except by observing the phenomena.

The profiles of velocity appear to be more or less gaussian. This cannot be strictly true of the gaussian profile at large values of r where, in the case of the axisymmetric jet at any rate, $\overline{w}$ is identically zero when $r > R$. Perhaps a single wavelength of a cosine-shaped curve (which is very similar to a gaussian curve except at the ends) is a more accurate representation, but people prefer the gaussian formula because it is familiar and easier to handle mathematically. In any case the profiles measured for a plume and a jet are sufficiently alike for it to be assumed that in equation (8.11.1), m takes the same value for both. It is very much easier to observe n (because it is equal to z/R) and it does appear to take more or less the same value for plumes and jets. Therefore (8.11.5 and 8.11.6) imply that the entrainment ratio for a plume is equal to $5/3$ times its value for a jet, and it appears that buoyancy is a direct cause of increased entrainment through the creation of vorticity. This conclusion depends (through 8.11.3) on the definition of W (in 8.11.1) and that is why the equality of values of m in the two cases is important. Some treatments have been based on the assumption that α is a kind of universal constant on the grounds that mixing is caused by the velocity difference between the jet and its environment, and it has not been noted that this requires a different value of either m or n for jets and plumes.

If, for the moment, (8.11.2) is regarded as being applicable to flux of pollution of concentration b in the case of a jet with no buoyancy (it being noted that buoyancy and pollution are both carried with the fluid and are diluted identically so that (8.11.2) may refer to either quantity in a plume), then to make any progress with the theory of a buoyant jet we need to assume that (8.11.2) applies throughout with the same value of k. At present there is no good evidence as to whether this is a correct assumption, but its consequences are of some interest.

The vertical rate of change of momentum flux is equal to the buoyancy, thus

$$\frac{\mathrm{d}}{\mathrm{d}z}(mW^2R^2) = gBR^2, \tag{8.11.7}$$

and we can write equality because we have defined the quantities for this purpose. The buoyancy flux is constant so that by (8.11.2)

$$WgBR^2 = \frac{1}{k}\Phi = \text{constant}. \tag{8.11.8}$$

Therefore (8.11.7) becomes

$$\frac{\mathrm{d}}{\mathrm{d}z}\left(mW^2\frac{z^2}{n^2}\right) = \frac{\Phi}{kW} = \frac{\Phi z}{k(W^2z^2)^{1/2}} \tag{8.11.9}$$

This has solution

$$\tfrac{3}{2}(W^2z^2)^{3/2} = \frac{n^2\Phi}{km}\tfrac{1}{2}(z^2+c^2) \tag{8.11.10}$$

where c is a constant of integration. It can be reshaped as

$$W^3 = C^3\left(\frac{1}{z} + \frac{c^2}{z^3}\right), \qquad (8.11.11)$$

where

$$C^3 = \frac{3n^2}{4km}\Phi. \qquad (8.11.12)$$

Equations (8.11.3 and 8.11.4) give α in the form

$$\alpha = \frac{1}{n}\left(1 - \frac{1}{6}\frac{z^2 + 3c^2}{z^2 + c^2}\right). \qquad (8.11.13)$$

For the jet stage of the phenomenon where $z \ll c$ we find

$$W = Cc^{2/3}z^{-1}, \qquad \alpha = \frac{1}{2n}; \qquad (8.11.14)$$

while for the plume stage where $z \gg c$

$$W = Cz^{-1/3}, \qquad \alpha = \frac{5}{6n}. \qquad (8.11.15)$$

The buoyancy may be derived from (8.11.7) and is

$$gB = \frac{n^2\Phi}{kz(z^2 + c^2)^{1/3}}, \qquad (8.11.16)$$

which is proportional to z^{-1} and $z^{-5/3}$ in the jet and plume stages respectively.

These formulas can be applied straightforwardly in the case where the momentum is in excess of that of a plume of the given buoyancy, the transition from a jet to a plume taking place at heights near $z = c$.

If a jet has negative buoyancy, (8.11.12) shows that C^3 is negative, which means that for W to be positive in (8.11.11) c^2 must be negative, and the range of z for which the plume rises is restricted. If we replace c^2 by $-c'^2$, we see that the velocity vanishes at $z = c'$, after which it must fall back towards the source and the analysis is of doubtful meaning. The entrainment ratio is

$$\alpha = \frac{1}{n}\left(1 + \frac{1}{6}\frac{3c'^2 - z^2}{c'^2 - z^2}\right), \qquad (8.11.17)$$

which is the same as for a jet when $z \ll c'$, but which decreases with z and becomes zero when $z^2 = \tfrac{3}{5}c'^2$. This expresses literally the idea that, when positive, buoyancy increases mixing, and therefore when negative reduces it to zero before the upward velocity falls to zero. The model is unrealistic between this level and $z = c'$ because 'detrainment' is required to keep the shape conical while it is being decelerated.

Above $z = c'$ the implication of the model is that, where the buoyancy is greater than in a plume of the given velocity, entrainment is increased above the value for a plume. The entrainment is infinite at $z = c'$ where $W = 0$ because it is implied that mixing occurs on account of the buoyancy even when there is no velocity relative to the surroundings. But this conclusion is questioned in Section 8.13 on the grounds that it is not correct to assume that $R \propto z$.

8.12 SIMILARITY IN STRATIFIED SURROUNDINGS

In the cases so far considered it appears that the rates of change of all quantities depend only on local conditions, that is on the values of the various quantities involved at the particular value of z. This has sometimes been taken to imply that, because the theory accords so well with observations, the rates of change do actually depend only on local values. All that has happened is that the similarity ensures they can be expressed that way, although the mechanics are such that the motion depends very much on what is going on above and below as well as on what is happening at the same level.

When the surroundings are stratified according to

$$\beta = \frac{1}{\rho}\frac{\partial \rho}{\partial z} \tag{8.12.1}$$

in the incompressible case, the equations for the momentum and buoyancy fluxes in the more general case than that covered by (8.11.7 and 8.11.8) are

$$m \frac{\mathrm{d}}{\mathrm{d}z}(W^2 R^2) = gBR^2 \tag{8.12.2}$$

and

$$k \frac{\mathrm{d}}{\mathrm{d}z}(WR^2 B) = -WR^2 \beta \tag{8.12.3}$$

The last of these equations expresses the fact that they buoyancy is reduced by an upward displacement in a stratified environment. These become

$$\frac{\mathrm{d}R}{\mathrm{d}z} + \frac{R}{W}\frac{\mathrm{d}W}{\mathrm{d}z} = \frac{RgB}{2mW^2} = \frac{\mathrm{F}}{2m}, \tag{8.12.4}$$

and

$$\frac{\mathrm{d}R}{\mathrm{d}z} + \frac{3R}{W}\frac{\mathrm{d}W}{\mathrm{d}z} = -\frac{R\beta}{kB} - \frac{R}{\mathrm{F}}\frac{\mathrm{dF}}{\mathrm{d}z}$$

$$= -\frac{\mathrm{S}}{k\mathrm{F}} - \frac{R}{\mathrm{F}}\frac{\mathrm{dF}}{\mathrm{d}z} \tag{8.12.5}$$

where

$$\mathrm{F} = gBR/W^2 \tag{8.12.6}$$

is a Froude number relating the buoyancy and inertia forces,

and

$$\mathrm{S} = g\beta R^2/W^2 \tag{8.12.7}$$

is a stability number relating the static stability to inertia forces. We then obtain

$$\frac{\mathrm{d}R}{\mathrm{d}z} = \frac{1}{n} = \frac{3\mathrm{F}}{4m} + \frac{R}{2\mathrm{F}}\frac{\mathrm{dF}}{\mathrm{d}z} + \frac{\mathrm{S}}{2k\mathrm{F}}, \tag{8.12.8}$$

and from the continuity equation in the form (8.11.4) we get

$$\alpha = \frac{\mathrm{d}R}{\mathrm{d}z} + \frac{R}{2W}\frac{\mathrm{d}W}{\mathrm{d}z}$$

$$= \frac{5\mathrm{F}}{8m} + \frac{R}{4\mathrm{F}}\frac{\mathrm{dF}}{\mathrm{d}z} + \frac{\mathrm{S}}{4k\mathrm{F}}$$

$$= \frac{1}{2n} + \frac{\mathrm{F}}{4m}. \tag{8.12.9}$$

The particular case of no stratification is recovered by putting β and S equal to zero. If F is then constant, and if it is not zero (8.12.6) corresponds to (8.10.17); so $\frac{1}{n} = \frac{3\mathrm{F}}{4m}$, and $\alpha = \frac{1}{2n} + \frac{1}{3n}$ for a plume. If F is zero we simply get $\alpha = \frac{1}{2n}$ for a jet. If n is constant when β is zero, (8.12.8) can be solved to give

$$\mathrm{F} = \frac{4m}{3n}\frac{z^2}{z^2 + c^2} \tag{8.12.10}$$

which gives the variation of F in the case of 8.11.

In practice plumes depart very little from the conical shape in stratified surroundings (except close to the level at which they cease to rise), and it is probably a realistic assumption to put n constant. We have to assume that m and k are also constant in order to make any progress with the analysis. This implies constancy of the profiles of mean and fluctuating velocity and buoyancy.

If the additional assumption is made that S is constant then (8.12.8) may be written as

$$-2n\mathrm{d}z/z \;=\; \mathrm{dF}\bigg/\left(\frac{3\mathrm{F}^2}{4m} - \frac{\mathrm{F}}{n} + \frac{\mathrm{S}}{2k}\right) \tag{8.12.11}$$

and integrated to give

$$\mathrm{F} \;=\; \frac{2m}{3n}\left(\sigma + 1 - \frac{2\sigma c^{2\sigma}}{z^{2\sigma} + c^{2\sigma}}\right) \tag{8.12.12}$$

where

$$\sigma^2 \;=\; 1 - \frac{3n^2\mathrm{S}}{2mk}\,. \tag{8.12.13}$$

The supposition that S is constant implies a relationship between W and β through (8.12.7), and is very artificial. (8.12.12) then gives B as a function of z through (8.12.6); also it reduces to (8.12.10) when β, and consequently S, is zero and $\sigma = 1$.

W is then given by integration of (8.12.4) in the form

$$W^3 \;=\; \mathrm{C}^3 z^{-\sigma-2}(z^{2\sigma} + c^{2\sigma}) \tag{8.12.14}$$

which is (8.11.7) when $\sigma = 1$.

The constant of integration c is given by putting in some known set of values denoted by suffix 1; thus

$$\frac{gB_1R_1}{W_1^{\,2}} \;=\; \mathrm{F}_1 \;=\; \frac{2m}{3n}\left(\sigma + 1 - \frac{2\sigma c^{2\sigma}}{z_1^{\,2\sigma} + c^{2\sigma}}\right), \tag{8.12.15}$$

and C is then determined by (8.12.14) in the form

$$\mathrm{C}^3 \;=\; \frac{W_1^{\,3} z_1^{\,\sigma+2}}{2\sigma c^{2\sigma}}\left(\frac{2m}{3n}(\sigma + 1) - \mathrm{F}_1\right). \tag{8.12.16}$$

This special case has been given because it is susceptible to complete solution; but is is utterly artificial in requiring S to be a constant, although this is not impossible in reality of course, and might be set up in a laboratory. It also possesses similarity.

A special case of this last result is that in which F is also a constant, and represents a plume possessing similarity in stratified surroundings. Then, with $\sigma = 0$, we have

$$\frac{gBR}{W^2} = \mathrm{F} = \frac{2m}{3n},$$

$$\frac{g\beta R^2}{W^2} = \mathrm{S} = \frac{2mk}{3n^2}, \tag{8.12.17}$$

and

$$W^3 = \mathrm{C}^3 z^{-2}$$

as a solution, which is possible only if the initial conditions make c equal to zero.

These similarity solutions do not actually make the assumption that everything is determined at the particular value of z, but mould the surroundings and initial conditions so that this is so. On the other hand, if we make the assumption (which is not justified by any observations and is clearly wrong in cases where there is any sort of overshoot), that everything is determined at the local level only so that the profiles across the plume are fixed and m and k are constants, Equations (8.12.2 and 8.12.3) may be combined to give

$$\frac{\mathrm{d}}{\mathrm{d}z}\left(W\frac{\mathrm{d}}{\mathrm{d}z}(W^2 z^2)\right) = -\frac{1}{mk} g\beta W z^2 \tag{8.12.18}$$

which can be integrated numerically to give W as a function of z if β is given as a function of z. B is then given by (8.12.2), and α by (8.12.9).

The assumptions that are required to validate such a numerical solution are such that the much simpler approach of Chapter 10 probably gives more enlightenment. In any case, it is difficult to imagine a case other than a sinister military one in which this kind of theory would be useful before the event. Afterwards events speak more eloquently than such predictions, as when unintended fires occur.

More recently, however, a method of mining the metalliferous deposits on the bed of the Red Sea requires a slurry of unwanted material to be returned to the sea; the slurry is denser than the sea water which is stratified slightly except close to the bed where the salinity is much greater. These formulas are applicable in the region of slight stratification.

Special solutions of (8.12.18) can easily be found. For example if

$$\beta = \frac{mk}{W}\sum_{\lambda}\frac{a_\lambda}{z^\lambda} \tag{8.12.19}$$

the solution is

$$W^3 = \tfrac{3}{2} z^{-3}\left(k_0 + \tfrac{1}{2} k_2 z^2 + \sum_{\lambda} \frac{a_\lambda}{(3-\lambda)(5-\lambda)} z^{5-\lambda}\right) \qquad (8.12.20)$$

where k_0 and k_2 are constants of integration related to the conditions imposed. This is more general than the solutions in which W^3 is related to only one or two powers of z, but it is relevant to problems only in that it can be used to exemplify conical plumes in stratified surroundings, making all the necessary assumptions described above, because β has not been allowed to be specified independently of W.

8.13 LAMINAR PLUMES

In Section 8.11 we discussed at length the case of a plume starting with an excess of momentum. The analysis was unrealistic for the case when it starts with an excess of buoyancy, and required 'detrainment'. An approach more in accord with observation in this second case is to assume that when possessed of excessive buoyancy a plume accelerates and becomes narrower. The flow then remains more or less uniform across the whole cross section and is bounded by a free stream surface (see Fig. 8.13.i) on which the pressure is the same as the outside hydrostatic value, so that the exterior fluid is not disturbed.

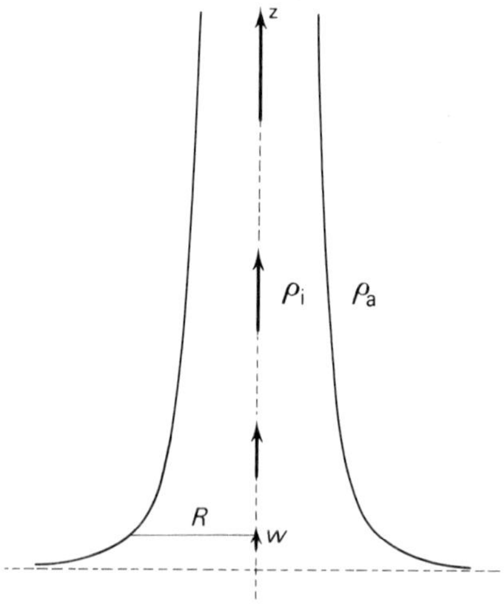

Fig. 8.13.i Coordinates for an accelerating laminar plume.

If the interior and exterior values are denoted respectively by suffixes i and a, and p_0 is the pressure at some starting level

$$p_a = p_0 - g\rho_a z \; . \tag{8.13.1}$$

Since the flux of volume up the plume, with uniform velocity W over a cross section having radius R, is constant

$$\pi R^2 W = \text{const} = Q \, , \tag{8.13.2}$$

and the pressure inside is given by

$$p_i = p_0 - g\rho_i z - \tfrac{1}{2}\rho_i W^2 \, , \tag{8.13.3}$$

assuming that z is measured from where $W = 0$. On the boundary $p_a = p_i$ and so

$$W^2 = 2gz \frac{\rho_a - \rho_i}{\rho_i} = 2gBz \, , \tag{8.13.4}$$

where

$$B = \frac{\rho_a - \rho_i}{\rho_i} \; . \tag{8.13.5}$$

Consequently, according to (8.13.2)

$$R^4 = \frac{Q^2}{\pi^2 W^2} = \frac{Q^2}{2\pi^2 gBz} \tag{8.13.6}$$

and

$$w \propto R^{-2}, \qquad R \propto z^{-1/4} . \tag{8.13.7}$$

Thus the buoyant fluid accelerates away from its source as a contracting stream. The buoyancy flux is constant and the momentum flux increases until it reaches the value appropriate to a turbulent axisymmetric plume of its own size, after which it becomes a plume. Figure 8.9.i is a picture of the beginning of the turbulence in the form of meanderings of the plume in a case where viscosity probably prevents the occurrence of much smaller eddies. As in the case of all our other motion patterns, compact bodies of the buoyant fluid are stretched out and carried first to one side then another and the eddies grow in size further along the plume.

Motion of this kind could well take place in calm conditions inside a cooling tower. The phenomenon is not to be confused with the vena contracta (see section 2.4) which is a phenomenon due to the curvature of the streamlines as a fluid emerges from an orifice into a region of constant pressure. This case is gravity dominated, and the curvature of the streamlines is of no importance.

8.14 RECIRCULATING JETS

The phenomenon of recirculation may be illustrated by considering a jet confined in a parallel sided tube. If a jet of cross section area A_2 and mean speed U_2 enters a tube of identical fluid moving with speed U_1 across the area A_1 surrounding the jet, and some distance downstream the flow across area $A_1 + A_2$ has average speed U_3, since the flux of material is constant we must have

$$A_1 U_1 + A_2 U_2 = (A_1 + A_2) U_3. \tag{8.14.1}$$

There is a rise in pressure Δp from the point of emergence of the jet to a point downstream where the two streams are thoroughly mixed equal to the decrease in momentum flux, so that

$$A_1 U_1^2 + A_2 U_2^2 = (A_1 + A_2)\left(\frac{A_1 U_1 + A_2 U_2}{A_1 + A_2}\right)^2 + \frac{\Delta p}{\rho}(A_1 + A_2). \tag{8.14.2}$$

Thus
$$\Delta p = \rho A_1 A_2 \left(\frac{U_1 - U_2}{A_1 + A_2}\right)^2 > 0 \tag{8.14.3}$$

if $U_1 \neq U_2$.

This rise in pressure will produce stagnation of the surrounding fluid if

$$\Delta p > \tfrac{1}{2}\rho U_1^2 \tag{8.14.4}$$

i.e. if
$$\frac{U_2 - U_1}{U_1} > \frac{A_1 + A_2}{2A_1 A_2}, \quad \text{or} \; \frac{U_2}{U_1} > \frac{A_1^2 + A_2^2}{2A_1 A_2}, \tag{8.14.5}$$

which is obviously possible if U_2 is large enough. The right hand side of (8.14.5) is a minimum when $A_1 = A_2$, and then stagnation in the outer stream occurs if $U_2 > U_1$. The flow separates from the wall at the stagnation point and a region of recirculation surrounds the jet from that point on until it has filled the tube. The conditions for recirculation are very easy to establish in practice.

Experimentally a steady symmetrical flow like that illustrated in Fig. 8.14.i is difficult to establish in two dimensions because transverse uniformity, i.e. strictly symmetrical motion, is difficult to set up and the jet tends to attach itself to one wall. In the axisymmetric case the flow fluctuates but the recirculation is easy to establish.

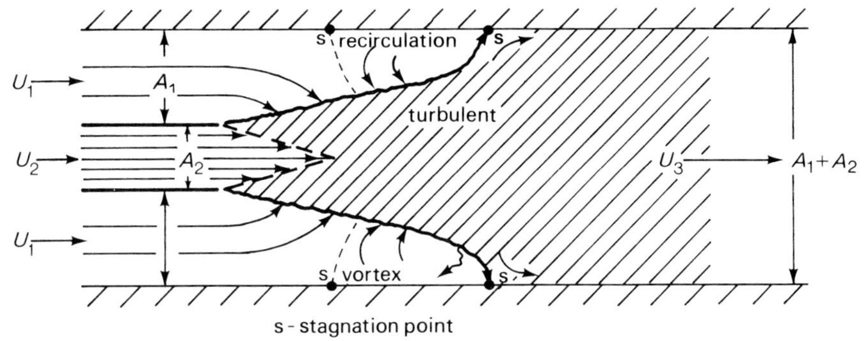

Fig. 8.14.i Enclosed jet with recirculation: a laminar jet entering an annular laminar stream becomes turbulent with a potential core diminishing to nothing at about 5 nozzle diameters downstream. The surrounding fluid stagnates at a point S before the turbulent cone reaches the wall, and a recirculation vortex is established.

8.15 SUMMARY OF EXPERIMENTAL VALUES

$$z = nR, \qquad w = C(gBR)^{1/2} = C'(gBz_c)^{1/2},$$

$$\text{Vol} = mR^3 \text{ (axisymmetric)} = m'{z_c}^3$$

$$= mR^2 \text{ (cylindrical)}$$

where z is either the height of the cap (z_c) or the distance up the axis, w is the velocity of the cap or a representative velocity, B is the overall mean buoyancy or buoyancy at height z, R is the maximum radius of half width. F is the buoyancy flux and $\dot{M}$ the momentum flux at the source. Also $w \propto z^{\alpha}$, $B \propto z^{\beta}$, $\sigma \propto z^{\beta}$, $z \propto t^{\gamma}$, and $\text{Re} \propto z^{\delta}$. When $\delta > 0$ the motion becomes turbulent, when $\delta < 0$ it becomes viscous-dominated with increasing z or t. σ is the concentration of pollution and is proportional to buoyancy. In a plume B_{max} and W_{max} are the maximum values for a given z. The values in brackets are extreme measured values. Suffixes m and f refer to mean and fluctuating components respectively.

Axisymmetric	n	θ	C	C′	m	m'	α	β	γ	δ	
Puff	(1.7) 4(6.3)				2.4		−3	−3	¼	−2	
Thermal	(2) 4 (7.3)		1.3		3		−1	−3	½	0	
Jet	4.76	11.8° ± 0.3					−1	−1		0	$\dot{M}_f = 0.74\dot{M}_m$
Plume	5.51	10.4° ± 0.6	$w_{\max} = 1.4(gB_{\max}z)^{1/2}$ $= 4.7^{-1/3}F^{1/3} \cong 3\overline{w}$		$gB_{\max} = 11z^{-5/3}F^{2/3}$ $\cong 3\,gB$		−⅓	−5/3		⅔	$F_f = 0.4F_m$
New Jet	4.76					.033	−1	−1	½	0	
New Plume	5.51		.04	.02		.025	−⅓	−5/3	¾	⅔	
Cylindrical											
Puff	(2.6) 4 (5.7)						−2	−2	⅓	−1	
Thermal	(1.3) 2 (3.2)		1.0		(1.8) 2.4 (3.0)		−½	−2	⅔	½	
Jet							−½	−½		1	
Plume							0	−1		1	
New Jet							−½	−½	⅔	0	
New Plume							0	−1	1	1	

Chapter 9

BUOYANT CONVECTION IN THE DRY ATMOSPHERE

9.1 ATMOSPHERIC CONVECTION; REGIMES AND OBSERVATIONS

While convection is occurring there are great variations in temperature horizontally. Because thermals entrain outside air into their centres the air with the greatest updraught does not have any significant buoyancy, but is surrounded by a ring of air that has the maximum buoyancy. Outside that there may be air that is temporarily descending at the maximum rate in the neighbourhood, but this may be on the outer reaches of a thermal and have a positive temperature anomaly. In between thermals the air may be slightly stably stratified and very slowly subsiding, being warmed as a result even though the warm air rising from below is not actually being mixed with it.

A sample taken by a rising balloon will differ according to whether the balloon rises up in a thermal or instead rises through the slowly subsiding surroundings.

Because of great variations in the nature of the surface the temperature close to the ground varies accordingly. Thus the temperature over a dry street may be 5°C or more warmer than that over a grassy field nearby, and a quanity such as the maximum daytime shade temperature is very much characteristic of the surface.

The situations we shall be discussing will therefore have features extremely dependent on whether we are concerned mainly with the different temperature of different parcels of air of different origin, or with the vertical profile of the horizontally averaged conditions. Even when we are discussing the latter, the properties will depend very much on the existence of inhomogeneities.

Near the ground the lapse rate may become very high, and in the bottom inch or two, where mirages occur on a hot day, convection occurs on a small scale, and is inhibited by viscosity and radiation. Rather higher up we may enter a similarity regime where these physical properties of the fluid have no influence. As we approach the condensation level, the influence of clouds becomes dominant, and their effects will be discussed in Chapter 11.

The rate at which the bottom temperature rises makes a big difference to the geometry of convection. It is very different over grassland, desert, or sea, and in the morning, afternoon, or evening. The reader must be warned not to assume that the conclusions reached by any of the treatments now following are applicable in particular cases. It is for each observer, paractitioner, consultant, or researcher to determine for himself what regime he is actually concerned with. Many discussions have been frustrated by assertion and counter assertion concerning the geometry of convection, whether it is composed of plumes or thermals. Clearly it could look like either if the scope of the observation is restricted, but the fitting of a model to observations is no guarantee that the model is correct.

9.2 SLOW CONVECTION: THE RAYLEIGH NUMBER

Near the rigid boundary at the bottom of the atmosphere the motion is inhibited by viscosity. Equally it is inhibited by thermal conductivity, which conducts the heat upwards without motion when the boundary is hot, and also reduces the horizontal gradients of temperature without which vertical motion would not occur. When vertical motion does occur it is resisted by viscosity away from the boundary because it produces stresses which resist up and down motion side by side. If the up and down motions are moved further apart the horizontal temperature gradients are reduced and the horizontal motion is increased, and this is resisted by viscosity near the boundary. Above the up and down motion there must be other horizontal motion in the opposite direction, and so a cellular structure is completed. The cells are very roughly square in cross section for otherwise the resistance to motion just mentioned would be greater. The precise shape depends on whether they are square or hexagonal in plan, and on what conditions occur at the upper and lower boundaries. There is a considerable literature about the properties of cellular convection between horizontal boundaries or fluid surfaces but it is not relevant to the atmosphere except in the bottom inch or so. Even there most theories are irrelevent because the radiative heat flux far exceeds the convective heat flux in that layer when the surface is hot. Also the conduction conditions beneath the surface have a big effect on the geometry of the convection above.

The driving force of this laminar convection is gravity operating on the relative density anomaly $\Delta\rho/\rho$. If we imagine that there is a particular density gradient at which motion begins, then a length h, representing the depth of the layer, must be involved. If $\Delta\rho$ is associated with the buoyancy anomaly resulting from a vertical displacement of a particle at distance h, then the critical gradient is increased by the conductivity, κ, and the viscosity, ν, which must enter the equations in an identical manner. To obtain the dimensions of a gradient of buoyancy (accel/length) the relationship must take the form

$$\frac{g\Delta\rho}{h\rho} = \mathrm{Ra}\frac{\kappa\nu}{h^4} \tag{9.2.1}$$

where Ra is non-dimensional and is called the Rayleigh number. Its value is determined by what phenomenon is under consideration. For the onset of motion with rigid perfectly conducting boundaries the classical analysis gives $\mathrm{Ra} = 27\pi^4/4$, which is of the order of 10^3. Because of its dependence on h^3 we deduce that very large gradients may be set up in shallow layers close to the heated boundary. In practice the maximum temperature exceeds the critical value and heat is also lost upwards by radiation. Even in relatively dry climates this radiation is mainly absorbed by the water vapour in the lowest 2 metres of air. For this reason the molecular coefficients κ and ν are not relevant to the geometry of convection in the atmosphere, and it is no use pretending that an eddy conductivity and viscosity will do instead. This is because it is the eddies themselves that are under investigation which cause the transfers of heat and momentum: there is no uniformity of such coefficients over the space occupied by the individual eddies, and the eddy transfer coefficient concept is useless in this context.

The theory of laminar cellular convection is a good example of inbreeding of ideas. All the interesting problems have been generated as a result of laboratory attempts to demonstrate the validity of a theory which was designed to be applied to the laboratory experiments: it has little direct application elsewhere. In particular the direction of the cellular motion, in so far as it has meaning in a cellular pattern (which in some cases it does not), has been explained in terms of the variation of viscosity and conductivity with temperature. In the case of the atmosphere the direction is always determined by the fact that parcels of air break out of the very hot layer near the hot ground and travel upwards. They entrain environmental fluid by mixing as they go, and rise penetratively through the environment. Various suggestions have been put forward that these parcels are plumes near the ground (within a very few metres or less) and become more like thermals higher, up, probably passing through a new though temporary plume stage before again rising as a thermal. As we shall see, the actual configuration depends on the rate of heating and the nature of the surface.

9.3 SIMILAR FULLY TURBULENT CONVECTION

A theoretical state which has attracted much interest is that in which the only dimension determining the parcel size, or geometrical scale of the buoyant convection, is distance above the ground. The analysis proceeds without reference to the geometry as follows: The vertical velocity, w, is given by

$$w = C(gBz)^{1/2} \tag{9.3.1}$$

while

$$B \propto \frac{z}{\theta}\frac{\partial\theta}{\partial z} \tag{9.3.2}$$

because $\Delta\theta = z\partial\theta/\partial z$ for adiabatic vertical displacements and $B = \Delta\theta/\theta$. θ is the potential temperature. If we now assume geometrical similarity the same fraction of any horizontal surface is occupied by upcurrents and if the flux of buoyancy is the same at all heights so that the heat fed in at the bottom passes out at the top, we must have that

$$wB = \text{const} \tag{9.3.3}$$

or

$$\frac{1}{\theta}\frac{\partial\theta}{\partial z} \propto z^{-4/3} \tag{9.3.4}$$

If the whole layer is being warmed up slowly then the heat flux decreases linearly and we have an expression of the form $C_1 - C_2 z$ in the place of the constant in (9.3.3). Then

$$\frac{1}{\theta}\frac{\partial\theta}{\partial z} \propto \left(\frac{C_1 - C_2 z}{z^4}\right)^{1/3}. \tag{9.3.5}$$

The similarity has now been lost because an additional height has been introduced, namely that at which the buoyancy flux goes to zero, and the convection does not exist above it. But this makes very little difference to the result (9.3.4) near the bottom boundary. Indeed, with the kind of measurements that are available in the bottom half of the layer it would be very difficult to detect any departure from (9.3.4) unless there was a large rate of change of the situation in time. The well-known result is obtained by noting that $1/\theta$ varies much less than $\partial\theta/\partial z$ and so, roughly, (9.3.4) is

$$\partial\theta/\partial z \propto z^{-4/3}, \qquad B \propto z^{-1/3}, \qquad w \propto z^{1/3}. \tag{9.3.6}$$

The significant point about this result is that it has been obtained without reference to the geometry of the convection, and solely on the basis of a simple dimensional argument with generates (9.3.1 and 9.3.2). The argument has been presented in a variety of forms dependent on a supposed geometry, but even if (9.3.4) is observed to be true in any particular case nothing at all is proved about the geometry.

If the convection consists of thermals, plumes or new plumes then we must suppose that the number of them in unit volume decreases so that the same fraction of a horizontal surface at any height may be occupied by up currents. This can only be achieved by their amalgamation when the cones along which they rise overlap. In consequence of this amalgamation the vertical velocity increases like $z^{1/3}$ instead of decreasing as z increases, as it must in any isolated thermal, plume, or new plume. This is because the buoyancy decreases more slowly when some of the entrained fluid contains buoyancy.

9.4 SIMILAR CONVECTION WITH A SUBSIDING UNSTRATIFIED ENVIRONMENT

The following treatment is taken from Scorer (1969).

If σ is the fraction of a horizontal area occupied by upward motion, and w and w_e are the vertical velocities of the upcurrents and their environment respectively, if also there is no upward transfer of volume (i.e. no total horizontal convergence, neglecting the density anomalies in this context),

$$\sigma w + (1-\sigma)w_e = 0 . \tag{9.4.1}$$

If $\bar{\tau}$ is the mean value of τ, $= \theta^{-1}$, at any height, and τ' and τ_e' are the anomalies of θ^{-1} in the upcurrents and the environment, then

$$\bar{\tau} = \sigma(\bar{\tau}-\tau') + (1-\sigma)(\bar{\tau}-\tau_e') ,$$

or

$$\sigma\tau' + (1-\sigma)\tau_e' = 0 , \tag{9.4.2}$$

where by definition τ' is positive, and so τ_e' is negative. For a liquid we would replace τ by ρ.

The buoyancy flux, omitting the factor g, is

$$F = \sigma w\tau' + (1-\sigma)w_e\tau_e' = \frac{\sigma}{1-\sigma}w\tau'. \tag{9.4.3}$$

Also

$$B = \tau'/\bar{\tau} \tag{9.4.4}$$

and by the formula (9.3.1), which holds for all configurations of the convective parcels,

$$w\tau' = \mathrm{C}\left(g\frac{\tau'}{\tau}r\right)^{1/2} = \mathrm{C}(g\tau'^3 z/n\bar{\tau})^{1/2} \tag{9.4.5}$$

where C and n are the values appropriate to whatever the parcel configuration may be, as described in Chapter 8. Since

$$\bar{\tau} = \tau_e - \tau_e' = \tau_e + \frac{\sigma}{1-\sigma}\tau' , \tag{9.4.6}$$

where τ_e is the constant value of τ in the adiabatically subsiding unstratified environment, ignoring the region close to the ground in which τ' is comparable with τ, we may put $\bar{\tau} \triangleq \tau_e$, so that (9.4.3) and (9.4.5) give

$$F = \frac{\sigma C}{1-\sigma}\left(\frac{g}{n\tau_e}\tau'^3 z\right)^{1/2} . \tag{9.4.7}$$

Thus if F is independent of z, we must have

$$\tau' \propto z^{-1/3} \tag{9.4.8}$$

as in (9.3.6). If we differentiate (9.4.6) we get

$$\frac{\partial\bar{\tau}}{\partial z} = \frac{\sigma}{1-\sigma}\frac{\partial\tau'}{\partial z} \tag{9.4.9}$$

or

$$-\frac{\sigma\tau'}{1-\sigma}\bigg/\frac{\partial\bar{\tau}}{\partial z} = -\tau'\bigg/\frac{\partial\tau'}{\partial z} = 3z . \tag{9.4.10}$$

This result is required in Section 9.6. It leads to

$$\frac{\partial\tau'}{\partial z} \propto \frac{\partial\bar{\tau}}{\partial z} \propto z^{-4/3} \tag{9.4.11}$$

which is equivalent to (9.3.4), and (9.3.6) follows.

It is important to note that the total environment penetrated by the thermals is slightly unstably stratified. Thermals, with a decreasing excess temperature, undergo a more ‘unstable’ lapse rate. The subsiding surroundings have a netural lapse rate: sampling must be made representative to obtain the lapse rate of the total environment. If in an actual case the surroundings were slightly stably stratified, the sinking motion would cause an apparent warming. Indeed this warming is very real in practice, but this similarity analysis is no longer strictly applicable, for an extra linear dimension is introduced defining the stratification.

9.5 RISE OF THERMALS IN WIND SHEAR

A thermal rising through wind shear $\partial U/\partial z$ experiences a horizontal acceleration relative to its environment in the negative direction of U of magnitude

$w\partial U/\partial z$. It also experiences an Archimedean acceleration of its environment downwards of magnitude gB. It is therefore subject to a total acceleration which is the vector sum of these two, of magnitude

$$g^*B = gB/\sin\theta = w\frac{\partial U}{\partial z}\Big/\cos\theta \qquad (9.5.1)$$

where θ is the inclination of the axis of the thermal to the vertical (Fig. 9.5.i).

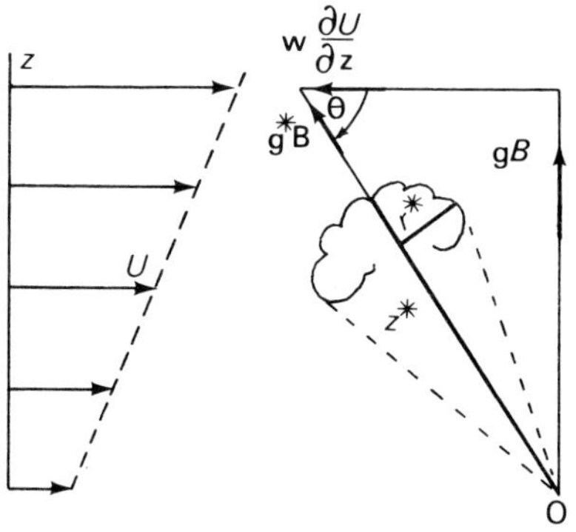

Fig. 9.5.i A thermal rising in wind shear represented by the velocity profile on the left experiences a horizontal acceleration of its environment equal to $w\partial U/\partial z$, and a downward acceleration of its environment of magnitude gB. These combine to produce a total acceleration g^*B at an angle θ to the horizontal, and its axis of symmetry is inclined accordingly.

Assuming the mechanics of the inclined thermal to be the same as that of a vertically accelerated one, its velocity w^* relative to the surrounding air is related to its vertical and relative horizontal velocities, w and u, by

$$w^* = \frac{w}{\sin\theta} = \frac{u}{\cos\theta} = \mathrm{C}(g^*Br^*)^{1/2} = \mathrm{C}\left(\frac{g}{\sin\theta}B\frac{r}{\sin\theta}\right)^{1/2} \qquad (9.5.2)$$

where r^* is its radius measured perpendicular to its axis of symmetry. From this

$$w = \mathrm{C}(gBr)^{1/2} \qquad (9.5.3)$$

and
$$\frac{u}{\partial U/\partial z} = \frac{w\cot\theta}{w^{-1}gB\cot\theta} = \mathrm{C}^2 r = \mathrm{C}^2 z/n \qquad (9.5.4)$$

The argument of this section is based on the properties of a thermal, and could not be repeated for a plume or a new plume except by means of much more extravagant assumptions. This is because, although the motion of the top of a new plume rising through wind shear would be much the same as for a thermal, the top would be carried downwind faster than the bottom and there would remain no axis of symmetry. We have to assume in this treatment that the variation in U across the depth of the thermal is not significant and is small compared with w, which requires that

$$r\frac{\partial U}{\partial z} << w \tag{9.5.5}$$

9.6 THE RATIO OF THE EDDY TRANSFER COEFFICIENTS FOR MOMENTUM AND HEAT

The previous sections provide the means to calculate the magnitudes of the rates of vertical transfer of horizontal momentum and heat or material. The transfer coefficient of buoyancy K_H, is related to the flux F given by (9.4.3), and the mean gradient (according to Fickian diffusion theory, or K-theory) by the equation which defines it, namely

$$F = \frac{\sigma}{1-\sigma} w\tau' = -K_H \frac{\partial \bar{\tau}}{\partial z}, \tag{9.6.1}$$

and so by (9.4.10)

$$K_H = -\frac{\sigma}{1-\sigma} w\tau' \bigg/ \frac{\partial \bar{\tau}}{\partial z} = 3wz\,. \tag{9.6.2}$$

By the definition of U as the mean wind, if u and u_e are the horizontal velocity anomalies of the thermals and of the environment, the equation analogous to (9.4.2) is

$$U = \sigma(U-u) + (1-\sigma)(U-u_e) \tag{9.6.3}$$

and so

$$\sigma u + (1-\sigma)u_e = 0. \tag{9.6.4}$$

By the definition of K_M, the vertical transfer coefficient of horizontal momentum,

$$-K_M \frac{\partial U}{\partial z} = -\sigma uw - (1-\sigma)u_e w_e \tag{9.6.5}$$

which is a measure of the Reynolds stress described in Section 7.3.

By means of (9.5.4)

$$K_M = \frac{\sigma}{1-\sigma}\frac{uw}{\partial U/\partial z} = \frac{\sigma}{1-\sigma}\frac{C^2}{n}wz\,. \tag{9.6.6}$$

Therefore, by (9.6.2),

$$\frac{K_M}{K_H} = \frac{\sigma C^2}{(1-\sigma)\,3n}\,. \tag{9.6.7}$$

This argument does not require that there be a very deep layer in which similarity exists, it is merely only applicable in such a layer. Even if the flux were not constant with height it would only lead to a small change in the factor 3 in (9.6.7) which comes from (9.4.10) and depends on τ_e being independent of z in (9.4.6). The numbers C^2 and n are characteristic of the nature of the convective parcels: if they are like thermals we may use (8.2.13) so that with $n = 4$ and $m = 3$ we have that

$$\frac{C^2}{n} \triangleq 0.4\,. \tag{9.6.8}$$

The value of σ is extremely uncertain, and probably varies with circumstances. Some radar plan pictures of thermals (e.g. Fig. 9.6.i) obtained by Konrad and Kropfli (1968) indicate values of σ between 0.1 and 0.6 which is probably excessive because the method collects thermals which exist in a finite layer and not at an infinitesimally thin surface. In a layer of cumulus cloud σ could be very occasionally as large as 0.5 or as small as 0.01: but the large values probably arise in cases where there is horizontal convergence and a non-zero value of the mean vertical velocity. In most circumstances we can probably be confident that $\sigma < 0.3$. If $\sigma = 0.1$ which seems a likely value immediately under cumulus,

$$\frac{K_M}{K_H} = 0.014. \tag{9.6.9}$$

If $\sigma = 0.5$,

$$\frac{K_M}{K_H} = 0.13 \tag{9.6.10}$$

which seems to be about the upper limit. Any larger value of the ratio, of the order of unity for example, must be associated with a different kind of eddy.

The cause of such a low value of the ratio was explained in Section 7.3. It is, simply, that buoyancy produces vorticity which is horizontal, and eddies

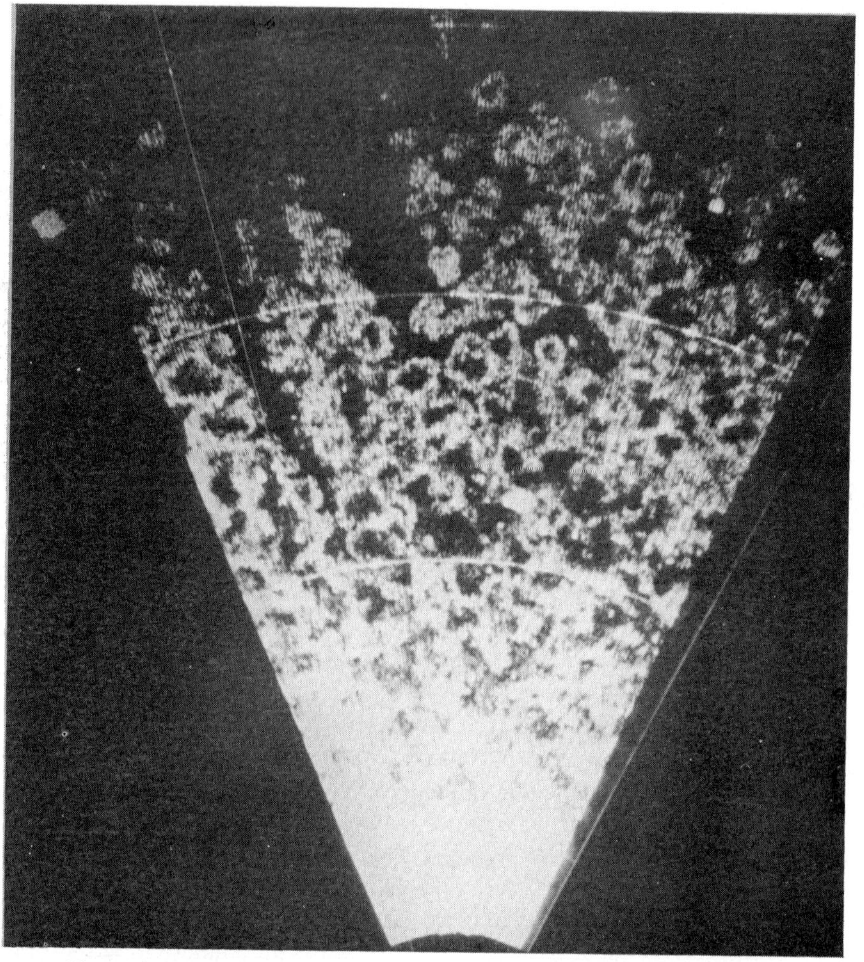

Fig. 9.6.i A radar photo of thermals by Konrad and Kropfli (1968) taken at 2° elevation with range markers at 5 n. mi intervals. The unsteadiness in time, variability with height, and dependence upon intensity make it difficult to find a representative value of σ even for a single occasion. This occasion was one of great convective intensity, although the fraction of the area occupied by active thermals is difficult to determine it does appear to decrease with height. See also Fig. 6.2.ii. [Photograph courtesy of Johns Hopkins University]

composed of horizontal vorticity are not stretched by shear of a horizontal wind. However, closed horizontal vortex lines transfer volume vertically and therefore very effectively transfer mass or buoyancy without producing a shear stress because they extract no energy from the mean shear. In Section 9.5 we saw that the tilting of the thermal axis by wind shear produces inclined vortex lines, and these are then stretched by the mean flow and extract energy from it. The small eddies in thermals continuously destroy the energy of the thermal so that a large proportion of this energy is of recent production. This means that a large proportion of the vorticity present is perpendicular to gravity which produced it.

At the other extreme, namely in a stably stratified fluid, as soon as a horizontal vortex line tilts the isopycnic surfaces out of horizontal, vorticity of opposite sign begins to be generated. This is simply to say that a stable stratification opposes penetrative vertical motion and puts it into reverse. Therefore if eddies are generated by some means the horizontal component of vorticity is destroyed, but the other component remains and can produce eddy stresses by becoming tilted in shear flow. Turbulence which produces vertical mass and heat transfer cannot persist with stable stratification: either the stratification or the penetrative motions are destroyed. The extreme case is when the only eddies are stable waves. Then horizontal momentum can be transferred (see Section 5.11) without any heat or mass transfer, and then

$$K_M/K_H = \infty . \tag{9.6.11}$$

In numerical weather forecasting the finite difference approximations inevitably introduce artificial smoothing equivalent to a diffusion mechanism. This is evidently sufficient to represent any vertical transfer of momentum, but not of heat and water vapour. In practical cases therefore K_M is put equal to zero while K_H is made infinite in situations in which convection is known to be active, namely when the lapse rate exceeds the wet adiabatic. K_H then refers to wet bulb potential temperature and humidity mixing ratio.

9.7 THE DOUBTFUL MEANING OF AN EDDY TRANSFER COEFFICIENT

The use of an eddy transfer coefficient is valid only if an appropriate gap in the spectrum of eddy sizes exists over which the sampling may be done (see Section 7.3). Moreover, it is not correct to suppose that a long enough sampling time will necessarily provide a meaningful parameter, even though the value does not depend on the sampling time. Atmospheric convection does not take place when the air is stably stratified, for example at night. The average lapse rate measured over a period of several days is usually stable: it never becomes very unstable because thermals start up and transfer heat upwards even when it is slightly unstable. Since there is no convective heat transfer downwards when it is stable, there is a net transfer of heat upwards by convection over a long time. The average heat transfer is up the average gradient and so the transfer coefficient is negative. This nonsense was remarked upon by Gold two generations ago, but his comment still remains unheeded by those devoted to finding out what the coefficient is. Its value is very large when convection is active, but zero when it is not: its average value is meaningless.

9.8 FORMS OF BUOYANT CONVECTION

It is so difficult to obtain slow convection, even in the laboratory, that it almost never occurs in the atmosphere. A slight increase in the heat supply causes the motion even in a viscous liquid to become unsteady with parcels rising from the heated boundary through the surroundings and flattening out at the top. The motion is therefore penetrative: in a shallow viscous layer in the laboratory the turbulent nature of atmospheric convection is not simulated because viscosity prevents the eddies which mix a thermal into its surroundings from operating.

(1) Streets and cells

Cloud streets may consist of very shallow clouds, as if they were merely the top of a motion that goes on below and is not significantly affected by the buoyancy effects of condensation. Although some cloud lines do contain large, even raining, clouds, here we are concerned with the small variety, exemplified in Fig. 9.8.i. Cloud streets occur over land temporarily in the morning, and

Fig. 9.8.i An aerial photograph of cloud streets over Oxfordshire.

occasionally in the evening, when the vigour of the convection is not great and is confined to a layer of more or less uniform depth. This influences the spacing of the clouds: they cannot be supported closer together because the upcurrents supporting them would overlap, and would leave no space for downcurrents even if they were initiated as tiny thermals at the ground, while if they were further apart fresh thermals would pop up in between. It does not require much sophistication to appreciate this argument, and if we allow about as much width at the top for downcurrents as upcurrents and assume that in rising from near the ground to the top of the layer the width of a thermal is about doubled, we find that with $n = 4$, as for thermals, this doubling of width occurs in an ascent equal to the width at the top. Thus the cell width is about twice its depth. Cloud streets usually are spaced about twice the cloud base height apart, and the inevitability of cells being about this wide in no way validates the argument sometimes advanced that because convection cells in a liquid of uniform viscosity also have roughly this shape, the mechanics in the atmosphere are analogous. Certainly the eddy mixing in the atmosphere varies greatly across the system. There have also been arguments put forward concerning the relevance of the Prandtl number to cell shape (the eddy Prandtl number is K_M/K_H for a uniformly turbulent layer). However we have no measurements which would give any idea of the horizontal transport of momentum and heat by eddies such as would be required if an analogy with the case of a viscous conducting liquid in the laboratory were to be made. Some people, however, prefer a wrong theory to an admission that we have no mathematical formulation of the detailed mechanics. This is silly because occasions differ so greatly in the atmosphere that a precise theory would be of no use, especially as an eddy transfer theory reveals nothing, being only a representation of a kinematical idea and not of basic dynamics.

Streets are common over the sea, particularly where the air stream is being slowly warmed, and the layer is of uniform depth over a large area. The warming is slow so that interactions between thermals can achieve a more or less steady state. More vigorous convection does not arrange itself in streets but clouds grow to a great variety of sizes. Vigorous convection in a layer capped by an inversion through which clouds do not penetrate is rare, but it usually generates a complete layer of anvil cloud anyway.

Streets are formed in a matter of 10–30 minutes and are arranged in rows when the clouds first appear so that their pattern already exists in the air motion below. This is seen when air moves from cool land to warm sea where the situation is suitable, or when the streets first form in the morning. The time is comparable with the ascent time for parcels through the layer. Thus the mechanics of streets cannot be connected with the earth's rotation, which requires a much larger fraction of a day to produce effects. One theory of streets supposes them to arise in the 'Ekman layer', which is a layer with a theoretical wind profile corresponding to a steady horizontal wind whose speed and direction are constant at each height over a large area and are determined by a balance of the pressure gradient force, the Coriolis force, and the shear stresses (see Section 4.10). This wind profile is unstable to longitudinal perturbations, which are said to correspond to the streets. The profile depends on a constant eddy momentum transfer coefficient, which does not exist when streets are present. Furthermore there is a large vertical exchange of fluid so that the supposed steady wind profile would not be present. Finally, of course, the air particles pass round the cells in a very small fraction of a day. Angell and Pack (1963 and 1967) made balloon measurements showing circulation times of the order of 30 mins, which agrees with deductions from observed air velocities in clouds or upcurrents in streets of the order of 1–10 m $\sec^{-1}$ found by glider pilots. This means that the Coriolis force is not a significant factor in the dynamics.

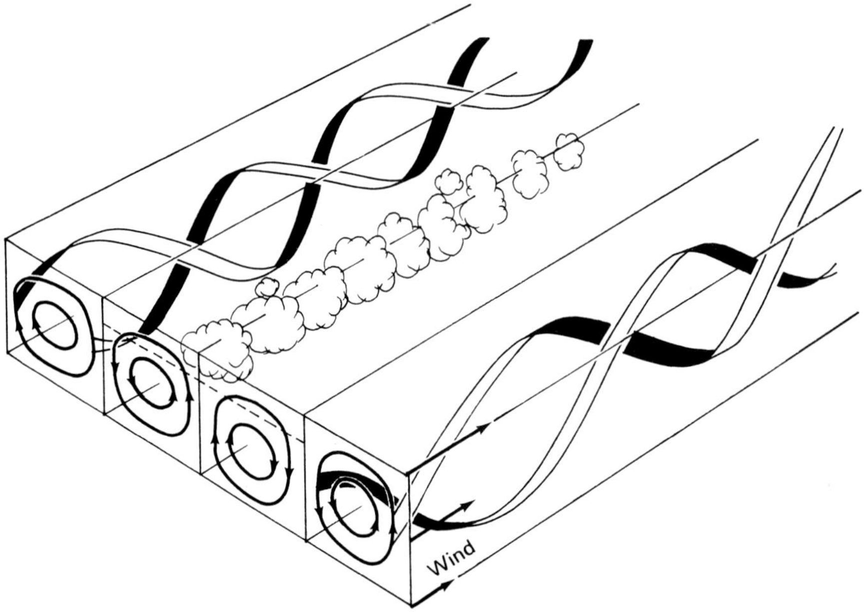

Fig. 9.8.ii Illustration of the helical average motion of particles in convection streets. Clouds may be formed at the top of the upcurrents either in the cells or penetrating a little above them: they are only represented in the middle upcurrent in this diagram. If the condensation level were at the dashed line and the flow were laminar a cloud layer would occupy the whole area above that level up to the top of the cells.

Streets appear to lie nearly along the wind: the motion of particles is helical (Fig. 9.8.ii), and it is generally thought that the arrangement in longitudinal rolls is caused by the drag of the ground. The difficulty with transverse rolls is that adjacent rolls would place the upcurrent of one beside the downcurrent of the next. This is because they would have to rotate in the direction of the vorticity in the shear produced by ground drag, whereas adjacent longitudinal rolls have opposite directions of rotation. In the free air, on the other hand, a shear layer can generate billows all with the same direction of rotation because the upper and lower surfaces of the overturning layer can be distorted to separate the rolls (Fig. 6.3.ii right of diag.). In layers of altocumulus arranged in long cells the motion is either billows of the kind described in Chapter 6, where the shear is large, or else the overturning is very slow, and is more usually arranged in squares or not arranged at all. This contradicts the old idea, drawn from laboratory experiments in smoke between sheets of glass in relative motion, that the cells are longitudinal when the shear is large, but transverse when it is small. There is no evidence that longitudinal rolls occur in thin layers of altocumulus at all, and a rigid boundary is probably a requirement for them to be produced.

It has often been suggested that the shadows of the cloud streets over land cause the heating of the ground to be arranged in lines, with the new upcurrents coming from the sunny areas. This could cause a migration of the lines across their length, as is often observed. This is less likely over the sea, where sunshine has little effect on the small time scale involved.

When there is no wind, and therefore no wind shear, small convection clouds are isolated and arranged with equal spacing unless the terrain is variegated.

(2) Plumes and lines

Solitary lines of cloud are often formed downwind of some obstacle or heat source. Figure 9.8.iii shows the kind of convective motion which is like a thermal taking off from a superadiabatic layer at the ground, part of which is pushed upwards by flow over a hump (see Section 3.6). Equally a heat source such as a large steel works where a hot slag heap is created, or a lake in a burst of cold air from polar regions in the autumn, may produce a line of convection clouds downwind. Oceanic islands may do the same, but more commonly they

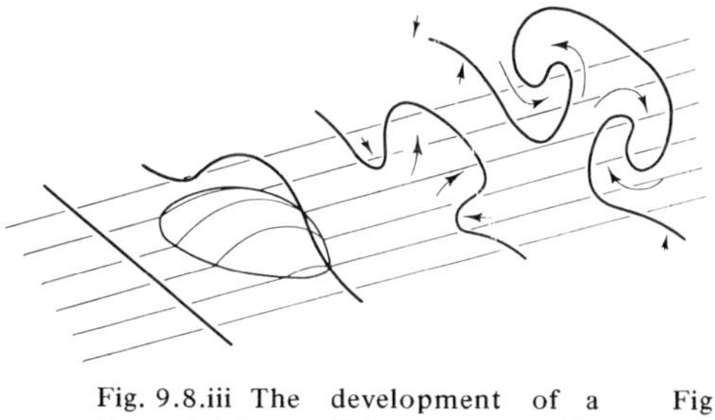

Fig. 9.8.iii The development of a thermal taking off from an unstable layer, initiated by a hump on the ground.

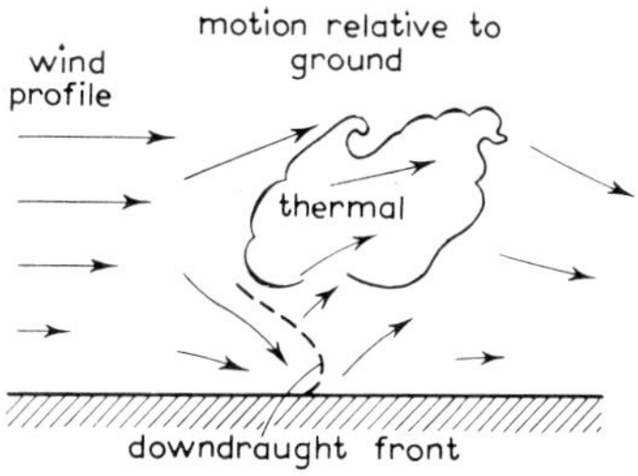

Fig. 9.8.iv When the wind increases with height the downdraught close to an upcurrent taking off from the found is moving faster than the warm air at the ground, and scoops it up, feeding the cumulus cloud with fresh thermals.

produce streamers of cirrus cloud: the low level air is slow moving so that the island in the sunshine is a continuous isolated heat source, and the tops of large anvil cumulus are carried away by a strong wind at say 10 km or higher. These streamers can be observed clearly by fixed satellite photographs and estimates of the wind made accordingly.

Sometimes it seems that a small cumulus cloud, which appears at the top of the ascent of a thermal and which owes its position to what goes on below, consists of a succession of thermals coming up from a source moving along more slowly than the cloud. Figure 9.8.iv shows how in wind shear the downcurrent coming in from above is predominantly from one side when warm air takes off from the superadiabatic layer at the ground. The shear causes the colder air descending to the ground to travel along scooping up the slower moving warmer air at the ground, and so sending up thermals from a source point which moves faster than the surface wind but slower than the speed of travel at the cloud level.

Clouds of this kind show the motion of a thermal rising in wind shear depicted in Fig. 9.5.i and Fig. 11.5.iv.

(3) Thermal sources

There are several well known sources of thermals, some already described in (2) above. Others are produced by variations in the nature of the ground surface. Standing ripe corn can collect a reservoir of warm air which is carried away by the arrival of a cool gust which bends the corn down. Green vegetation, by contrast, remains cool by transpiration, so that it is a source of latent rather than sensible heat. Bare dry earth, rock, concrete, or sand beside it may therefore initiate thermals. A favourite among glider pilots is the wind-shadow source: it is supposed that in the lee of a wood or cliff there is a body of slow moving air which gets hotter than the surrounding air because it does not have its heat diffused upwards by turbulence as much as in the case of fast moving air. Therefore from time to time it generates an accumulation of buoyancy which causes it to rise as a thermal.

(4) Rotating convection

The desert provides a very hot surface but one in which very little heat is stored. Infrared radiation from the surface is the agent of a large fraction of the heat loss, and this is absorbed and reemitted over a depth of a few metres mainly by the water vapour present. Consequently an unstable layer is built up, often to a depth of a few tens or even hundreds of metres, without there being any well marked thermal source at the ground. As soon as a small thermal rises, cooler air descends to the surface from higher up and the surface is cooled significantly as only the uppermost very few mm are hot. The air spaces between the sand particles make the sand layer a very effective insulator. Thus at a depth of a very few cm there is no perceptible diurnal variation of temperature in spite of variations of the order of 50°C at the surface. Consequently one thermal is not followed by others which feed it, and large thermals are inhibited.

In this situation any system able to sweep up heat from the ground as it travels over it can persist and perhaps grow. Such a system is the dust devil. The rotation must be initiated by a variation in the drag of the surface offered to the wind so that a horizontal gradient of wind is produced. Convergence taking place in this air produces vertical stretching of the vortex lines and an

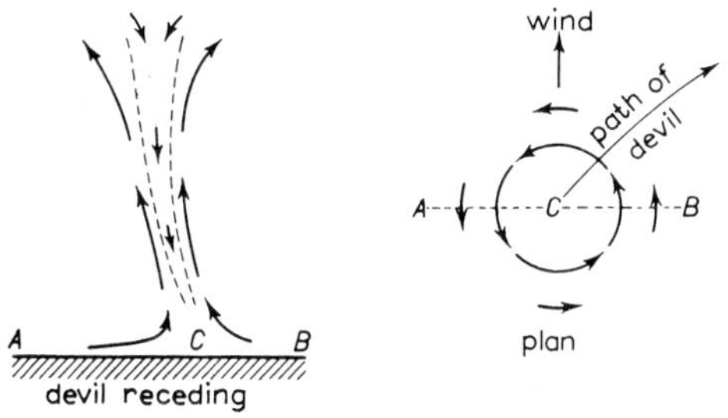

Fig. 9.8.v The motion in a dust devil in vertical section and plan. The base moves with the wind but also towards the forward moving side. There is sometimes a down current in the middle, especially when the system is new and not very tall.

intensification of the rotation. The result is the generation of secondary inflow as described in Section 3.8, so that heat from the surface over which the system travels is continuously fed into it, predominantly on the side where the circumferential velocity is in the direction of the wind. Thus a cyclonically rotating dust devil moves to the right of the wind (Fig. 9.8.v). A dust devil may remain stationary in a suitable position: one account described how a dust devil sat on the end of a causeway across a depression in the ground being built by lorry loads of sand brought to the point reached – only to be whisked away by the whirlwind. Swinn (see Willis, 1953, p. 78) gives a picturesque description of dust devils in Egypt, including their hollow appearance. This could be due to the centrifuging of particles from the centre, but it also appears that the air close to the axis has often been drawn into the vortex from the top, and is colder and free of dust. On the other hand, the dust is sometimes said to absorb sufficient sunshine to increase the temperature of the air containing it. This could be true of a dust devil a few hundred feet high, for then the dust would be airborne for several minutes, and the dust free air in the vortex core would be cooler.

Dust devils are understood in general terms but many details remain elusive and not susceptible to even the most sophisticated theory. There are great variations in the importance of friction, conduction, and the transfer of heat from the hot dust to air into which it is carried and from the shallow layer of hot air a few mm thick which is almost stagnant before the arrival of the winds of the vortex. Devils vary from a foot or two to several hundred metres in height, sometimes extending up to the condensation level and being capped by cloud (Ives 1947). Of course they are great fun, readily associated with spirits by primitive people, and equally liable to facile explanation in a technological age.

(5) Anabatic winds

This name is given to the winds which may be produced on a slope heated by sunshine. They are best developed when the air mass against the slope is stably stratified, because the warmed air is thereby prevented from rising far from the sloping ground and is warmed as it continues up the slope. Often the anabatic flow is stopped at the snow line, above which the surface reflects the

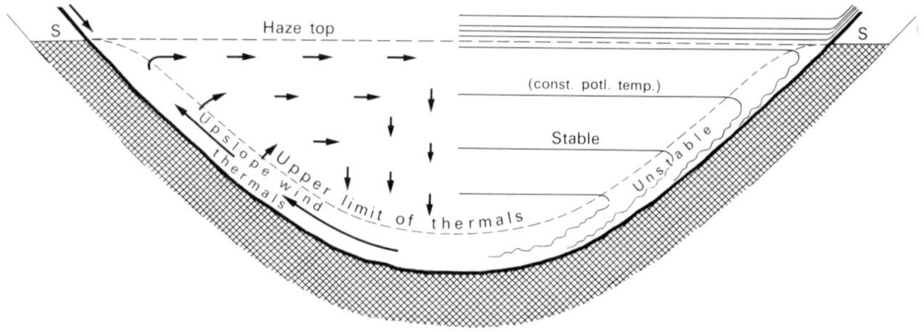

Fig. 9.8.vi The air motion (left) and stratification (right) in a valley with anabatic winds blowing up the sides. S denotes the snow line above which there is a shallow cold layer of air and possible a katabatic wind.

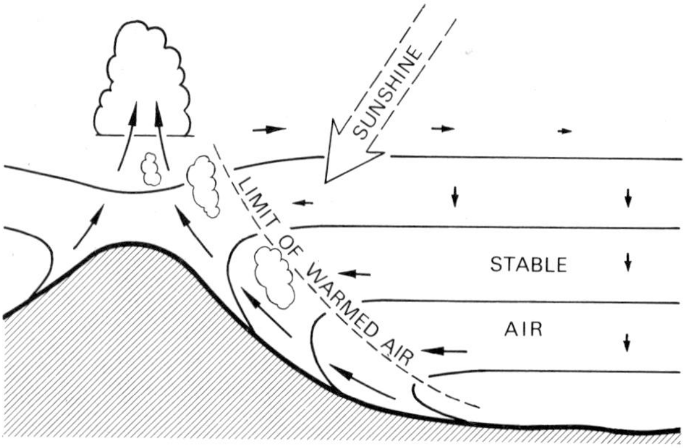

Fig. 9.8.vii Upslope winds feeding cumulus clouds over mountains when the air mass is slightly stable.

sunshine and remains colder than the adjacent air, perhaps with a katabatic flow on it. Haze generated in a valley travels up the slopes and spreads out into the stratified air mass at various levels according to the stratification (Fig. 9.8.vi). A well defined haze top is generated at the upper limit of the anabatic wind, which is often the general level of the mountain tops.

Fig 9.8.viii Cumulus cloud growing out of the anabatic wind up a mountain in Bohemia while the air in the valley remains stably stratified, seen from Milesovka by M. Koldovsky.

On occasions when there is no strong stratification above the mountain tops the anabatic flow feeds cumulus clouds above them (Fig. 9.8.vii and Fig. 9.8.viii). For more detailed treatment reference is made to *Clouds of the World* and *Air Pollution*. The more intrepid glider pilots who are prepared to fly close to a mountain slope have developed the art of soaring in anabatic winds.

(6) Effects of mechanical stirring: Föhn

When air flows over rough terrain and is mechanically stirred, the stirred layer acquires a uniform potential temperature. When the layer initially had a stable stratification, the effect is to cool the upper part of the layer and warm the lower part. This may sometimes explain why there is a higher air temperature at the ground on the lee side of hills than on the upwind side (Föhn effect), particularly if the terrain is rough enough to stir it over a depth of a few hundred metres. The most significant effect is to produce an inversion at the top of the stirred layer, which may then be the upper limit of convection on the lee side. The inversion may also be intensified by the presence of cloud produced at the level of maximum cooling: this cloud radiates into space in the infrared and is cooled thereby, giving rise to downward convection in the layer and a progressive cooling of it. This sometimes leads to a situation in which smog occurs. Again the reader is referred to *Air Pollution.*

9.9 SEA BREEZE FRONTS

Sea breezes are a commonplace of elementary geography and of the experience of coast dwellers. They have also been referred to for countless years, at inland places to which they sometimes penetrate, as the cause of a change of air mass during the afternoon or evening. That they arrive with a sudden drop in temperature has also been long known, but because knowledge of their upper air structure is lacking, this has not been thought of as being due to fronts.

When air over the land is warmed by morning convection, the penetration of sea air inland generates a temperature discontinuity as the cool air undercuts the warm. This discontinuity creates difficulty in any computer-based study using a grid of points of finite spacing, and the history of the subject contains several attempts at analytical representation of the steady state achieved after a long time. Because of mixing on the interface (which is rather like that on top of a thermal, with lobes of the advancing fluid entrapping the fluid into which it advances), the bulk parameters of the phenomena are the useful ones as in the case of thermals.

The speed, V, of advance of the front relative to the fluid it undercuts is related to the buoyancy, B $(= \Delta\rho/\rho)$, and the height, d, reached by the head of the advancing fluid before it sinks to a lower level further back (see Fig. 9.9.i) by

$$V = \mathrm{C}(gBd)^{1/2} \tag{9.9.1}$$

where C is found to be roughly about 2/3. V is defined as $U_w + U_c$, which is the sum of the speeds with which the warm and the cold air at some distance approach the front. The result (9.9.1) is true for laboratory experiments and sea breeze fronts in the atmosphere, mainly observed by Simpson (1969, 1972,

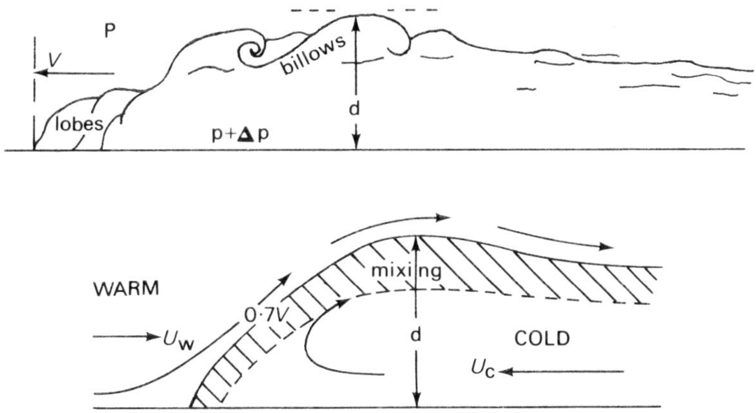

Fig. 9.9.i Schematic representation of a sea breeze front with density difference $\Delta\rho$, maximum height d, and speed V relative to the warm air into which it is moving. Mixing occurs at the front where warm air is entrapped in the lobes, and on the top where billow-like motions occur. The maximum upcurrent is of the order of $0.7\ V$.

1977) near Lasham using a glider. The cold air, having a higher humidity, was often clearly defined by its greater haziness.

The region of maximum upcurrent above the nose of magnitude about $0.7V$ is surprisingly narrow in the atmosphere, on many occasions giving lift to a glider only over a strip 100 m wide. This apparent effect may be due to the presence of irregularities on the advancing front in which mushroom shaped lobes of cold air entrap the warm air. The mixed zone is buoyant relative to the cold air. In model experiements the mixing looks very like billows, with a wavelength fractionally in excess of d, when seen in a vertical plane of the motion illuminated by light from a slit. The billows travel backwards on the head of the advancing current. Fig. 9.9.ii shows a model experiment of the phenomenon.

The sea breezes observed by Simpson typically had values of d equal to 700 m, $\Delta\rho$ corresponding to about 1°C temperature difference and speeds of about 3 m $\sec^{-1}$. Land breezes caused by the cooling of the land at night are much shallower, like katabatic winds, because the nocturnal cooling affects only a few metres when there is not much wind otherwise. When there is a significant wind the land breeze produces only a small effect. However land breezes can have significant consequences, notably by driving banks of fog out across the coast. These are observed early in the day and are sucked inland and evaporated during the morning. They are common on the coasts of the Western Mediterranean, near the Straits of Gibraltar particularly.

Similar density currents occur when an avalanche causes snow to be lifted from the surface and cool the air. Outflows from rain storms also have well defined fronts, and perhaps the most spectacular of these are the Haboobs, or

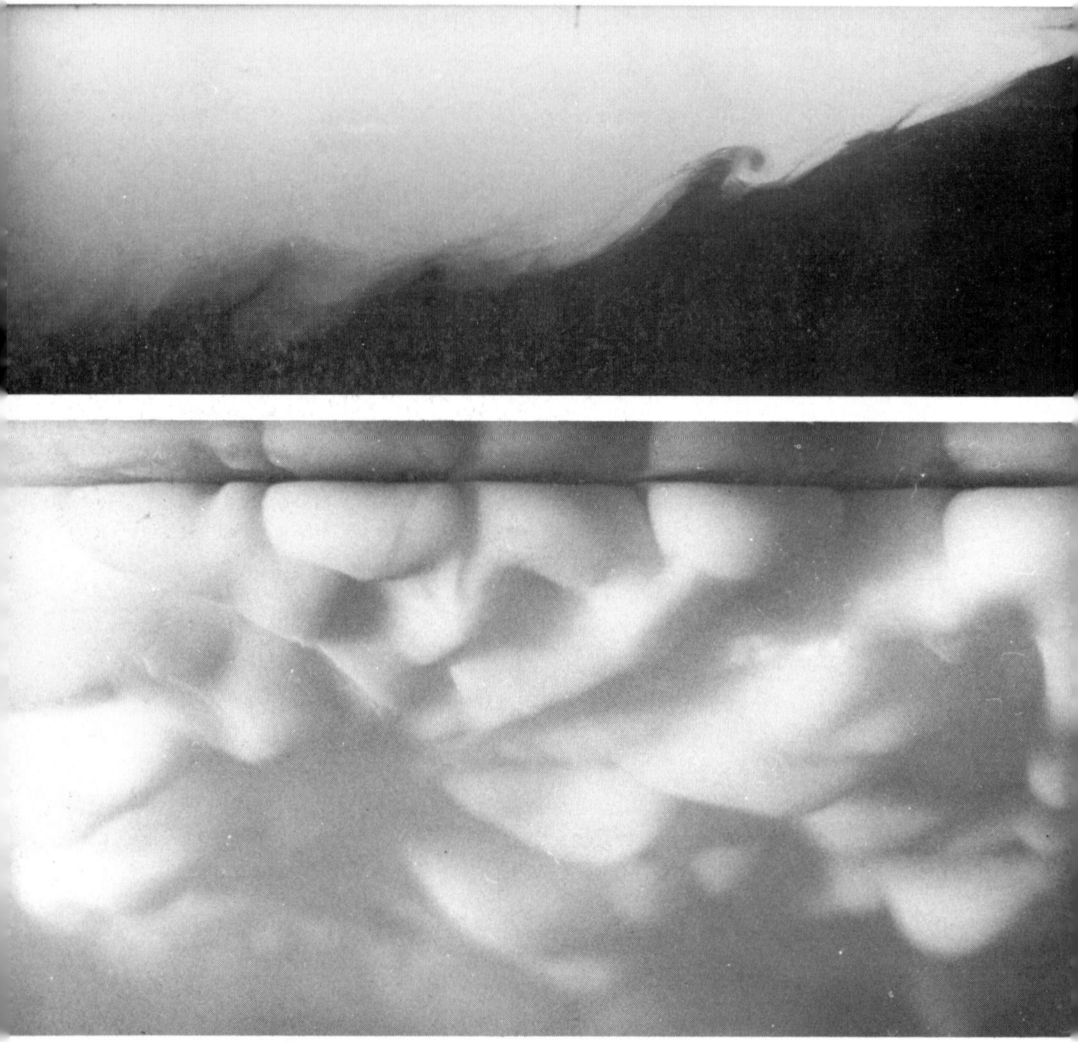

Fig. 9.9.ii A model experiment by H. O. Anwar. It differs from the experiments of Simpson (1969) in having lighter, whitened, fluid advancing along the free surface of a water tank. The views are from the side and front. Because of the absence of friction at the boundary there was less overhang than when the boundary is rigid (see Simpson 1972). In both views reflections can be seen in the surface.

[Photographs by courtesy of the Hydraulics Research Laboratory]

dust storms, of the Sudan. The rain storms which generate these maybe 100 miles away and may not be observed by those who experience the Haboobs.

The air particles flowing through the sea breeze front system pass through it in a small fraction of a day, the upcurrents frequently being around 3 m sec^{-1}

or larger. Consequently Coriolis forces play a negligible part in their detailed mechanics, although the direction of the air motion at the coast is seen to veer after a few hours of sea breeze. Thus the Formula (4.8.3) for the slope of a front given in Chapter 4 is irrelevent. Because of the motion inside the cold air the slope is not 60° as would be suggested by Fig. 2.4.ii(a). An overhanging nose is observed to varying extents and in model experiments is clearly associated with the viscous drag of the bottom boundary.

9.10 AIRCRAFT DOWNWASH AND CONTRAILS

In order to support its weight an aircraft leaves behind it a downwash which for practical purposes consists of a pair of equal and opposite vortices and a motion pattern as shown by the streamlines in the left half of Fig. 9.10.i. Relative to the vortices the motion is as shown in the right half of Fig. 9.10.i, where it is seen that a body of accompanying fluid exists which circulates around the vortices, and travels downwards with them. The surrounding fluid passes round as if the accompanying fluid were a solid body.

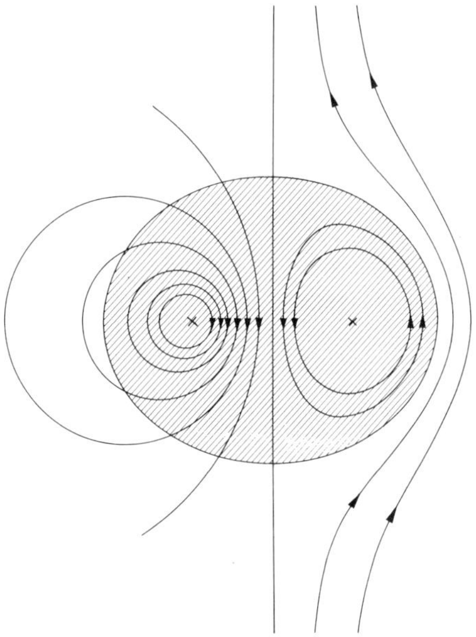

Fig. 9.10.i The streamlines of the motion produced by two parallel line vortices. On the right is the motion relative to the vortices, and on the left is the motion relative to the fluid at a great distance. The shaded area shows the fluid which travels with the vortices, called the 'accompanying fluid'.

In a neutrally stratified atmosphere these vortices may descend to the ground, where they come under the influence of their image vortices and this causes them to move apart close to the ground. But usually before that happens they are disrupted by ambient turbulence and by instabilities inherent in them. Trailing vortices usually contain an axial component of velocity which varies with distance from the axis, and this is destabilising according to Section 3.10, and as a result the vorticity is spread away from the axis. Furthermore, a vortex pair is unstable because if a sinusoidal disturbance is created it will grow. The most unstable such disturbance occurs in planes at 45° to the plane containing the vortices, so that the parts of the vortices which are brought closer together and displaced downwards are subject to an acceleration in that direction. This result is demonstrated by classical small perturbation methods, which usually make various simplifying assumptions about how to calculate the motion field due to the sinuous vortices. Because of the difficulties it is impossible to pursue the calculation when the disturbance becomes large.

The atmosphere above cloud base is stably stratified, and this means that the accompanying fluid acquires buoyancy when it has descended any distance through the air. As a consequence a density discontinuity is produced at the outer surface of the accompanying fluid and this generates a vortex sheet there with the direction of the vorticity as indicated in the left half of Fig. 9.10.ii

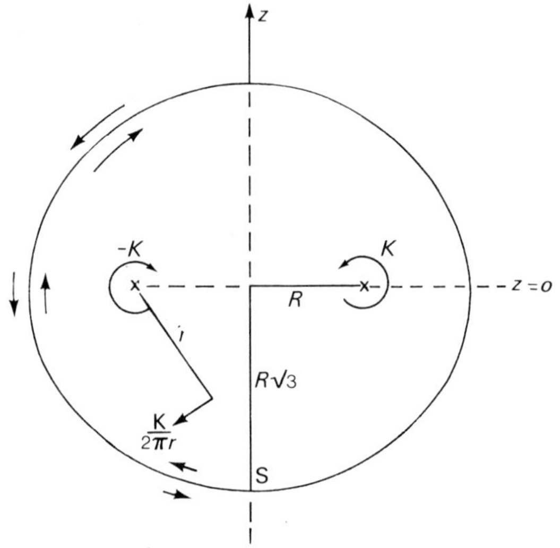

Fig. 9.10.ii Coordinates for a vortex pair having circulations $\pm K$, a distance $2R$ apart. S is the lower stagnation point on the boundary of the accompanying fluid: it travels downwards with the same speed as the vortices. The vorticity generated on the boundary by the density discontinuity increases as the air flows round from the lower to the upper stagnation point.

The magnitude of this vorticity increases from zero at the bottom to a maximum near the top as the fluid flows around this boundary having vorticity generated in it all the way.

After some time the amount of vorticity on the boundary has grown large enough to alter the flow pattern. If we apply equation (6.3.4), using (6.3.5), which are obtained on the assumption that the fluid acceleration is small compared with gravity, the vorticity is obtained by integration. Actually the assumption just mentioned is not all that restrictive because the greater part of the fluid acceleration is due to the curvature of the flow and is nearly normal to the isopycnics. Therefore it does not produce much vorticity. Thus we find that at height z measured from the level of the vortices the vorticity is

$$\eta = \frac{g\beta_0}{q_0}(z + \sqrt{3}R). \tag{9.10.1}$$

This is because it is zero at the bottom stagnation point, S, in Fig. 9.10.ii, where $z = -\sqrt{3}R$, and the downward velocity of the fluid is the same there as that of the vortices. The speed q is the speed along the boundary of the accompanying fluid relative to the vortices, each of which produces a circumferential velocity field of magnitude $K/2\pi r$ at distance r from itself.

We assume that approximately the value of $\Delta\rho$ across the boundary remains constant for a particle as it passes round: this is not strictly correct, partly because of the up and down motions induced in the outside fluid which is stratified. The amount of vorticity in the transition layer at the boundary is obtained by multiplying (9.10.1) by the thickness δ, so long as it is small.

Putting

$$\eta\delta = \mathrm{H} \tag{9.10.2}$$

and

$$\beta_0\delta = \frac{\Delta\rho}{\rho} \tag{9.10.3}$$

H is now the vorticity in unit length of the boundary and is

$$\mathrm{H} = \frac{g\Delta\rho}{q_0\rho}(z + \sqrt{3}R). \tag{9.10.4}$$

The velocity produced by this new vorticity is obtained by integrating its effect around the boundary. On the boundary z is proportional to R multiplied by a function of position which represents the geometrical shape of the system. The length of the boundary is likewise proportional to R. Before calculating it we consider its effect. The vorticity on the left half of the boundary produces a horizontal velocity at the left vortex which is greater in magnitude than and opposite in direction to that produced by the right half. The upper left quadrant has larger vorticity than corresponding points in the lower left quadrant, and since it has the effect of moving the left vortex to the right the net result will be to move the two vortices towards one another. As a consequence each will cause the other to descend more rapidly. The velocity due to the vorticity H is proportional to H/R, and when integrated round the boundary the total effect is proportional to RH/R. Conseqently

$$\frac{\mathrm{d}R}{\mathrm{d}t} \propto -\mathrm{H} \propto -g\frac{\Delta\rho}{\rho}\frac{R}{q_0} \propto -g\frac{\Delta\rho}{\rho}\frac{R^2}{K} \tag{9.10.5}$$

because the velocity q_0 on the boundary, being substantially that due to the vortices, is proportional to K/R.

If we now suppose there to be a stable stratification of the environment β, as the vortex system descends with velocity proportional to K/R, the density anomaly is given by

$$\frac{\mathrm{d}}{\mathrm{d}t}\left(g\frac{\Delta\rho}{\rho}\right) \propto g\beta\frac{K}{R}. \tag{9.10.6}$$

Then from (9.10.5) we have

$$g\beta\frac{K}{R} \propto \frac{\mathrm{d}}{\mathrm{d}t}\left(\frac{K}{R^2}\frac{\mathrm{d}R}{\mathrm{d}t}\right), \tag{9.10.7}$$

i.e.

$$g\beta\frac{1}{R} \propto \frac{\mathrm{d}^2}{\mathrm{d}t^2}\frac{1}{R},$$

so that

$$R \propto R_1\, e^{-\alpha\sqrt{g\beta}t}. \tag{9.10.8}$$

R_1 is the initial value of R and α is a constant determined by the geometry of the system, and we shall not be concerned to evaluate it here.

Clearly the system cannot last. As R is decreased the rate of descent increases until $\Delta\rho$ becomes so large that the boundary vorticity H begins to produce important effects to be discussed in a moment.

An alternative approach was made by Scorer and Davenport (1970). They followed a theme initiated by J. S. Turner in his study of buoyant vortex rings, and the reader is referred to his book *Buoyancy effects in fluids*. According to this approach the downward momentum of the vortex system is (see Lamb, 1932, Art. 152, Eq. 6) equal to $2\rho KR$. Since an upward buoyancy force is operating on the accompanying fluid this momentum is decreased. Since K is constant, R, must decrease. By ignoring the effect of the vorticity in the boundary on the momentum, this led to the conclusion that

$$R = R_1 \operatorname{sech} \frac{\sqrt{mg\beta}}{2\sqrt{2\pi}} t \tag{9.10.9}$$

which is similar to (9.10.8) except close to the start. The number m is defined so that mR^2 is the cross section area of the accompanying fluid, and it is equal to 11.5. We may therefore set α equal to $(11.5/8\pi)^{1/2}$, i.e. 0.68.

The effect of the vorticity in the boundary is to alter the position of the upper stagnation point and bring it inside the accompanying fluid. Consequently the fluid containing the new vorticity is removed and left behind, and the volume of the accompanying fluid decreases so that the vortices come closer together. The streamlines are shown in Fgi. 9.10.iii.

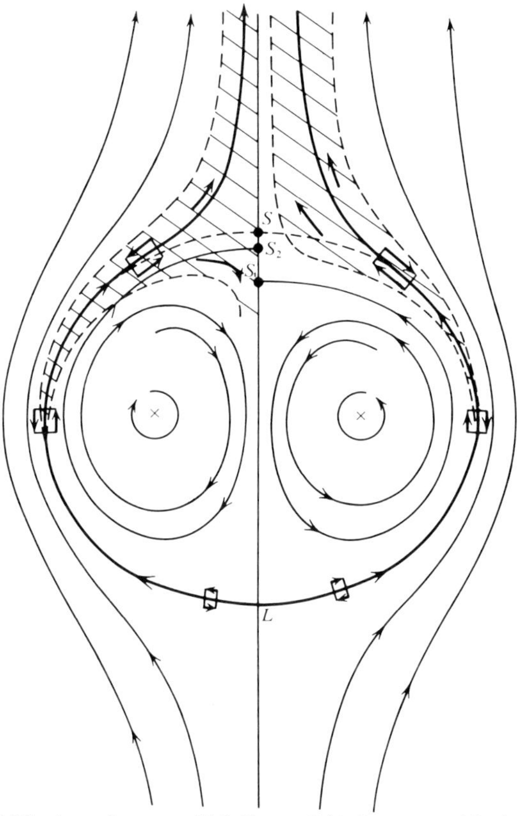

Fig. 9.10.iii The boundary on which the vorticity is generated is shown by the heavy line. This is detrained at the top of the accompanying fluid when the upper stagnation point is lowered from its starting position, S, to S_1. The shaded area represents what may be imagined to be the mixing region produced by instability on the vortex sheet at the boundary. Most of this is detrained e.g. in the right half of the diagram. But if the stagnation point rises to S_2 part of this mixed fluid may circulate down the centre of the accompanying fluid.

As the vorticity is fed towards S, the original upper stagnation point, it modifies the velocity field and carries itself away upwards as a thin double vortex layer. Here the total vorticity is zero and is soon cancelled out by the proximity of equal vorticity of opposite sign.

However, the vortex layer is unstable and as it becomes more intense it will break into billows and become thicker. Scorer and Davenport showed that although initially the stagnation point moved down to a position S_1 so that the mixed layer would all be fed way upwards, the stagnation point would move upwards again after a time which, with typical atmospheric stratification would be about two minutes. At the same time the vortex layer became thicker. In consequence some of the mixed fluid would then begin to be fed into the middle of the accompanying fluid, as indicated in the left half of Fig. 9.10.iii, with the stagnation point at S_2. As soon as this happens the motion in the accompanying fluid becomes unstable in the sense described in Section 3.11 and illustrated in Fig. 3.11.ii.

We now recall that the original vortex pair was unstable for sinusoidal disturbances even before buoyancy became important. Consequently small perturbations would grow. However, because of the decrease of R given in (9.10.8) and the consequent exponential increase in the vertical velocity of the vortices, $K/2\pi R$, this sort of disturbance will occur on account of the bouyancy forces alone, and it is impossible to distinguish the two effects simply by observation. The theory just outlined can be used to interpret the visible appearance of condensation trails behind aircraft. Because it involves initially a discussion of the physics of condensation, that part of the treatment is left until Section 11.3 and Section 11.4.

The dynamical problem has recently been treated more thoroughly by numerical methods by Hill (1975). She took the calculation as far as the beginning of the detrainment of the vortex sheet.

Chapter 10

DISPERSION OF POLLUTION

10.1 GENERAL PRINCIPLES

Many authors have been taken in by diffusion theory. Their approach has been to develop an analysis assuming that dispersion is diffusion, and that diffusion is best represented by a diffusion coefficient. The whole weight of conduction theory may then be deployed. The objection to this procedure is that we do not know what the diffusion coefficients ought to be, if only because they vary in time and space so enormously. In any case the assumption initially is that the motion which disperses the pollution is a property of the atmosphere and that the pollution is carried passively by it. This is not correct in situations where the pollution is emitted with a significant amount of buoyancy because in that case the buoyancy is the chief cause of dilution.

The theory of partly turbulent flow is fairly successful because Reynold's analogy, which assumes that momentum and material are diluted in the same way, is more or less in accordance with fact in such cases. However it can be completely wrong when applied to flow which is all turbulent, as the wind is, and to such entities as thermals, puffs and jets where the constant bulk parameters do not exist. For example, we showed in Section 9.6 that the ratio of the coefficients of vertical transfer of heat (or matter) and horizontal momentum by eddies can differ by a factor of 100 or more between cases. Many studies do not appreciate that it is easier to measure the distribution of pollution than to measure the eddy characteristics of the air motion and then predict the distribution of pollution by means of some theory. Since the theory is utterly dependent on the measurement of pollution for determining the coefficients, it is often a pointless exercise because it does not establish anything about mechanism. It does not even test the assumption that the transfer is due to the action of eddies: it merely calculates what the coefficients would have to have been if the observed distribution had been the result of some kind of eddy diffusion.

Of course the dispersion is indeed effected by motions which can be called eddies, and we may suppose that the sampling time or volume is large enough to engulf a large number of these eddies so that the statistical assumptions of the theory are valid. But that may well be useless because there is no limit to the time scale of fluctuations. In the observation of pollution we must recognise that the sampling time always affects the concentration measured, so that the assumption that eddy diffusion analysis is valid is simply incorrect.

When examined without preconceptions all diffusion theories are seen to do no more than specify coefficients of diffusion leading to an analysis and an answer which can then matched, usually only very crudely, to such observations as may be available. There is always a spread in the measurements so enormous that they cannot provide a test of any hypothesis. In any case they usually refer to a collection of occasions when the weather was rather different and the transport rates therefore different also. Most of the measurements are made on the ground downwind of a fairly isolated chimney, and theory is applied to the diffusion of pollution from an elevated point source. Since in most cases no measurements exist about the vertical distribution of pollution, the vertical diffusion rate is assumed to have a value which produces the right answer. Most such measurements have been made of SO_2 (Sulphur dioxide) without any knowledge of its rate of absorption at the ground, so that the whole exercise is a fantasy. Usually the answers are not silly but only because they cannot be. The pollution must go somewhere, so the simple principle of continuity means that, very roughly, its concentration must be estimated correctly to within a factor of 2 or 3 unless unrealistic assumptions are made about where it is. Since the observations vary by a similar factor anyway it is a case of 'you can't be wrong'!

The history of the subject shows that the laboratory approach greatly affected thinking. Occasions were chosen for study because of their supposed freedom from complications. It was always appreciated, at least by some, that the static stability and wind shear characteristics of an airstream could make a big difference to the mechanics of dilution of pollution, and many attempts have been made to introduce one or two additional parameters which might be adjusted to represent the various situations. Sutton's parameter 'n' and Pasquill's 'stability categories' are probably the best representatives of this technique, and many authors have mimicked with tedious monotony the routines which these two significant originators suggested, as if the existence of an analysis or a routine validated a procedure. The matter will not therefore be pursued further here. It should be noted, however, that any complication of any theory designed to take account of what are gaily called 'the meteorological parameters' cannot be better than very crude. This is because it must represent by one or two numbers the features of the profiles of several physical quantities and it cannot actually be of any use unless the meteorological parameters can be predicted. The whole procedure is therefore either dependent on provision of forecasts which cannot exist or upon the assumption that a climatology can be defined. Such a climatology must be measured over a period of several years, and when assessments come to be made the most useful climatology that is ever obtained is that of pollution itself.

This is because several decades are required to define normals of the atmospheric parameters, and during this time the sources of pollution undergo a great change. Furthermore the topography of the ground and the distribution of sources make it almost impossible to apply diffusion theories in practice. This has led to the invention of new theories which are crude in the extreme but realistic in their simplicity, namely pollution budget theories based on the misnamed 'airshed' concept. These are simple mathematical expressions of the continuity principle that the pollution remains close to the source, or it is carried into a neighbouring box of space by the wind, or else it may be diluted into a greater volume by upward mixing.

There is no point in pretending that there is any general theory of these calculations because they include whatever the author feels he needs or can have. They may include chemical reactions which vary with time of day if they are dealing with a 'Los Angeles smog' situation, and with absorption at the ground in circumstances dominated by SO_2 pollution. The sources and sinks of the variety of pollutants that may be considered depend on what inventories are available. The concepts of meteorology and fluid mechanics used are elementary in the extreme, but the computing techniques may be very sophisticated. This is typical of the 'indoor culture' which thinks that our brains rather than our fuel supply differentiates us from our less rich ancestors.

In this chapter we are concerned to gain a clearer idea of mechanisms, and to examine some of the techniques used to represent them. We are not so much concerned with offering ready made methods of calculating the dispersion of pollution as with examining the validity of ideas so that the student may more effectively discriminate between methods and have a clearer idea of what any calculation actually tells us. He is referred to my book *Air Pollution* for a fuller account of some of the meteorological phenomena which have an important effect on dispersion of pollution, and to Dr. Pasquill's *Atmospheric Diffusion* for an excellent comprehensive treatment.

It is not obvious to engineers and designers how little use diffusion theories have actually been. These theories have acted as a substitute for knowledge in many design exercises, and so they may said to be have been made use of. However they have been rather complicated, using impressive-looking formulas and requiring a considerable computer capacity, often leading to vast amounts of data logging and processing. All of this so befuddles the mind that only a few people have been able to keep sight of reality throughout the exercise. This makes them seem necessary because they appear to be doing what could not be done with pencil, paper, and a slide rule. When enough people have been persuaded of its value, the exercise assumes a prestige of its own without which any major undertaking involving chimneys would not seem complete.

At the bottom of it all is the simple concept of an isolated chimney standing on a plane surface in a steady uniform wind and emitting a known amount of pollution and buoyancy. The properties of this system will now be considered.

10.2 POLLUTION FROM AN ISOLATED STACK

The motion of the air out of doors is almost always at a high Reynolds number and low Mach number so that, other things being equal, the motion is similar whatever the wind speed unless the wind is light and variable. This means that the volume of air into which pollution is emitted from a stack is proportional to the wind speed U, and so the pollution concentration P is proportional to the emission rate Q and to U^{-1}.

The next step in the argument is deceptive in its simplicity: since eddies of all sizes are present in the airstream we have to consider what sizes are most important. An observer at distance d from the source is not concerned with eddies whose size is $d \times 10^{-3}$, for they merely cause a slight blurring of the outline of the plume. Equally eddies of size $d \times 10^3$ do not concern him for they merely cause a slow variation in what he calls the wind direction. The eddies with which he is chiefly concerned are those between $d \times 10^{-1}$ and $d \times 10$ in size. If we recognise that this is true at all distances and suppose that eddies of all sizes are present, we conclude that they must have the same effect on all scales. Thus the plume spreads out along a cone and the picture at distance nd is just the same as the picture at distance d enlarged n times. This picture of similarity at all distances leads to the conclusion that the concentration is proportional to b^{-2}, b being the width of the cone (Fig. 10.2.i). It does not even matter what the distance is, for whether it spreads out along a cone of wide or one of narrow angle, α, the dilution depends only on the width of the cone. Indeed this particular conclusion does not even depend on whether it is a cone or a paraboloid of revolution, which is what it would be under diffusion by a constant eddy diffusion coefficient in a uniform wind. This is because the concentration in the plume cross section is still proportional to b^{-2}.

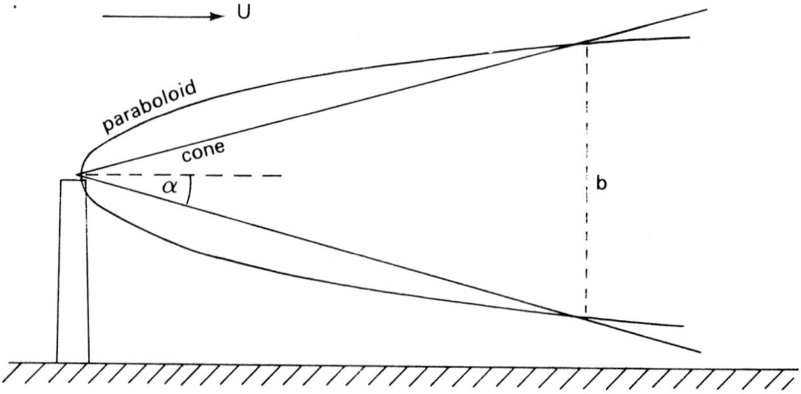

Fig. 10.2i The average pollution across a plume is the same when its cross section area is b^2 whether it has spread out along a cone or along a paraboloid. If the shape of the cross section is the same in case of different spreading rates a single linear dimension, b, is enough to specify the area.

The plume must impinge on the ground at some stage. If the pollution is absorbed by the surface it is leached out and the concentration becomes less than if the surface were not present. If it is not absorbed then all the pollution is concentrated in the air above as if there were another source at the image position of the actual source in the ground (Fig. 10.2.ii).

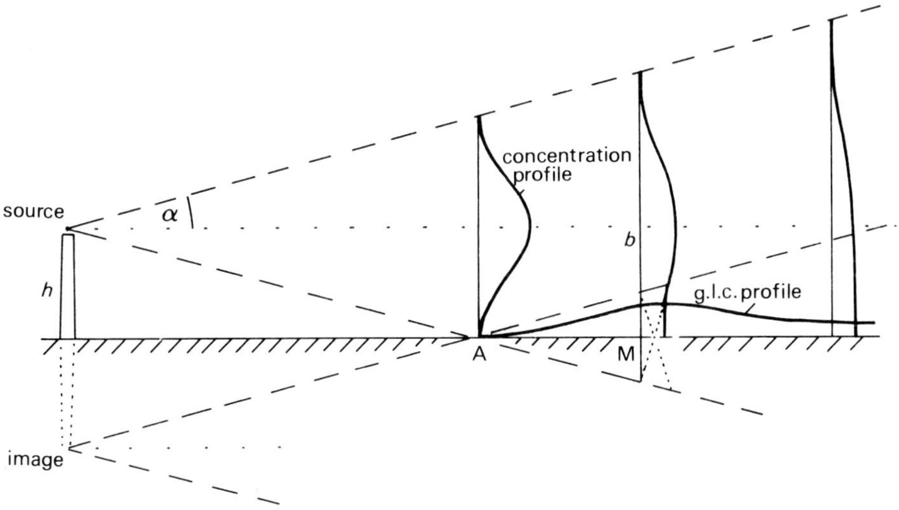

Fig. 10.2.ii If there is no absorption of pollution at the ground the concentration is as if there were an image source below ground. The profile across the plume is shown by the heavy line at three positions for conical spreading. At M the ground level concentration (g.l.c.) reaches a maximum, and is the sum of the two dotted profiles. The downwind g.l.c. profile is shown using a vertical coordinate for its magnitude.

At all points downstream of the point where the plume impinges on the ground the pollution has twice the concentration it would have at the same point in the cone in the absence of the ground. In the case of pollutants such as SO_2 there is a slow absorption on the ground or water surface, and the concentration in the air we would expect to measure is not far from what would be expected if the ground were absent. The value is very much dependent on the absorbing properties in the immediate neighbourhood of the measuring instrument. Perhaps the most important example is a hedge, which may quite easily reduce the SO_2 concentration in air passing through it by a factor of 2.

With such complications in mind we then note that upstream of A in Fig. 10.2.ii there is no pollution at the ground, while far downstream the spreading of the plume causes a decrease in the ground level concentration. There is therefore a maximum ground level concentration, $P_{\max}$, occuring at some point

M downwind of A, and its precise position depends on the distribution of the pollution across the cone. This distribution is to be assumed similar at all distances because of the similarity of the eddy behaviour mentioned above (but we shall have some more to say about this later), and the magnitude decreases as the plume widens. Fig. 10.2.ii shows the vertical concentraion profile in the vertical plane through the source, the heavy line indicating what it would be with complete reflection at the ground. Above the ground line in the diagram is a curve indicating the downwind profile of ground level concentration in this central plane of the plume, and Fig. 10.2.iii shows a map of the ground level concentration in the form of concentration contours.

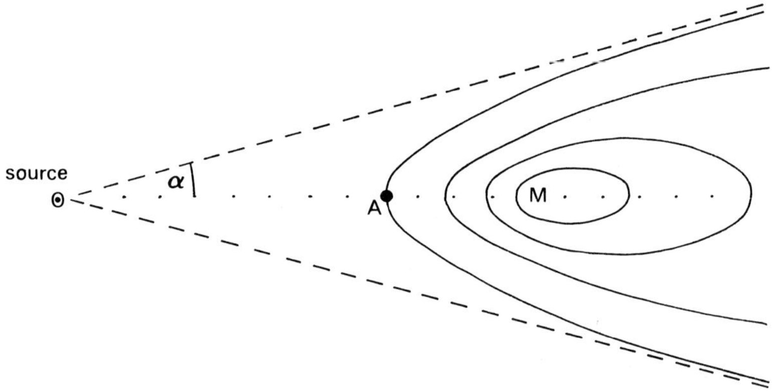

Fig. 10.2.iii A map of contours of constant concentration at ground level from a tall chimney. Outside the contours through A the concentration is everywhere zero, if the wind is steady, and the maximum occurs at M. The contour through A is asymptotic to lines at the angle of the cone, α.

The actual magnitude of P_{max}, at M, depends only on b, the width of the cone at M. b is a numerical multiple of H, the source height, the factor depending on the form of the concentration profile across the plume. In the diagrams above H is equal to h, the stack height and P_{max} is proportional to h^{-2}.

In practice many important sources of pollution are also sources of heat. The buoyancy causes the effluent gases to rise. Let us suppose that this rise can be represented by a height h', which we call the thermal rise. Far downwind the plume will behave as if it had come from a source somewhere in the neighbourhood of the actual source (Fig. 10.2.iv), and if we assume that the point M is far enough downwind for this to be true we may say simply that its height is equal to $h + h'$. We then arrive at the formula for the maximum ground level concentration:

$$P_{\max} = N \frac{Q}{U(h+h')^2} \qquad (10.2.1)$$

in which N is a numerical factor which is determined by the angle of the cone and the profile of concentration across it. Its value is around 0.15, but it should

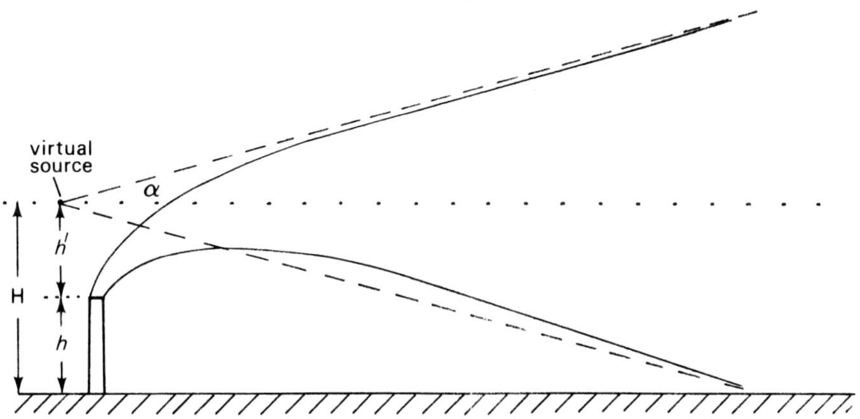

Fig. 10.2.iv By the time a buoyant plume is diffused down to the ground it is behaving more or less passivley, as if it had originated at a virtual source at height h' above the actual source without any buoyancy. In the diagram the passive spread is represented as conical.

be no cause for surprise if on particular occasions it is found to be as small as 0.05 or as large as 0.5, especially when the value of h' is open to argument. A simple estimate is obtained if we guess that the maximum ground level concentration occurs at a point where the concentration is about the average value in the plume cross-section and that the area of the cross-section is about 2/3 of the area of a circle of radius $3H/2$. The flux of pollution acrôss this area is Q, and is also equal to $\frac{2}{3}\pi\frac{9H^2}{4}UP_{\max}$, and from this we find that $N = 2/3\pi$. But, *caveat emptor*, this is not an assertion about the value of N other than it is probably about 0.2 $\overset{\times}{\div}$ 3.

The formula (10.2.1) is not useless; it is of great vaue in emphasising that ground level concentrations are much reduced by increasing the chimney height. However it must be emphasised that the advantage of a high emission point often lies just as much in the fact that when the air is stably stratified a higher chimney more often reaches above the stable layer thereby decreasing the number of occasions on which the plume is trapped below it, as illustrated in Fig. 10.2.v.

Fig. 10.2.v The tall chimney of an aluminium works in the valley of the Durance standing above the pollution accumulated at night from many small sources in the cold pool of air generated by katabatic flow down the mountain sides. The plumes escapes and rises under its buoyancy. The disc on the stack top is a device sometimes employed to prevent flagging when the efflux velocity is low.

10.3 JET AND THERMAL RISE

In the following discussion (based on Scorer 1959) there is an assumption that the effect of thermal rise of effluent gases on the ground level concentration can be correctly represented by adding a quantity h' to stack height h. Although this is a simple way of achieving some recognition of the effect of thermal rise in the formula, it has never been justified.

Fig. 10.3.i Bent-over buoyant plumes are bifurcated by the buoyancy-induced circulation which gives it the cross section of a 2D thermal.

Thus if the angle of spread of the cone is very wide the plume impinges on the ground fairly close to the chimney. This may cause the maximum ground level concentration to lie closer to the chimney than the distance at which the theoretical thermal rise is actually achieved, and then its full value should not be added to h in (10.2.1). Many suggestions have been made as to how h' should be defined, for example as 'the height at which the centre of the plume levels out'. This cannot be measured with any accuracy, and any such definition depends on how the height of the centre is measured. The cross-section of a buoyant plume is initially like that of a two-dimensional thermal (see Fig. 10.3.i), namely bifurcated (see Fig. 8.4.i), and there is no part of it which can be easily defined and observed other than the top and the bottom. Thus it becomes more like a non-buoyant plume passively dispersed by ambient eddies further downwind, and consequently there is no simple definition which can be used in practice for the determination of h'. Our purpose here is to discuss what h' must depend upon, and since we shall discover that it depends on Q and U, it is clear that the simplicity of our formula (10.2.1) is about to disappear. The temptation to disappear with it and become engulfed in a welter of academic and useless analysis has to be studiously resisted. In a more complicated situation we do not benefit from having a more complicated formula: in the case of partly turbulent flow the complexity is overcome by a contentment with simplicity. No advantage has been gained by complexity in the history of plume rise formulas. Simple formulas can guide policy: complicated ones disappear into a computer and then judgements are obfuscated.

The effluent emerges from the chimney as a buoyant jet, with a velocity given by (8.11.11), in which the constants C and c determine whether it is behaving like a plume or a jet or something in between. They are determined by the buoyancy and momentum fluxes at the stack mouth.

We recognise that the horizontal wind will cause the effluent to become bent over, and we assume this to occur when the vertical velocity w is equal to μU, where U is the wind speed and μ is a numerical factor, to be determined from observation, about equal to unity. When the plume is bent over we imagine that it has taken up the speed of the wind and that a vertical cross-section of it behaves like a cross-section of a two-dimensional (cylindrical) thermal or puff.

The thermal rise will be complete when the vertical velocity due to the buoyancy has been reduced to the order of magnitude of the ambient eddy velocities, which we represent by λU, λ being typically of the order of 0.1 in a wind over not very rough ground and of neutral static stability. λ could be much larger than this if there are large obstacles (such as exist in an urban built-up area) to create large eddies or when thermal convection is active in sunshine. In this latter case λ could easily be 0.4, or even as large as 1.0 when convection is vigorous in a light wind. At the other extreme λ could be as low as 0.001 when a chimney stands well above very smooth ground or when the lower layers are very stably stratified, as they often are at or after dusk on a clear evening. When the ambient turbulence is reduced to negligible proportions in this way the thermal rise is restricted only by the stratification of the environment, and for reasons explained in Section 10.9 and Chapter 11 this means that the plume will normally ascend up to the level of the base of convection cloud.

To illustrate the method to be employed we shall take first the simplest case of zero buoyancy. Suffix 0 indicates values at the orifice, whose radius is a, σ is the pollution concentration, while suffix 1 denotes values at the height where the jet becomes bent over and $w = \mu U$. In the bent over part the height, denoted by y, is measured from a virtual origin at the height where the cylindrical puff which is equivalent to the cross-section of the bent over plume would have originated. As mentioned in Section 8.8 (see also Section 8.15) the typical value of n, is about 4 and so the width, r, of the bent over part is given by

$$y = 4r. \qquad (10.3.1)$$

In the vertical part, by (8.9.11),

$$w = w_0 z_0/z, \qquad w_1 = \mu U = w_0 z_0/z_1,$$

$$\sigma = \sigma_0 z_0/z, \qquad \sigma_1 = \sigma_0 z_0/z_1 = \sigma_0 \mu U/w_0, \tag{10.3.2}$$

$$z_1 = z_0 w_0/\mu U.$$

The flux of pollution is

$$F = \sigma_0 w_0 \pi a^2. \tag{10.3.3}$$

In the bent over part the cross-section area (see Section 8.15) is taken to be $2.4r^2$ and so the pollution flux, when the effluent is travelling with the wind U, is

$$F = 2.4r^2 \, \sigma \, U. \tag{10.3.4}$$

Using (10.3.1 and 10.3.3) we therefore deduce that

$$2.4\left(\frac{y_1}{4}\right)^2 \frac{\sigma_0 \mu U}{w_0} U = \sigma_0 w_0 \pi a^2,$$

or

$$y_1 = 4\left(\frac{\pi}{2.4\mu}\right)^{1/2} \frac{w_0 a}{U}. \tag{10.3.5}$$

Since σ and w are proportional throughout, the above argument is the same if applied to the conservation of upward momentum.

In the bent over part, according to (8.8.5)

$$w = w_1 y_1^2/y^2, \tag{10.3.6}$$

so that the height y' at which the bent-over jet becomes passive, where $w = \lambda U$, is

$$y' = y_1 (w_1/\lambda U)^{1/2}$$

$$= 4\left(\frac{\pi}{2.4\lambda}\right)^{1/2} \frac{w_0 a}{U} \cong 4.5 \, w_0 a/\lambda^{1/2} U. \tag{10.3.7}$$

Taking $z_0 = 5a$, i.e. $n = 5$ for a jet, the rise from the orifice before the plume becomes passive is equal to

$$h' = y' - y_1 + z_1 - z_0$$

$$= a\left(\frac{w_0}{U}\left(\frac{4.5}{\lambda^{1/2}} - \frac{4.5}{\mu^{1/2}} + \frac{5}{\mu}\right) - 5\right). \qquad (10.3.8)$$

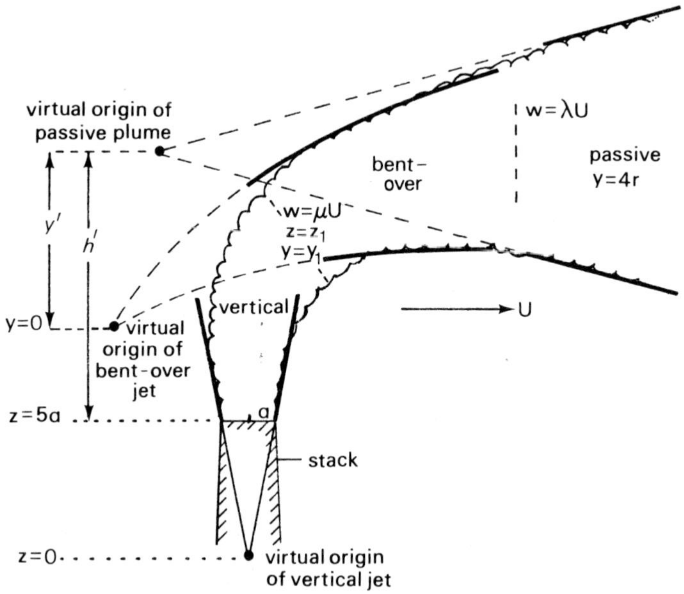

Fig. 10.3.ii Stages in the dispersion of a bent-over jet. In the jet stage the virtual origin is within the stack. It becomes bent-over after the point where $w = \mu U$, at which $z = z_1$ and $y = y_1$. It behaves like a 2D puff until the ambient eddies take over when w has decreased to λU, and has virtual origin at height $y = 0$. When $w < \lambda U$ it behaves like a passive plume, diffusing along a cone originating at height h' above the stack top, where $y = y'$.

This formula is irrelevant if $w_0 < U$, for the plume is bent over from the orifice. In discussing it therefore it is to be assumed that h' is positive, presumably because $w_0 > U$, μ is about unity and λ is much less. If the orifice is narrowed while the volume flux is unaltered we have that

$$w_0 a^2 = \text{constant} \qquad (10.3.9)$$

so that a decrease in a is equivalent to an increase in w_0. It is most effective when λ is much smaller than μ, which is in conditions of low ambient turbulence. In this most effective case the **jet rise** is

$$h_j' \cong 4.5 \frac{w_0 a}{U\lambda^{1/2}} \tag{10.3.10}$$

and so the effect of a large efflux velocity is to lift the effluent by jet rise a few orifice diameters above the stack top. This is an effective method of avoiding the capture of effluent by the eddies in the lee of the stack. When such capture occurs it is called 'flagging' because the plume looks like a flag on the stack (Fig. 10.3.iii). It is sometimes also called 'downwash.'

Fig. 10.3.iii A 'flagging' plume. The efflux velocity is not large enough compared with the wind to prevent smoke entering the wake of the chimney. When this happens frequently the chimney becomes blackened on the outside and the effective height of the stack is reduced.

At greater distances than the chimney height from the chimney, assuming there is no flagging, the most important effect is the rise due to buoyancy after bending over has occurred. Cases in which the other smaller factors are included were discussed by Scorer (1959) but we now calculate the dominant component of h', namely y', for the case of significant buoyancy. This is the **thermal rise.** In this case, instead of (10.3.1) we have roughly, according to Richards's measurements for a 2D thermal referred to in Chapter 8, that in the bent-over plume

$$y \cong 2r, \tag{10.3.11}$$

$$w = 1.0(gBr)^{1/2}, \tag{10.3.12}$$

and the buoyancy flux is $F = 2.4r^2 gBU.$ (10.3.13)

If we put $w = \lambda U$, $y = y'$ and eliminate B and $\boldsymbol{r}$ we obtain the thermal rise as

$$h_b' \cong y' = \frac{2F}{2.4\lambda^2 U^3} \cong 0.8F/\lambda^2 U^3. \tag{10.3.14}$$

In view of the roughness of most of the approximations it is scarcely worth including the values of y_1, z_1 and z_0 in this formula. Since at the orifice

$$F = \pi a^2 w_0 g B_0 \tag{10.3.15}$$

the value of h' given by (10.3.14) exceeds that given by (10.3.10) in the ratio

$$\frac{h_b'}{h_j'} = \frac{0.8\pi a g B_0}{4.5\lambda^{3/2} U^2}, \tag{10.3.16}$$

and the thermal rise is much more significant than any jet rise if

$$gB_0 >> 3\lambda^{3/2} U^2/a. \tag{10.3.17}$$

The most significant feature of these formulas is their extreme dependence on λ, which is probably the least well known factor. This is not a criticism of the formulas: it merely indicates that the plume rise h' is very dependent on the level of ambient turbulence. This fact cannot be altered by any other theory except one which seeks to define h' without reference to λ.

In (10.3.12) h' is seen to be proportional to F, the buoyancy flux. Measurements made by Lucas and others (1963) indicate that for large chimneys the dependence is more nearly like $F^{1/4}$. But that could be due to their assumption that h' depends on a power of U different from U^3 and also to the fact that taller chimneys have larger values of F and may have a stronger wind to contend with. Furthermore, the definition of h' may interfere: for example, if it is defined as the height reached by a buoyant plume at a given distance downwind or given time from efflux, then according to (8.8.5) we have for a two-dimensional thermal

$$z \propto t^{2/3}, \tag{10.3.18}$$

while distance downwind is given by

$$x = Ut. \tag{10.3.19}$$

Therefore the height h' attained at $t = T$ and $X = UT$ is

$$h' \propto T^{2/3} \propto X^{2/3} U^{-2/3} \tag{10.3.20}$$

which is a very different power of U from that in (10.3.14). Thus there have never been adequate measurements to check any proposed formula for h', and certainly none which have been able to eliminate the dependence of the results on the unknown power of U and on the way in which h' is defined. In this treatment the dependence upon λ is very great, which is to be expected. In practice λ is likely to decrease with height, and this again emphasises the importance of tall stacks, Somewhat different powers of U can be obtained by other models of a buoyant plume.

The Formula (10.3.14) indicates that if $B_0 = 0.3$, $g = 10^3$, $w_0 = 2 \times 10^3$, and $a = 300$ CGS (which is typical for a power station) when $\lambda = 10^{-2}$ and $U = 3 \times 10^2$ (reasonable for a clear evening when the ground has begun to cool) we find that $h' = 170$ km. On the other hand when $\lambda = 10^{-1}$, $h' = 1.7$ km.

In either case the plume rise is very large, but if $U = 10^3$ CGS (10m sec^{-1}) the results are about 4.6 km and 46 m respectively. These answers are quite realistic except that a neutrally stratified atmosphere of sufficient depth for it to happen does not exist in all cases.

In practice, however, the plume is not diffused downwards by small eddies, but by large ones. Hence all theories which give an important place to a definite value of h' must be viewed with the greatest scepticism.

Bosanquet pointed out that if $h' \propto U^{-1}$, which it was according to one of his suppositions, then when the wind was such that the maximum ground level concentration according to (10.2.1) took its greatest value, the value of h' was h. Thus a chimney emitting buoyant effluent could always be safely thought of as being twice as high as it actually was. This ludicrous result is not so much wrong as irrelevant to reality, because it is eddies large compared with plume width which effectively bring the effluent to the ground. At very large distances the chimney height makes no difference to the pollution concentration, especially if its upward diffusion is limited by a stable layer, which to a large extent it usually is at cloud base.

Furthermore, if $h' \propto U^{-3}$, then Bosanquet's result is that when the wind is such as to maximise P_{max}, $h' = h/5$, which means that we can safely always add on $h/5$ to a stack height in calculations of ground level pollution. This is a very different result from Bosanquet's original one and is more realistic, but still silly because both results are clearly false for infinitesimal buoyancy. This shows that the functional form of (10.2.1) is inadequate for representing thermal rise.

The upward momentum of the effluent at efflux is w_0 per unit mass. The upward momentum acquired by buoyancy gB_0 during time t after efflux is gB_0t. The buoyancy effects are therefore more important than any effects due to efflux velocity from time

$$t = w_0/gB_0 \text{ onwards.} \tag{10.3.21}$$

For an excess temperature of around 100°C, B_0 is of the order of 1/3. For an efflux velocity of the order of 10 m sec^{-1}, the buoyancy effects are dominant from about 3 sec after efflux, or from 30 m downwind from the stack in a 10 m sec^{-1} wind. For large stacks buoyancy is clearly the only significant contributor to plume rise.

10.4 NEGATIVE PLUME RISE: EFFECT OF LIQUID WATER CONTENT

If a plume contains liquid water at efflux and it is evaporated subsequently as a result of mixing into the surroundings, the latent heat of evaporation is lost, and a negative buoyancy may be created. The plume then sinks to the ground, and behaves as if it were travelling with negative buoyancy (see Fig. 10.4.i) from some imaginary source higher than the actual chimney top. Figure 10.4.ii gives the cooling effect. The isopleths give the temperature deficit which the effluent would have to have at this imaginary emission point in order to have the same temperature deficit as the actual plume after the water has evaporated. More practically it may by regarded as the factor by which the plume will have been diluted into the surroundings when it has a temperature deficit of 1°C.

Usually the liquid water is present because of a gas washing operation. Here water is evaporated into the gases, but further cooling occurs by the loss of some sensible heat to the water. This emerges from the bottom of the washing chamber warmer than it went in and there is a further loss through the walls of the stack. The result is the condensation of a cloud of droplets. In the diagram T_1 is the temperature at which the gases are saturated and T_2 is the temperature

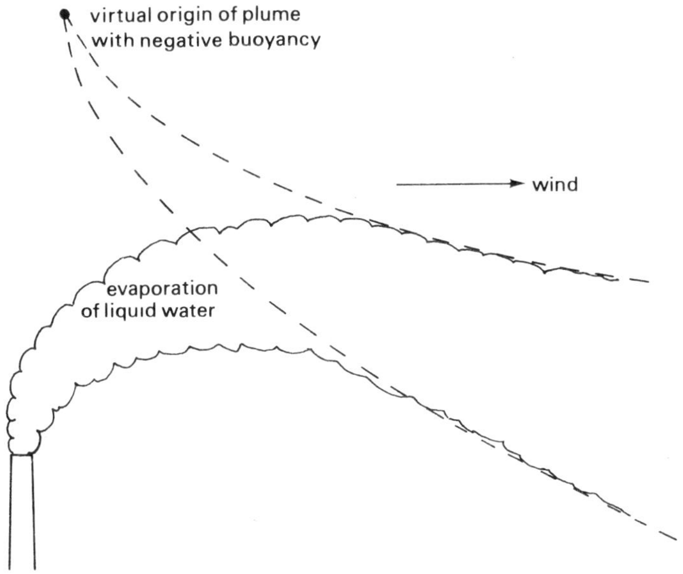

Fig. 10.4.i When a plume acquires negative buoyancy as a result of the evaporation of liquid water contained in it at efflux it behaves as if it were descending from a virtual origin some distance above the chimney top. This behaviour is common for plumes from which pollution has been removed by wet gas washing.

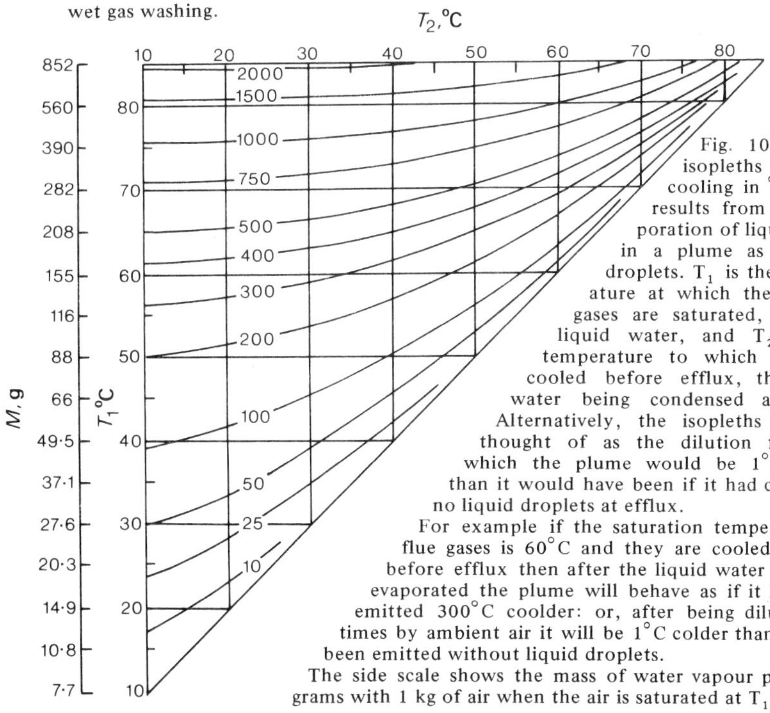

Fig. 10.4.ii The isopleths show the cooling in °C which results from the evaporation of liquid water in a plume as airborne droplets. T_1 is the temperature at which the effluent gases are saturated, with no liquid water, and T_2 is the temperature to which they are cooled before efflux, the liquid water being condensed as cloud. Alternatively, the isopleths may be thought of as the dilution factor at which the plume would be 1°C colder than it would have been if it had contained no liquid droplets at efflux.

For example if the saturation temperature of flue gases is 60°C and they are cooled to 33°C before efflux then after the liquid water has been evaporated the plume will behave as if it had been emitted 300°C coolder: or, after being diluted 300 times by ambient air it will be 1°C colder than if it had been emitted without liquid droplets.

The side scale shows the mass of water vapour present in grams with 1 kg of air when the air is saturated at T_1.

to which they are cooled before efflux. The side scale shows the mass of water present as vapour in 1 kg of air when the gases are saturated at T_1 from which the mass condensed on being cooled to T_2 may be deduced. Typically T_2 is about 25°C and T_1 about 40°C, and the cooling has the same effect as if the effluent had emerged about 80°C colder than the environment, implying that $B_0 \triangleq -1/3$. No surprise is warranted if such a plume descends to the ground within 5-10 stack heights of the stack, or even nearer in light winds.

10.5 PLUMES IN STAGNANT AIR UNDER AN INVERSION

We are concerned here with the change in temperature at any height in plumes with a vertical axis in calm air due to a reduction of the orifice size. The chief effect is to increase the efflux velocity, which increases the rate of dilution because the jet is now everywhere nearer to its virtual origin. Eventually it becomes a plume, according to the arguments of Section 8.11, in which the relationships between w, B, and height z above the origin are given by (8.11.11 and 8.11.16) so that the plume is the same regardless of the duration of the jet stage lower down. The only difference is that the apex of the cone along which the effluent rises has been raised from $5a$ below the chimney top to $5a'$ below it, a' being the new reduced radius. This is unlikely to have any significance as far as penetrating inversions is concerned, and the effect could equally be obtained by simply increasing the stack height by $5(a - a')$. This would not increase the fan power required to drive the flue gases.

10.6 COLD INFLOW

If a bottle of water is held upside down it empties and air flows in but if a downward-pointing tap is turned on the efflux velocity of the water is usually too large to allow air to penetrate into the pipe. Therefore at some value of the efflux velocity as it is decreased the air must begin to flow in. The same phenomenon occurs at a chimney mouth: when gas of large buoyancy is emerging with a rather small velocity, cold air from outside can penetrate down the stack. The cold inflowing air is mixed with the rising warm gas as it penetrates down, and eventually it reaches its maximum penetration beyond which its much reduced negative buoyancy is too small to carry it against the upcurrent.

The phenomenon was studied by Jorg and Scorer (1967), who found that the critical mean efflux velocity below which inflow occured depended not only on the buoyancy of the effluent but also on its velocity profile. If the velocity in the boundary layer was increased by flattening the profile for a given volume flux, the buoyancy difference needed for inflow was increased. No non-dimensional relationships were obtainable for this reason, but it was clear that in turbulent flow the diameter of the orifice was not generally a determining factor, although the thickness of the boundary layer was.

The distance d penetrated into a tube by exterior fluid was found roughly to fit the formula

$$\frac{Bg\nu}{(d^2/D^2 + 8)^3 V^3} = 10^{-6} \tag{10.6.1}$$

where D is the tube diameter, V the mean velocity far inside, B the buoyancy, and ν the kinematic viscosity. Although this provides an order of magnitude for the critical velocity at which d is zero for a given B, and provides also an idea of the depth of penetration, it has considerable limitations in the spread of values upon which it is based. This is because there are many extraneous influences such as tube interior roughness to be reckoned with. In practice there is nearly always a cross wind, and then the experiments showed that the inflow usually took place at the sides rather than at the up- or downwind points of the orifice.

The problem has been a real one. Oil fired furnaces used to produce black acid smuts, particularly when starting up from cold and with metal stacks. On low load the efflux velocity was reduced in greater proportion than the buoyancy, and so (10.6.1) shows that penetration is more likely. Also, when penetration occurs in cold weather it can be enhanced by cooling through the stack wall. The result was that the chimney wall was cooled below the acid dewpoint (H_2SO_4) and black carbon deposited on the wall soaked up the acid. When the efflux velocity was then increased particles a few mm in size were carried out by the airstream. The concentrated sulphuric acid burnt holes in clothing, in the paint of car roofs, and in other objects. The remedy has been to maintain sufficient efflux velocity and to insulate the chimney so that the temperature is kept above the acid dewpoint. Various bevelled tops have also been found to reduce the likelihood of cold inflow.

Although cold inflow does sometimes occur in cooling towers, the more likely phenomenon is separation of the outflow from the wall at the level of smallest diameter in a hyperboloidal tower. Actually very little is known about inflow and possible penetration to the bottom in cooling towers. However it is probably not a serious problem anyway because of the fairly uniform distribution of the heat source horizontally across the bottom in a conventional wet cooling tower. Thus the heat source distribution prevents a large fraction of the volume becoming cold.

The picture in Fig. 10.4.iii is of the experiment from which (10.1.6) was derived.

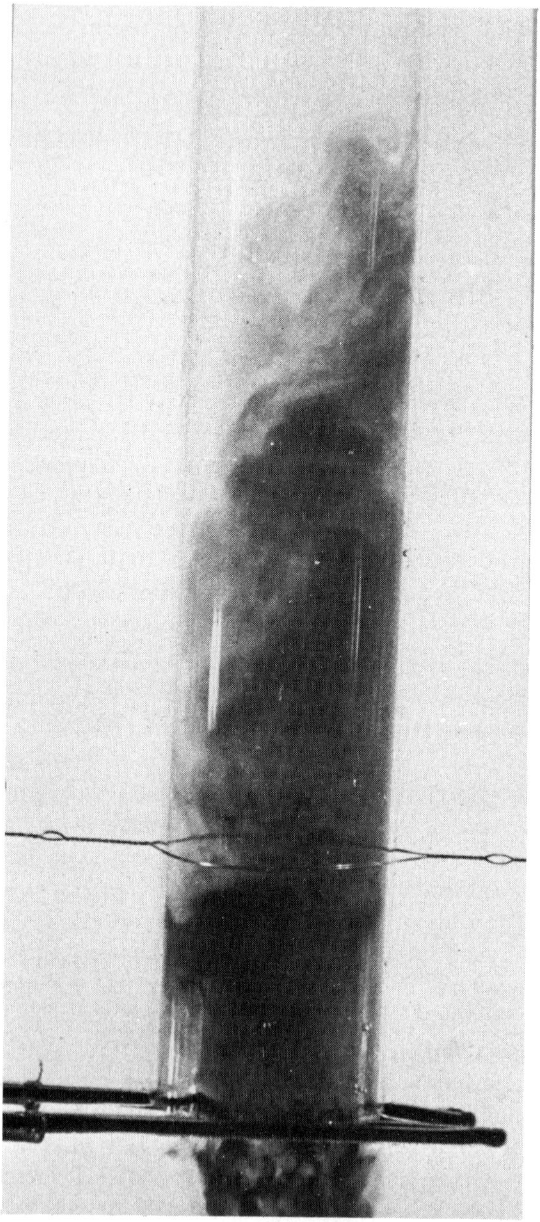

Fig. 10.6.i A model experiment of cold inflow by Jorg (Jorg and Scorer (1967)). Salt water is descending the glass tube (8.5 cm outside diameter) which is immersed in a tank of fresh water. Around the open bottom of the tube are six jets of dye. Any inflow of the buoyant fresh water up the tube carries dye with it, and the penetration can be observed. In this case it is a little more than 4 tube diameters.

10.7 THE SHAPE OF CHIMNEY PLUMES

Any observer soon notices that there are several different shapes of plume. Several authors have sought to list the distinct shapes by their appearance, on the following lines (see Fig. 10.7.i):

looping	having large sinuosities: dispersion by eddies large compared with plume width.
coning	having a steadily widening boundary: dispersion by eddies of size smaller than the plume width.
fanning	dispersing horizontally more than vertically: plume achieves equilibrium level in a stably stratified airstream soon after emission.
fumigating	dispersing downwards but not upwards: emission below an inversion which is not penetrated.
lofting	dispersing upwards but not downwards: stack height sufficient to place effluent above an inversion.
thermalling	the plume breaks up into distinct masses indicating that the buoyancy from the chimney is initiating thermals or being entrained into natural ones.

There are other features such as:

flagging (downwash)	effluent entrained into the eddies in the lee of the chimney.
downdraught	effluent entrained from time to time in the lee of the building associated with the chimney. This is common with dwelling houses.
puffing	regular formation of cauliflower-like lumps at the upwind edge of the chimney mouth. These usually have a periodicity of 1 to 5 puffs per second.
bifurcation	this is due to the entrainment of clear air up the middle of a plume bent over by the wind so as to make it have a cross-section like a cylindrical thermal of puff.

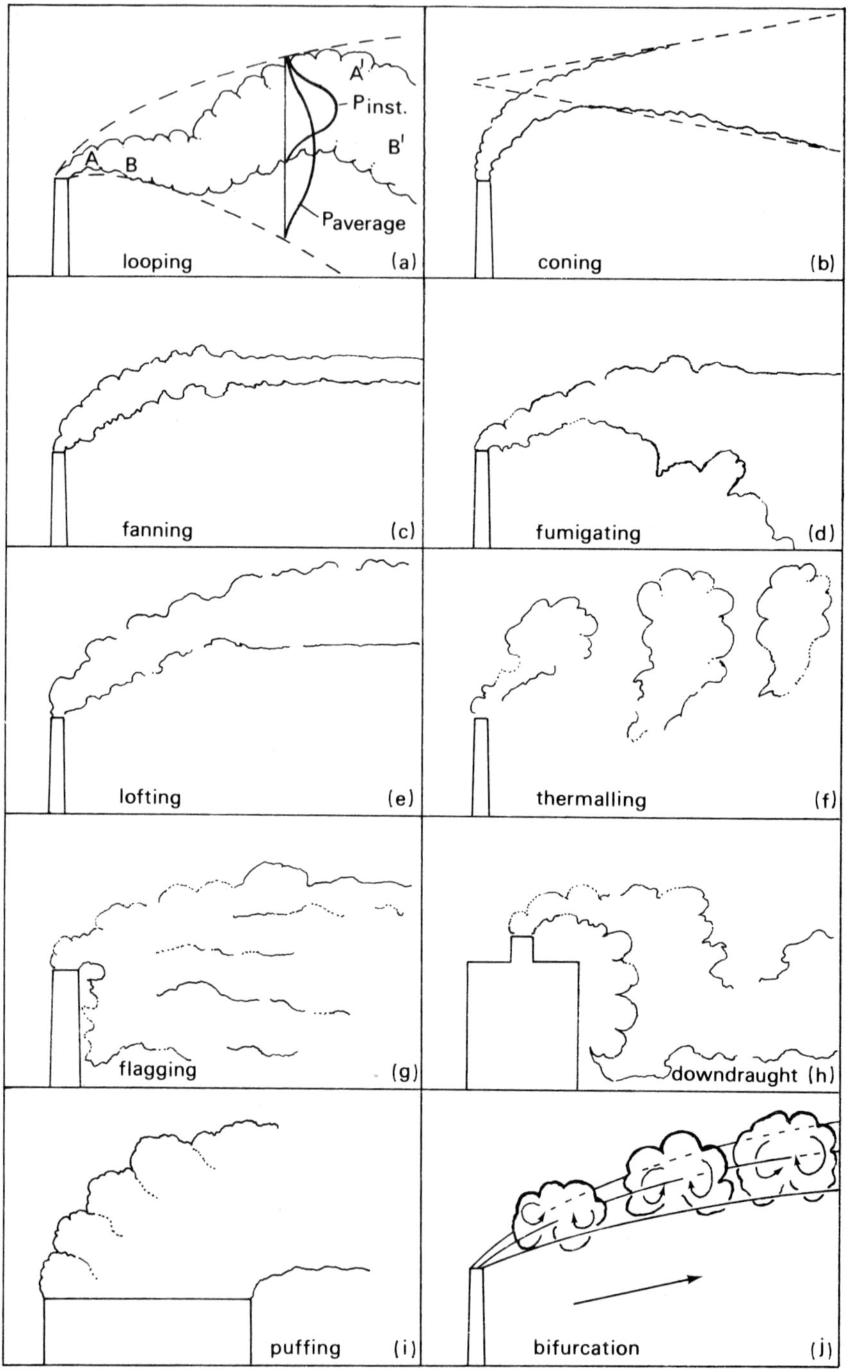

A
B
A'
B'
Pinst.
Paverage
looping (a)
coning (b)
fanning (c)
fumigating (d)
lofting (e)
thermalling (f)
flagging (g)
downdraught (h)
puffing (i)
bifurcation (j)

There now follow comments on these various patterns.

The large eddies which cause a fairly narrow plume to wander about within a bent-over cone may have been shed by large buildings or hills, or generated by thermal convection in sunshine. As far as any formula for dispersion is concerned such a plume is treated as if it were distributed over the **bent-over cone** which envelopes it. The particles of the plume do not travel on sinuous tracks, but the particle at A travels to A′, and B to B′ (Fig. 10.7.i); thus the sinuosities move along with the wind, and the sinuous plume itself is widened more slowly by smaller eddies. There is an absence of eddy sizes in between, and this is usually because the smaller eddies are produced by the buoyancy of the plume or by small roughnesses on the ground, while the larger one are produced by thermals or are shed from large obstacles.

A **coning plume** is usually **bifurcated** and rising under its own buoyancy, and it appears to have the shape of a bent-over cone only when seen from the side. From beneath it is usually bifurcated very clearly (Fig. 10.3.i). Indeed a plume seen from below to be bifurcated is certainly rising because of its buoyancy, and the eddies which mix it into the surroundings are due to the buoyancy and are not present in the ambient air.

If a buoyant plume is emitted into an airstream which is strongly stratified it starts off bifurcated (and coning) but soon reaches its equilibrium level and then travels horizontally. The stable stratification is most common in rather light winds whose horizontal direction may vary, and then the plume appears to wander horizontally. This does not always happen, so that a behaviour appropriately called **fanning** is rather rare (Fig. 10.7.ii).

Fig. 10.7.i (preceeding page) (a) *Looping:* the sinuosites are large compared with the width of the instantaneous plume, whose profile contains a greater maximum concentration than the longer term average across the region swept by the sinuosities.

(b) *Coning:* this is typical of a buoyant plume growing by mixing in surroundings with little turbulence.

(c) *Fanning:* in side view the buoyant plume is seen quickly to reach its equilibrium level in a stably stratified environment. It may often be seen to wander horizontally if the wind is light and variable. See Fig. 10.7.ii.

(d) *Fumigating:* this occurs when the plume is trapped in a well mixed layer below a very stable layer. If a fanning plume is dispersed downwards as a result of thermal convection in the sunshine the ground below is said to be fumigated.

(e) *Lofting:* in this case the plume asends by its own buoyancy to above a stable layer which limits its downward diffusion.

(f) *Thermalling:* when thermal convection is active the plume may be broken into lumps by the individual thermals, whose buoyancy dominates that of the plume.

(g) *Flagging:* the effluent is fed into the turbulent wake of the stack and descends several diameters on the lee side.

(h) *Downdraught:* in this case the plume is fed into eddies behind a building on which the chimeny stands.

(i) *Puffing:* on close examination a periodic turbulent fluctuation may be seen on the upwind lip of the stack.

(j) *Bifurcation:* a buoyant plume, here sketched in perspective, has a crosssection like that of a thermal. Coning plumes are usually bifurcated.

Fig. 10.7.ii The plume from the copper mine at Mt. Isa, Queensland, seen early in the morining. The mixing ceases when the plume settles at its equilibrium level. In the distance the plume can be seen just above the skyline. Fumigation occurred about an hour later at the airport.

When the wind is very light a plume may rise as a vertical hot plume and spread out horizontally on reaching its equilibrium level (Fig. 10.7.iii). The top of the plume usually overshoots a short distance, and because there is often a wind discontinuity at a very stable layer the plume is carried away as thin layer of polluted air.

When a polluted layer is formed and thermal convection begins in the morning causing **fumigation**, the pollution is sensed at ground level over a wide area at about the same time. For this reason the source may not easily be identified, especially if the pollution has accumulated during the hours of darkness. This mechanism was first identified by Hewson when a smelter at Trail (B.C.) caused pollution to be observed about 50 miles down the valley of the Columbia river in the morning without it having been observed previously at the ground in between. The rise in pollution levels at the ground may be quite sudden because the morning convection may produce anabatic winds up the side of the valley.

Fig. 10.7.iii The plume from the old power station at Ironbridge. The shear in the flow above the stagnant valley air can be seen in the overshooting elements. This, like Fig. 10.7.ii was followed by fumigation.

These spread out above the polluted layer and displace it downwards in the centre of the valley without any dilution.

When a plume is emitted above an inversion, as if often the case when there is a cold layer of air at the ground in the evening or at night, the ambient turbulence is usually very slight. Then the plume rises under its own buoyancy, with bifurcation, until it reaches the cloud base or a stable layer. **Lofting**, according to the scheme in Fig. 10.7.i, is rare because it includes situations in which the plume is dispersed upwards but not downwards by ambient turbulence. Moreover turbulence greater than that due to the buoyancy of the plume itself is rare above a very stable layer because the stability is seldom zero there.

Thermalling occurs when thermals are being generated by sunshine on the ground, the thermals being the dominant form of eddy. A plume of large buoyancy initiates its own thermals and so all the pollution is likely to be carried upwards while dispersion downwards is negligible.

Flagging is a feature mainly of plumes of low buoyancy and low efflux velocity. In the case of some chimneys, vortices with vertical axes are shed from the lee side and pollution can sometimes be seen to be entrained into the centre of these and carried away downwind. It is important only when a significant fraction is drawn down the chimney most of the time. In the case shown in Fig. 10.3.iii the chimney was blackened down to about half its height. Some architectural designs cause flagging, but the problem of teaching architects that a chimney should be designed to emit the pollution cleanly and clear of the building has been alleviated in the last two decades. During these years some architects have seen their job, at last, as taking account of these phenomena, and the problem of architects has become a problem for architects.

Downdraughts can be avoided by making the chimney stand sufficiently clear of the building. Many chimneys fail to emit so as to ensure that pollution is clear of all eddies due to separation of the airflow from salient edges on the building. The reader is referred to the chapter on separation in *Air Pollution* (Scorer 1968) for a fuller discussion. A special case is that of the plume from a power station employing gas washing where the descent of the effluent to the ground was clearly due to the negative buoyancy. The old rule of thumb that a chimney should be 2½ times as high as the building on which it stands is based on nothing except its acceptance at a time when no other guidelines were offered. Sir David Brunt told me that he invented it on the basis of some observations of balloon trajectories over a hill, in which they seemed to be unaffected if they were more than 2½ times as high as its top. That, of course, was a quite irrelevant observation, and even generalisation from it about flow over hills is incorrect. The 2½ times rule is irrelevant for buildings which are tower shaped, and wind tunnel experiments are then recommended: but it is important that a wide variety of turbulence characteristics should be tried out in the wind tunnel airstream, otherwise possible natural events may not be produced in the model. However one should not take too seriously the problem of how the turbulence is produced because the variety of eddy forms, including cold downdraughts, is so great in nature that they cannot all be simulated by one device. It is important to experiment with a variety of obstacles, some moving, upstream of the working section in order to find out what might possibly happen if the model were built on the full scale.

Puffing is a phenomenon of little importance in practice and its causes have not been studied, although guesses at its cause have been made. **Bifurcation** has been treated elsewhere in Chapter 8; it is a consequence of the upward momentum of the effluent gases.

10.8 SAMPLING OF POLLUTION

The weather and associated physical parameters, namely wind speed and direction, temperature, and humidity, are varying enough to make it difficult to define them without reference to some standard instrument. An objective definition of any usefulness does not exist because the volume of air sampled and the duration of the sampling are finite, and anyway if small samples are taken it is not useful to know the details of the variations from one very small sample to the next in any particular case. Therefore most definitions state the manner in which the air shall be sampled, and, as we have discussed in Chapter 7, there is no definition that does not to some extent, and often significantly, depend on the volume of air sampled.

Pollution introduces further variations of greater magnitude on smaller time and distance scales. This is because of our nearness to sources which increase the pollution content by several orders of magnitude. Furthermore sinks of many forms of pollution exist both in natural and artificial surroundings.

We therefore find it useful to define a **background level of pollution**, originating from distant sources and not varying over small distances in the air because it has been well mixed, except near sinks. (For example, the SO_2 concentration is decreased in some cases by a factor of 2 or 3 when the air passes through a hedge or well wooded patch of ground). In addition we have **pollution due to sources near at hand** which may arrive in whiffs. For example, the pollution due to a sinuous plume may be absent at one moment and at its maximum 30 seconds later. Thus the average concentration over a time long enough to include many sinuosities would show a profile like P_{av} in Fig. 10.7.i(a), while if taken in a few seconds the profile would be like P_{inst}.

In practice therefore if the averaging time is to bear any relation to concentrations expected according to a dispersion formula it should be several times the time taken for a particle to travel with the wind to the point of observation. At the same time it has also to be noted that the concentration in whiffs could well be up to 10 times the average, the duration of the whiffs being determined by distance from the source. If any visible plume is observed the nature of these variations can be seen on any actual occasion. On the other hand, the average for 24 hours, a month, or a year, may easily be around a fifth, a hundredth, or a five hundredth of the hourly or three minute average. This is mainly because of the distribution around all points of the compass and the inclusion of many cases in which there is no pollution near the stack because of lofting *et cetera*. These multiples vary from case to case, and make difficult any attempt to maximise the pollution emitted into the environment. That policy is sometimes advocated on the quite spurious supposition that there exists a level up to which the pollution concentration may be raised without any harm, but beyond which harm begins. Such a level cannot, of course, ever be discovered, and in any case the continual variations would be a factor determining its value. Monitoring the actual level is a very crude business unless a large number of samples is used, and then the quantity of information obtained is oppressively voluminous unless handled by a computer, in which case intelligent treatment of it is generally precluded. There is no substitute for lively and perceptive observation using a few monitoring instruments only, for visual observation of pollution incidents and behaviour of visible plumes, and for an intelligent application of simple principles of natural aerodynamics.

If any specific sampling time were to be stated here, experience shows that it would be quoted as my recommendation, whereas my recommendation is that anyone who cannot make a suggestion as to sampling time in any particular situation is not fit to advise on the matter.

10.9 DISPERSION OF NATURAL POLLUTION

There is plenty of natural pollution originating from fires and desert dust, and it is carried over great distances. Like background pollution it soon becomes well dispersed into a certain part of the air, but not all. Forest-fire smoke or smoke produced by agriculture is usually produced at a time of year when there is thermal convection from the ground. If it is produced in the morning it is mixed well into the convection layer below cloud base. If it is generated in the evening it may rise to the top of the convection layer and stay there overnight because the layers below are stably stratified: it will then be fumigated down into the whole layer the next morning.

The air above cloud base is stably stratified except when saturated, and pollution is only carried up with cumulus cloud towers. When the cloud evaporates, which is typically within 30 minutes or so, a tower of higher humidity and greater pollution remains. This is usually sheared over within an hour or so into a thin layer (see Fig. 11.5.vii) which is not easily identified as the product of a cumulus cloud and of thermal convection. The towers of smoke can easily be seen immediately after the cumulus cloud has evaporated if the air below cloud base is very much more polluted than the air above.

Chapter 11

CLOUDS AND FALLOUT

11.1 BASIC PRINCIPLES

First, it is helpful to state certain simplifications of the actual complexities which we shall assume. We do this because any departure from these simplifications is negligible compared with the inaccuracies present in our knowledge of the more important factors. We may understand the thermodynamics of condensation physics with great precision, and we may have equations of motion which are equally accurate, but measurements are always crude and so we can never apply the equations with less crudity. There are therefore some mechanisms which can always be ignored in practice. On the other hand, some are basic:

(1) Condensation

There are always enough nuclei present in the atmosphere for condensation of water vapour to take place in the form of droplets as soon as the air is cooled below its dew point. The actual number and their chemical nature makes a difference to the sizes of the droplets, but there are always enough for a cloud to be visible. Although in some cases hygroscopic particles grow large enough to reduce the visibility when the relative humidity is between 70 and 100 per cent, there is no effect of cloud particle size on the motion until rain or freezing occurs. Any degree of supersaturation that might occur is too small to measure in practice and has no measurable or observable effect on the buoyancy or motion except soon after glaciation (see Sections 11.1(7), 11.2(6) and 11.6(4) below).

(2) Radiation

Clouds, like snow on the ground, radiate as black bodies. If a cloud is too thick to see through its upper surface radiates to space in those long infrared wavelengths that are not absorbed by the air above. The water vapour absorbs in a very complicated way, some wavelengths being almost completely absorbed by two metres of air while at other wavelengths the air is virtually transparent.

Consequently cloud is cooling very significantly at its upper surface and is losing heat at around the same rate as the ground on a clear night when dew and frost occur.

On the other hand clouds, like snow, are warmed only to a negligible extent by sunshine. Almost all the visible radiation and neighbouring wavelengths spanning the incoming solar energy are reflected and scattered by water droplets In so far as clouds absorb more of this radiation that the clear air it is due to absorption not by the droplets themselves but by the water vapour in the air between them. This is because multiple scattering of the sun's rays by clouds produces a very long path length for light before it is finally directed out of the cloud. Incoming sunlight is either reflected back into space or is transmitted ultimately to the ground: the number of times a ray is scattered before leaving a cloud can vary from one or two in very thin cloud to thousands in a deep cloud. The amount of heat absorbed by a deep cloud is not zero, but it is small compared with the latent heat of condensation released or absorbed as a result of up and down motions small in magnitude compared with those which are operating in a cloud anyway.

A cloud radiates in the long (infrared) wavelengths downwards. Thus, since its temperature is quite close to that of the ground below compared with the temperature of outer space (which is virtually absolute zero), a cover of even thin cloud serves to send back to earth a large proportion of the long wave radiation from the ground. Consequently, clouds very greatly reduce the incidence of frost and fog at night. They also reduce very considerably the heat received from outer space, and cloud amount is the greatest single factor, followed by snow cover, in reducing the amount of sunshine absorbed by the earth. At the same time they reduce the outgoing radiation by being at a lower temperature than the ground, and the radiation is proportional to the 4th power of the absolute temperature.

(3) Cloud formation

Except for cases where water vapour is introduced from sources at high temperature, such as geysers, aircraft exhaust, and industrial chimneys, the only mechanisms producing cloud are adiabatic cooling due to ascent of the air or proximity to a cold surface causing fog formation. Cloud form may be modified by motion which occurs partly as a result of the radiative heat losses of the cloud, and the buoyancy forces which result. Such cloud as is formed by mixing of two unsaturated air masses of different temperature is secondary in the sense that the latent heat of condensation is of no importance in producing motion. Thus billows sometimes produce cloud by such mixing, and fog is often formed by the mixing of cold air near the ground with warmer air above which has not been cooled directly by the bround. By comparison with the heat subsequently lost rather slowly by radiation, the latent heat released by the condensation of such cloud is negligible. Compared with the imprecision in our knowledge of the other more important mechanisms, it is certainly irrelevent in forecasting, or even explaining, the cloud.

(4) Rain formation

Rain formation is very much dependent on the nature of the nuclei on which condensation occurs. A few large (i.e. hygroscopic) nuclei, 100μ or larger in diameter, among a multitude of much smaller ones will generate rain quickly by collision of droplets of different sizes having different fall speeds (see Section 2.10). If there is a narrow dropsize spectrum (all very nearly the same size) the collision mechanism takes very much longer, especially if the droplets are all well below 20μ in size. See also Section 11.6(4).

(5) Electrical effects

Although thunder and lightning make a great show, the amount of energy involved in such displays is very small compared with that involved in the mechanics of the storm. A hailstorm which does not generate lightning is not measurably different from one which does, and although thunderstorms are usually large that is not becuase of the lightning. Electrical forces do not have any effect comparable with buoyancy forces, and are due to temperature gradients within ice crystals and effects of particle collision. Electrical energy is a trivial by-product of the buoyancy forces, and the fields are produced by a gravitational separation of ice crystals and hailstones. The small crystals are carried up in the updraught, acquiring a positive charge from larger hailstones falling much faster. The hailstones are warmer than the ice crystals because they are collecting supercooled droplets which release their latent heat of fusion as they freeze onto the hailstones. The ice crystals bounce off, but a charge separation takes place because of the temperature gradient.

The older parts of a glaciated cloud contain very few unfrozen particles, so that the temperature differences are reduced there. The charge separation takes place mainly in the newer, more vigorous updraughts where a large amount of supercooled water exists. The rapid generation of kinetic energy by buoyancy forces in new cloud cells completely dominates any electrical forces on the scale of the air motion. It used to be thought that when a lightning discharge took place a large electric field was suddenly removed, and that this released a gush of rain or hail previously held in suspension by electrostatic forces. In fact the charge separation occurs because the hail has already fallen, and naturally it would be expected to reach the ground soon after a lightning flash, especially from a new cloud cell.

(6) Cloud droplet temperature

Unless otherwise stated, the temperature of water droplets and ice crystals is very nearly the same as that of the air through which they are falling. Hailstones are an exception of practical importance: they are warmer than the air when they are capturing supercooled droplets inside a cloud, and because of their great fall speed they may arrive at the ground much colder than the air into which they have descended.

Similarly, large rain drops may be cooler than the air, but that is mainly because they take up the wet bulb temperature on falling through the air (see Section 11.2(5) below).

(7) Glaciation and supercooling

When air is cooled to its dewpoint condensation occurs either onto a solid surface or as cloud. On further cooling the droplets do not freeze at 0°C but may remain liquid and supercooled down to −40°C. Below this temperature freezing is spontaneous unless there is a high concentration of a dissolved substance such as a sulphate or chloride in the droplet.

At temperatures between 0° and −40°C the time for which a droplet may remain unfrozen depends on the droplet size and the temperature. A cloud of small droplets (less than, say, 5μ diameter) may remain unfrozen for many hours at −30°C, while raindrops may freeze in a few seconds in a cumulus cloud at −10°C. Thus cumulus clouds glaciate at much higher temperatures than thin layer clouds. The generation of rain is hastened by a strong updraught and a rapid condensation of liquid water. This is especially so when a few large nuclei are present to provide a wide spectrum of drop sizes, for the drops then collide and coalesce with one another more readily.

Just as the chance of the formation of an incipient crystal is greatly increased by a droplet becoming larger, so it is increased by a decrease in temperature. For practical purposes glaciation is so rapid below −40°C that it takes place immediately even in the smallest droplets.

(8) Latent heat

The amount of heat liberated by condensation varies slightly with temperature, and is so large as to be a dominant factor in cloud dynamics. Water has a larger latent heat than most commonly occurring vapours, and condensation of only 5 to 10 per cent of the vapour is enough to warm the air by 1°C.

Glaciation of liquid produces only about one sixth as much latent heat as condensation of vapour to liquid water, and is therefore only a modifying influence. The chief, and by far the more important, effect of glaciation is to delay evaporation and so prevent the loss of latent heat of condensation.

11.2 THE T-ϕGRAM

In order to represent the state of the atmosphere as revealed by a sounding, two coordinates are required. In the T-ϕgram these are temperature, T, and entropy. Since entropy is proportional to the logarithm of the potential temperature, θ, the vertical scale is a lograithmic one marked out with values of θ. Horizontally we have a linear scale of temperature. A T-ϕgram is printed inside the front cover of this book.

(1) Isobars

By (1.6.3) $\theta \propto p^{\frac{1}{\gamma} - 1}$ at a constant temperature, and so the vertical scale is a logarthmic one for pressure, but with pressure decreasing upwards. The scales may be chosen so that the isobars are inclined at about 45° to the axes, and because we are operating far from the zero of potential temperature the logarithmic scale is not far from linear. As a consequence the isobars are only slightly curved (see Fig. 11.2.i).

(2) Dry adiabatics

If the air is neutrally stratified the potential temperature is the same at all heights, and so a sounding is represented by a line along θ = const., i.e. a

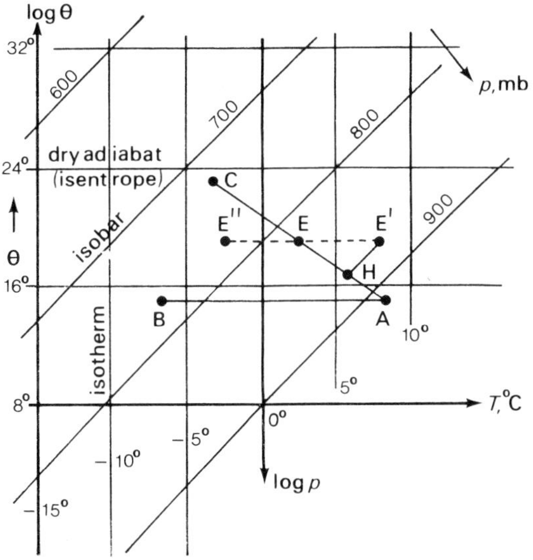

Fig. 11.2.i Part of a T-ϕgram showing isotherms at 5°C intervals and lines of constant potential temperature θ at 4°C intervals. $T = \theta$ on the 1000 mb isobar; the isobars are shown at 100 mb intervals from 900 to 600 mb. A horizontal line such as AB is a dry adiabatic, and neutral stratification on a sounding. AC represents stable stratification, or θ increasing with height (decreasing p).

horizontal line. This is called a **dry adiabatic** because it is the line followed by a particle displaced vertically to a different pressure adiabatically without condensation. Thus in Fig. 11.2.i the line AB represents the sounding of a neutral layer of the atmosphere. AC represents a stably stratified layer of the atmosphere because the potential temperature increases upwards (i.e. as pressure decreases and height increases). When a sounding is made the instruments record the pressure and temperature, and these two quantities are therefore used as the coordinates for plotting a sounding on the diagram.

Some meteorological services use a very similar diagram in which the coordinates are T and $\log p$. This is drawn with the axes skew so as to look like the T-ϕgram, and the only difference is that the isobars are straight and the dry adiabatics slightly curved.

A parcel of air displaced downwards adiabatically from E to E′ in the environment AC would be warmer than its new surroundings at H, and would therefore experience a restoring force towards E. If it were displaced upwards adiabatically to E″ it would be colder and denser than its new surroundings and would again suffer a restoring force. The air layer AC is therefore stable.

(3) Humidity mixing ratio: condensation level

This ratio is defined as the number of grams of water mixed with 1 kg of dry air. This is for practical purposes, the same as the ratio of the partial vapour pressure of the water vapour to the total pressure ÷ 1000. The physical information about water vapour is placed on the diagram for saturated air in the form of lines indicating the **saturation mixing ratio.** These lines are inclined to the left of the vertical as indicated in Fig. 11.2.ii. A parcel of air whose temperature and pressure placed it at A in Fig. 11.2.ii might have a huminity mixing ratio of, say, 3g/kg. If it were lifted adiabatically to C, which is on the 3g/kg line, it would become saturated. If the saturation mixing ratio at A were 5g/kg the **relative humidity** of the air when at A would be 3/5, or 60 per cent. When lifted to C it would be 100 per cent, and the altitude of C is called the **condensation level** of the parcel.

(4) Wet Adiabatics

If the parcel of air now at C were lifted further to the isobar through C′ in Fig. 11.2.ii, some vapour would be condensed out as liquid. Thus instead of cooling dry adiabatically to the point C′ it would be warmed by the latent heat released and would be at S, where less would be condensed out.

The line CS is part of the **wet adiabatic** through C. It is calculated assuming that the liquid water is removed as it is condensed. This is done because the amount present would depend on the level at which the parcel first became saturated, but no very large error results from this. If the liquid water remains with the air, the air cools more slowly on ascent because it is bringing liquid

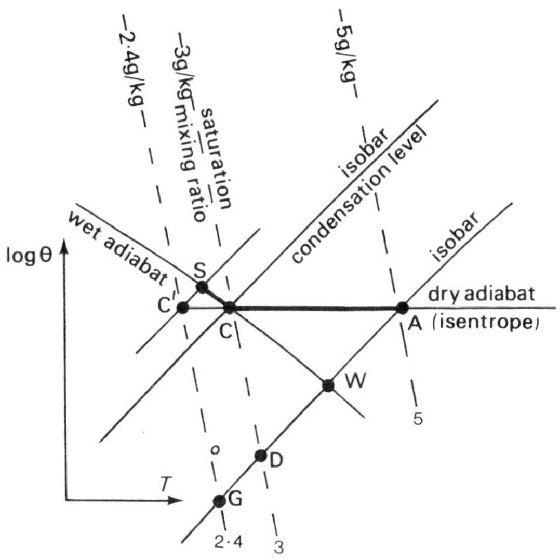

Fig. 11.2.ii The lines of constant saturation humidity mixing ratio are inclined upwards towards the left on the *T-ϕ*gram. They are marked in g per kg by means of dashed lines. If a parcel of air lifted from A adiabatically becomes saturated at C its mixing ratio is 3g/kg. The height, or pressure, at C is called the condensation level. If lifted further the parcel follows the wet adiabatic up to S.

At the pressure of A, air having its condensation level at C has its wet bulb temperature at W and its dewpoint at D.

water with it which also has to be cooled. Strictly speaking if air is brought down a wet adiabatic the water to be evaporated into it must be supplied at the same temperature as the air. In the figure the saturation mixing ratio line through S is 2.7 g/kg, so that 0.3g of liquid water would be condensed out in rising from C to S.

(5) Normands' Theorem: wet blub and dew point

If we now draw the isobar through A to meet the wet adiabatic through C, the condensation level, at W, W represents the temperature to which the air would be cooled if water were evaporated into it in descending from C to the pressure of A, instead of descending to the temperature of A. A wet bulb is cooled by evaporation, and because the temperature of the thermometer bulb is steady all the heat for the evaporation comes from the air, the only difference being that in the case of a wet bulb the water is evaporated at the wet bulb temperature and not at the variable temperature along CW. Nevertheless, the difference is small and for practical purposes W may be taken as the wet bulb temperature. Errors occur due to the conduction of heat down the thermometer stem to the wet bulb from the dry end: theoretically therefore the whole thermometer should be enclosed in a wet surface. The best way to minimise this effect is to increase the rate of evaporation by increasing the ventilation speed of the wet bulb, so that the total heat lost is greater, and insulating the rest of the thermometer in an outer casing of glass.

The constant mixing ratio line through C meets the isobar through A in D, which is the **dew point**, namely the temperature to which the air must be cooled at constant pressure to become saturated. **Normand's theorem** states that the dry adiabatic through the dry bulb temperature meets the wet adiabatic through the wet bulb temperature and the mixing ratio line through the dew point in a point at the condensation level.

If the air is cooled to G, in Fig. 11.2.ii, it can only contain 2.4g/kg as water vapour. Thus 0.6g must be condensed out either as dew on the cold surface producing the cooling or, if the cooling is largely by radiation loss from the air (mainly, from the water vapour in it) to the cold surface, in the air as fog. In the former case some of the air is cooled well below the dewpoint and dew is deposited when the wind is light. If the air becomes stirred up in the morning the cooling is spread over a greater depth and fog is formed by the mechanism of mixing two parcels at different temperatures (see Section 11.3).

(6) Frost Point: ice evaporation level

The vapour pressure over ice is less than that over a water surface at the same temperature, which means that the mixing ratio lines for saturation over ice differ from those for water by an amount which increases as the temperature falls below 0°C. As the air is cooled on the isobar (i.e. *in situ*) through A the frost point F is reached before the dew point D (in Fig. 11.2.iii). If the air is cooled by lifting along the dry adiabatic no condensation occurs until the water condensation level is reached at C.

If a finite amount of water were condensed out by further ascent to S and then the condensed water were frozen, first the latent heat of fusion would be liberated and then further sublimation of water vapour on to the ice crystals would take place. This would cause further warming until the air achieved the state represented by T which is on the saturated adiabatic line representing ice saturation with the same total water content as at C, but with ice instead of liquid water.

If the air is now caused to descend from T, remaining saturated for ice by the ice crystals evaporating into it, the ice will not all be evaporated until E is reached. This is the point where the ice adiabatic through T meets the dry adiabatic through A, and it lies on the ice saturation mixing ratio line through F. The saturation mixing ratio line for water through C and D meets that for ice through E and F on the 0°C isotherm. Likewise the wet adiabatic through C (and S) meets the ice adiabatic through E (and T) on the 0°C isotherm. If at the

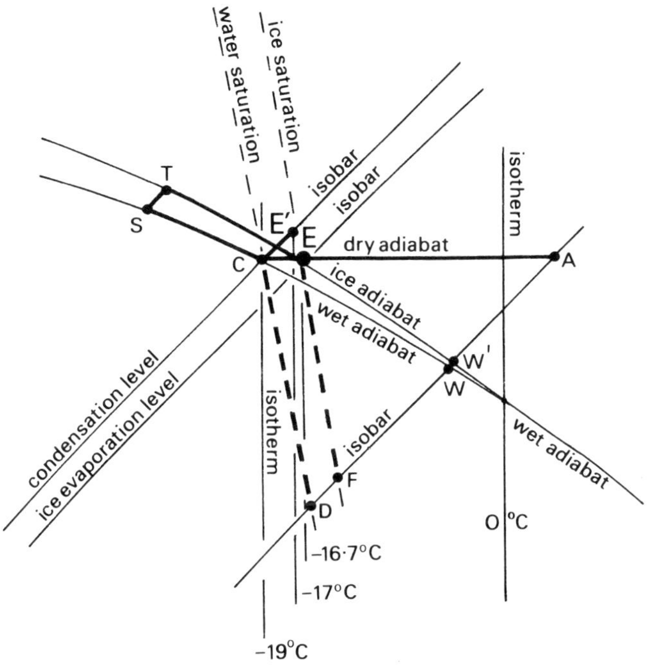

Fig. 11.2.iii The water and ice saturated adiabatic lines for an air parcel colder than 0°C meet on the 0°C isotherm. If the parcel originates at A and has condensation level at C its wet bulb (supercooled) and dewpoint temperatures are at W and D. But if the bulb is ice covered the 'wet' bulb temperature is W′ and the frost point, at which frost begins to form on a cooled surface, is at F. Strictly F should be measured at the temperature at which frost begins to evaporate from a surface as it is warmed, because sometimes the first deposition of ice on a clean surface may occur at a temperature below that of F. If condensed water remains supercooled the parcel may be on the wet adiabat CS, but if the droplets freeze the parcel moves to TE, the ice adiabat for the same total water content. Saturation for ice at that water content is on the ice saturation line through E which meets the water saturation line through C on the 0°C isotherm, off the bottom of the diagram. Because condensation to ice does not occur out of clear air when it is cooled by ascent, the isobar through E is a pressure level called the ice evaporation level.

pressure of A the water on the wet bulb is frozen, the temperature it will record is given by W′ on the isobar through A and the ice adiabatic through E.

E is not said to be at the ice condensation level because ice crystals do not form out of the vapour as the pressure falls. But if a frozen cloud is present, at T say, it will not disappear by evaporation when the air descends until it reaches E. This is therefore called **ice evaporation level.**

The difference between the ice and water adiabatics and saturation mixing ratio lines can be expressed simply by the difference between the frost point and dew point at the same pressure for different values of the dew point. This is given in table 11.2.I, in which T_f and T_d are the frost point and dew point respectively. Thus if the dew point is −19°C, the frost point is −17°C; air which at −19°C condensed cloud which then froze would not have it evaporated until it had descended rather more than 200 metres lower than the condensation level. For this case the isobar through C meets the ice saturation line in E′, which is 2° warmer. Thus E is about 2.3° warmer than C and is 230 m lower. There are thus below the −19°C isotherm 230 m of air where an ice cloud would persist whose particles might grow, but where a water cloud would evaporate.

Table 11.2.I

T_d, °C	0	−4.4	−9	−14	−19.4	−24.5	−32.2	−40.4
$T_f - T_d$, °C	0	0.5	1	1.5	2	2.5	3	3.5

(7) Virtual temperature

Water vapour has a density 0.622 times that of air at the same temperature and pressure. Thus if air is 1 per cent water vapour, i.e. has a mixing ratio of 10g/kg, its density is reduced to 99.622 per cent of the density of dry air. At a temperature of 270°K, a parcel of air having 1g/kg more water vapour than another but having the same density must be 0.0388 per cent cooler, i.e. 0.11°C cooler. Roughly a 1g/kg difference in water content is equivalent to 0.1°C temperature difference.

The **virtual temperature** of air is the temperature of dry air with the same pressure and density. Meteorologically this is an effect of negligible importance, but it can have important effects on the refractive index, which is influenced both by humidity and temperature (Scorer 1961). Thus if the humidity is not uniform because of the irregular evaporation of cloud, scattering of radio waves and the refraction of light rays as seen in a powerful telescope may occur even though there are no differences in density present. Such a distribution of refractive index used to be ascribed to turbulence when it was first observed by means of radar scattering. In air of uniform composition the velocities required to produce the pressure fluctuations needed to produce the same scattering were so large as to be very unlikely. This humidity effect is a more likely explanation, and the phenomenon was given the name ‘fossil turbulence’ because it indicated where there had been some motion.

(8) Conservative properties of air

If a parcel of air is not mixed with others and loses no heat by radiation its potential temperature is conserved. If condensation occurs it moves along a wet adiabatic. If cloud condenses and some of the liquid water falls out and it then descends below the point where all the cloud evaporates, the same wet adiabatic remains characteristic of it. If on its descent water which was not there before is evaporated into it, its wet bulb remains unchanged, just as it remains unchanged if water is evaporated into it at constant pressure. Thus only mixing, conduction, or radiation can alter the wet adiabatic on which the parcel will move. This line can be identified by the temperature at which it crosses the 1000 mb isobar, giving the wet bulb temperature in the parcel at 1000 mb. This conservation property is called the **wet bulb potential temperature** of the air.

A complete T-ϕgram is displayed in the front of this book and as the discussion proceeds may be referred to to give the physical magnitudes which actually occur with various condensation and vertical motions described.

11.3 CONTRAIL PHYSICS: MINTRA

When two parcels of unsaturated air mix they end up at a temperature between their original temperatures according to the proportions in which they are mixed. Aircraft exhaust gradually mixes with an increasing volume of surrounding air until the mixture rises to an equilibrium level at which no density difference remains. It will always find such a level if no liquid water is present, but if some remains as water droplets the heat lost by their evaporation causes density differences leading to further mixing. Thus water cloud is always in a state of mixing and evaporation at the edges where it is in contact with unsaturated air at the same level.

The saturation mixing ratio may be drawn as a function of temperature for water and ice at different pressures. Figures 11.3.i, ii, and iii show this for three different pressures, namely 550, 400 and 250 mb. The aircraft exhaust is very hot and has a very large mixing ratio, and so the point representing it on the diagrams is indicated only in approximate direction by a tangent to the water saturation curve. If the ambient air is represented by a point above this tangent but below the water saturation curve in the shaded area to the left of the point of contact, then at some stage during the dilution of the exhaust the point representing the mixture will lie above the water saturation curve and cloud will be formed. If the ambient air point lies below the ice saturation curves in the half-shaded area, the cloud will soon evaporate. If it lies between the water and ice saturation curves, in the fully shaded area, the cloud will persist if it is frozen. It will certainly be frozen if its temperature is below -40°C; otherwise it may or may not be, according to the nature of the nuclei present. Usually it will not be frozen, because the mixing occurs in too short a time for the cloud to freeze on its own: there are so many nuclei in the exhaust that the cloud consists of a very large number of extremely small particles less than 1μ in diameter.

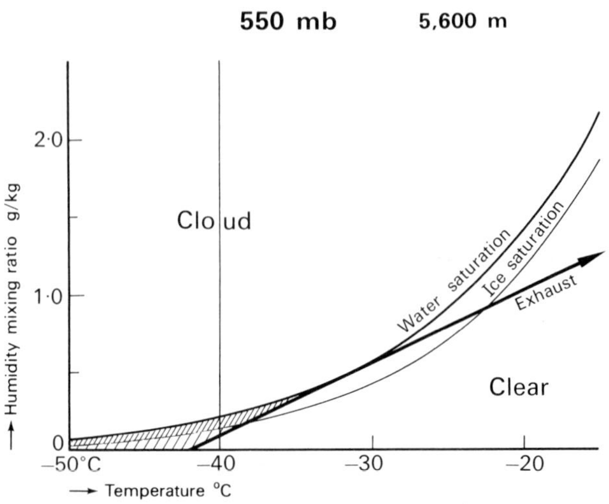

Fig. 11.3.i The water and ice saturation curves for air at 550 mb which is at a height of roughly 5600 m. Typical aircraft exhaust lies far off the diagram.

If the point representing the state of the ambient air lies in the shaded region the mixture will be saturated at some point as the exhaust is diluted by mixing, when the point representing the mixture lies above the water saturation curve.

If the ambient air point lies in the doubly shaded region and the cloud generated when mixing produces supersaturation becomes frozen it will not evaporate however much the mixture is further diluted with ambient air because it lies above the ice saturation line, i.e. it is above its ice evaporation level.

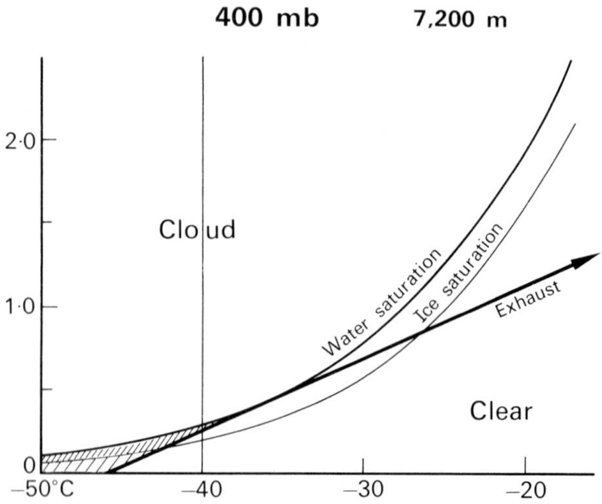

Fig. 11.3.ii The same as 11.3.i but for 400 mb, roughly 7200 m.

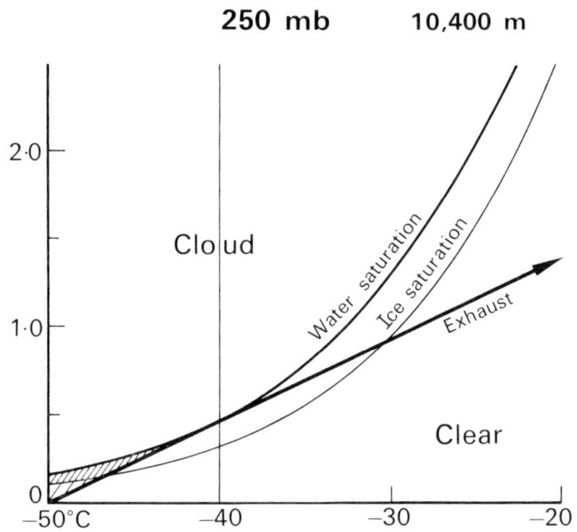

Fig. 11.3.iii The same as 11.3.i but for 250 mb, roughly 10,400 m.

In general the air at 550 and 400 mb is not at a temperature below −40°C and so contrails, besides being very rare, do not persist. But above about 350 mb (i.e. at lower pressures) temperatures below −40°C are quite common. Hence trails are formed if the ambient air is between the water saturation curve and the tangent, and they will persist if it is above the ice saturation curve. This information is contained in Fig. 11.3.iv, which is a *T*-ϕgram turned so that the isobars are approximately horizontal. If the ambient air point is in the lower right unshaded half of the diagram no trails are formed. If it is in the shaded area trails are possible, but will only occur for certain if the air is colder than the 'mintra dry' line. This indicates the lowest altitude on each isotherm at which trails will form even when the ambient air has zero humidity. If the air is saturated they will form at all altitudes above (at all pressures below) those on the 'mintra sat' line for each isotherm. This line depends on the proportion of hydrogen in the fuel, and so the line for different fuels is shown. The line appropriate to jet aircraft burning kerosene is taken as the boundary of the shaded area. The fuel $C_{16}H_{12}$ has been suggested as a possibility for military purposes to avoid making trails. The lower wedge marked 'water trails possible' is the part of the 'trails possible' area warmer than −40°C.

The wiggly boundary indicates the region in which the tropopause is usually found in mid latitudes, roughly between 300 and 200 mb. Above the tropopause the air is usually too dry for trails to persist and so they are called semi-persistent. Below the tropopause the air is frequently saturated for ice and so persistent trails may form.

The standard ICAO atmosphere is also shown with a tropopause at about −57°C and 235 mb, and this is below the 'mintra' from 360 mb down to the ground. Trails are unlikely very far from this line because the atmosphere does not depart from it by more than about 20°C most of the time.

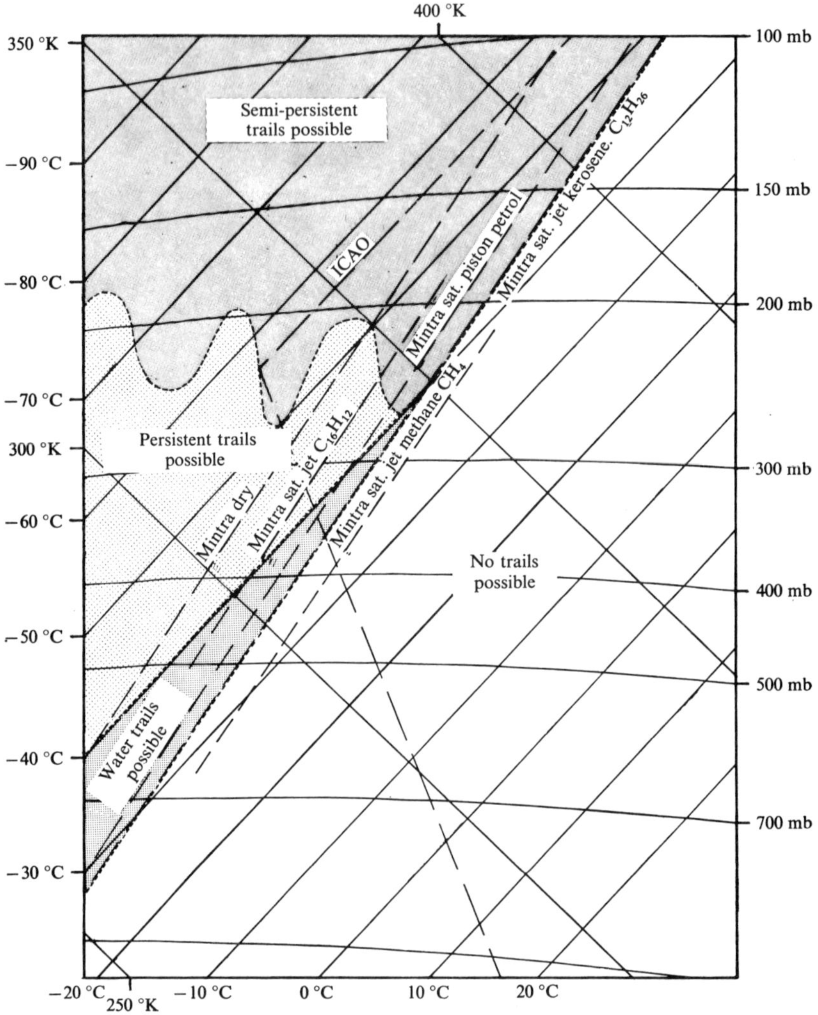

Fig. 11.3.iv A T-ϕgram showing the properties of contrails. The dry adiabatics are the diagonals marked in °K. The isotherms are marked on the left and at the bottom in °C. The diagram indicates the properties of trails according to the position of the point representing the ambient air and the nature of the fuel. The 'mintra' is the line below which no trails are formed, and this is at the lowest altitude for methane which has the highest hydrogen content of the fuels considered. The region in which trails are possible for typical kerosene jet aircraft is shaded. Mintra lines are drawn for saturated and for dry air. At temperatures above −40°C unfrozen trails are possible. The position of the tropopause above which the humidity is much less is shown as a wavy boundary. Below it persistent trails are possible because it may be above the ice evaporation level. Above the tropopause it is usually too dry for trails to persist. The ICAO standard atmosphere is represented by the dashed line. An actual sounding is likely to lie within 10° of this over much of the Earth, and so it may be seen that persistent trails are most likely about 50 mb on either side of 300 mb. This figure together with Figs. 11.3.i–iii are reproduced from Scorer and Davenport (1970).

11.4 CONTRAIL DYNAMICS

When trails form they first appear about 30 m behind the engine. Fig. 11.4.i shows the motion streamlines inside the accompanying fluid of the trailing vortices (see Section 9.10). In the centre we see a puff indicating a cross-section

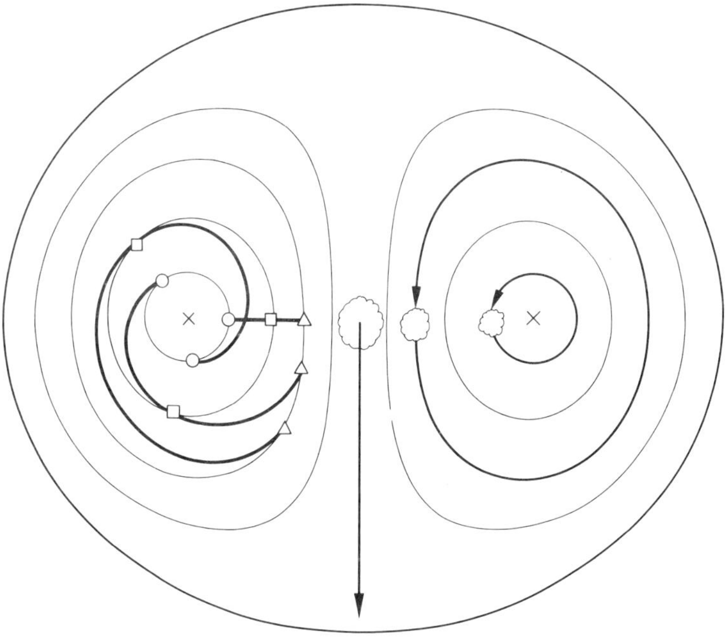

Fig.11.4.i The streamlines of the flow inside the accompanying fluid of a vortex pair which is approximately the velocity pattern in the cross section of an aircraft wake initially. The vortices are marked X. In the right half are shown the tracks taken by the exhaust trails from outboard engines, and in the centre of the track taken by the exhaust from a central engine. On the left are shown the successive simultaneous positions of three particles initially in a horizontal line, together with successive positions of the line of particles joining them.

of the trail formed behind a single central engine. To the right we have two puffs indicating sections of the starboard trails of a 4-engined aircraft. In the left half are shown three positions of a line of particles originally in line with the trailing edge of the wing. While the outer of the three particles makes more than one rotation round the vortex the two inner ones make two thirds and one eighth of a rotation respectively. Consequently, if viewed from below, the two inner trails from a 4-engined aircraft will pass beneath the outer trails to a position outside them. All the time some diffusion and shearing takes place and they may no longer be distinguishable as four separate trails after about 20 seconds.

Since the trail from an outer engine is often drawn close to the centre of the vortex on each side it usually persists longer because the pressure is lower there, perhaps by as much as 50 mb. About 30 seconds behind the aircraft therefore the trail often has the appearance of two tubes of cloud, in which the flow appears to be non-turbulent. Much of the cloud from the inner engines, the amount depending on their position, passes close to the outside of the accompanying fluid and is detrained above the upper stagnation point as described in Fig. 9.10.iv. It is therefore seen as a vertical curtain of cloud stretching up to the level at which it was first detrained (Fig. 11.4.ii).

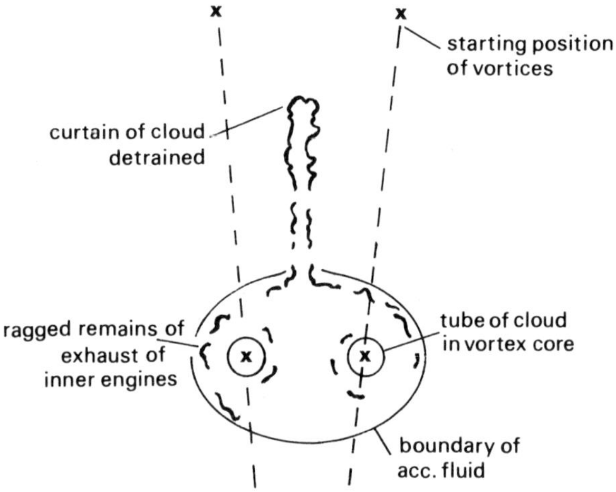

Fig. 11.4.ii The shape of a contrail cloud 30 seconds or more behind a 4-engined aircraft. A three engined aircraft will probably show a more distanct curtain of cloud with a sharp upper edge because the trial from a central engine is more quickly detrained at the top of the accompanying fluid than that from an inner outboard engine.

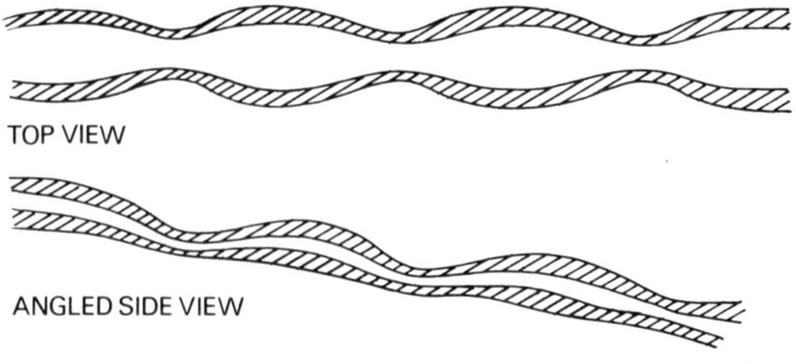

Fig. 11.4.iii Top view and angled view from slightly higher on one side of the tube trails contained in the vortex cores. The parts which are closer together have descended further than those that are further apart.

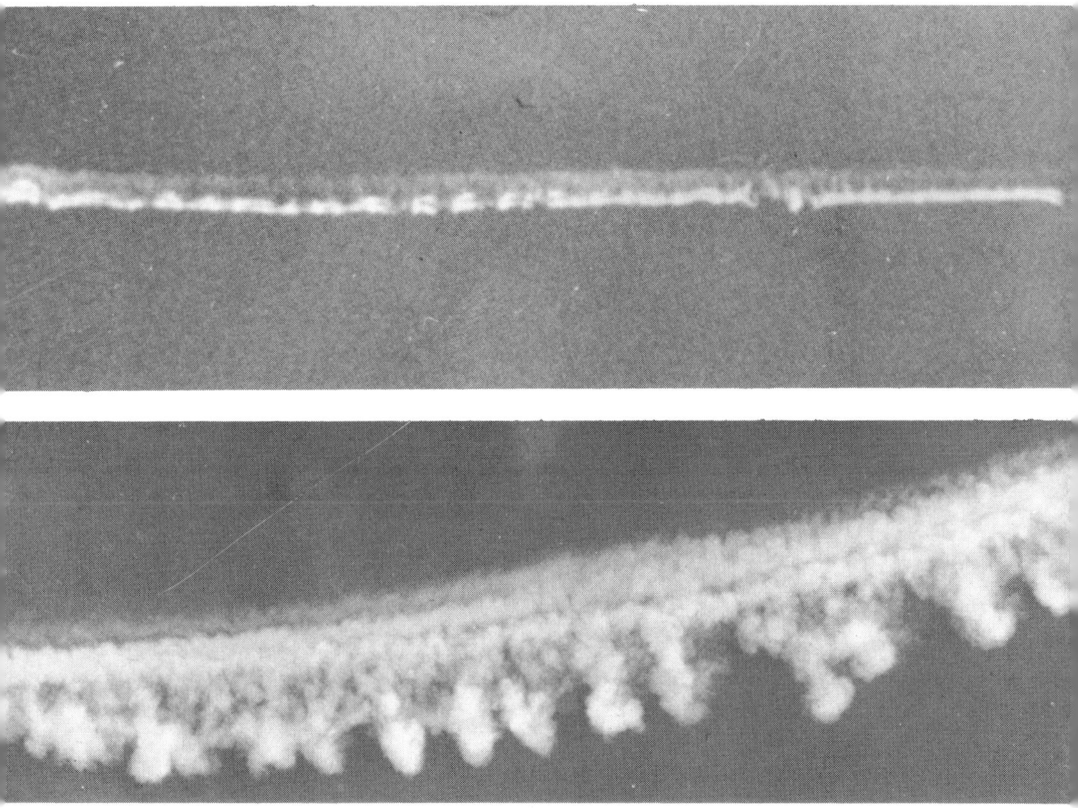

Fig. 11.4.iv (upper) Side view of a trail from a 3-engined plane showing the curtain of cloud mainly from the central engine and the tube clouds in the vortex cores which are denser because of the lower pressure. The tubes are beginning to break into loops.

(lower) A view from below of an aircraft trail from 4-engined aircraft on a turn. The downwash now has an outward component, which makes the shape visible from vertically below. The blobs are where the vortices have burst and the tube-like clouds in the vortices diluted. The blobs occur on the parts of the trails that are bent downwards (see Fig. 11.4.iii).

The vortex pair is an unstable configuration because it is accelerating downwards and any part that is slightly ahead of another part will appear as a downward and inward distortion of the trail (Fig. 11.4.iii). If there is a trail from a central engine it will appear in this motion pattern as a blob of the trail pushed downwards (Fig. 11.4.iv). However if only two trails are left in the cores of the vortices, where they are distorted downwards they quickly become unstable because of the entrainment of mixed air from the outer boundary of the accompanying fluid down the trail's centre, as described in Section 9.10.

If an aircraft flies above a layer of cloud its downwash produces a clear lane in it (Fig. 11.4.v), and if this happens to be so placed that the blobs reach to the bottom of the layer a series of holes is made in it (Fig. 11.4.vi).

Fig. 11.4.v A photo of the wake of an aircraft as seen from the aircraft, when flying just above a layer of cloud. [Photograph by V. G. Plank].

The spacing of the loops seen when two cloud tubes only exist (Fig. 11.4.vii) is obviously related to the width, but it is not very regular and one wavelength is not picked out. Obviously if two blobs are closer together than about four trail widths there is no chance of another developing between them, whereas at 12 widths apart there is room for another disturbance to grow. Where the instability occurs the contra-rotating vortices cancel one another and the vortices that remain join across the ends, which are sometimes visible in condensation cloud. Fig. 11.4.vii shows smoke trails from the wingtips of a Comet aircraft at 10,000 ft, where the air is stably stratified. The bursting of the vortices appears to occur after about a minute, as would be expected if it is occasioned by the rise of the upper stagnation point in the accompanying fluid so that some exterior fluid is entrained down the centre. When the instability occurs the rotation is rapidly reduced, the central pressure rises, the tubes of cloud disappear, and rapid mixing occurs. In Fig. 11.4.vii the smoke is rapidly diffused. Where the vortices burst they cancel one another so that the loose ends of the vortices join up because vortex lines cannot end in the fluid. An example of different appearance is given in Fig. 11.4.viii.

Fig. 11.4.vi The upper picture shows a distrail beginning to appear where an aircraft on a turn flew just in or above the top of a cloud. At the top of the picture the cloud is glaciated and the trail has intensified the cloud.

The lower picture taken 4 minutes later shows the holes due to the downwash blobs made in a thicker piece of the same cloud.

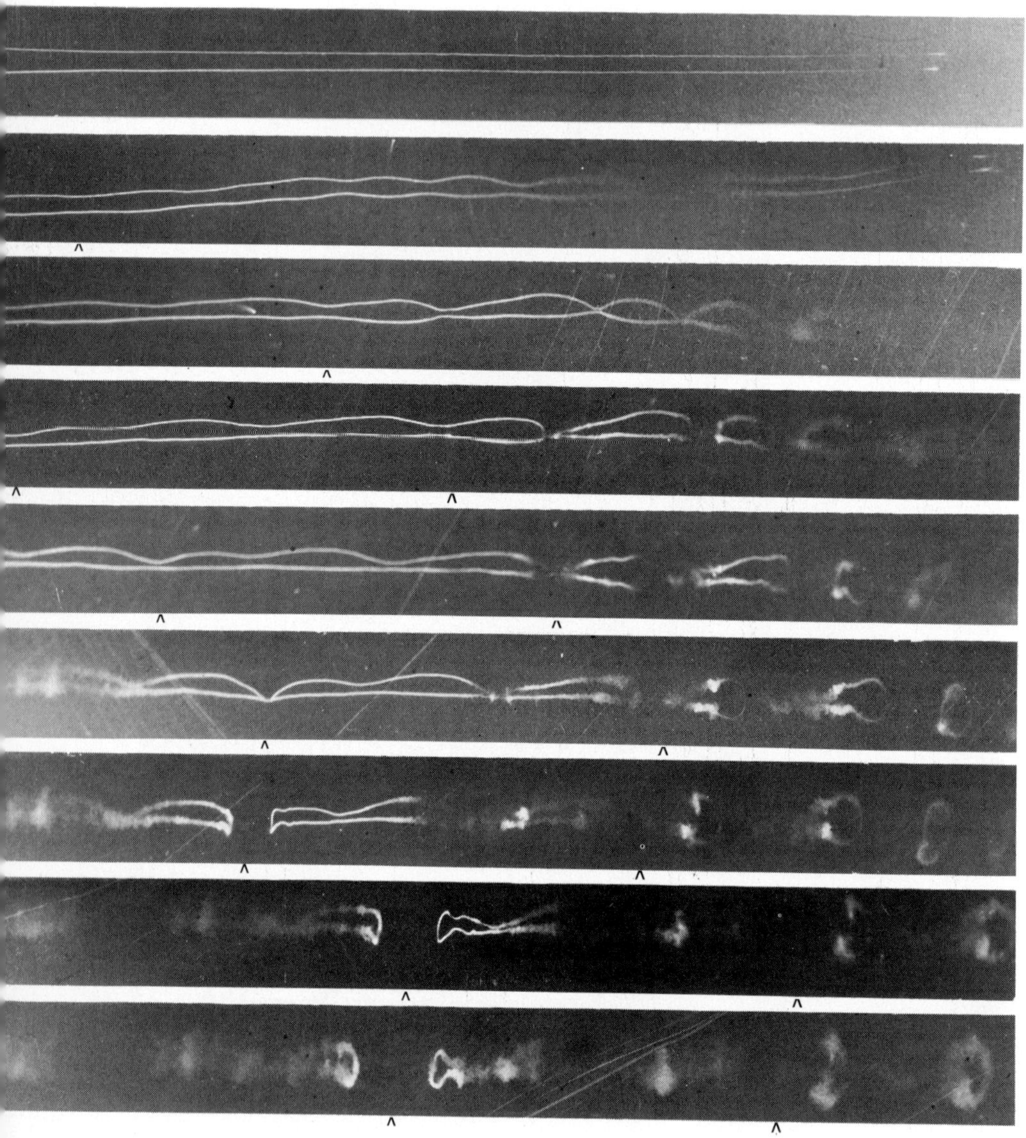

Fig. 11.4.vii Smoke trails from the wingtips of a Comet aircraft flying at 10,000 ft photographed from below at the times given in seconds after the passage of the aircraft. The arrows identify the same points of the trail, one of which is beyond the left end of the picture at 69 secs.

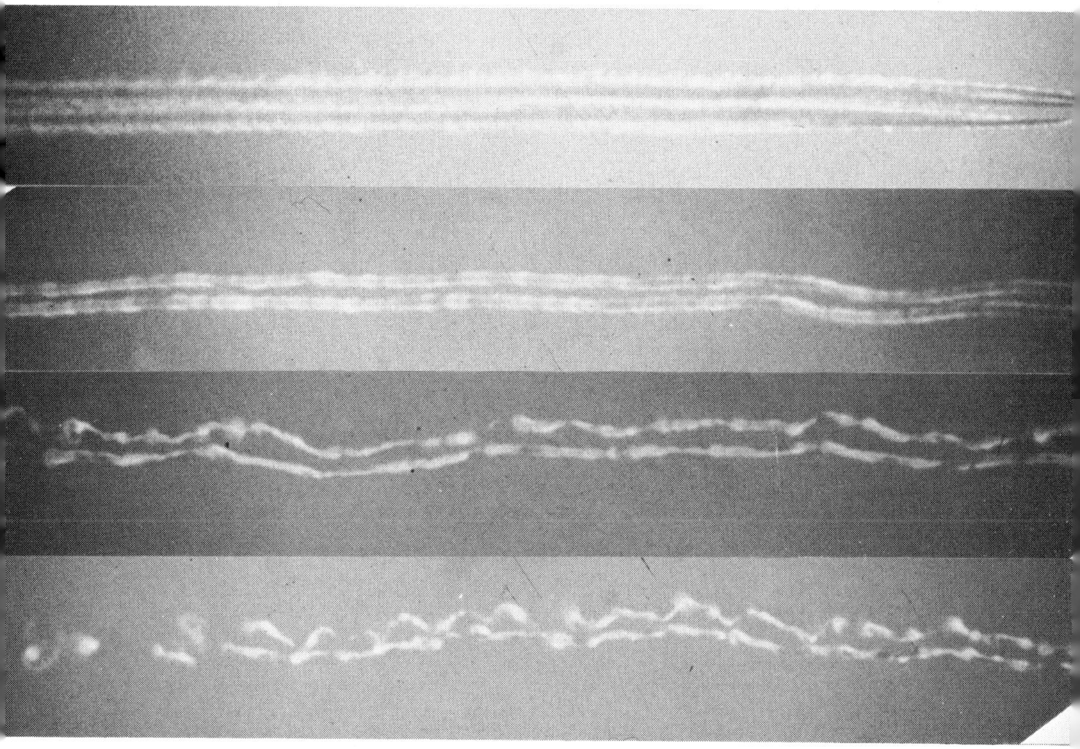

Fig. 11.4.viii Four successive pictures of an aircraft trail from a 4-engined aircraft. In the first the downwash which rotates the inner engine trails around the outer can be seen to produce a different appearance of the two pairs when seen at an angle of elevation of about 30°.

In the later pictures only the trails from the outer engines remain, surrounding the vortex cores. In the fourth picture the near (upper) trail appears more distorted than the far one because of the angle of view from below.

11.5 CUMULUS DYNAMICS

Convection carries parcels of air to the condensation level, whereupon the heat liberated by condensation becomes the dominant source of buoyancy. Although some rain falls to the ground, most cumulus clouds are evaporated by mixing into the surroundings into which they ascend, and so the heat lost when evaporation occurs is also an important feature of the mechanics. As soon as the role of water becomes important the mathematical simplicity of the situation vanishes, and vain attempts have been made by many authors to recapture the similarity love of the honeymoon period with partly turbulent flow. But the problem possesses an essential elusiveness which charts, computers, and all the time in the world cannot portray better than the clouds themselves. The system is essentially very complex and is continuously reacting with everything that is going on both close at hand and many thousands of miles away. It is part of what meteorologists naively call 'the general circulation', meaning something that can only be studied with the aid of all the world's meteorological services and the world's biggest computer. However it is still elusive in the sense that nothing very useful can be done for the man in the field, that is, the field which grows the food; not the one turned into a research laboratory. The world's scientific effort has been diverted by the so-called 'rich' nations to the facilitation of their narrow purposes. Meanwhile the so-called 'poor' nations have been taught to value an air line more than an agriculture and ecology matched to their weather and long term climate. Consequently the approach to agricultural needs has been through attempts to make rain and hurry the irrigation plans, as if some purpose were served by supplying the needs created by the previous technological effort. We, however, shall look simply at the cumulus clouds to get the feel of how they continuously circulate material through the atmosphere, creating a beautiful variety of weather in which man may evolve the variety of cultures which is the treasure of his civilisation.

(1) Erosion of thermals

A sounding of the air is depicted in Fig. 11.5.i as the line QLR on the *T*-ϕgram. Suppose a thermal source produces a warm parcel at A slightly warmer

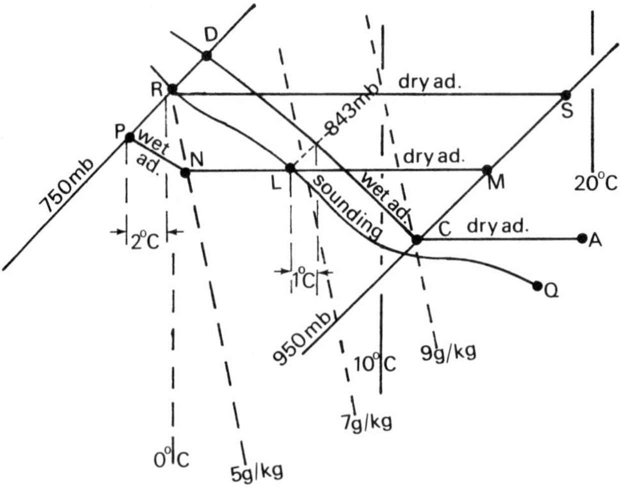

Fig. 11.5.i The ambient air is represented by the sounding QLR. A warm parcel rising without significant dilution from A would pass along ACD if it became saturated at D. If it is mixed in equal proportions with the environment at R with 20 per cent humidity the mixture would be at P, M being the mid-point of CS. This parcel at P will find its equilibrium level at L. (See text for details).

than the average ground temperature at Q, and that it rises to its condensation level at C along the dry adiabatic, and then up through the middle of a cumulus cloud along the wet adiabatic to D. All the way up it is warmer than the surroundings. It now emerges to the exterior of the cloud and mixes with the surroundings at R, which are unsaturated and, we suppose, have a mixing ratio of 1g/kg. The consequences of the mixing may be discovered by imagining the parcel and the surroundings to be brought down to the condensation level, so that the parcel is back at C and the surroundings are at S, on the dry adiabatic through R.

If the two are mixed in equal quantities the mixture will be at the mid point M, of SC, and the mixing ratio will be the mean of 9 and 1, namely 5g/kg. If we now take the mixture back to its proper level it will go dry adiabatically to N, on the 5g/kg saturation line, and then wet adiabatically to P. There we find that it is colder by 2°C than the surroundings, which are at R, whereas the surroundings at L were 1°C cooler. If this parcel is now left to find its equilibrium level this will be at L, where it has the same temperature as the sounding.

It is thus demonstrated by an example, and others may be multiplied at will on the T-ϕgram, that at the top of a cumulus cloud we may generate parcels which are much colder than the surroundings, especially when they are dry (this case had an environment with a 20 per cent relative humidity). The downdraughts which occur as a consequence can easily be seen by a careful cloud watcher and are a common experience of glider pilots.

In order for a cloud to rise far into dry surroundings it must be very large so that parcels may rise to its top without being subject to evaporation, and it is most likely to do this if it is ascending along a path already moistened by previous thermals. This happens best over a good thermal source such as a mountain side where there is almost no shear or general wind to carry away the towers of air moistened by previous thermals.

The intensity of the mixing in a thermal is shown in Fig. 11.5.ii, which is a sequence showing the successive positions of a blob of surroudnings which is

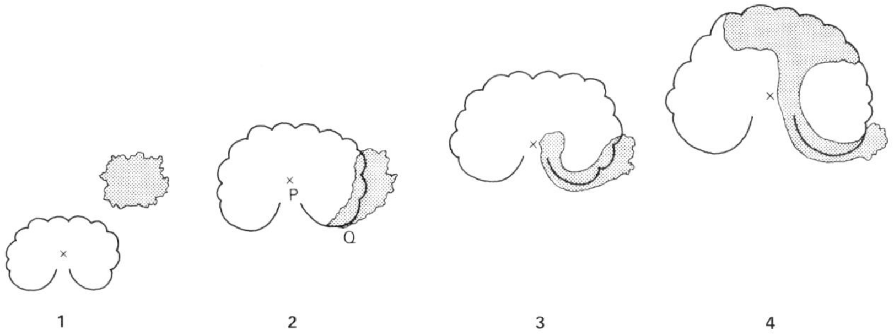

Fig. 11.5.ii Successive positions of a blob of exterior air entrained into a rising thermal. After position 4 it would be spread around all sides.

entrained into a thermal. After the fourth position the material will be circulated around the thermal on all sides of the axis so that the blob will be completely mixed into the thermal as it rises about 1½ diameters. When the entrained blob is exterior dry air it is clear that the thermal cannot rise very far before all the cloud in it is evaporated, and Fig. 11.5.iii shows a sequence in the life of a cloud tower. The sharp outline on the top, which is characteristic of

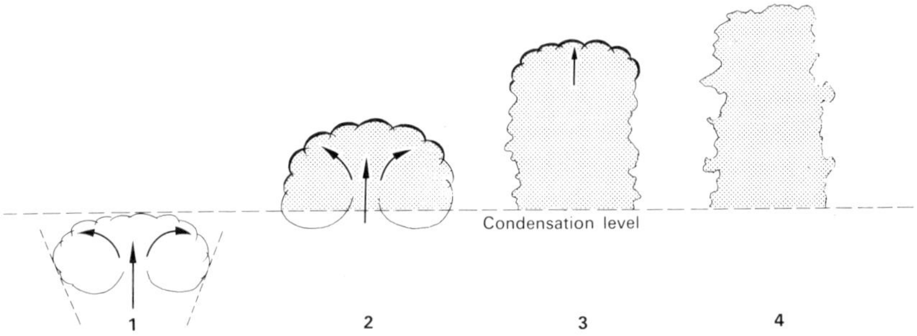

Fig. 11.5.iii Successive positions of a thermal rising above the condensation level. After the final position shown the tower would soon evaporate sinking down as it does so.

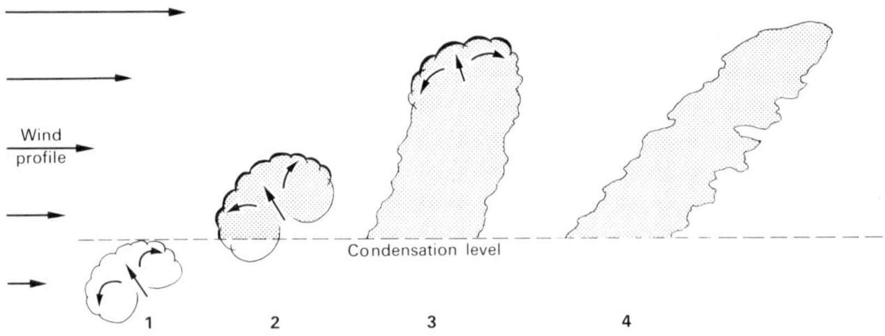

Fig. 11.5.iv Successive positions of a cloud tower rising in wind shear. Although the tower leans over in the direction of the shear the axis of the thermal is tilted in the opposite direction while it is rising.

of growing cumulus, is where fresh interior saturated air is rising to the surface. Because of the rapid rate of ascent there is plenty of water to form a droplet on every nucleus so that the cloud is very dense, and has a visibility of only a metre or less. That is why its outline is sharp. After position 4 the remains of the tower find their equilibrium level in the stably stratified surroundings and gradually evaporate, generating small motions as they do so because of the cooling by evaporation, until there is no cloud left. All that remains is a moistened tower of clear air.

If the thermal rises through wind shear (Fig. 11.5.iv) its axis is tilted in the opposite direction to the tilting of the tower of residual cloud. The erosion mechanism is easily demonstrated in the laboratory, by allowing a thermal to

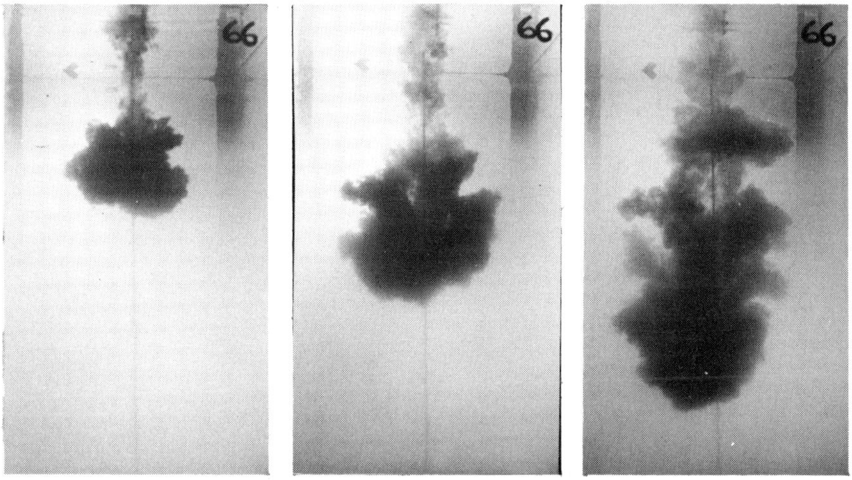

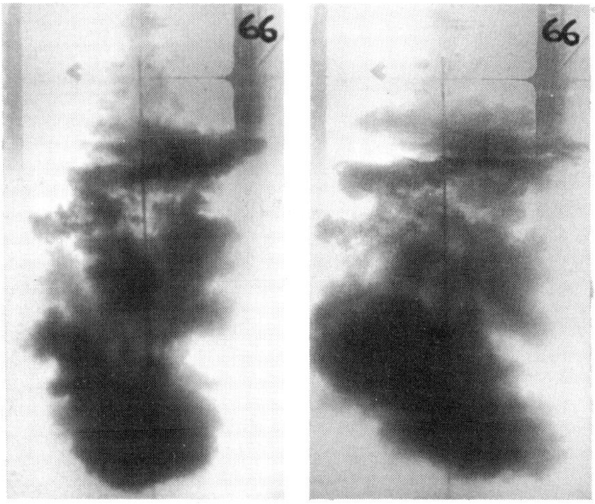

Fig. 11.5.v A model of a cloud tower rising in stratified surroundings. A thermal composed of dyed salt water is releasted at the top of a tank stratified by increasing the salt concentration towards the bottom. A tower of eroded material is left behind, each part at its own equilibrium level in the stratified surroundings. The front of the tower slightly overshoots its final equilibrium level.

sink through a stratified mass of water. In Fig. 11.5.v the upper part of the tank has no stratification and the thermal behaves like those in Fig. 8.2.i. As soon as it meets the stratification it leaves a trail of mixed material where each bit comes to rest at the level of its own density. The last picture shows the thermal, which has not grown since it entered the stratified layers, flattening out slightly above the level to which it overshot. Clouds like this tower are common in the trade winds and other oceanic areas, as shown in Fig. 11.5.vi, see Malkus and Scorer (1955).

Fig. 11.5.vi Trade wind cumulus of the castellatus type leaning over in wind shear. The wind is from the ENE and decreasing in strength with height above the cloud base. Often these clouds are arranged in streets [Photograph by J. S. Malkus]

When moistened towers of air are subjected to shear for several hours, the top may be hundreds of miles from the base and so towers become thin layers of high humidity with less humid layers in between (Fig. 11.5.vii). When these are made visible in a wave cloud we get the 'pile of plates' depicted in Fig. 11.5.viii. If a cumulus rises rather slowly in wind shear, the new growth

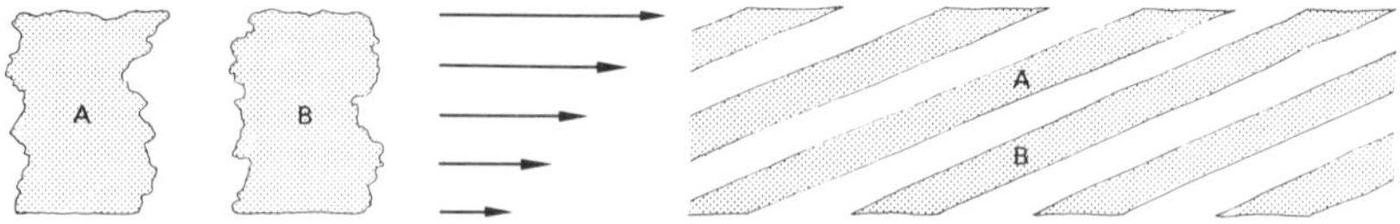

Fig. 11.5.vii When towers of moistened air are sheared over, after a few hours they take the form of layers of alternately high and low humidity. After a few days storm areas may produce the same effect, and large areas of sky with laminations of moist air are created.

Fig. 11.5.viii This wave cloud seen at the edge of the steep Rockies at Boulder, Coloradio, shows the laminations of humidity produced by shear. This type of wave cloud is often called 'a pile of plates', particularly when the cloud is isolated. [Photograph by J. P. Lodge]

takes place on the up-shear side because the thermals are usually rising in succession from a moving source, as depicted in Fig. 9.8.v. The evaporation and sinking, which occurs in air mixed with a high proportion of surroundings and travelling nearly with the wind at that level, produces downdraughts on the down-shear side (Fig. 11.5.ix).

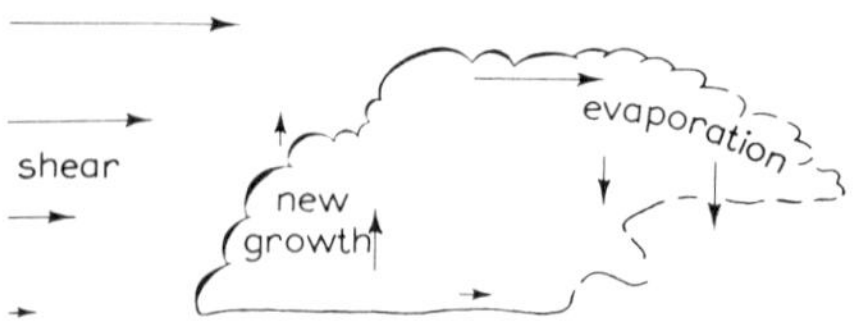

Fig. 11.5.ix In wind shear the new growth of a cumulus comes up on the up-shear side from the slower moving layers below. The evaporation and down-draughts caused by it occur on the down-shear side.

(2) Cumulus and inversions: pileus; cooling and warming

When a thermal arrives at an inversion it is flattened out as at a ceiling, although it may lift it temporarily by its upward momentum. It may therefore form a moist layer beneath the inversion, and may or may not be visible as cloud. In Fig. 11.5.x a thermal is depicted as displacing the moist layer upwards

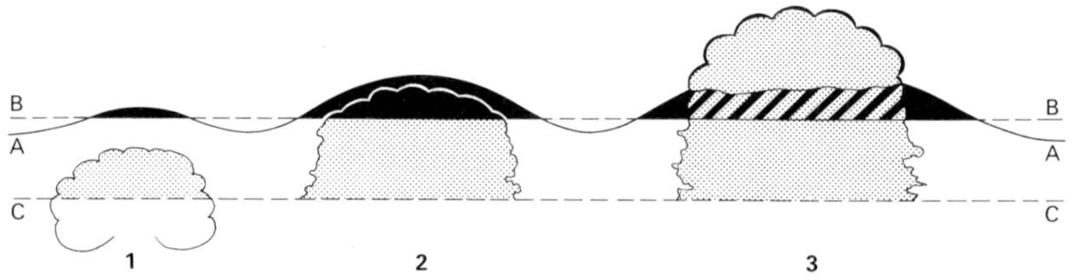

Fig. 11.5.x Possible successive positions of a pileus (cap) cloud and the cumulus (thermal) rising beneath it. If it penetrates, the pileus becomes a 'scarf' cloud, but is soon evaporated in the downdraughts on the outside of the cumulus. C and B are the condensation levels of the cumulus and pileus forming air respectively. The thermal may be invisible if B and C coincide. A is the position to which the stable layer just below B is distorted by the thermal.

above its condensation level to form a pileus (cap) cloud which may or may not be subsequently penetrated. If the thermal is a cumulus cloud, the inversion being well above its condensation level, the pileus may appear above the top of the cumulus and may or may not be subsequently penetrated. The pileus appears temporarily because it sinks back to its own level: sometimes the thermal producing it is not visible if the pileus is at the condensation level for surface air.

Less obvious is the fact that cumulus may create inversions. We note that at the base of a cumulus cloud there is a flux of heat, and buoyancy, upwards. At the top there is evaporation of cloud at the highest level reached, and this may cause cooling if the surroundings are dry enough. Somewhere in the middle of such an average cloud is a level where the flux of heat as buoyancy is exactly equal to the flux of negative latent heat in the form of the liquid water droplets moving upwards. Somewhere a little higher up is a level at which the cooling effect predominates. We speak of the 'average' cloud because when the sky has widespread scattered cumulus each cloud passes through a life cycle of 10–45 minutes, and some are growing bigger while others are rapidly disappearing by evaporation. In Fig. 11.5.xi we imagine the sounding on the T-ϕgram to be

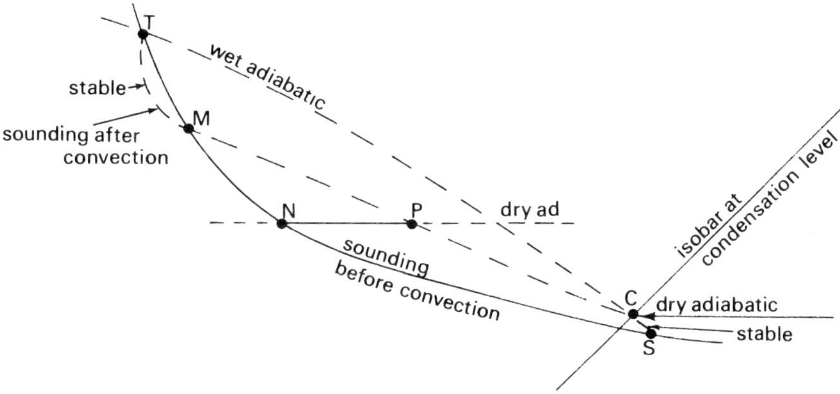

Fig. 11.5.xi The initial sounding SNMT is modified by convection from below to SPMT. C is the point of condensation of thermals from below. Stable layers are generated between S and C, and between M and T by the convection.

initially as indicated. As thermals rise they mix with the surroundings and therefore acquire a temperature somewhere between the wet adiabatic through their point of condensation, C, which they would follow if unmixed, and the environment. The evaporative cooling causes them to acquire the same temperature as the environment at some level M.

Some thermals rise higher than M and then fall back below that level as a result of the cooling. Others reach the level of M with less mixing and have a higher temperature than the average and therefore more liquid water. All thermals overshoot their equilibrium and mix with air higher up, and therefore cool it and cause it to sink; but it cannot be caused to sink below C even in an extreme case. Thermals above M cause a cooling of the air. None can possibly rise above T, where the wet adiabatic through C cuts the sounding, except by overshooting after rising unmixed. Since mixing always occurs, T is a good estimate of the highest level to be reached by any thermal. The environment is gradually cooled between T and M. Below M it is warmed by the subsidence necessary to compensate for the ascending thermals until the thermals can no longer rise through it, and so the sounding is modified to the dashed line after some time of convection. Thus an inversion is produced below T, and subsidence warming is represented by the sinking of the air at N along the dry adiabatic to P.

Below the condensation level the air is more nearly neutral, and may be slowly warmed as described in Section 9.4 (end). The air just above the cloud base is stably stratified, and so the subsidence of air through the condensation level which compensates for the upward flow of air in the form of thermals generates a layer called the **sub-cloud inversion**, occupying the layer CS.

This sub-cloud inversion is intensified if there is also subsidence at that level due to large scale motion of the air. On the other hand, if there is large scale horizontal convergence, as over land when there are sea breezes blowing, the lifting of the whole air mass cools all the air above cloud base. This is because it is stably stratified there, which makes it possible for convection to continue even though the temperature at the ground is not increasing.

Thus the existence of convection clouds may be an indication of horizontal convergence as much as of an increasing surface temperature. Equally, warming is as often due to subsidence as to heating by thermals. This topic is discussed at length by Fraser (1968).

In a burst of cold air from the north the whole air mass is rapidly warmed by subsidence. Thus clouds which reached up to 8 or 9 km above the sea off the west of Scotland may be replaced by clouds up to only 3 or 4 km off the coast of Portugal, even though the sea is warmer there, and by a layer of strato-cumulus below an intense inversion at 1 km at the Canary Islands. It then travels across the Atlantic ocean as a Trade Wind, approaching the equator slowly and passing over ever warmer sea. The convection penetrates higher and higher as the sea temperature rises, but the cumulus clouds are continuously evaporated into surroundings which are subsiding and becoming drier above the inversion which the thermals penetrate more and more. Gradually the inversion is lifted higher by the penetration of the still warmer cumulus, and by the time the air arrives in the Caribbean it has reached around 3 km. All this time the air above the inversion into which it is being mixed is coming from the west.

What happens next depends on the time of year. In autumn the sea is very warm and the cumulus begin to penetrate the inversion and generate showers which may soon lead to hurricane formation (see Sections 11.10 and 4.9). Otherwise widespread showers occur from time to time. The air has been transformed from a shallow cool layer capped by warm dry air as it entered the trade winds into a warm moist layer. This is then turned northwards to become the warm sectors of the frontal cyclones of middle latitudes. Such are the intermittent and seasonally variable mechanisms of the general circulation of the atmosphere the world over. Their importance in determining how much subsidence or general convergence occurs in any area where cumulus may exist is overwhelming, and in no part of the atmosphere can the behaviour of cumulus be fully understood without reference to them.

By rising through a stable environment cumulus clouds warm not only the air with which they mix but all the air in between which subsides to compensate for their updraughts as well. Thus the air over a narrow sea may be warmed by subsidence compensating for convection over the surrounding land, and this may prevent the formation of the sub-cloud inversion over the land. Because it has had the latent heat of condensation of all the cumulus clouds which appear during the day added to it, the air over the land acts as if it were all warmer on the average than the air over the sea during the daytime. It is as if each cumulus, as it rises, sends out a wave, which lowers the altitude of all the surrounding air over an ever increasing radius, just as the level of a lake is lowered by an outward travelling wave when a boat is lifted out of it. All the waves from an area of cumulus production add together to cause subsidence in neighbouring areas not growing cumulus. The wave travels at a speed typical of lee waves, which is typical of the wind but may be slightly faster, up to say 60 kt.

In the evening when cumulus clouds evaporate, the air mass which con-

Fig. 11.5.xii Castellatus (turret) clouds sprouting from thin cloud produced by convergence and slow ascent. When they evaporate very quickly they are called floccus (fleece).

tained them is cooled, but not by as much as the daytime warming if some rain has fallen. This cooling sends a wave travelling outwards and causes the lifting of areas of air, over water for example, where there had been no daytime cumulus, see Scorer (1955). Sea surrounded by land may therefore appear to be producing cumulus in the evening, but they are castellatus and do not have roots in thermals generated over warm sea, and the clouds have nothing to do with any land breeze which may occur.

Cumulus generated in air in which there is convergence and upward motion begin when some condensation occurs. If the sounding slopes more than the wet adiabatic, thermals may sprout spontaneously from this cloud. Such cumulus is called castellatus (turrent cloud), or floccus (fleece) when it evaporates very quickly because of the dryness of the environment (see Fig. 11.5.xii). Many trade wind cumulus clouds (Fig. 11.5.vi) are castellatus. They have no roots in identifiable thermals rising through the sub-cloud layer, but derive almost all their buoyancy from condensation.

Recently satellite pictures have shown how cloud over narrow seas such as the eastern Mediterranean comes and goes in response to the cumulus convection over the surrounding land. For example, the morning disappearance moves across the sea from the coast at about 30 kt on some occasions. Wind and wind shear together with the stability modify the way waves may travel. For example, it may be deduced from the theory in Chapter 5 that waves can more readily travel against the wind shear than with it.

11.6 THE DYNAMICS OF FALLOUT

We shall not here be concerned in much detail with the mechanics whereby a large number of cloud particles formed by condensation from vapour are collected into a smaller number of larger particles. In size these will range from those large enough to have fall speeds of the order of 1 m sec^{-1} or more, to hailstones as big as grapefruit with speeds slightly greater than the largest up-draughts ever occuring in hailstorms, namely 30–35 m sec^{-1}.

In the early 1960's the name fallout was popularly used to mean radio-active fallout because large nuclear explosions were then introducing it into the atmosphere. Even so, it was not used to refer primarily to the radioactive debris falling in the neighbourhood of the explosions, but to the much smaller total mass of very small particles carried into the stratosphere by the thermals resulting from the release of nuclear energy. Those particles were not fallout, being too small to have a fall speed comparable with the vertical motions of the air. They were nearly all removed from the stratosphere by vertical motions which brought them into the troposphere, soon after which they were brought down in rain: they were therefore fallout only parasitically. They were, however, removed from the stratosphere more quickly than was expected according to incorrect assumptions about vertical motion in the stratosphere, and the assumption that their actual speed of sink through the air was most revelent.

We use the term fallout for rain, hail, snow and so on, namely for water particles which have a fall speed causing consequences to the air motion or to the appearance of cloud. Thus fallout leaves behind a large amount of heat that appeared when it was condensed, and perhaps was frozen. Fallout can cause downdraughts by applying its weight to air through which it falls and by cooling that air by evaporation into it; and it can alter the appearance of clouds by the falling motion.

(1) Fallout fronts

Simple experiments to reproduce the effects of the weight of the particles may be done using clouds of air bubbles in water. The downdraught is the inte-

Fig. 11.6.i Successive positions of a cloud of air bubbles released from small holes. The cloud produces a motion in the water rather like that of a thermal, but the bubbles are rising through the water at about a third of the speed of advance of the cloud.

Fig. 11.6.ii A cloud of air bubbles rising out of a polluted part of the water tank dragging the fluid behind it. A second cloud of air bubbles may readily be made to rise through the middle of one released just before it, whereupon the first then rises through the second.

grated effect of the wakes of the falling particles. At their terminal velocity their weight is added to the air. The particles tend to accumulate at the bottom, i.e. the front, of the downdraught, where they are falling into slower moving air, and the motion is rather like a thermal or new plume with the downdraught, spreading out sideways at the front. Fig. 11.6.i shows four successive positions of a cloud of air bubbles released into water. In the first they are still coming from the array of small holes; in the second the last few to emerge are rising rapidly up the axis of the motion system. By the fourth position a more or less steady state has been achieved in which they occupy a ring shaped volume with very few near the axis, rather like a thermal. The second picture (Fig. 11.6.ii) shows how, if a cloud of bubbles is allowed to rise out of a layer of cloudy water into clear water, it drags the cloudy water after it, but moving in advance with the rising speed of the individual bubbles. At this stage the rate of rise of the cloud was between two and three times the rate of rise of the individual bubbles.

With a large rate of rainfall, say 5 cm hr^{-1}, which would usually be maintained for only a few minutes at most even in a heavy shower, with a rain drop fall speed of 12 m sec^{-1} (8 relative fall speed + 4 for the downdraught) the amount of water in a cubic metre of air is only about 1 gram, and its weight is equivalent to cooling the air by about 0.3°C. With three times this value obtained by suitable suppositions, we may conjure up a situation in which the weight of water may be equivalent to a cooling of about 1°C. This is actually about the same order of magnitude as the weight of water condensed in a dense cloud. On the other hand, differences of 2-3°C in the temperature of the air at the same level in a vigorous cloud are likely on the grounds that the wet adiabatic for a rising parcel departs from the temperature sounding by that amount.

A more important effect or rain, particularly when it is not composed of particularly large drops, is to cool the air below cloud (or beside cloud, if rain falls out of an overhanging piece of cloud) down towards its wet bulb temperature by evaporation. This cooling increases from zero at cloud base at about 4°C 1 km below, if the air below cloud has a uniform mixing ratio. If it is drier the cooling quite possibly may be up to 10°C, and therefore the effect of evaporation is likely to be much more important than the effect of added weight of water. Furthermore, a downdraught of cold air will continue to drain outwards from a storm and to flow down sloping land far from the storm long after the rain has fallen out of it.

Such outflows from storms proceed as fronts which have much in common with sea breeze fronts (see Section 9.9). They can be seen on radar because of the temperature discontinuity, and sometimes because of the birds soaring in the upcurrent ahead of it, for this often contains insects scooped up in the warm air ahead of the front. Haboobs are such outflows, in which the cold air is filled with sand raised by the strong wind and by intense convection due to the hot desert underlying it. On the other hand, the outflow from a shower over the sea may prevent convection reaching up to the condensation level for many minutes. Thus, just after a shower cloud has evaporated there may be an area surrounding its former position without any clouds at all.

(2) Föhn fallacies

It is almost always stated that when the air on the lee side of a mountain is found to be warmer than on the windward side, it is because of the latent heat of the rain which falls out on the mountain. This explanation is usually incorrect. We imagine a mountain range reaching to 700 mb (about 3000 m) above sea level, with a cloud base at 900 mb (about 1000 m) on the windward side at A′, and suppose that the cloud base on the lee side is at 700 mb. This is as high as it can possibly be in this situation (Fig. 11.6.iii). From the T-ϕgram (Fig. 11.6.iv) we deduce that if the temperature at A′ at 900 mb on the upwind

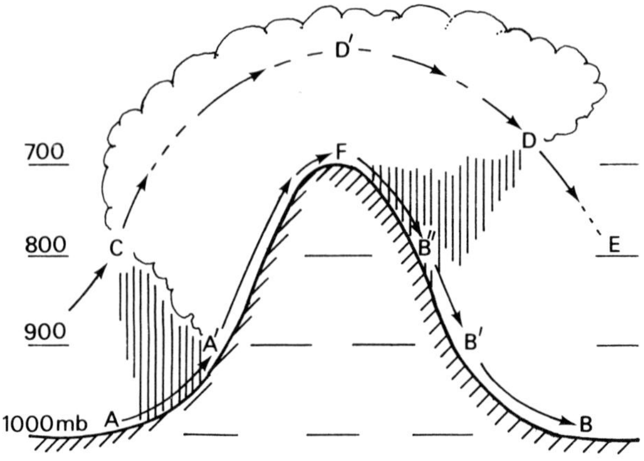

Fig. 11.6.iii Illustration of possible cooling mechanisms in air passing over a mountain range with cloud and rain. The air at B′ is found in Fig. 11.6.iv to be about 8.5°C warmer than the air at A′ if the cloud base is at 900 and 700 mb on the upwind and lee sides. Rain alters this conclusion if evaporation occurs below cloud. Other possibilities are described in the text.

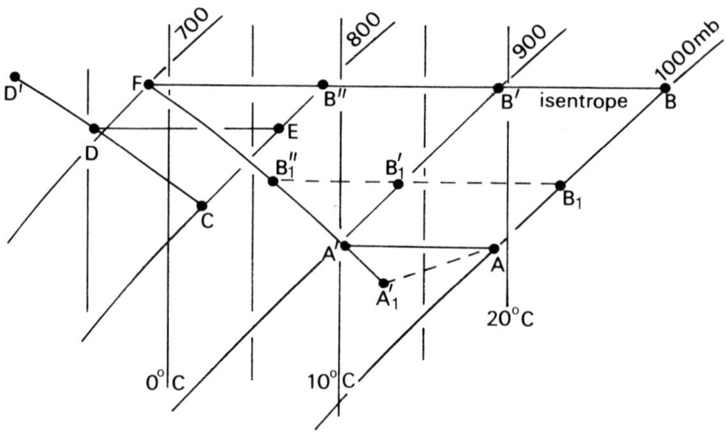

Fig. 11.6.iv A T-ϕgram representing the possible temperature changes in air passing over a mountain range, illustrated in Fig. 11.6.iii.

side is 10°C, using the wet adiabatic to F at 700 mb and a dry adiabatic to return to B′ at 900 mb the air will be at 18.5°C at B′ on the lee side. Thus a warming of the order of 8.5°C is explained. Such an increase is more easily explained by the blocking mechanism in Section 5.18(7) and Fig. 5.18.xii, and this is more likely because the warming if often observed when there is no cloud at the mountain top. The mixing caused by turbulence may also be effective as described in Section 9.8(6), where the effect of stirring a stably stratified layer is to warm the lower half of it and make the lapse rate dry adiabatic throughout its depth.

One must also be wary of explanations in terms of rain because before the air ascends through the cloud base on the upwind side it may have rain falling through it between A and A′ which cools it and lowers the cloud base to A_1'. The suffix 1 is used to indicate that rain has been evaporated into the air at some time. Unless there is a corresponding increase in the rain falling from the air close to the mountain surface on the upwind side, the cloud base on the lee side will be lower and the parcel will return via B_1'' and B_1'. In that case the warming above the temperature at A is roughly half that previously calculated. Likewise rain falling through the air on the lee side would have the effect of bringing the air down the wet adiabatic from F, with the same consequence. Furthermore, when the cloud does engulf the top of the mountain and the air flows up the front face, it is usually observed that the cloud base on the lee side is lower close to the mountain, at B″ for example, than at D. This indicates that on the lowest streamline the air does not have as much water rained out as on higher streamlines which reached a greater altitude in the cloud.

The column of air AC is not unstable, and will normally be slightly stably stratified. In that case C lines on a warmer isentrope than A and is above AA′ on the T-ϕgram (Fig. 11.6.iv). The air moving along the track CD′DE is represented on the T-ϕgram, where it is seen that unless D lies above the line BF, E will be on a colder isentrope than B, and the column of air BE will be unstably stratified. This is not observed, which means that in its passage over the mountain the cloud base must be raised on the streamline CE by more than it is raised on the streamline AB, unless the column AC is very stably stratified. If it is observed that the cloud base does satisfy this condition, it is more probable that this is because the air below the mountain top does not pass over it. Moreover, because of blocking the air above the top descends on the lee side and evaporates any cloud that may be in it. This evaporation has nothing to do with rain on the mountain. Furthermore, wave theory shows that the clouds do not necessarily disappear on the lee side of a hill (see Fig. 5.16.iii), and Föhn and Chinook warm winds are common with no rain on the mountains. Chinooks in their early stages take-off from the ground not far from the lee of the mountain. It seems probable that, since they descend from a high level because of blocking of the lower layers, the lee wave is of large amplitude and a rotor is present which is filled with cold air not yet displaced from near the foot of the mountain. On such occasions the ground may still be snow covered at the end of the winter on the lee side.

Thus the commonly employed explanation is a facile one which depends on the cloud engulfing the mountain top with rain on the upwind side only. It may occasionally be correct, but blocking is the more common cause of the warming. Perhaps the most spectacular case of warming is provided by the Berg winds of Natal: on arrival at Durban the air sometimes has a potential temperature indicating that it has descended from higher than the mountain tops (3600 m, 150 miles away) with no cloud present. Mechanical mixing of a stably stratified layer probably contributes to the high surface temperature in this case.

(3) Dynamics of mamma

If cloud thermals spread out at a stable layer the lower surface (i.e. the base) of this 'anvil' cloud is a discontinuity of humidity mixing ratio. This surface is caused to sink, perhaps by the ascent of further thermals into the space between the inversion and the cloud base, or because the air is moving over unlevel ground. Then the air above cloud base will sink along the wet adiabatic and will quickly become colder than the air below, which sinks along the dry adiabatic. If the base sinks 100 m a temperature difference of about ½°C is produced, because the wet adiabatic corresponds to around ½°C for 100 m over a large part of the T-ϕgram. If the lapse rate in the air both above and below cloud base was the wet adiabatic (which would be typical) before the subsidence took place, a layer in each mass equal in thickness to the extent of the subsidence can be involved in the instability which follows. It takes the form of a wrinkling of the surface as illustrated in Fig. 11.6.v and vi. Further pictures are to be found in Scorer (1958).

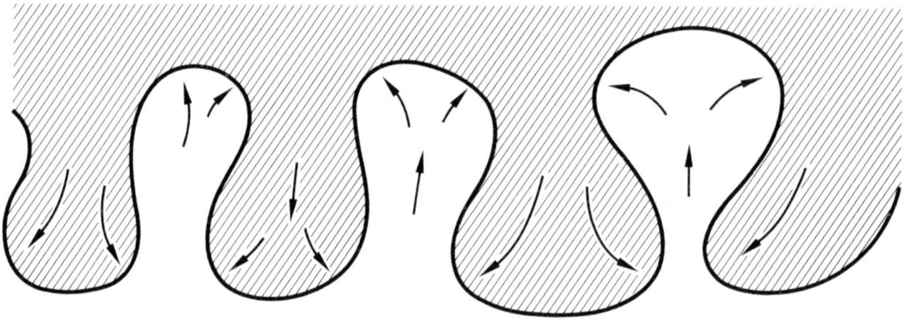

Fig. 11.6.v The motion pattern at the base of an anvil cloud made unstable by subsidence. Such a cloud is shown in Fig. 11.6.vi.

The upper surface of a cloudy thermal has a typical cumulus, or cauliflower, appearance, with a sharp outline occasioned by the large number of nuclei forming droplets in the stronger upcurrent. By contrast the base of the downward thermals, or **mamma** (breast-like pendulous clouds), are smoothly and less sharply outlined. This is because the smaller droplets are quickly evaporated in the downward motion and so the visibility is greatly increased to several metres. When illuminated by the setting sun shining on the cloud base the mamma are bright, but when the sun shines on the top of the layer the bright spaces between often show the cloud to be very thin.

Fig. 11.6.vi This cloud is first formed by the spreading out of a large cumulus at a stable layer. The cumulonibus anvil cloud is typical and spectacular, but similar motions occur when smaller cumulus spread out in the same way at their tops.

As the parent cloud grows and the anvil thickens, the anvil base descends and an unstable stratification is produced there by the subsidence. The mamma (breast) clouds are the result. The instability may be enhanced by fallout which cools the air just below the mamma and increases the down-draught there. Sometimes the lower boundary of the mamma is double, the lower boundary being that of the fallout, the upper that of the cloud.

In this picture the parent cloud is beyond the lower left corner. [Photograph by M. Koldovsky]

When snow falls, which it does with a much smaller speed than rain, the outline of the fallout base is often quite sharp. The snow not only drags the air down with it but also cools it, and these effects cause mamma on the fallout base, for example when a snow shower or warm front snow fall begins. Many cumulonimbus anvil clouds have copious mamma, which differ from those on the stratocumulus anvil clouds just mentioned in being due to a fallout of ice crystals into air which is unsaturated and therefore cooled by the fallout.

A common form of mamma is on the trail of fallout from a high level castellatus cloud which becomes glaciated (see Section 11.6(4) below) forming dense trails. If the air into which the crystals fall is supersaturated for ice the crystals grow, and this is quite likely to happen just below castellatus where air at the base is saturated for water. If there is a slight shear the trails are carried away to one side, and then when they reach an unsaturated layer they begin to produce mamma by evaporation (see Fig. 11.6.vii).

Fig. 11.6.vii The base of the ice trails formed from castellatus which became glaciated have fallout mamma. The flat tops of the clouds are near to the condensation level of the original castellatus. The water cloud has evaporated but the ice fallout from them has grown, down as far as the ice evaporation level. It then falls into air unsaturated for ice, the shear drawing the trails of falling particles to the left. The evaporation then cools the air causing descending motion.

(4) Glaciated fallout

The mechanism proposed by Bergeron and Findeisen whereby fallout particles originate as ice crystals in a cloud of supercooled droplets is well exemplified by fallout from medium level clouds referred to in the preceding paragraph. If all the particles are glaciated, for example because the temperature is below −40°C, they will all remain small. The cirrus cloud so generated will then show some feeble effects of fallout, probably due to the slight growth of the particles in supersaturated air after they become frozen. At the other extreme we have floccus (Fig. 11.5.xiii), castellatus clouds which evaporate quickly because the surrounding air is very dry and it is too warm for much glaciation (e.g. −5° to −20°C). In between (say from −25 to −35°C) we may have various degrees of partial glaciation where a few ice particles grow big enough to fall out. The remainder evaporate, partly because of the presence of ice among them and partly at the edges of the cloud where the surrounding air is unsaturated.

More spectacular is the behaviour of a layer of supercooled cloud when glaciation is induced at one or two points and the temperature is too high (say $-20°C$) for any spontaneous glaciation in the tiny particles present. The glaciation may be started by ice crystals falling from some cloud frozen at a higher level, or by the flight of an aircraft. This may freeze some cloud droplets on contact and scatter crystals in its trail or its exhaust may contain freezing nuclei. Glaciation is perhaps started by deliberate artificial seeding with dry ice pellets (solid CO_2 sublimes at $-72°C$, so the pellets leave a trail of air temporarily cooled below $-40°C$) or silver iodide. It spreads because droplets freeze first on the outside and then fragment when the inside freezes and expands accordingly. The freezing also causes the release of latent heat and thereby sets up mixing motions which spread the frozen particles among the surrounding unglaciated cloud. As a consequence the area of glaciation increases and forms a fallout hole, illustrated in Fig. 11.6.viii. The total amount of fallout in such circumstances is small, since it comes from a thin layer; nevertheless the spectacular nature of the phenomenon has often been misused in arguments in favour of rain-making programs. Fallout of this kind usually evaporates long before reaching the ground.

Fig. 11.6.viii A fallout hole in a layer of a cloud of supercooled water droplets caused by local glaciation. The ice crystals fall out because they grow in size while they are in the cloud where the air is saturated for water, but usually evaporate in the drier air below. [Photograph by H. H. Samuelson]

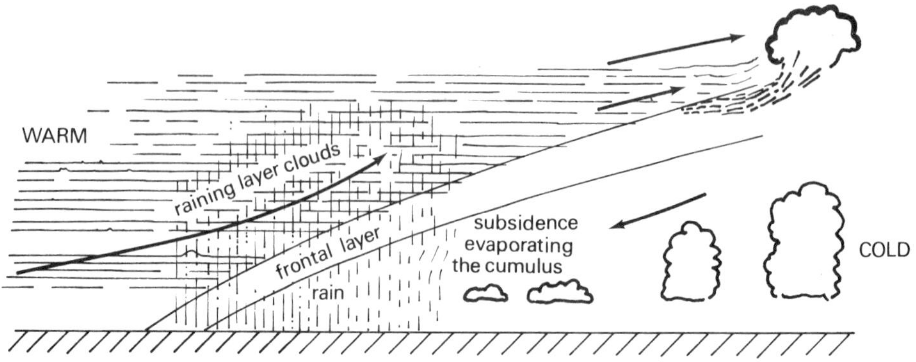

Fig. 11.6.ix A cross section of an advancing warm front. The raining layer clouds are preceded by lines of cirrus often with fallstreaks oriented by the shear. Beneath the frontal zone the cloud is evaporated by the subsidence.

Fig. 11.6.x An approaching cyclone. The lines of cirrus at the top are moving along their lines from the NW in the jet stream. The direct circulation (see section 4.8) is made visible by the fallstreaks pointing inwards towards the warm air. Wave clouds in the middle level are made likely by the increase of wind with height. The low level wind is seen in the smoke to be from the SW (see Fig. 4.8.iii). [Photograph by F. H. Ludlam]

The cirrus clouds which can be seen in the jet stream ahead of an advancing warm front often produce falling rails of great length because of the large horizontal and vertical extent of the supersaturated air in that situation. The direct circulation characteristic of a developing storm can then be seen in the sky. The orientation of the front is revealed by the long lines of fibrous cloud moving rapidly along their length with the thermal wind. The air at high levels is moving outwards and upwards from the storm; that at low levels is moving inwards and downwards, thereby decreasing the amount of low and middle level cloud beneath the increasing cirrus. The fall streaks lie across the lines to indicate the direct circulation, and when they do not show this feature the front may be expected to be less active, and the wind and rain reduced. This visible feature was remarked upon by Bewsey and the situation is indicated in Fig. 11.6.ix.

11.7 MISCELLANEOUS CLOUD FORMS

For a fuller treatment and illustrations the reader is referred to *Clouds of the World*. Here were are concerned primarily with a few mechanisms which can produce an enormous variety of cloud forms according to the circumstances. **Scud** is fragmentary cloud below the main cloud base in rainy weather. It is formed when a thermal rises in air into which water has been evaporated from rain or wet ground, thereby lowering the condensation level. **Steaming fog** is produced when cold air passes over warm water, and is often seen in the autumn downstream of power stations which warm a river used for cooling. It is the same as bathroom cloud, and is produced by the mixing of air saturated at the water surface with cooler air above to produce condensation, as in a contrail.

It is unusual for layer clouds to remain formless for long. The top of a layer is losing heat by radiation to space both day and night, unless there is another layer above radiating downwards at its own temperature: consequently downward convection from the top surface goes on all the time and both layers of cloud and fogs have a cellular structure in otherwise calm conditions. This is particularly true of sea fog which, contrary to popular supposition, is not simply due to the sea being cold. Sea fog is in fact being warmed by the sea all the time, even in cool sea areas famous for their sea fog. In many cases the fog is first formed over the land and drifts out to sea where a steady state is reached with an upward flux of heat. Land fog shows cellular structure less often because the ground is often cooled at night to a rather low temperature before the fog is formed. Although the cells can be seen in the structure of these thin layers of cloud, the cellular motion is not easily detected because it is very slow.

Much more rapid is the formation of billow clouds which make visible the motion described in Chapter 6 without significantly affecting it. The buoyancy forces due to condensation are too small to influence this motion, which has large vorticity already, and radiation takes too long to produce significant temperature differences. The shape of clouds in billows depends on their position in the motion pattern, and the various possibilities are illustrated in Fig. 11.7.i. Not all these forms are visible at the same time, and because the pattern is always developing many forms are transient. If, as is very common, the billows are generated in a wave (see Section 6.3) the full development of the overturning may be stopped when the air passes the wave crest and vorticity of opposite sign begins to be generated. Thus billow clouds of the kind depicted in the early stages of billow growth may actually persist and may not be transformed into the later forms (Fig. 11.7.ii).

Fig. 11.7.i The shaded areas illustrate the various cloud shapes which may appear within the billow motion discussed in Chapter 6. The progressive development of the billows is sketched from left to right. Most of the forms occur in typical billow manner, that is several parallel elements together regularly spaced. However subsequent billow motions may generate several criss-cross patterns in the clouds they produce, because their orientations may differ (see Fig. 11.7.ii).

Since the clouds represent a redistribution of humidity they may remain as a fossil pattern if evaporation by subsidence occurs as it does on the downwind side of a wave. Thus billows or cells formed in one wave cloud may appear already formed at the upwind side of the cloud in the next wave. The fossil pattern (see 11.2(7)) may remain for a very long time to appear a day later when cloud is reformed at a front. Very often on first appearance layer clouds at a warm front have a cell structure much larger than can be accounted for by their own thickness, and so reveal humidity patterns caused by convection many hours before.

Wave clouds may take the form of smooth lens-shaped clouds formed when the sharp top of a well mixed layer, which was the upper limit of convection somewhere upstream, is carried above its condensation level. The bases are then flat, while billows are sometimes formed on their upper surface. On the other hand, waves may be visible from below simply because of a few holes in a layer of cloud in the wave troughs. If the layer was formed by the spreading out of cumulus tops at an inversion, the base may be sharply defined by a discontinuity in humidity and may therefore be arched in waves.

Waves often produce similar effects to lifting and subsidence of a layer. Thus castellatus may be initiated by waves, and may extend in lines downstream from the point at which condensation occurs. On the other hand the instability may be less in magnitidue and merely break the first wave cloud into a cellular structure, which evaporates at the downwind edge of the wave cloud and reappears at the upwind edge of the next.

Fig. 11.7.ii Billows of different orientations and wavelength superimposed in a layer of cloud, together with tiny cumulus created by static instability, caused by radiative heat loss or ascent of the layer.

Waves may occur at very high altitudes, in which case the temperature may be below −40°C and the droplets freeze almost immediately. If the air remains above the ice evaporation level a trail of cirrus may extend for many miles downwind as **orographic cirrus.** These trails have much in common with the trail of ice cloud from the anvil top of a cumulus situated over an oceanic island or over isolated mountain in a dry country. The source of thermals is fixed, and often the low level wind is light, but at cirrus levels there may be a strong wind continuously carrying away the newly formed anvil cloud.

11.8 SELF PROPAGATING STORMS

An air mass ready for shower formation often has warm moist air near the ground with much drier air aloft. Isolated thermals growing up into the dry layer are quickly evaporated. Large clouds with a long enough life to generate rain are not formed, especially if there is wind shear. Wind shear turns the humidified towers of air due to evaporating clouds into thin layers (see Fig. 11.5.vii) which scarcely reduce the evaporation of cloudy thermals rising through them (see Ludlam, 1966).

Evaporation is greatly reduced by glaciation, and this is probably a more important consequence of glaciation than the extra 80 calories per gm released as latent heat. It also causes the onset of rain by the Bergeron-Findeisen mechanism before evaporation occurs, thereby further retaining the latent heat of condensation produced when the rain falls out as liquid. The following very much simplified description is based on work of Ludlam (1963) since taken up by Newton (1966) and others. It describes a simple two dimensional flow

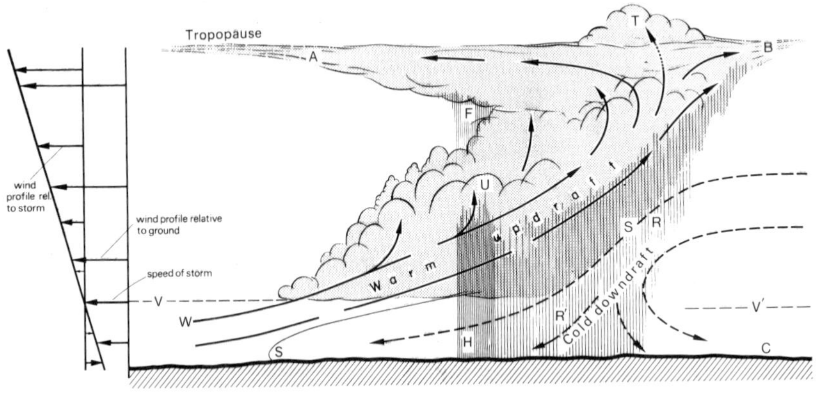

Fig. 11.8.i A cross-section of a travelling shower cloud showing features essential for self-propagation. This 'machine' processes the air, replacing the moist air ahead of it with a carpet of cooler air behind made moist by the rain. S is the front of the advancing cold air and is experienced as a sudden squall when it arrives. It may spread far ahead of the storm in some circumstances.

pattern which can cause a shower to proceed through the air mass, processing the air as it goes by carrying up the warm air and releasing latent heat which is retained because of rain falling out. At the same time the shear of the wind ensures that the cloud leans over, and because the rain (R) falls out of what was the side of the cloud it falls into the dry air above cloud base and cools it. This cooling is much more than results from the evaporation of cloud because rain (R′) continues to fall into the air until it reaches the ground. Because it comes from a level of stronger wind the downdraught spreads rapidly along the ground, shovelling up the warm moist air into the cloud so as to continue the whole process. The motion is shown in Fig. 11.8.i.

At the highest levels, which may be just beneath the tropopause, the rising air begins to take up the speed of its surroundings and is carried forward as a great anvil cloud (A). Some of the very rapidly rising air may also spread backwards at the tropopause (B), and intermittent towers (T) penetrate several

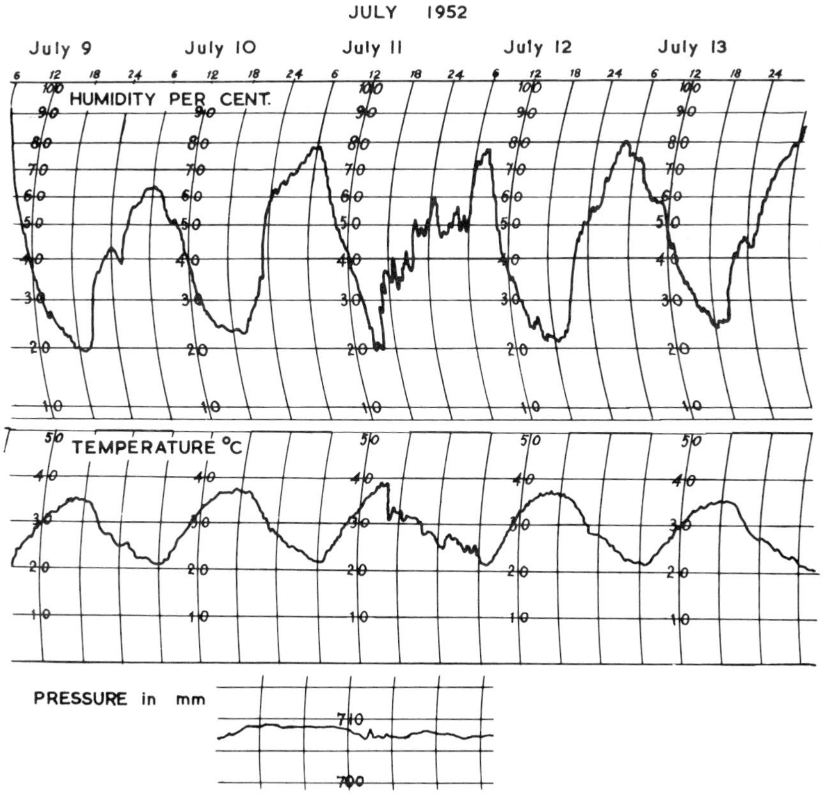

Fig. 11.8.ii Humidity, temperature and pressure traces at Madrid showing the sudden arrival of cold air from a storm at about 16 h on 11 July 1952. On the previous and following two days no showers occured in the area so that the air mass remained unchanged.

hundred metres into the stratosphere. Ice fallout from the upper forward leaning part (F) may initiate hail (H) in the strong upcurrents (U) boiling up faster than the average in the sloping updraught.

The cold air (S) spreading forwards on the ground may move far ahead of the storm and initiate new towers of cloud separated from the main storm so that the system may proceed in jerks. Furthermore it is not a two-dimensional system, and in reality many complications produce local variations, as do features of the topography. An example of cold outflow ahead of a storm is shown in Fig. 11.8.ii, taken from Scorer (1953). The traces show the fall in temperature, rise in humidity and sudden rise in pressure.

The warm air (W) ahead of the storm may be beneath a sub-cloud inversion (V) and after the storm has passed it may leave a lower inversion (V′) topping the air that has been brought down by cooling and laid out as a carpet (C) behind the storm.

The changes in surface wind can be quite variable. The warm air may or may not be moving towards the storm relative to the ground. The cold air may or

Fig. 11.8.iii A storm with the anvil streching predominantly backwards with most new growth on the right side.

may not arrive as a squall, and if the front has a lobed structure such as is indicated in Fig. 9.9.ii, the squall will vary very much along the front in strength and direction. This is usual, and the arrival of the squall before the rain is usually accompanied by slamming of doors and windows, flapping of washing on the line, flying of loose sheets of corregated iron, clanging of dustbin lids, raising of dust and leaves, and so on. In the rain there may be an abundance of scud. Above the squall may be seen a variety of soaring birds reluctant to descend into the turbulent cold air but at the same time enjoying the updraught. Figures 11.8.iii, iv, and v are examples of such storms.

Mathematical representation of this situation is not easy, and requires that the parameters be chosen to produce the assumed pattern. This represents the essential difficulty with all complicated motion patterns involving several mechanisms which are not easily quantified: the form of the motion is never discovered by the calculation, which is, in effect, only bodged up to tell us what we already knew. No new laws of physics or dynamics are revealed by the study of these phenomena, only new combinations or complications.

Fig. 11.8.iv A storm seen as the rain approaches, advancing from the left. [Photograph by Betsy Woodward]

Fig. 11.8.v The cloud at the squall front of an advancing storm showing the typical ragged base on the boundary of the warm and cold air. Above is the anvil cloud. [Photo by M. Koldovsky]

11.9 THE ROLE OF CLOUDS IN FRONTAL CYCLONES

It appears from many computer studies of the mechanics of cyclones that fronts may be generated in them without the intervention of condensation and the release of latent heat. On the other hand the starting point, namely a stably stratified atmosphere with horizontal temperature gradients, could not be generated without clouds. The lapse rate up to cloud base is dry adiabatic most of the time, except for occasional shallow layers near the ground. Above the cloud base the air is always stably stratified except in cloud, and convection clouds maintain this situation. The result is that flow up frontal surfaces may remain concentrated in the frontal cloud systems, and does not extend far away from them as it otherwise would in a stably stratified air mass. This means that the direct circulations which generate winds and rain are confined mainly to the neighbourhood of fronts and last much longer than if this were not so because the general characteristics of the contrasting air masses are preserved. Computer forecasts spread the effects of fronts over a wide zone because discontinuties cannot be handled with a course grid of points. They are interpreted by inserting fronts in the appropriate positions and using knowledge of their properties to invent details of weather which should follow from the calculated movement of fronts and overall development.

If the earth had no clouds, cyclones would have a much shorter life history.

11.10 TROPICAL CYCLONES, TORNADOES, AND WATERSPOUTS

Because of their small size and great intensity, particular tropical cyclones are about the most difficult phenomenon to handle by computer. There is very little in the soundings over the tropical oceans which enables forecasts to predict precisely where convection will produce rain and whether it will be produced by a few large storms or several small ones. The criterion seems to be that large ones are likely if they can be sustained, and the subsidence which must take place elsewhere to compensate for the rising air in the storms warms the rest of the air mass. A new situation is created when a revolving system is set up.

When a revolving mass of air is situated over sea which has a more or less uniform temperature of, say, 28°C over a large area, the deficit of pressure at the centre may be as much as 60 mb. Air moving towards the centre may be cooled by 5°C or so by adiabatic expansion, and is therefore warmed by contact with the sea, which thus becomes a source of buoyancy *because* of the rotation. The sea also becomes a source of latent heat because water is evaporated from its surface. It will be seen from the T-ϕgram that at 950 mb an increase in temperature of 5°C at 25°C raises the saturation mixing ratio by about 7g/kg^{-1}, and that when air rises wet adiabatically to 500 mb it condenses out roughly 15g/kg^{-1}. At these warm temperatures therefore, there is a very large supply of heat by condensation, and the turbulent nature of the air close to the sea surface due to the strong wind together with the copious spray ensure effective evaporation from the sea. Rainfall under the ring of convection cloud around the eye may commonly be of the order of 12 cm per hour, or around 3 m a day.

In order to have a pressure 50 mb lower than in the surrounding air it would be necessary to warm the air up to 500 mb by 10 per cent of its absolute temperature, namely by about 30°C. This clearly cannot be achieved by the heating from the sea referred to above which causes at most about one sixth of this temperature increase. Furthermore, if a parcel of air were to move 50 mb down the horizontal pressure gradient without retardation it would attain a speed of about 100 m sec^{-1} ($\Delta p = \frac{1}{2}\rho q^2$, and so $50 \times 10^3 \text{d}y \text{ cm}^{-2} = \frac{1}{2} \times 1.3 \times 10^3 \text{g} (100 \text{ cm})^{-3} q^2$, giving $q = 1.14 \times 10^4 \text{ cm sec}^{-1}$). Consequently air moving spirally inwards at the surface feeds the upcurrents and does not reach the centre of the storm. Instead air is sucked down dry adiabatically from the stratosphere and the tropopause is brought down to the sea surface. By descending dry from 300 mb and −50°C to 950 mb the air would attain a temperature of about 40°C and so the funnel of stratospheric air provides the low density which causes the low pressure at the centre.

Many small vortices are observed by ships and aircraft, some having winds of 30 m sec^{-1} or more, but the vortices do not endure because they are not big enough to suck down the stratosphere to the surface and maintain a low enough pressure to cool the air enough to make the sea a local heat source. The initi-

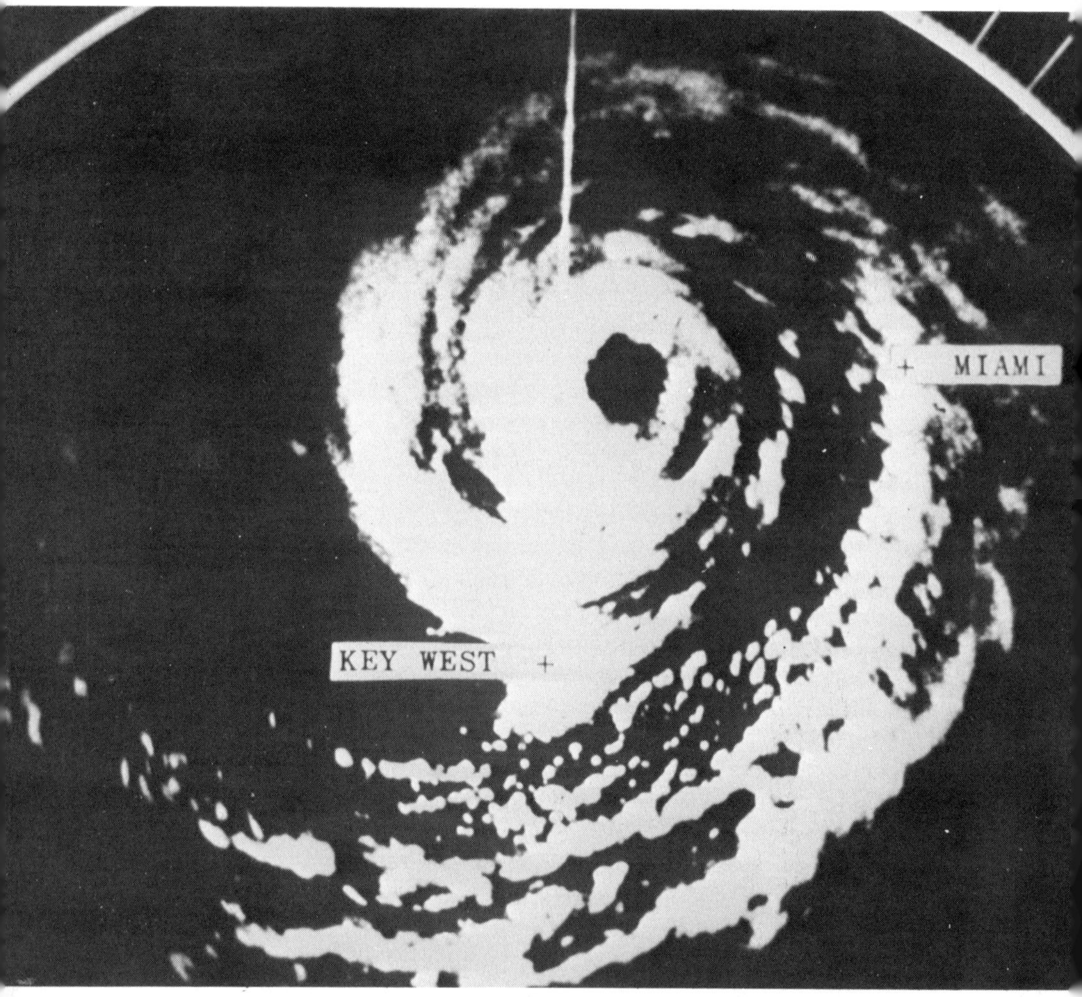

Fig. 11.10.i Plan of hurricane Donna seen by Key West Radar. These may be interpreted by reference to the cloud cross sections in Figs. 11.10.ii and iii. The radar shows principally the cumulus which are raining. The largest are those surrounding the eye. The distance from Key West to Miami is about 220 km. The vertical section (Fig. 11.10.ii) was made about eight hours earlier through the eye, when it was about 30 km wide. Heights are given in thousands of feet.

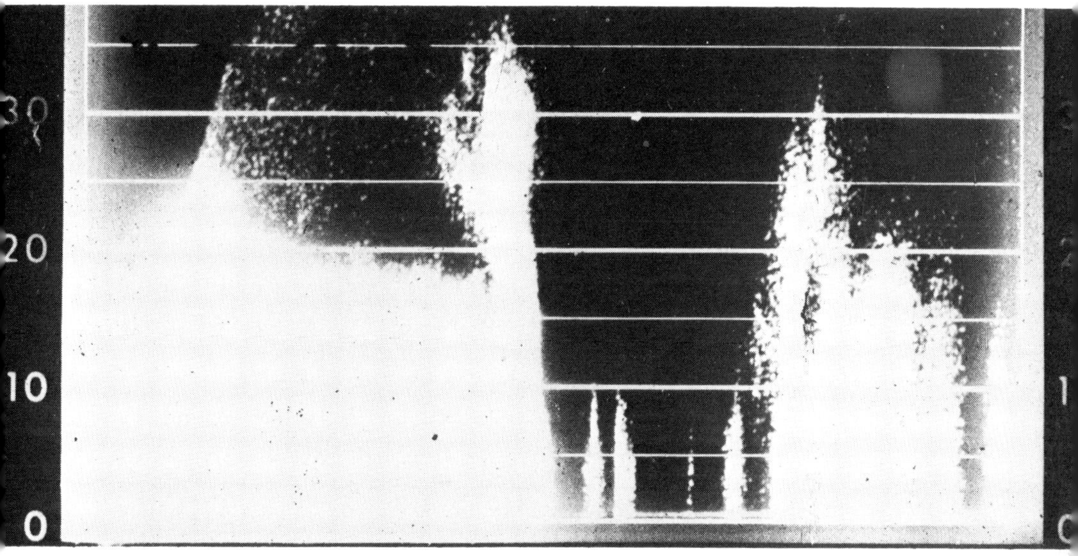

Fig.11.10.ii

ation of a hurricane simply requires the assembly of sufficient vorticity in one small area, and this is a chance phenomenon in an area where the instability can be relieved by a large number of scattered shower clouds.

A mature hurricane is not symmetrical but has more clouds in the quadrant towards which it is advancing. It moves roughly with the wind at 3000 m, i.e. about 700 mb (see Chapter 4). It has spiral bands of cloud, shown in Fig. 11.10.i and Fig. 11.10.ii, which are radar pictures of a Caribbean hurricane. The bands move around with the wind, and are not streamlines of the flow. A vertical cross section is drawn in Fig. 11.10.iii, in which the vertical scale is exaggerated and the clouds are therefore stretched vertically. We next draw attention to motion associated with clouds.

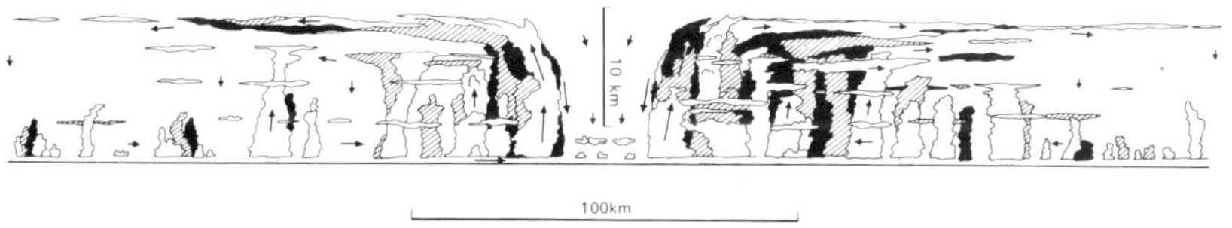

Fig.11.10.iii A schematic cross-section of a hurricane, with typical dimensions, through the eye and the lines of shower cloud such as are illustrated in Fig. 11.10.i. The motion indicated by the arrows has the circumferential motion described in section 4.9 superimposed.

Descending, very dry, stratospheric air lies against the wall of cloud surrounding the eye. The liquid water content of the wall cloud is about 10g m^{-3}.

The mixing of these two utterly different masses of air would cause a cooling of 20°C if the water were evaporated at the level of mixing. Consequently mixing causes a powerful downdraught, of the order of 30 m sec^{-1} at times, which fills the bottom of the eye with much moister air than that of the stratosphere. Sometimes,but not always, there is a 'hub cloud' at about 1 km from the sea in the eye.

The drag of the sea reduces the wind at the surface so that instead of being proportional to r^{-1} as in a frictionless vortex, it is observed to be proportional to $r^{-1/2}$ roughly in a typical hurricane. In the eye the wind is calm near the centre. The drag is communicated upwards by the upcurrents, which are largely composed of retarded air, so that as it spreads out aloft (usually with a circular canopy of cirrus) it soon acquires a rotation speed slower than the earth and rotates anticyclonically. Mariners have long used this and other features to guide them as to the whereabouts of the storm centre when they are on its periphery. Storm centres move typically at 10 to 15 kt and may therefore be avoided by most ships with good navigation and reading of the sky. At a radius of 300 km the vertical component of the rotation of the earth at latitude 19° (where $\cos\theta = \pi^{-1}$) causes a circumferential velocity of 25 km hr^{-1} = 13.5 kt. A circumferential wind of 19 kt is therefore an actual speed of 32.5 kt in space. If the speed relative to the earth increases like $r^{-1/2}$ this leads to a wind of 60 kt near the wall of the eye at 30 km radius, which is about 61 kt in space. This decreases to 6 kt in space at a radius of 300 km which is a speed of −7.5 kt relative to the ground at this radius, a significant anticyclonic rotation. At 600 km this circumferential velocity is increased in magnitude to $-(13.5 \times 2 - 61 \div \frac{600}{30}) \triangleq -24$ kt.

A hurricane is subjected to more surface friction over land so that the circumferential speeds around the centre are soon significantly decreased. But, more important, the source of latent heat is removed, the rate of rainfall is decreased, and the storm quickly loses its ferocity. The worst effects from the human viewpoint occur in the first few hours after it arrives near the coast.

The cold outflow from the downdraughts created in raining cumulus clouds is often responsible for very intense local updraughts. If the air lifted has vertical vorticity the rapid generation of buoyancy in the cloud creates low pressure and convergence at the cloud base similar to that into the base of a thermal. The stretching of the vortex lines then intensifies the rotation and prolongs the life of the low pressure centre. Convergence is therefore induced in the air below, and provided that there is vorticity there also, the vortex is extended towards the ground.

The sudden growth of cumulus towers in these circumstances produces funnel clouds which gradually extend down towards the ground, and have a much prolonged life if they reach it. Over the sea these funnel clouds are called waterspouts, and they are fairly common under showers near the coast over warm seas. Many spouts last only a few minutes and do not reach down to the sea. However the vortex may reach the sea even if the base of the funnel cloud, which is approximately an isobaric surface, does not, and then the spray generated looks like a dust deveil (see Section 9.8(4)) rising upwards from the surface.

Over land, because of the much more varied nature of the surface, vorticity is greater in the lower layers. Thus when a hurricane passes over land it may produce many more funnel clouds, called tornadoes over land, than it does over the sea. This is further enhanced by the fact that the surface air under a hurricane is now much drier than it is over the warm sea. In consequence, when a downdraught occurs it may be much more intense than one over the sea because of the greater evaporation of rain in it (see Section 11.6(1)). This produces a correspondingly greater supply of air into the cloud producing the funnel and makes it possible that the downdraught front may generate a velocity discontinuity which is a much more intense supply of vorticity than is found over the sea.

Tornadoes seem to be most intense when the air above cloud base is drier than usual, because then the number of cumulus clouds is decreased and any rapidly growing one is more intense as the lapse rate above cloud base is steeper. Outside hurricanes, tornadoes are most intense at minor cold fronts produced by storms. The base of the funnel cloud is sometimes truncated by evaporation near the ground because of warming due to friction in the intense wind.

When a tornado begins to become intense it soon occupies the centre of the storm cloud because of the rapid cloud growth above it. This means that earlier tornadoes of the same storm have ceased to grow and their tops are sheared over so that the vortex becomes stretched and inclined. This is called the rope stage of a tornado because it becomes a long, tortuous tube of cloud.

For further discussion see *Clouds of the World*, Chapter 14.

11.11 ATMOSPHERIC TIDES

The gravitational attraction of the sun and moon produce a tide in the atmosphere whose magnitude is of the order of 10^{-2} mb in low latitudes. There is observed to be another tide nearly 100 times as great with a period of exactly 12 solar hours, i.e. semidiurnal. An example is shown in Fig. 11.11.i. It is thought to be of thermal origin because the heating and cooling due to day and night is not a pure sinusoidal input with a period of 24 hrs but has a very substantial 12 hr component. It also happens that the atmosphere has a natural period of oscillation of about 12 hrs for waves whose crests are along lines of longitude. Consequently, on a rotating earth these wave crests remain along longitudes fixed in relation to the sun at about 10 and 22 hr solar time. This quasi-resonance was demonstrated Weekes and Wilkes (1947), who suggested

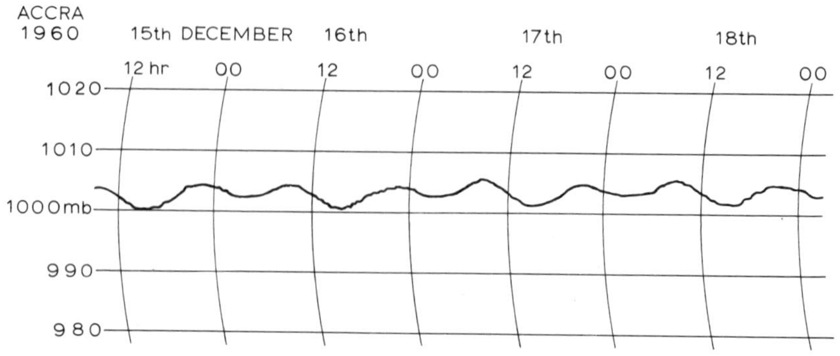

Fig. 11.11.i A typical tropical barogram showing the pressure variations due to the solar semi-diurnal tide in the atmosphere.

that the resonance was so sharp that it amplified the sun's gravitational tide but not the moon's which had a period about 50 min longer. Such a large (× 100) amplification is unlikely because of the variation of winds and height of the tropopause with latitude, longitude, and time. These variations cause variations in the speed of propagation of the tidal wave. The energy input from atmospheric temperature variations is not easy to estimate, and is not thought by some authors to be large enough to generate the observed tide without resonance. Besides the incidence of sunlight and the nocturnal heat loss by radiation observed in air near the surface, there is a large daily input of heat due to the diurnal production of clouds and rain followed by cooling due to evaporation of clouds several hours later. The quasi-resonance smoothes out the variations of input from day to day, but the oscillations do show some variation with the weather. Although some authors think that the heat input in the ozone layer around 40 km above the ground is a major cause of the tide, this conclusion is based on an underestimation of the dissipation and the input from clouds and rain.

A most interesting aspect of this tide is that the sun's gravitational field must produce an accleration of the earth's rotation because the wave crest occurs about 2 hrs before noon and midnight. With an amplitude in the tropics of about 2 mb, which is equivalent to a mass of about 2 cm of water, the mass of air displaced by the air tide could possibly equal that of the ocean displaced by its tide. This should be averaged over the whole earth, for the ocean tide manifests itself only when it enters shallow water at the coast, which occupies perhaps 0.5 per cent of the oceanic area, and a 2 cm tide there may be magnified to 10 m. On reflection at the coast the oceanic tide is put out of phase and therefore over a large area produces a couple retarding the earth's rotation in the gravitational field which generates it. This is that of the moon primarily, but also that of the sun.

Consequently, because of the quasi-resonance of the atmospheric tide, the slowing down of the earth's rotation by the ocean is cancelled out by the acceleration of the thermally induced air tide by the sun's gravitation. This means that the couple exerted by the moon, which increases the moon's angular momentum about the earth is probably about equal and opposite to that of the sun which lengthens the year. The kinetic energy gained by the moon in its orbit round the earth does not come, ultimately, from the rotation of the earth about its axis, but from the rotation of the earth around the sun, and it increases the length of the year, not that of the day. The changes in rotation speed are so small that they are unlikely to be measured directly except over many decades. Moreover, since there are changes in the rotation of the solid earth due to variations in the zonal wind, it is not simply a question of obtaining accurate measurements to confirm or refute the theory that tidal resonance preserves the length of the day. Climatic change could be the cause of any observed variations.

11.12 CLOUDS AND CLIMATIC CHANGE

The mean temperature of the atmosphere is determined by the total incoming and outgoing radiation. Changes are buffered by the large thermal capacity of the ocean. Much has been written about the increased greenhouse effect due to the increase in the proportion of carbon dioxide in the air. The burning of forests and fossil fuels will continue to cause an increase, and it has been estimated that, other things being equal, the increase since 1850 has raised the world temeperature by 0.5°C, and will cause a further increase of around 2°C in the coming century if the use of fuel "follows the expected trends".

In spite of great variations in the power of the sun there has been a large part of the earth's surface at a temperature near enough to 10°C for it to be very suitable for life for 3000 million years. How the earth has avoided much larger variations in climate than have actually occurred is a much more important problem because the stabilising mechanisms could overwhelm any effect due to activities of mankind.

The most obvious factor, and probably by far the most important, is the cloud amount. This is because changes of the magnitude which are observed to occur cause changes in the earth's albedo (the fraction of sunshine which is reflected back into space) which must dominate the radiation budget. Cloud amount changes from hour to hour, and with the seasons and over the decades. If there were not a powerful stabilising mechanism the temperature would have been observed to vary in historical times by much larger amounts than are known to have taken place. A general rise in temperature increases the amount of water vapour in the air and thereby increases the global average amount of cloud accordingly and more sunshine is reflected.

The only effective agents of climatic change are those which interfere with the mechanisms determining cloud amount. Even those which directly alter the cloud amount are of little consequence: for example if the amount of cirrus were substantially increased by aircraft contrails it would merely cause the mechanisms to generate less, and the effect would be negligible.

The most important influence is nowadays thought to be the precession of the earth's axis with a period of about 45,000 yards. At present the earth is nearest to the sun in the northern winter, but when it is farthest from the sun in that season the northern winters become more severe. This causes a decrease in the cloud amount over the greatest land masses in winter when the effect is least because sunshine is feeble. Furthermore the snow cover of the land hemisphere is increased which has the same effect as an increase of cloud amount.

This very brief reminder of the overwhelming importance of cloud amount, which is determined on a very short time scale of a few hours, in stabilising and determining the average world temperature, which varies with the seasons and over much longer time scales, is needed because of the popularity of many preposterous and fanciful theories about man's inadvertent or even deliberate influence on climate.

Unfortunately, the smallness of the details of cloud cover and its rapid changes make it very difficult to observe and quantify any measure of the global effects even with the present satellites, some of whose remarkable photographs are reproduced in this book. It will be many years before records have been obtained which give us a basis for quantitative assessment of the other natural and man-made influences which are rather small by comparison.

Chapter 12

THE AERODYNAMIC ENVIRONMENT OF LIFE

12.1 HOW THE PHENOMENA ARE STUDIED

We are in an era in which expertise in a branch of science seems to require a knowledge so detailed that it is very difficult for anyone to argue with the expert. If one takes issue with a specialist one is liable to be slapped down with statements which appear to disprove any contrary suggestion. There is a real danger that the narrowness of the specialist's imagination has excluded from his consideration ideas which are obvious to an untrained observer. Thus in Exodus X 13 it may be read, 'And the Lord brought an East Wind upon the land all that day, and all that night; and when it was morning the East wind brought the locusts', and in verse 19, 'And the Lord turned an exceeding strong West wind, which took up the locusts and drove them into the Red Sea and there remained not one locust in all the borders of Egypt'. However, for a long time it appeared to a trained biologist within a swarm that the locusts were in purposefully directed flight, and to them the problem was to find out where they were making for. We shall see in Section 12.6 that their flight is not purposeful in direction and that they are carried by the wind on their great migrations, as Moses's 'scriptwriter' indicated; yet there was great resistance to this idea from biologists because it degraded the status of the locust, which had been thought to act purposefully. Even today many biologists have ridiculously grandiose ideas of a bird's power of migration and prefer to look to such mechanisms as a map of the heavens incorporated from birth in the bird's mind and to ability to detect the earth's magnetic field. There is, of course, no physiological evidence for either of these suggestions, and it would seem simple common sense to gain first an understanding of how species which live an important part of their lives airborne react to the air motion. Scientists appear to be so badly trained these days that their imaginations seem able to work only on concepts of laboratory origin.

We must avoid asking such questions as 'how do birds navigate?' before we are sure that they do. Life would be simpler if they had ways of travelling about with the seasons which did not involve any intellectual activity whatever (like our breathing; which we can, but don't usually, control). The mode of success requiring the minimum of biological activity is likely to be the one practiced because it is the most reliable. We start, therefore, with the assumption that airborne life forms have very simple aeronautical behaviour which only involve the sense mechanisms to the minimum.

We assume that the pattern of behaviour must favour successfully the survival of the species, and that a simple pattern is more likely than a complicated one involving overt purpose.

Without any disrespect intended to that great enterprise 'Nessie', it is appropriate to use the Loch Ness Monster as a generic term (in the manner of 'boycott', 'meander', 'cardigan' etc.) indicating a marvellous being used to explain a large number of different phenomena which have come to be erroneously attributed to the same cause. Such explanations must, of necessity, be more marvellous than any of the explanations offered for the separate phenomena. Thus the Monster must be able to take the appearance of decomposing vegetation rising to the surface, motor boats, whirlwinds, fish shoals, and probably many more quite separate events. It cannot be like a motorboat because it it is like the rotting weed; it cannot be like rotting weed because it moves like a whirlwind; and of course it cannot be like any other animal or it would have been caught long ago. Thus UFO's are a Loch Ness Monster, for they are like planets, like aeroplane lights, like geese or patches of clear air turbulence (to radar), like various meteorological optical phenomena (e.g. rings round a sub-moon or sub-sun due to multiple reflections on almost horizontal surfaces of ice crystals in tenuous clouds), and like mirages, together with a few other illusions and exaggerations. UFO's are required to be the single explanation of several unrelated phenomena which are only thrown together by being observed at a time when science fiction propagates certain ideas.

Such marvellous explanations are often taken seriously and the public as well as the scientists have become educated to expect instant answers. If I say that it is contrary to experimental physics to suppose that birds navigate by means of the magnetic field I receive the retort, 'well how do they do it, then?'. The situation is so complex in the great outdoors that laboratory concepts are often quite inappropriate. At the same time a long exposure to the medium can often help in finding simple explanations through experience gained living in it. In this respect the present generation, living mostly in towns and effectively protected from the weather even when outside towns, is the worst equipped for many centuries to learn what is going on. The stars are effectively hidden by city lights, and we have no need to know them. A few meteorologists, astronomers and naturalists are exceptions in modern society. Most biologists seem to go out of doors with laboratory blinkers on. So, of course, do many engineers, physicists and chemists. Thus a radar engineer who obtained an ordinary photograph of the melting level which he had previously only 'seen' on the radar as a 'bright band' (it does not appear that way optically), called it a visible example of the 'bright band'. (See *Clouds of the World*, Chapters 2.2.1 and 4.3.7, Scorer,

1972, or *Cloud Studies in Colour*, 20, Scorer and Wexler, 1967.)

There have been many eminent biologists in the past who have believed that a bird or a fish experiences a force equal to that on an object held fixed, relative to the ground, in a moving stream of air or water (see Ackworth, 1955). This is the same as saying an airship would feel an excess pressure on one side it if were travelling at right angles to the wind, and indeed the director of the film *Round the World in 80 Days* depicted the air as streaming past the gondola of a balloon.

In the light of these introductory remarks it will be appreciated that where the mechanisms to be described in this chapter give purely physical and mechanical explanations of phenomena involving life forms whose behaviour has been 'explained' in the past as due to biological, sensory, behaviour there is likely still to remain a residue of scepticism in some quarters previously committed to the search for such explanations.

12.2 APHIDS

Greenfly, blackfly, greyfly, and a few other insects are covered by this name. During the summer they live on growing green vegetation in colonies consisting only of females which reproduce numerous female young without sex. The production continues as long as the vegetation survives. During this time, when the air is warm enough members of a colony mature enough to fly away do so each morning. Each afternoon there is another flight, this time of those which have become mature enough since sunrise. They are carried away by the air motion, and may ascend to heights of several km according to what natural convection currents they may enter by chance. They are more likely to be carried upwards if they begin to fly only when the temperature reaches a certain minimum.

Aphids are thereby dispersed over a wide area and on descending to ground level they fly around, with an air speed usually less than the average wind speed which has transported them at higher altitudes, until they find a suitable host plant on which to start a new colony. During the day they have been found by traps carried on aircraft and balloons to decrease in numbers logarithmically with height, as would be expected on the assumption that they are carried up by thermal convection. However at night and in the early morning they are often found in layers in the air (Johnson, 1957, and Johnson *et al*, 1962). These are presumed to be at temperature inversions: as the aphids descend the air becomes warmer, except at an inversion. Here it is supposed that they fly upwards into warmer air as soon as they sense a significantly colder mass of air, causing a rapid change of around 2°C or more. Equally, when in the warm air near the ground they fly about but then stop propelling themselves in order to descend once carried to much colder levels by convection currents.

As the days lengthen in the autumn, males are produced and copulation leads to the production of eggs which survive the winter and hatch out the following spring.

Dragonflies have also been observed to make use of upcurrents for soaring. They find the aerial plankton on which they feed more numerous in asending air (see Scorer, 1954) and soar only over fixed objects such as walls from which the sunshine causes continuous currents of rising air (Hankin, 1913).

12.3 SWALLOWS AND CUCKOOS

These two species are characteristic of migrant birds and are very easily recognised by their manner of flight and call respectively. Swallows can be observed soaring in thermal upcurrents, these being the parts of the air which carry up the insects, such as aphids, on which they feed. It has been objected that they are not 'soaring' birds, but this does not mean they they do not gain considerable advantage from the upward motion of the air in which they find their food. Indeed, it is probable that they could not sustain such prolonged flight otherwise. Some species, such as swifts, travel hundreds of miles in upcurrents in cold fronts, feeding on the insects rising in warm air shovelled up by the advancing cold air. They have been observed also in sea breeze fronts and in outflow fronts from rainstorms (Simpson, 1967, Berson and Simpson, 1971, Harper, 1958).

Since significant numbers of swallows and cuckoos do not simply travel through Britain *en route* to countries beyond, their arrival and departure here is easily observed to accompany winds in their direction of travel. Therefore it is likely that they perform most of their migrations by moving with the wind. Other species are known to fly in large flocks by night on migration flights of considerable distance. The effect of wind and weather on these is described by Williamson (1969). The discovery that birds were responsible for strong echoes at cloudless cold fronts was made by Harper (1958), (see also Harper 1957 and 1959).

The philosophy behind these studies is that the birds' behaviour is more likely to be made up of instinctive patterns of flight which exploit the air motion than of purposeful activities which are required to operate from time to time in order to be successful.

12.4 VULTURES, BUZZARDS AND OTHER BIRDS OF PREY

A most interesting study of the flight of vultures and similar birds in India was made by Hankin (1913). He was of the opinion that soaring birds gained energy from sunshine on their wings, by a mechanism unknown, and that they were not using upcurrents except in obvious cases where the air flowed upwards over an obstacle such as a hill, a building, or a ship. In retrospect it is difficult to understand why we had to wait for Wolf Hirth to demonstrate in 1928 that thermals were quite adequate to raise the height of a glider, for nowadays the imagination of someone watching soaring birds circling immediately fills in the picture with a rising mass of warm air. Indeed Hankin's writing suggests that even in his time it was widely believed that birds soared in upcurrents. The fact that the obvious was not accepted as the correct explanation suggests that scientific thinking can produce a very obtuse wit. But then, Hankin was a chemist.

Much of the detailed behaviour of soaring birds confirms the opinion that thermal convection consists of thermals, sometimes in complicated relationship with one another, and shaped more like new plumes in their early stages near the ground (see Chapter 9). A very brief summary of Hankin's observations was given by Scorer (1954), at a time before the significance of the early experiments on thermals was fully appreciated and the concept of bubbles was still prevalent; nevertheless those notes may be stimulating to readers interested in intelligent observation of soaring birds.

The most important point which emerges from the phenomenon of bird soaring is that certain species, such as vultures in desert areas, are completely dependent on soaring. Like buzzards and eagles they may live in mountainous areas where slope soaring is readily available on most days. Similarly hawks, kestrels, ravens, crows, and of course many coastal birds also exploit lift in upslope air streams to remain airborne without wing flapping. Hankin noted that near Agra vultures were only seen flapping extensively when returning home to roost in the evening when thermals were failing. In the morning vultures were seen sampling the air from time to time to see how good the thermals were, and only becoming airborne in large numbers when some of them had established that it was worthwhile. Indeed, some glider pilots even suggested that the vultures, just noted as sampling the air, had the ability to 'trigger off' thermals in air that was becoming unstable. Some of the larger soaring birds emigrate from areas in which thermals are not reliable in the winter, while the others which remain scarcely take to the air during that season.

There are no vultures in Australia, which is at first sight surprising because it is a continent of varying degrees of desert. However, it suffers from long periods of drought as a whole and so does not contain enough variety of climate to provide a reliable base for vultures to exist in when most of the continent is arid. There are two factors militating against the establishment of vultures there. The first is that soaring birds of prey and feeders on carrion do not have the wing power which would enable them to cross the ocean to Australia, and so a colony cannot, except artificially, be established from an outside population. The winds do not carry birds towards Australia from Asia. The second factor is the rapidity with which the well established fly population of Australia can consume any carrion available. Flies can survive a peiod of drought in the form of dehydrated eggs and can multiply very rapidly in a few days if the opportunity arises. In order to reduce the number of flies it might be worth considering the introduction of vultures. However, they would have to be fed artificially through periods of drought and the numbers kept up in order to be effective in stealing the flies' food if a situation arose favourable for an increase in flies. There is, by contrast, a viable but small population of herons and shags dependent on fish in the few permanent saline lakes among the mountain ranges of the Centre.

The wing form of birds dependent on thermal soaring, or slope soaring in small spaces, is shown in Fig. 12.4.i. This shape has several advantages. First, the

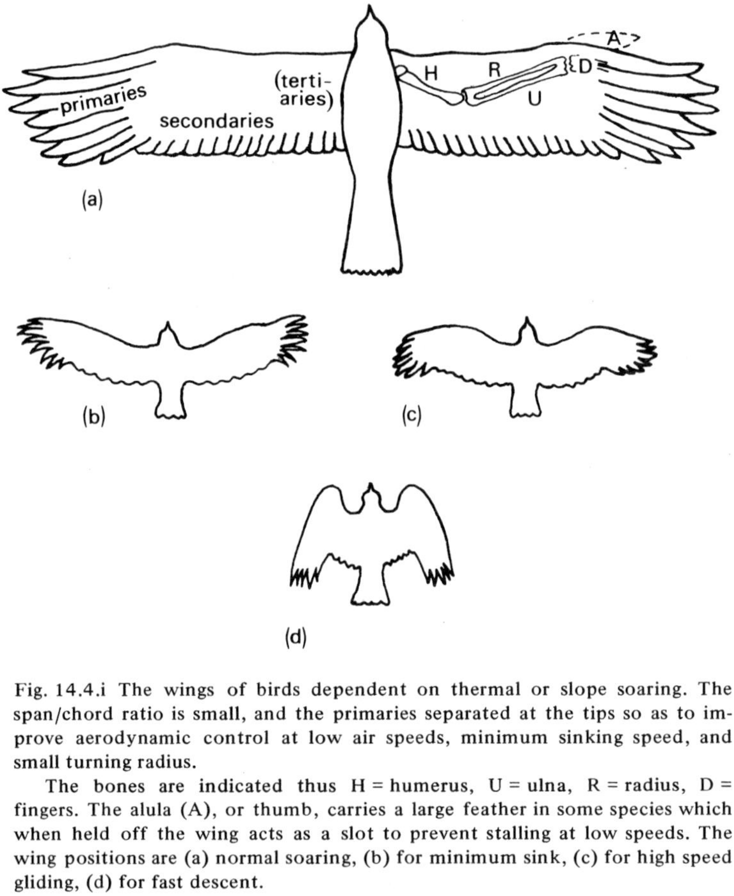

Fig. 14.4.i The wings of birds dependent on thermal or slope soaring. The span/chord ratio is small, and the primaries separated at the tips so as to improve aerodynamic control at low air speeds, minimum sinking speed, and small turning radius.

The bones are indicated thus H = humerus, U = ulna, R = radius, D = fingers. The alula (A), or thumb, carries a large feather in some species which when held off the wing acts as a slot to prevent stalling at low speeds. The wing positions are (a) normal soaring, (b) for minimum sink, (c) for high speed gliding, (d) for fast descent.

curvature of the section containing the secondaries is increased when the primaries are extended fully forwards. This increases the wing area and reduces the best flying speed (i.e. the speed at which the rate of descent relative to the air is a minimum). Secondly primary feathers at the tip are spaced apart and provide boundary layer control as in a slotted wing, and the alula (thumb) appears to serve the same purpose in some large species. These effects reduce the stalling speed, especially on the inside wing during circling in small thermals or turning in confined upslope winds when a larger wing span would be an embarrassment. When the primaries are closed together and turned backwards, thereby reducing the curvature of the primary section, very much higher airspeeds are possible. This is especially so in diving downwards without loss of control, or in flying fast upwind between thermals, which is necessary to avoid being carried progressively downwind while thermal soaring.

Hankin's description of soaring flight as he observed it is spoiled to some extent by his conviction that soaring was not due to the air having an upward component of velocity. He says that he offers no explanations, which is true, but he was determined to disprove the orthodox, and very sensible, explanation of the day. He makes a Loch Ness monster out of soaring. The upward motion of the air can be of several kinds due to different causes. By making a single phenomenon of it he proves, by instances, that each of the suggested origins of the upcurrents is incorrect because he can find cases to which they are each in turn inapplicable. In some cases he even provides good evidence from observation of clouds that the soaring was due to upcurrents, but does not appreciate that he does so. This is because in these cases he is seeking to prove that the upcurrents cannot be of small dimension, as suggested by one contemporary.

12.5 ALBATROSSES, AND DYNAMIC SOARING

If for the moment we ignore rate of sink of a gliding bird on the grounds that any method of gaining height can be used to overcome this, we may think of the problem as finding a way of increasing the bird's air speed. If there were a vertical discontinuity in the wind a bird could fly across it repeatedly, always in the direction which would cause an increase in air speed each time. In Fig. 12.5.i we imagine two air currents, side by side, in opposite directions. A

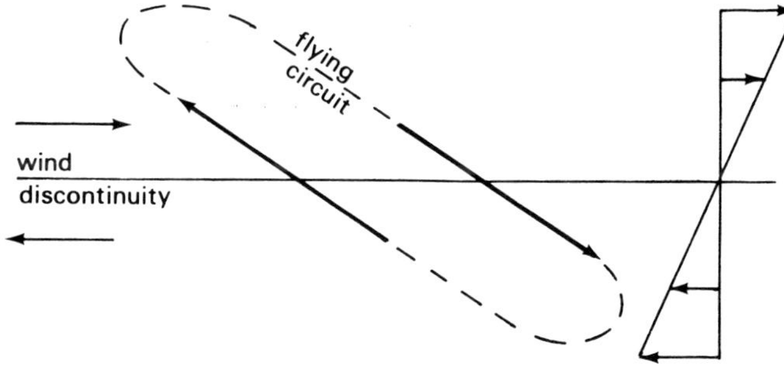

Fig. 12.5.i By flying the circuit shown across a wind discontinuity a bird repeatedly increases its air speed where the arrows are thickened. A wind gradient as indicated on the right will support unpowered flight in the same way.

bird crossing the interface in the directions shown by the heavy arrows in the presence of the wind discontinuity given on the left, experiences an increase in airspeed which can be used up to gain height. It is only necessary to complete the circuit along the dashed part of the track to continue to ascend up the discontinuity.

If we now imagine the diagram to be a vertical section of one airstream superposed on another, the bird will convert as much gravitational potential energy into kinetic energy on the descending arm as it converts the other way on the ascending arm, and so flight can be supported indefinitely if the discontinuity is sufficient. If the discontinuity is replaced by a wind gradient, as shown on the right of the diagram, the same circuit will, with sufficient wind gradient, support flight without power indefinitely.

If there is a wind gradient $\mathrm{d}v/\mathrm{d}z$, and the vertical component of the bird's velocity is w, on ascending the upward arm the bird's airspeed, u, will be increased at the rate $w\,\mathrm{d}v/\mathrm{d}z$. Its kinetic energy will therefore increase at the rate

$$\frac{\mathrm{d}}{\mathrm{d}t}\frac{1}{2}u^2 = u\frac{\mathrm{d}u}{\mathrm{d}t} = u\,w\frac{\mathrm{d}v}{\mathrm{d}z}, \qquad (12.5.1)$$

while potential energy is stored at the rate gw, and must therefore be subtracted from the kinetic energy. The bird will be able to ascend this arm without loss of airspeed (neglecting the aerodynamic drag) if

$$g\,w < u\,w\frac{\mathrm{d}v}{\mathrm{d}z},$$

i.e., if

$$\frac{\mathrm{d}v}{\mathrm{d}z} > \frac{g}{u}. \qquad (12.5.2)$$

Clearly, a bird with a large airspeed, u, will be most likely to satisfy this criterion. On the downward arm of the circuit the potential energy will be converted back into kinetic energy and the loss to aerodynamic drag made up. Consequently the downslope arm of the circuit will be the most prolonged so as to minimise the loss due to drag while gaining height.

A tactic often observed to be employed by birds is to fly into a region of increasing air speed, gaining height rapidly, then to turn quickly and descend near to the surface with a greatly enhanced airspeed. This is used up as required, for instance to search the sea surface for edible material. The sudden upward leap into the wind increase and rapid descent is then repeated.

The rapid ascent is often made by passing over the crest of a wave from the slow moving air on the lee side into the larger velocity at the wave crest. This is particularly advantageous if the flow separates at the wave crest, often made

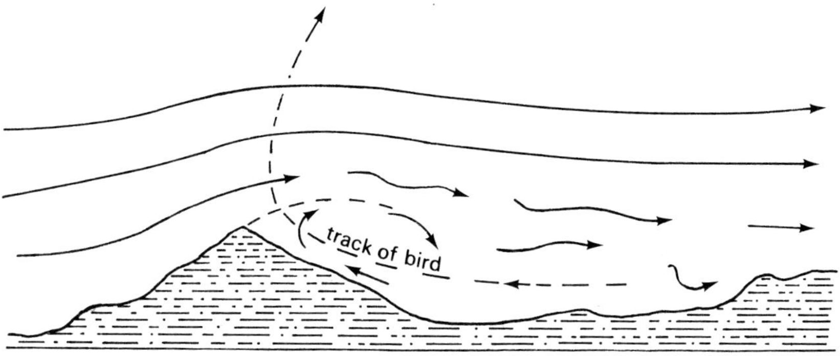

Fig. 12.5.ii Birds exploit the large wind gradient behind the crest of a cusped wave to gain airspeed and height. This is most effective when the waves move against the wind, as may be the case when the waves are produced by a ship.

visible by the motion of spray blown off at a cusped crest (Fig. 12.5.ii). Birds following a ship may use the wave pattern of the ship's wake to proceed upwind with the ship. This is more effective than using waves raised by the wind because the greater speed of the waves relative to the air makes separation at their cusped crests more likely to occur. Birds undoubtedly develop an unconscious skill from experience in performing these soaring feats, and skill is required since the waves are moving relative to the sea. Birds have the advantage that mistakes can be rectified by wing flapping or settling on the water; dynamic soaring is unlikely to be practical in gliders because of the difficulty of learning the art without frequent accidents.

If the airflow over wave crests is not used, the wind gradient required can be guessed from (12.5.2). Thus if the bird's airspeed is 30 m $\sec^{-1}$ (about 60 kt), since $g = 9.81$ m $\sec^{-2}$ a wind gradient of about $\frac{1}{3}$ m $\sec^{-1}$ per metre is required, and this is only likely to be found in the lowest 10 metres of the air over the sea.

Albatrosses have long (almost two dimensional) wings which minimise the aerodynamic drag at high speeds (Fig. 12.5.iii) and have much higher airspeeds

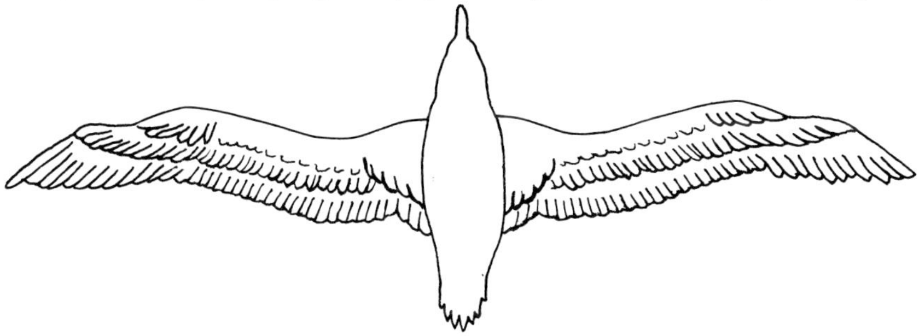

Fig. 12.5.iii Albatross wings have a small chord/span ratio and a shape more suitable for high speed flight at minimum sink, which is required for dynamic soaring.

than birds which soar in upslope winds or thermals. The largest birds have difficulty in flapping flight close to the sea surface and cannot take off from it unless the wind is strong, i.e. comparable with their best airspeed. The great southern albatross breeds on remote oceanic islands where it can get airborne by running down a steep wind-facing slope. These birds do not inhabit the northern hemisphere because, as was pointed out above, the non-ascending part of the flight circuit in dynamic soaring will carry the bird downwind unless a special effort is made to direct it upwind. I have observed smaller abatrosses and similar sea birds travelling with the ocean liners *Georgic* and *Fairstar* at speeds of 15-20 kt in still air or against the wind, but they do not seem able to keep level with the Queens travelling at around 29 kt by any means other than exploting upsloping air motion produced by the ship. The consequences of this limitation is that birds living by dynamic soaring alone must drift downwind, and that means having to travel round the earth in the direction of the wind without crossing land. This can only be done in the southern hemisphere westerlies. Large albatrosses cannot survive on a continental coast, nor in an ocean subject to frequent or prolonged periods of light wind in which they could not soar. The northern ocean westerlies have anticyclonic periods in which dynamic soaring would fail. The land obstacles in the southern oceans, notably New Zealand, South Africa, and the tip of South America and South Georgia, are easily circumvented.

Flying fishes, which remain airborne for up to half a mile, may use dynamic soaring, although this seems unlikely because their main object is to escape from predators. In such cases unidirectional travel would be preferred. They could however use the dynamic soaring routine while heading nearly across the wind all the time, and I have seen this technique used by gannets. On the other hand, it is said that flying fishes dip their tails into the water surface while airborne and propel themselves by means of sideways oscillations of the the fin (Hankin, *loc. cit.*).

12.6 DESERT LOCUSTS

Desert locusts occur in two forms, **solitary** and **gregarious**. In the solitary form they are not usually a serious pest. The gregarious form is physiologically different and individuals can be changed into the gregarious form by being crowded together. The existence of the solitary form ensures the persistence of the species during periods when there are no swarms of the gregarious locust. When there are numerous swarms travelling about over a large area there is said to be a plague of desert locusts.

A plague begins when there are rainstorms over a desert area, such as Southern Arabia, provided that there are enough locusts there to start with. The swarms occur only as a result of rain on the desert, which is a rare phenomenon, and they are of interest because they have thrown light on how a species which spends an important part of its life flying can exploit the air motions in a simple manner.

In order to produce rain the air must ascend and that means that there must be horizontal convergence. If locusts flew at a constant height in a random manner they would be diffused by their random motion but aggregated by the converging air motion. At cold fronts the convergence is very large because the warm air is lifted bodily off the ground by the cold air and so locusts in the warm air mass may be collected from a very large area as in a shovel.

Swarms lay fields of eggs at a depth of around 15 cm in sandy ground after rain and provided they remain moist the eggs hatch out in just under three weeks. If the rains are not sufficient the eggs, which are dehydrated, will not hatch out but will remain buried until sufficient rain occurs. When the rain is sufficient it produces a luscious growth of vegetation which provides the hatching swarm with food, for which there are usually no competitors.

The insects which emerge are small and wingless, and are called **hoppers**. They aggregate together in **hopper bands** which **march** forward with a circulatory motion continaully bringing fresh hoppers to the front, especially when those at the front halt to feed. Almost everything edible in the way of the marching bands is eaten. In some areas where sufficient human labour is available, trenches are dug in which the hoppers are buried as soon as they march in. Another technique is to put down poisoned bait in front of them.

The most effective method of destroying hopper bands so far used has been to cover the vegetation with pesticide by **drift spraying**. The technique with a closely growing crop has usually been to spray insecticide from the air in calm weather so that the spray settles on to the crop by gravity. This is not fully satisfactory because of the sheltering of the lower parts of the crops by the upper. It is a very wasteful technique in desert areas because most of the insecticide falls on the ground and it has the serious limitation that it can only be carried out in suitable wind conditions, which do not obtain most of the time. In drift spraying a moderate wind is used to carry the spray horizontally across the ground, and the vegetation is covered by a cloud of spray droplets which are kept airborne for a few km by the turbulence but are caught on parts of the vegetation with small geometry by impact. Even if the hoppers have to march a mile or two they will eventually consume a lethal dose of insecticide. Since insecticides which are very persistent can be used, a barrier to their progress can be set down in the desert.

The locust swarms reach the flying stage in about 6 weeks. They then fly by day, provided the temperature is warm enough, and settle at night. A large flying swarm may cover a few square miles and have as big an appetite as 100,000 people. The destructive power of a swarm has been recognised since civilisation began, and the Old Testament recognised several elementary attributes. For instance, unlike ants and bees they have no leader although they look like a well ordered army (Proverbs 30.27), and they remain settled in cold weather, flying to an unknown destination when it is warm (Nahum 3.17).

It used to be thought by ecologists that the swarms were aggregated by topographical features or vegetation; thus the early arrivals would stay around while others arrived to concentrate the swarm. There is no evidence of this sort of effect, quite the contrary, for they depart from an area of food to one of desert if the wind carries them that way.

Rainey demonstrated that swarms move with the wind, any apparent departure from this behaviour being of the order of the uncertainty in knowledge of the wind vector at the swarm. The time at which swarms take off in the morning and settle in the evening is not usually known accurately, and uncertainty about the flying time is increased because under a swarm there is usually a large number of settled locusts. The swarm therefore travels somewhat slower than the wind except in very warm conditions. There is circulatory motion (Fig. 12.6.i) with those at higher altitude travelling faster both because of the greater wind strength and also because locusts tend to fly upwind when near the ground, especially when taking off or landing. There tends to be a greater accumulation of locusts at the front of the swarm (Fig. 12.6.ii), which can add to the mass of the air to such an extent as to produce a significant

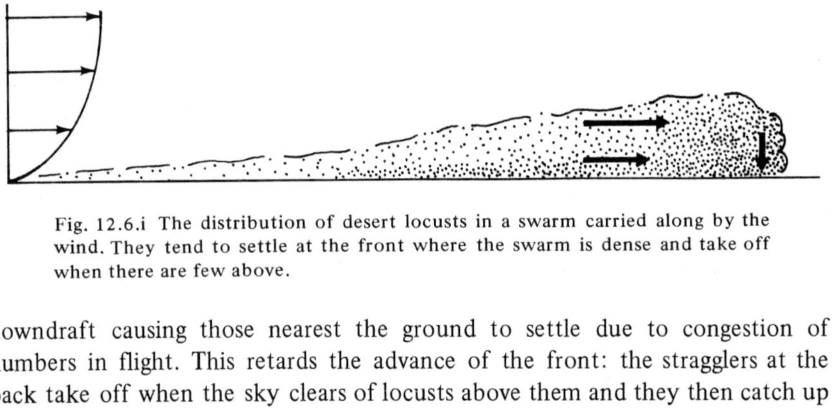

Fig. 12.6.i The distribution of desert locusts in a swarm carried along by the wind. They tend to settle at the front where the swarm is dense and take off when there are few above.

downdraft causing those nearest the ground to settle due to congestion of numbers in flight. This retards the advance of the front: the stragglers at the back take off when the sky clears of locusts above them and they then catch up with the swarm, probably by dispersing in the warm sunshine to a greater altitude than the majority.

At the edge of the swarm it appears, from considerable observational evidence, that when a group of locusts finds itself flying outwards into unoccupied air it turns round and flies towards the greater concentration. This skin effect is not like surface tension, which tends always to reduce the surface area. If a downdraft produced a hole in the middle of a big swarm those on the boundary of the hole would fly away from it so as to increase its size and eventually, perhaps, divide the swarm. It does, however have the affect of preventing the random flight of the locusts from dispersing the swarm by slowly decreasing its concentration at the edge. A swarm is considered dense if it has a

Fig. 12.6.ii A small low flying swarm, about 1 km^2, near Wajir, Kenya. [Photograph by H. J. Sayer]

concentration of more than one locust, per $metre^3$, or has more than 100 locusts above a square metre of ground.

In situations in which a swarm remains compact, smoke or any other airborne material carried more or less passively by the air (e.g. water vapour, odour, pollen, aphids) is dispersed. Even when the dispersion is confined to the air between the ground and a stable layer, the air motion spreads smoke out horizontally. Locusts differ in that they tend to remain in the air either at a more or less constant height or in the lowest layers. When convection is strong over a considerable depth locusts are carried up in columns, but, as explained in Section 8.5, even in that case they may often fall back to a lower level in high concentrations. Locusts are not observed above the levels reached by thermal convection except that when there is no convection they may nevertheless fly, in which case they remain close to the ground. If they were merely to partake of the horizontal component of the air motion and remain at a constant height they would be concentrated at the base of the upcurrents. If a downcurrent were then to occur, unless it scored a direct hit on the centre of an aggregation found in the base of a previous upcurrent, the concentration of locusts would simply be transported bodily into the base of another upcurrent. Dispersion like smoke would therefore not occur.

A large swarm viewed from the air appears more like a network, with high concentrations in the threads and low concentrations between. This structure is to some extent illusory because the locusts are most easily seen when sunlight is reflected to the observer from their shiny wings. In a swarm groups of locust are seen flying in random directions, the local unaminity of direction serving to avoid collision. However the direction of flight relative to the air often changes with height in a deep swarm. Consequently some parts of the swarm are more easily seen than others, and become visible or invisible quite quickly to an observer circling around the swarm in an aircraft.

The small scale air motion and random flight are thus prevented by simple mechanisms from dispersing the swarm. A swarm may be divided by airflow around a mountain when it does not pass over the top, and may even reform beyond an isolated mountain which divides it. Swarms do not fly over mountains except when the air flows over the top. If there is a convergence on one side (as on the eastern slopes of the mountains of Eritrea) the swarms are concentrated against the mountain by their reluctance to be carried away from the ground (Fig. 12.6.iii).

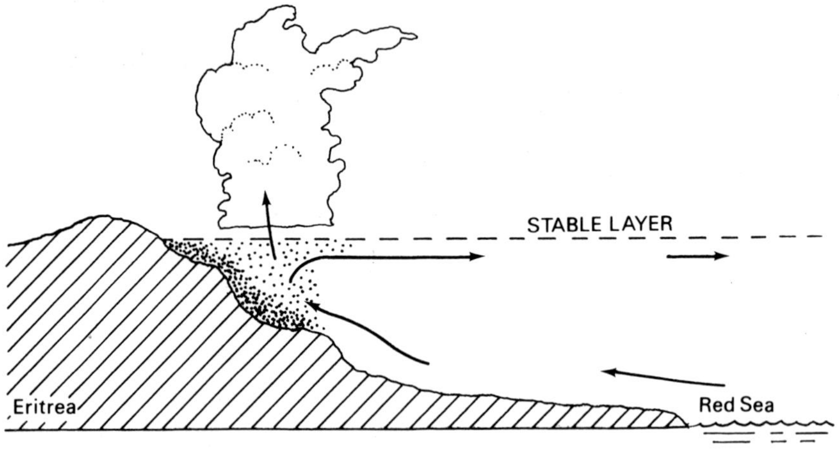

Fig. 12.6.iii Swarms are sometimes accumulated under showers on a mountain range. They avoid being carried up into the shower or away beneath the inversion.

Often there are showers in the airflow from the Red Sea and then locusts would avoid, if possible, being carried up into the clouds, which would be located above the upslope wind. There is often a layer of cloud beneath the stable layer, which is less pronounced if there are showers. The storms may be above the mountain tops, in which case the air flow would carry locusts there.

It is not certain what mechanism causes locusts to resist being carried upwards but there are several obvious possibilities – sight of the ground or sense of movement relative to it, or avoidance of the lower temperature at greater altitude. This could be achieved either by deliberate powered flight to lower levels in an upcurrent, or by an automatic cessation of wing beating when a temperature is reached at which the locusts would not have taken off from the ground. These mechanisms can be represented mathematically by imagining particles to be carried with the horizontal air velocity but having a downward velocity relative to the air which is a simple function of height above the ground (or above a level of desired flight) or a function of the upward velocity of the air. Cold fronts can concentrate widely dispersed insects as they lift warm air off the ground. Very large concentrations of locusts have been observed near Hargeisha in Somalia when air currents from the north and south east converge at a front.

The rate of change of insect concentration σ following the insects which have a velocity $\mathbf{v}$ relative to the air which has a velocity $\mathbf{u}$ is

$$\frac{\mathrm{D}\sigma}{\mathrm{D}t} = \frac{\partial\sigma}{\partial t} + (\mathbf{v} + \mathbf{u}).\mathbf{grad}\,\sigma = \frac{\partial\sigma}{\partial t} + \mathrm{div}\,\sigma(\mathbf{u} + \mathbf{v}) - \sigma\,\mathrm{div}\,\mathbf{v} \qquad (12.6.1)$$

if $\mathrm{div}\,\mathbf{u} = 0$. If the concentration is steady in a region the first two terms on the right are zero and the equation tells us the rate of change of σ experienced by an insect. It is positive if $\mathrm{div}\,\mathbf{v}$ is negative. If the horizontal air velocity of the insects is small or random the only component which can be effective in bringing the insects towards higher concentrations is the vertical velocity v_z, and so if $\partial v_z/\partial z$ is negative the regions of high concentration are maintained even when the wind is from a direction where the concentration is lower. In practical terms, if the downward velocity of the insects relative to the air increases upwards conditions favour aggregation of swarms in a region towards which the wind blows.

Small scale details such as fronts may thus produce and maintain very large concentrations of locusts. Cochemé (1960, see Rainey, 1963) performed some simple calculations and found that on the larger scale also particles following only the horizontal component of air velocity could increase their concentration by a factor of 10^3 in a few days. Furthermore, he showed that in particular cases air from a widely dispersed area of Southern Arabia and East Africa, where gregarious locusts had been observed, would be concentrated in areas of Baluchistan where large swarms were observed 2 or 3 weeks later. This converging motion produces the rain which the locusts exploit to perpetuate the species.

Even the long distance 'migrations' of desert locusts serve the same purpose of carrying swarms towards the convergence zones which produce rain in the desert. Such rain is very unreliable, and does not support human settlement except at oases or other places where the water is stored underground; but desert locusts have evolved a flight behaviour to exploit the opportunities provided by the 'random' and sparse rainfall.

Many other species such as the swarming weaver bird (Quelea Quelea) or locust bird, and many varieties of moth (e.g. army worm) and grasshopper probably have a similar survival mechanism. Indeed the understanding of the life style of locusts has opened up quite new possibilities in the control of insect pests. The development of this theme is clear in the works of Rainey (1951, 1963, 1969, 1973).

12.7 CONTROL OF PESTS THROUGH SWARMS

It is conventional to think that the best way to protect a crop from a pest is to spray the crop with pesticide. This may often be done too late, and since we know that insects are carried over large distances by air motion, to spray the crop certainly has a negligible effect on the population reserves of the pest elsewhere. Joyce *et al.* (1970) found that when farmers in a large part of Java sprayed at the same time as one another at intervals of 2 or 3 weeks, the effect was greater than if they sprayed at different times (which had made much more economic sense to them in terms of utilisation of aerial or ground based spray teams). This was because when the pest population is reduced over a large area it takes longer to recover, particularly if the territory is an island with only infrequent communication with other population reserves. On the other hand, a small area can be quickly reinfested from outside.

In the case of crop spraying, the proportion of spray which actually enters the pest is about one part in $10^{8\pm2}$. This is very wasteful and means that almost all the pesticide is harmful environmental pollution. Efficiency is usually measured in terms of the proportion of the crop surface covered by insecticide, or the proportion of the total insecticide that is deposited either on the crop or within the field containing the crop and not carried across its boundary: in such terms efficiences between 0.1 and 0.6 can be claimed, but this is a very misleading measure of efficiency. One of the most successful uses of insecticide was a case in which about half of the insecticide sprayed into a swarm of desert locusts was taken up by the flying locusts: the rest was deposited on the ground over a wide area at very low concentrations. Since many of the locusts received a dose considerably more than the minimum lethal one, the proportion of insecticide which actually killed insects in this case where the swarm was totally destroyed was about one part in 5. According to this measure the case mentioned is about the best ever achieved with any pest having an airborne phase.

It is therefore important to discover whether the swarm-forming mechanisms operate for many insects. Clearly they are not important for aphids, but for any species which propagates entirely as a result of copulation with egg and grub phases, aggregation into swarms is important for mating. Using an aircraft equipped with doppler radar navigation and insect trapping devices in the Sudan, Joyce (1973) has found that the concentration of insects is greatest by far, in small scale regions of horizontal wind convergence. It is probable that most of the infestations take place when a swarm created at convergence zone passes over the crop. There is a great deal of folk lore about the arrival of dense aggregations of moths, locusts, or other insects such as ladybirds, some of which assumes that the swarm was hatched out locally from eggs buried 7 years previously. The lore suggests that the storm which accompanied their appearance caused the outbreak. It is much more likely that the storm and the concentration of insects were both the result of the horizontal convergence of air.

Locust swarms have no natural enemies other than starvation, cold or disease. Predators such as storks can make no impression on their vast numbers. No grounded animal can concern itself with them because of their unpredictable and unreliable movements. Although they succumb to certain viruses in high humidities, and do not survive in well vegetated tropical areas for that reason, the concentration for breeding into a region which cannot be predicted by predators has great survival value for a species. All that is required for this to happen is that the insect should be inclined to fly at a constant height above the ground or at a constant temperature, and wait for its mates to accumulate around it at a suitable breeding ground.

The fact that desert locusts have two forms and that these were thought for some time to be different species obscured the relationship between solitary lucusts and the creation of populations by swarms. Many insects found in swarms may depend on swarms almost entirely for their reproduction. A species which could propagate that way would have many survival advantages against both predators and mankind. Small insects are not easy to observe in swarms, and moreover many moths probably only swarm at night or at dusk, so it requires radar to seek them out. Birds which feed on airborne insects (e.g. swallows, swifts) certainly benefit from insect swarms in cold fronts, including sea breezes and other micro-fronts.

G. W. Schaefer (1976) has shown how wing beat and breathing frequencies can be used to identify insect species by radar, and thus find swarms of particular species. Many swarms seem to occupy thin layers in the atmosphere, and this apparently horizontal arrangement of the high concentrations could be the result of the shearing over of more compact volumes containing the insects. If the layers are bounded below by colder air, the insects' reaction could well be to fly upwards if they find themselves sinking downward into the cold air, thus preserving the high concentration at the bottom of the warm air. Already successful operations against the spruce budworm moth have used knowledge of these mechanisms.

Early studies of bird migration caused it to be supposed that birds had wonderful powers of navigation. Insects were presumed, on the other hand, to be dispersed by air motion, their reproduction rate being so high as to make this an advantage. Such simple hypotheses created problems of how a moth found its mate and how birds navigated, and quite miraculous powers have been ascribed to them as a result. Actually we see that no intelligence is required. It is not even an advantage for a bird species to be very successful at navigation, because that would prevent it from exploiting the possibility of colonising new territory made suitable by long term (even decades) changes in climate. In a successful year parent birds would not wish their larger than usual broods to accompany them on migration but would more likely instruct them to 'get lost'. Through the work of Williamson (*loc. cit.*) it is now known that birds too are subject to wind trends in a very important way: when making an ocean crossing they appear to choose to fly downwind if their stellar means of navigation are obscured by cloud. This would be the best tactic for making the quickest land fall, but equally it would often lead the birds hundreds or thousands of miles from their normally expected destination. Such almost random flights are necessary for the preservation of a species existing in a continually changing climate and in competition with other species of continually changing numbers.

At the time of writing our knowledge of the movements of low intensity lines of convergence, of the kind found by Joyce, is small because no meteorological observing system has been designed to reveal them. The best prospects are certainly through the use of radar, which can see the insects, and it is probable that improved understanding of microfronts will come from the study of insects (Scorer 1974). That will probably provide the economic incentive for more detailed work. Meanwhile, knowledge of the mechanics of sea breeze fronts is the best guide.

Epilogue

MAKING PEACE WITH NATURE

We have moved out of an era during which mankind has believed that he was destined ultimately to have dominion over the earth. We move, after the industrial revolution, into a time in which it seemed that this moment of triumph had arrived and that for his own sake, if for no other, mankind would have to plan and control the world like a garden.

But a garden, like a city, or a drawing room, is possible only because of a bigger environment from which resources are taken and opportunities grasped and into which by-products are dispersed. For the whole world there is no outer environment accessible to us except as observers. Our technological success seemed potentially unbounded when we were only sampling the possibilities; but success and excess have changed the scene, and the rest of nature is either made sour by our destructiveness or has retalliated resiliently by evolving socially as quickly as mankind and biologically much more rapidly. Meanwhile we destroy for ever some of the pinnacles of evolution by our mechanisation and artificial political boundaries and our own overpopulation. Perhaps the art of nomadism and many minority cultures which we call primitive have a level of sophistication which we have hidden from overselves, and perhaps some of the species threatened by extinction as a result of man's dominance have an invaluable part to play in the totality of life on Earth!

We may find ourselves. being utterly simple minded. For example the attempts to defend irrigated territories against the quelea by extermination of their vast flocks may be very ill-advised if only because of the danger and waste of the poisons used. There are many species whose populations grow explosively in the semi-arid regions when occasional rainstorms appear, and irrigation may be the direct cause of their increase. Their populations are normally controlled by scarcity of food in the wild, so that one swarm or flock exterminated by man leaves food for others who would have died.

We have tended to be simple-minded about how things are controlled in nature. When we interfere with a major mechanism, minor ones emerge and take a major role: mistakenly we call them side-effects. More humility, and less impatience to come to conclusions, is required. Even our use of words misleads us as, for example, when we speak of 'pests' and imagine that harm to a pest is necessarily good. Often a 'pesticide' has been less lethal to the pest, which has seemed to thrive, than to its predators or parasites, which has formerly kept it in control, and with whom we would have done better to make friends.

The era from which we are now emerging was not only the era of the explosion of knowledge, but has brought also an explosion of ignorance: ignorance of factors which we have ourselves made newly important. We need an era of consolidation of knowledge, but instead we have a plethora of information. Consolidation means separating out and understanding the best from the flood of scientific literature, nearly all of which is evanescent. It is my modest hope that this book makes a contribution towards clarifying and ordering our growing knowledge of our atmospheric environment.

QUESTIONS AND (PERHAPS QUESTIONABLE) STATEMENTS FOR DISCUSSION

1. A cupful of air poured upwards in water forms a spherical cap bubble. How does a cupful of water poured downwards in air differ? What about a cupful of mercury poured into water?
2. Could a glider with an average sinking speed of 1 m sec^{-1} stay aloft by dynamic soaring in wind gradients that actually occur?
3. According to Bernoulli's theorem the wind speed is greater where the pressure is less: therefore the wind increases with height.
4. Where the wind increases with height, the high velocity at the chimney top causes a suction, according to Bernoulli's principle.
5. A hydrogen filled balloon is less likely than a soap bubble to make contact with a building in a wind.
6. At the point of intersection of two mutually perpendicular lines of fluid particles the sum of their rates of rotation in their plane is the same as for any other two such lines in the same plane.
7. Why is it usually inconveniently windy around the base of tall buildings, and can architects do anything about it?
8. How can you prevent yourself from falling over backwards, without moving your feet, by whirling your arms?
9. How can a cat land on its feet if dropped from rest with its feet upwards? (no air resistance). Try standing on a turntable and attempt to turn round.
10. If viscosity is suddenly increased in a short section of the flow around two parallel line vortices so that they cancel one another, what will happen to the 'loose ends' of the vortices on either side of this section? What about the loops found in contrails?
11. If silt collects where the flow of a river is slowest why doesn't it collect on the outside of bends where the pressure is greater and therefore the velocity less?
12. Can deposits of leaves, drifts of snow, accumulations of dust and silt, and so on be explained on the basis that deposition occurs where the velocity is least; and can scouring be attributed to high fluid velocity?

13. A basin of water is at rest. The basin is then set in rotation. It is observed that the fluid at the surface moves inwards towards the axis and yet the surface becomes hollowed out there. What sort of magic is this?
When the fluid is all in rotation the basin is stopped rotating. The hollow in the upper surface fills up but the fluid moves away from the centre!
14. A basin of water is in solid rotation. In the absence of air resistance a ping pong ball floating on the surface will move towards the middle; but if weighted so that it lies just totally immersed it moves away from the middle. In the latter case can the object be made not to move outwards, or perhaps even move inwards?
15. Can a tree derive any advantages from catching cloud water droplets on its leaves? What sort of leaves are best for this purpose?
16. Why is accretion of particles a maximum but condensation of vapour a minimum at a forward stagnation point? What kinds of ice accretion illustrate this?
17. If the pull on the rope of a flag is increased when the flag flaps how can animals propel themselves forwards by propagating a wave motion backwards? Is the swimming of plaice aided by the proximity of the sea bed?
18. Can one expect to release heavy spheres, e.g. ball bearings, into treacle so that collisions occur? Can they be made to collide with rising air bubbles, or bubbles with one another?
19. Why does the air behind a car, as made visible by exhaust smoke, appear to move as if there were two trailing vortices as behind an aircraft? Does this imply that there is a lift force on the car, and if so is it desirable?
20. A rigid circular disc with a hole in the middle is placed close to a flat wall and air is blown through the hole towards the wall. What is the force produced on the wall? What is the force due to a jet of air directed from a nozzle (without the disc) towards the wall? Devise a domestic experiment whereby a piece of paper is lifted from a table by blowing air downwards on to it.
21. Why do some orifices produce turbulent puffs while others produce ring vortices with laminar flow?
22. Two jam jars with tightly sealed lids are placed together on a smooth table, one completely filled with water, the other filled except for one or two cc of air. They are held firmly in the hand and moved around rapidly in a circle, for a second or so, with the base sliding on the table. In one the water is set in rotation, in the other not. The direction is seen (by tracer particles) to be that of circular motion of the hand, which is opposite to the direction of rotation about its centre of a ball placed in a similarly moved empty jar. How is the vorticity fed into the water in one case and how does such a small volume of air alter this effect?
23. Separation occurs in regions where the boundary layer is thickening downstream. Why then does the point of separation move forward if the

boundary layer is made thinner by increasing the velocity?

24. If the turbulence of the atmosphere is diffusive how is it that water evaporated from a large area of ocean can be concentrated in a small area and create a flood on land? Do not showers contradict the second law of thermodynamics?
25. In turbulent motion eddies are by definition, randomly oriented. Therefore as many are being compressed as are being stretched; as many are losing kinetic energy as are gaining it, and there is no net transfer of energy from the mean motion, and no work is done against eddy stresses, i.e. there are no such stresses. Equally there is no net transfer of energy between eddies of different sizes in the spectrum. What other properties of turbulence can be discovered from such definitions?
26. Is any purpose served by defining any kind of irrotational motion as turbulent?
27. Show that in a fluid in which the velocity and temperature gradients are determined by viscosity and thermal conductivity (e.g. Hawthorne's hot hump experiment) the Richardson number alone cannot provide a criterion for the onset of turbulence.
28. The flow of energy from large to small eddies is caused by the removal of energy from the small eddies by viscosity.
29. The flow of energy down the size scale to smaller eddies is caused by the eddy stresses, due to the small eddies in the organised motion of the larger ones.
30. How does turbulence occurring in stably stratified layers generate layers of greater static stability than existed before?
31. In what circumstances can thermal convection or mechanically generated turbulence increase the static stability at the top of the mixed layer?
32. The Richardson number is useless as a criterion for the onset of turbulence because it is easier to discover whether turbulence is present than to measure the Richardson number.
33. The Richardson number is a local number and therefore cannot provide a criterion for the onset of global turbulence.
34. Why do trees 'withered' by the prevailing wind assume 'aerofoil' shapes?
35. Would a shape good at catching airborne particles in a horizontal wind have a large viscous or a large form drag? Would large form drag be associated with better capture of falling particles by larger falling bodies?
36. Why is an anabatic wind stronger and shallower when the air above the lower ground is stably stratified?
37. How is stably stratified air warmed by thermally driven upcurrents while remaining stable (a) when clouds are present (b) when no clouds are present?
38. How is the sub-cloud stable layer related to the amount of (a) sunshine (b) moonshine?

39. If a plume from a fixed source of heat is in the form of a cone with semi-angle 5° it must be in an upcurrent.
40. Could a machine fixed to the ground in the bottom 100 m of a tornado extract energy released by buoyancy forces at 1000 m.
41. Show that the mean streamlines up a vertical jet follow lines for which $r \propto z^{1/2}$. What is the corresponding curve for a two-dimensional jet and for a plume?
42. Compare the efficacy of the following methods of introducing heat into a room:

    A fan heater blowing hot air across the floor
    A convection heater with a small high temperature source
    A convection heater with a large low temperature source
    A ventilation system which extracts air at floor level, warms it, and blows it in at high level
    A similar system which extracts it at high level and blows it in at a low level
    A heated floor
    A heated ceiling
    A hot source with a reflector to beam the radiation

    Which of these systems generates draught, cold feet, hot head; and which is most economical or efficient for particular purposes? Which is the most versatile?
43. How many non-dimensional numbers may be generated from the following factors controlling sonic gravity waves in the atmosphere: $g\beta$ in the stratosphere, $g\beta$ in the troposphere, c (velocity of sound) at the ground, h the height of the tropopause, σ the frequency of the source?
44. Why are wave clouds more common in Greenland and Tierra del Fuego than in Venezuela?
45. Why does the wind blow uphill more often than downhill?
46. In what circumstances would it be expected that the waves revealed by a radio-sonde balloon would decrease in frequency as it rises?
47. Why does rain fall in streaks, and not in horizontal sheets which would result from the following drops catching up the leading ones by falling in their wakes (downdraught)?
48. The air over a desert has 10 per cent humidity. The air over an oasis has 50 per cent humidity; show that unless it is more than 1.5°C cooler it will rise?
49. Show that saturated thermals can penetrate the sub-cloud stable layer in surroundings whose humidity is 50 per cent and whose temperature is 7°C if they are up to 0.3°C cooler, even without any upward momentum.
50. Why should lee waves have a greater amplitude at a level at which smooth-topped clouds are formed?

51. The radius r, of a streamer of pollution with no buoyancy released with zero velocity in a uniform wind is related to the distance, x, downwind according to $r \propto x^{1/3}$ except close to the source.

52. *'You'll find it's already stirred'*

53. Draw the stairs correctly, as viewed from about 3 m above the floor.

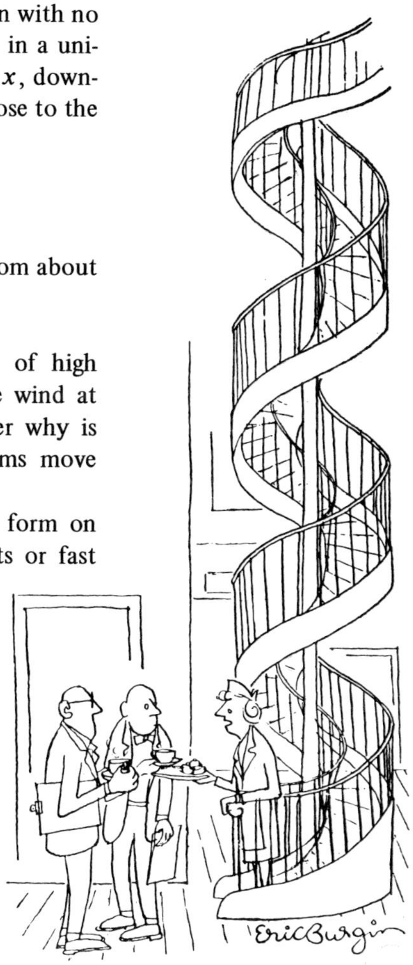

Reproduced by permission of PUNCH.

54. In what circumstances can a ridge of high pressure move backwards against the wind at all heights? (In view of your answer why is it often asserted that pressure systems move with the wind at an upper level?

55. Why are depressions more likely to form on the slow moving parts of cold fronts or fast moving parts of warm fronts than elsewhere?

56. Cyclogenesis is more likely on the left of a jet exit and on the right of a jet entrance.

57. Under what circumstances would there be divergence at the ground from a centre of low pressure?

58. Why is a tilted inversion a wind discontinuity? In what circumstances is the wind change vector (a) along the contours of the inversion (b) perpendicular to them?

59. Under what circumstances can the wind speed be equal to the geostrophic value when the isobars are curved?

60. A horizontal wind increases with height. It is rotating with the earth and therefore has some vertical vortex lines. These lines will be sheared over so that they have a horizontal (secondary) component which will make the upper level flow move to the right and the low level flow move to the left. How, then, might unidirectional shear flow persist?

61. In axial flow machinery, such as a turbine, there may be a significant obstacle on the axis. If there is a swirl in the flow a separation bubble is more likely to occur at the downstream end of this obstacle than if there

is no swirl.

62. If the updraught of a cumulonimbus shower be regarded as an obstacle to the flow at a high level the vortex lines are wrapped around it and will produce downflow into the cold downdraught. Likewise the cold nose will collect vortex lines which feed the updraught. Is this a significant effect.

63. Why does the airflow not disperse insect swarms, although it might divide large swarms?

64. Why does the wind tend to pile wind blown material, such as sand, into heaps rather than disperse it uniformly like smoke? How small do particles have to be to be dispersed? A larger barkhan dune is likely to be composed of larger particles than a small one?

65. Why does the chopping of vortices by the fan of a large by-pass ratio aero-engine emit noise? What might be the origin of vortices in the ambient air?

66. Why is the noise generated by the mechanism of question 65 greater when the engine is not moving relative to the ground (e.g. on testbed or start of runway) but reduced when flying above about 50 m from the ground?

67. The photograph shows water flowing past a circular cylinder in a channel 4 cm deep. The tracer is $KMnO_4$. Identify aspects of the flow at different depths.

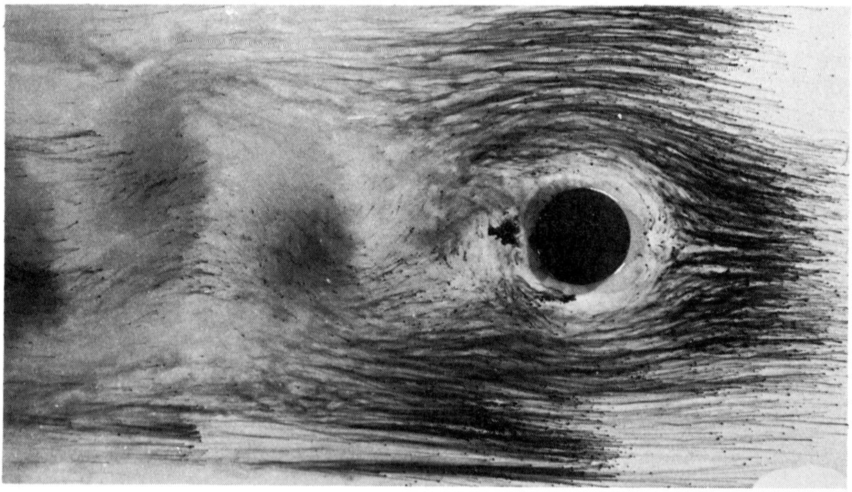

Fig. 67

68. Identify desert dust, low cloud, high cloud, a vortex within a vortex. Find Agadir, Canary Islands. (Satellite picture 5th March 1977)

Fig. 68

69. Close up of part of area of Fig. 68 on 3rd March 1977. Red rain fell in West Scotland a few days later. Note low cloud shadows; explain the 'ship wave patterns'.

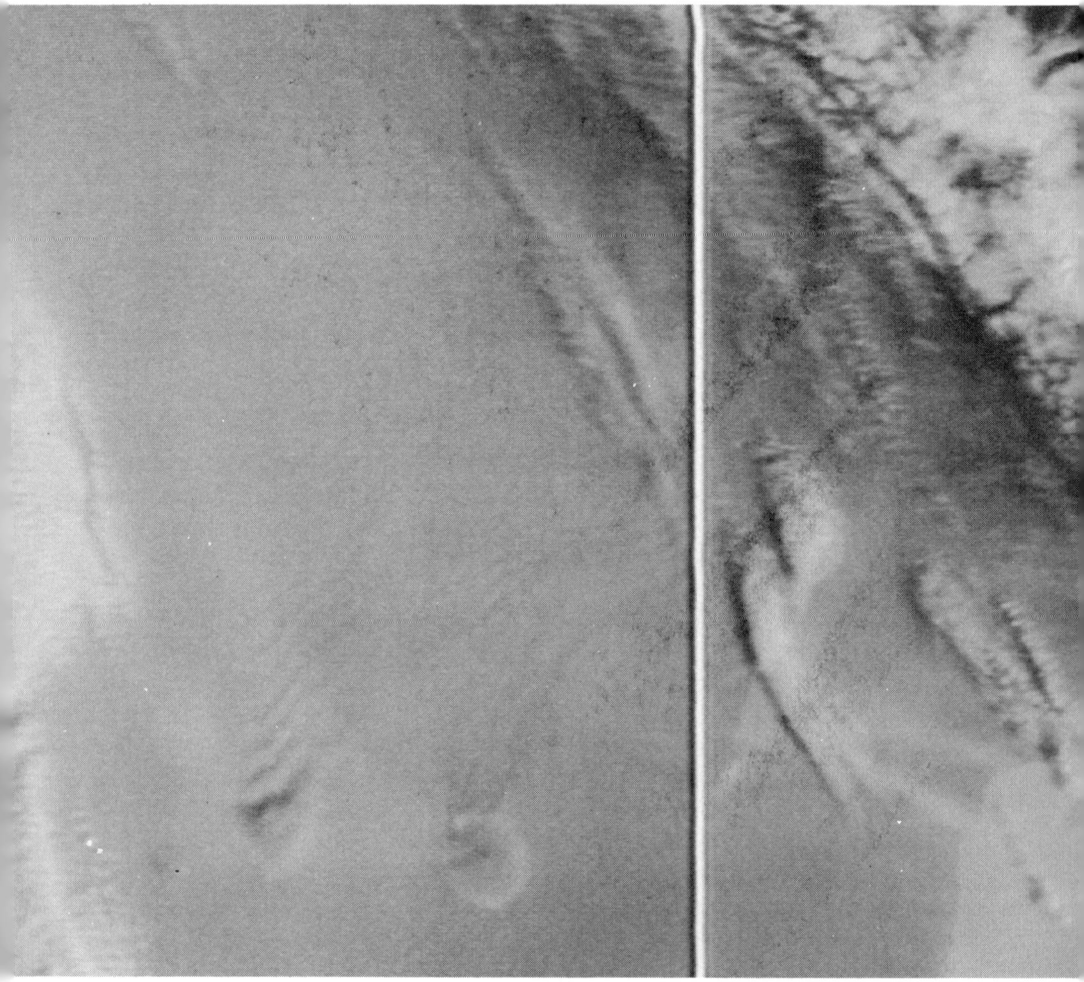

Fig. 69

70. Wind from ocean ice (infrared satellite picture 26th March 1977). Find streaming fog, cloud streets, vortex street (Jan Mayen) which is lopsided, orographic cirrus, snow line and coastline of Iceland, Greenland coast.

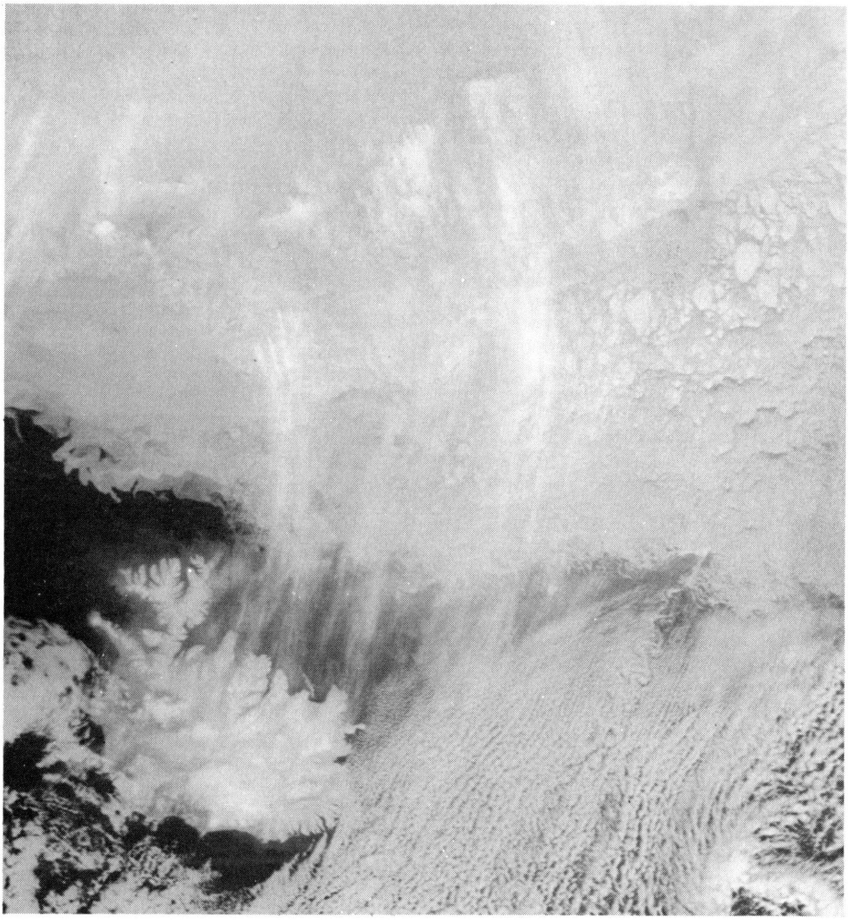

Fig. 70

71. Locate the land by the cloud forms (infrared satellite picture 9th September 1976). What is the wind direction?

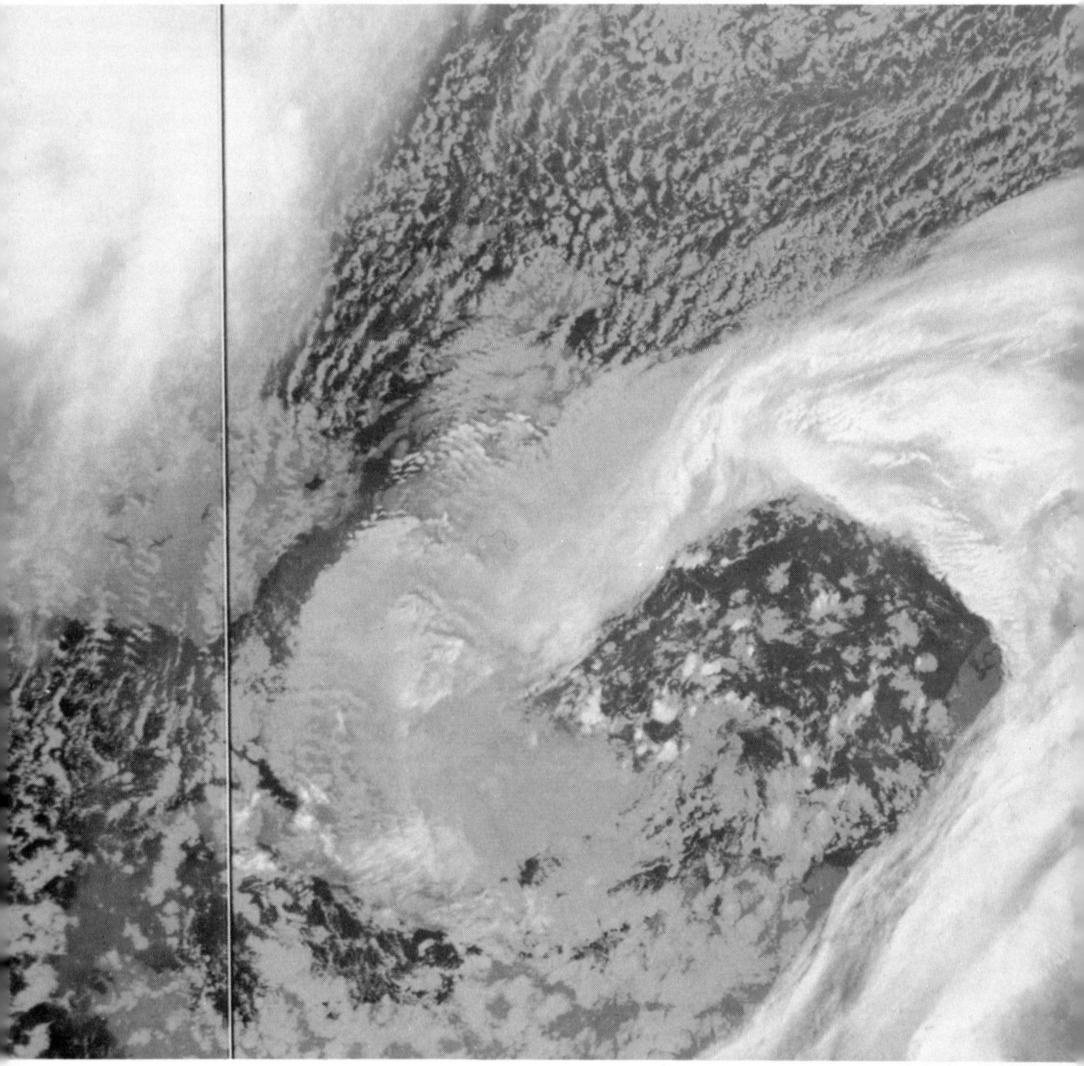

Fig. 71

72. Natural and man-made clouds: cirrus, altocummulus, cumulus, lee waves (Peterhead), contrails, industrial haze (Scot. Lowlands; Nottingham). What else? (Infrared satellite picture 0940GMT 13th August 1976).

Fig. 72

73. This photograph was obtained in the same channel as Fig. 67. Explain the details.

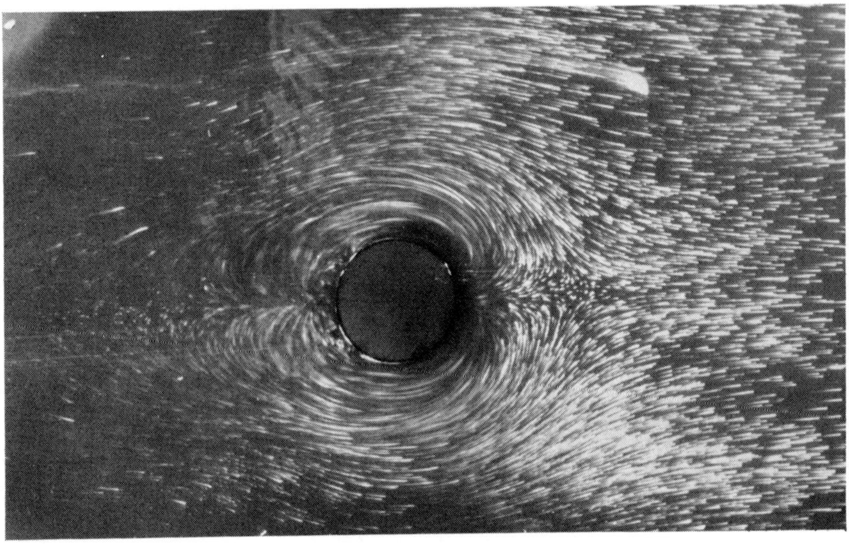

Fig. 73

References

Chapter 2

Lamb, Sir Horace, 1932, *Hydrodynamics* (Cambridge University Press).

Chapter 3

Dean, W. R., 1927, *Phil. Mag.*, **4**, 208.

Eustice, J., 1925, *Engineering*, issue Nov. 13.

Görtler, H., 1959, *Ingenieur Archiv.*, **28**, 71.

Hawthorne, W. R., 1951, *Proc. Roy. Soc. A*, **206**, 374.

Hawthorne, W. R. and Martin, M. E., 1951, *Proc. Roy. Soc. A*, **232**, 184.

McConologue, D. J. and Srivastava, R. S., 1968, *Proc. Roy. Soc. A*, **307**, 37.

Scorer, R. S., 1967, *J. Inst. Maths. Applic.*, **3**, 250.

Scorer, R. S. and Wilson, S. D. R., 1963, *Quart. J. R. Met. Soc.*, **89**, 532.

Squire, H. B. and Winter, K. G., 1951, *J. Aero Sci.*, **18**, 271.

Chapter 4

Brunt, D., 1939, *Physical and Dynamical Meteorology, 2nd Ed.*, (Cambridge University Press).

Scorer, R. S., 1972, *Clouds of the World* (David and Charles, Newton Abbot; Lothian Pub. Co. Melbourne; Stackpole, Harrisburg).

Tannehill, I. R., 1956, *Hurricanes, their nature and history, 9th Ed.*, (Princeton University Press).

Chapter 5

Astapowitsch, I. S., 1934, *Quart. J. Roy. Met. Soc.*, **60**, 493.

Banks, W. H. H., Drazin, P. G. and Zaturska, M. B., 1976, *J. Fluid Mech.*, **75**, 149.

Berkshire, F. H., 1975, *Pageoph.*, **113**, 561.

Berkshire, F. H. and Warren, F. W. G., 1970, *Quart. J. Roy. Met. Soc.*, **96**, 50.

Booker, J. R. and Bretherton, F. P., 1967, *J. Fluid. Mech.*, **27**, 513.

Drazin, P. G. and Howard, L. N., 1966, *Adv. Appl. Mech.*, **9**, 1–89.

Förchtgott, J., 1949, *Letectvi*, **25**, 350.

Förchtgott, J., 1950, *Met. Zpravy* No. 3–4, 72 (translated in *Gliding,* **2**, No. 4, 147).
Gerbier, N. and Berenger, M., 1961, *Quart. J. Roy. Met. Soc.,* **87**, 13.
Lamb, Sir Horace, 1932, *Hydrodynamics* (Cambridge University Press).
Lee, G. H. and Newmark, O. W., 1952, *Met. Mag.,* **81**, 307.
Long, R. R., 1953, *Tellus,* **5**, 42.
1955, *Ibid,* **7**, 341.
Manley, G., 1945, *Quart. J. Roy. Met. Soc.,* **71**, 197.
Queney, P., 1947, *Misc. Rep. Dept. Met. Univ. Chicago No. 23.*
1948, *Bull. Amer. Met. Soc.,* **29**, 16.
1955, *Tellus,* **3**, 367.
Roper, R. D., 1952, *Quart. J. Roy. Met. Soc.,* **78**, 415.
Sawyer, J. S., 1960, *Quart. J. Roy. Met. Soc.,* **86**, 326.
Scorer, R. S., 1949, *Ibid,* **75**, 41.
1950 (a), *Quart. J. Math. Appl. Mech.,* **3**, 107.
1950 (b), *Proc. Roy. Soc. A,* **201**, 138.
1951 (a), *Quart. J. Roy. Met. Soc.,* **77**, 76.
1951 (b), *Gliding,* **2**, 119.
1952 (a), *Sci. Prog., No. 159,* 466.
1952 (b), *Quart. J. Roy. Met. Soc.,* **78**, 76.
1953 (a), *Gliding,* **4**, 61.
1953 (b), *Quart. J. Roy. Met. Soc.,* **79**, 70.
1955, *Ibid,* **81**, 340.
1956, *Ibid,* **82**, 75.
Scorer, R. S. and Klieforth, H., 1959, *Ibid,* **85**, 131.
Scorer, R. S. and Wilkinson, M., 1956, *Ibid,* **82**, 419.
Wallington, C. E., 1958, *Ibid,* **84**, 428.
Warren, F. W. G., 1961, *J. Fluid Mech.,* **10**, 584.
Whipple, F. J. W., 1930, *Quart. J. Roy. Met. Soc.,* **56**, 287.
1935, *Ibid,* **61**, 285.
Yamamoto, R., 1957, *J. Met. Soc. Japan, Ser. II,* **35**, 32.
Yih, C.-S., 1960, *J. Fluid Mech.,* **9**, 161.
1967, *Ibid,* **29**, 539.

The reader is referred to two excellent reviews, both entitled *The Airflow over Mountains* (World Meteorological Organisation) viz:
Queney, P. (Chairman), Corby, G. A., Gerbier, N., Koschmieder, H., and Zierep, J., 1960, *Tech. Note No. 34 WMO, No. 98,* TP 43.
Nichols, J. M., 1973, *Tech. Note No. 127, WMO, No. 355.*
The form of clouds in mountain waves is extensively illustrated in
Scorer, R. S., 1972, *Clouds of the World,* Chapter 5.
A stimulating treatment of waves in stratified fluids will be found in
Yih, C.-S., 1969, *Mechanics of Non-homogeneous Fluids,* Chapters 1–3.

Chapter 6

Drazin, P. G., 1957, *J. Fluid Mech.*, **4**, 214.
1961, *Ibid,* **10**, 571.
1970, *Ibid,* **42**, 321.
1974, *Ibid,* **65**, 781.

Drazin, P. G.and Howard, L. N., 1961, *Proc. Amer. Soc. Civ. Eng. (J. Eng. Mech. Div.),* **87**, 101.

Hicks, J. J. and Angell, J. K., 1968, *J. App. Met.,* **7**, 114.

Howard, L. N., 1961, *J. Fluid Mech.,* **10**, 509.

Lamb, Sir Horace, 1932, *Hydrodynamics* (Cambridge University Press).

Miles, J. W., 1958, *J. Fluid Mech.,* **4**, 538.
1959, *Ibid,* **6**, 583.

Scorer, R. S., 1951, *Quart. J. Roy. Met. Soc.,* **77**, 235.
1961, *Paper 7 in RAE Sympos. – Atmos. Turb. in relation to Aircraft Op.* (HMSO).
1969, *Radio Science,* **4**, 1299.
1971, *Arch. Met. Geoph. Biokl. Ser. A,* **20**, 1.
1972, *Clouds of the World,* Chapter 6.

Thorpe, S. A., 1968, *J. Fluid Mech.,* **32**, 693.

Warren, F. W. G., 1968, *Quart. J. Mech. and Appl. Math.,* **21**, 433.
1974, *J. Fluid. Mech.,* **68**, 413.

Woods, J. D., 1968, *J. Fluid. Mech.,* **32**, 791.

Woods, J. D., and Wiley, R. L., 1972, *Deep Sea Res.,* **19**, 87.

See also:

Ludlam, F. H., 1967, *Quart. J. Roy. Met. Soc.,* **93**, 419.

Browning, K. A., 1971, *Ibid.,* **97**, 283.

Chapter 7

Batchelor, G. K., 1956, *Theory of Homogenous Turbulence*, (Cambridge University Press).

Brunt, D., 1939, *Physical and Dynamical Meteorology,* (Cambridge University Press).

Hinze, J. O., 1959, *Turbulence,* (McGraw-Hill).

Howard, L. N., 1961, *J. Fluid Mech.,* **10**, 509.

Lamb, Sir Horace, 1895 and 1932, *Hydrodynamics,* (Cambridge University Press).

McEwan, A. D., 1976, *Nature,* **260**, 126.

Saffman, P. G., 1962, *Quart. J. Roy. Met. Soc.,* **88**, 382.

Sutton, O. G., 1962, *The Challenge of the Atmosphere,* (Hutchinson).

Chapter 8

Davies, R. M. and Taylor, G. I., 1950, *Proc. Roy. Soc. A,* **200**, 375.

Goodhart, H. C. N., 1958, *Swiss. Aero. Revue,* OSTIV Section, **6**, 324.

Malkus, J. S. and Scorer R. S., 1955, *J. Met.,* **12**, 43.

Richards, J. M., 1961, *J. Fluid Mech.*, **11**, 369.
1963 (a), *Int. J. Air & Wat. Poll.*, **7**, 17.
1963 (b), *Quart. J. Roy. Met. Soc.*, **89**, 254.
1965, *J. Fluid Mech.*, **21**, 97.
1968, *Ibid*, **32**, 681.
1970, *Quart. J. Roy. Met. Soc.*, **96**, 702.
Ricou, F. P., and Spalding, D. B., 1961, *J. Fluid Mech.*, **11**, 21.
Rouse, H., Yih, C.-S., and Humphreys, H. W., 1952, *Tellus*, **4**, 201.
Schmidt, W., 1941, *Z.A.M.M.*, **21**, 265.
Scorer, R. S., 1957, *J. Fluid Mech.*, **2**, 583.
Spackman, E. A., 1964, *DIC Thesis*, Imperial College Department of Mathematics.
Woodward, B., 1959, *Quart. J. Roy. Met. Soc.*, **85**, 144.

Chapter 9

Angell, J. K. and Pack, D. H., 1963, *Monthly Weather Rev.*, **91**, 583.
1967, *Physics of Fluids – Suppl. Boundary layers and turbulence*, S 226.
Ives, R. L., 1947, *Bull. Amer. Met. Soc.*, **28**, 168.
Konrad, T. G. and Kropfli, R. A., 1968, *Proc. 13th Radar Meteor. Conf.*, Amer. Met. Soc., 262.
Lamb, Sir Horace, 1932, *Hydrodynamics*, (Cambridge University Press).
Scorer, R. S., 1968, *Air Pollution*, (Pergamon Press).
1969, *Arch. Met. Geoph. Biokl. Ser. A*, **18**, 249.
1972, *Clouds of the World*, (David and Charles).
Scorer, R. S. and Davenport, J. L., 1970, *J. Fluid Mech.*, **43**, 451.
Simpson, J. E., 1969, *Quart. J. Roy. Met. Soc.*, **95**, 758.
1972, *J. Fluid Mech.*, **53**, 759.
Simpson, J. E., Mansfield, D. A. and Milford, J. R., 1977, *Quart. J. Roy. Met. Soc.*, **103**, 47.
Turner, J. S., 1973, *Buoyancy Effects in Fluids*, (Cambridge University Press).
Willis, P. A., 1953, *On Being a Bird*, pp. 78-86. 1977 (David & Charles).

Chapter 10

Jorg, O. and Scorer, R. S., 1967, *Atmos. Environ.*, **1**, 645.
Lucas, D. H., Moore, D. J. and Spurr, G., 1963, *Int. J. Air. and Water Poll.*, **7**, 473.
Pasquill, F., 1974, *Atmospheric Diffusion, 2nd Ed.*, (Ellis Horwood, Ltd.).
R. S. Scorer, 1959, *Int. J. Air. Poll.*, **1**, 198.
Scorer, R. S., 1968, *Air Pollution*, (Pergamon Press).

Chapter 11

Browning, K. A. and Foote, G. B., 1976, *Quart. J. Roy. Met. Soc.*, **102**, 499.

Fraser, A. B., 1968, *Quart. J. R. Met. Soc.*, **94**, 71.
Ludlam, F. H., 1963, *Severe Local Storms: a review*, Met. Monographs, **5**, 1.
1966, *Tellus*, **18**, 687.
1976, *Bull. Amer. Met. Soc.*, **57**, 774.
Malkus, J. S. and Scorer, R. S., 1955, *J. Met.*, **12**, 43.
Newton, C. W., 1966, *Tellus*, **18**, 699.
Scorer, R. S., 1953, *Weather*, **8**, No. 7.
1955, *Arch. Met. Geoph. Biokl.*, **8**, 25.
1958, *Sci. Prog.* No. 181, 75.
1964, *Bull. Astr.*, XXIV Fasc. 2, 216.
1972, *Clouds of the World*, (David and Charles).
Scorer, R. S. and Davenport, L. J., 1970, *J. Fluid Mech.*, **43**, 451.

Chapter 12

Ackworth, B., 1955, *Bird and Butterfly Mysteries*, (Eyre and Spottiswoode).
Berson, F. A. and Simpson, K. G., 1971, *Weather*, **26**, 23.
Cochemé, J. and Scorer, R. S., *inter al.*, 1965, "Meteorology and the Desert Locust', *WMO/FAO. Seminar. WMO Tech. Note No. 69.*
Hankin, E. H., 1913, *Animal Flight: a Record of Observation*, (Iliffe).
Harper, W. G., 1957, *Nature*, **180**, 847.
1958, *Proc. Roy. Soc. B.*, **149**, 484.
1959, *Ibis*, **101**, 201.
Johnson, C. G., 1957, *Quart. J. Roy. Met. Soc.*, **83**, 194.
Johnson, C. G., Taylor, R. R. and Southwood, T. R. E., 1962, *J. Anim. Ecol.*, **31**, 373.
Joyce, R. J. V., 1973, *PANS*, **19**, 62.
Joyce, R. J. V., Marmol, L. C., Lucken, J., Bale, E. and Quantick, R., 1970, *PANS*, **16**, 309.
Rainey, R. C., 1951, *Nature*, **168**, 1057.
1963, 'Meteorology and the Migration of Desert Locusts', *WMO Tech. Note No. 54.*
1969, *Quart. J. Roy. Met. Soc.*, **95**, 424.
1973, *Weather*, **28**, 224.
Schaefer, G. W., 1976, *Insect Flight*, (Editor R. C. Rainey), Ch. 8, *Radar Observations of Insect Flight*, (Blackwell).
Scorer, R. S., 1972, *Clouds of the World*, (David and Charles).
1974, *Chemistry and Industry*, 2 March, 181.
Scorer, R. S. and Wexler, H., 1967, *Cloud Studies in Colour*, (Pergamon Press).
Simpson, J. E., 1967, *British Birds*, **60**, 225.
Williamson, K., 1965, *Fair Isle and its Birds*, (Oliver and Boyd).
1969, *Quart. J. Roy. Met. Soc.*, **95**, 414.

Index

Mathematical Symbols

A few symbols which occur only in one section have not been included: their meaning is clear from the context.

a	radius of sphere (2.2.7); chimney radius; particle dimension (2.8.2); tangential acceleration (3.1.12)
a	non-divergent vector (1.4.5)
b	horizontal oabstacle dimension (5.13); plume width (10.2)
b	unit binormal (3.1.6)
B	buoyancy number (2.12.1) (8.2.1) (8.2.7) (8.13.5)
B	Bernoulli vector $\mathbf{t} \times \mathbf{w}$
C	univeral dimensionless constant (8.2.1), (9.9.1); dimensional constant (8.11.12) (8.12.14)
C_D	drag coefficient (2.10.1)
c	velocity of sound (1.8.8); transition height (8.11.10) (8.12.12)
d	body dimension (2.8.1); density current depth (9.9.1)
D	drag force (2.8.2)
E	turbulent energy (7.1.1)
e_{ij}	rate of strain (1.3); $e_{r\theta}$ (7.9.17)
f	Coriolis parameter (Chs. 5, 6); f_3 (Ch. 4)
f	fluid acceleration = $D\mathbf{v}/Dt$ (1.4.1) (3.1.9); Earth's rotation (4.1.1)
F, F_i	body or total force
F	Froude number (2.12.3) (8.12.6), Rossby number (5.8.14)
F	buoyancy or pollution flux (8.10.18) (9.4.3) (10.3.3)
g	gravity, magnitude g.
g*	apparent gravity (4.1.6)
h	depth (2.5.1); representative height (2.12.2); isobaric contour height (4.4.1); layer depth (Chs. 5, 6); chimney height (10.2); h' = thermal rise (10.2)
H	Bernoulli constant gH (1.7.4) or (1.7.9); layer depth (6.2); effective chimney height (10.2)
j	unit vector across flow (5.1.6)
k	line vortex strength (2.5.3) (7.8.2); wave number (Chs. 5, 6); k* lee wave number; dimensional constant (8.2.9); momentum factor (8.7.1) (8.11.2)

k_0 isentropic constant (1.6.2) (1.8.1) (5.1.6)
K eddy transfer coefficient (4.11.1); vortex strength (9.10.5)
K_M momentum transfer coefficient (9.6.5)
K_H heat transfer coefficient (9.6.1)
$\mathbf{l}$ unit vector (1.1.1)
ℓ Scorer parameter (5.9.2); mixing length (7.3.i)
L $= \ln(\frac{1}{2}\rho q^2)$ (3.2.6)
m source strength (2.2.1); volume factor (8.2.1) (8.11.2)
M Mach number = U/c
$\dot{M}$ momentum flux (8.9.20)
n widening factor (8.2.6)
$\mathbf{n}$ unit normal (3.1.3)
N plume dilution factor (10.2.1)
p pressure, p_m mean pressure (1.3.18); lee wave parameter (5.13.20)
p_{ij} stress (1.3)
P pollution concentration (10.2)
q fluid speed (1.2.5)
Q volume or pollution flux (8.13.2) (10.2)
r radial coordinate
$\mathbf{r}$ position vector (4.1)
R gas constant (1.6.1); radius of earth (Ch. 4); residue (5.13); radius or width of thermal, puff, jet, plume (Ch. 8); vortex pair (Ch. 9)
Ra Rayleigh number (9.2.1)
Re Reynolds number (2.8.1) (7.3.11)
Ri Richardson number (2.12.4) (6.1.2) (7.5.3)
$\mathbf{R}$ $= \frac{1}{\rho}\,\mathbf{grad}\,\rho$ or $\frac{1}{\tau}\,\mathbf{grad}\,\tau$ (1.4.1)
s distance along trajectory
S Strouhal number (2.11.1); stability number (8.12.7)
S stream stratification (5.4.5); stratification factor (8.4.1)
t time
$\mathbf{t}$ unit tangent (3.1.1) $= \mathbf{v}/q$
T absolute temperature (1.6.1); kinetic energy (1.1.10)
u x‾ or r‾ velocity component velocity perturbation (5.8) (Ch. 6)
$\mathbf{u}$ velocity vector, magnitude u
U inflow velocity (8.9.5); stream velocity or wind speed
v $y-$ or $\theta-$ velocity component
$\mathbf{v}$ velocity $= (u, v, w)$ in Cartesian or cylindrical polar coordinates
$\mathbf{v}_a$ ageostrophic component of velocity (4.3.6)
$\mathbf{v}_g$ geostrophic component of velocity (4.3.4)
$\mathbf{v}_i$ isallobaric wind (4.3.11)
V transverse wind (5.8); speed of front (9.9.1)
w vertical velocity component

w_c vertical velocity of cap of thermal
W vertical velocity (8.11.1)
x horizontal streamwise Cartesian coordinate
$\mathbf{x}$ position vector
y horizontal transverse Cartesian coordinate
Y yield factor (8.4.2)
z vertical or axial coordinate

α angle of deflection of flow (2.4.1) (3.6.1); inclination of front (4.8.1); velocity stratification (shear) (5.4.4); entrainment ratio (8.11.3); plume cone angle (10.2)
β generally static stability (1.8.2) (5.2.4) (8.12.1); angle of helix (3.10–12); β-plane parameter (5.20)
γ ratio of specific heats (1.6.2), helical angle of vortex line (3.10.ii)
Γ adiabatic lapse rate (1.8.3)
δ boundary layer thickness (2.7.1)
ϵ energy dissipation rate (7.3.12)
ζ vertical displacement (1.8.5) (5.4.6) (Ch. 6); vertical or z-component of vorticity (5.20.1)
η y-component of vorticity; = vorticity in 2D motion (Chs. 5, 6, 7); magnitude of shear
H vorticity density (9.10.4)
θ potential temperature (1.6.3); angle of bend (Ch. 3); colatitude (3.10) (Ch. 4), spherical polar coordinates; inclination of thermal (9.5.1)
κ Hele-Shaw porosity (2.6.6); trajectory curvature (3.1.3); von Karman's constant (7.3)
λ dynamic viscosity (1.3.7 and 23); stratification (3.6.2); turbulent intensit intensity (10.3.7)
μ dynamic viscosity (1.3.7 and 23); doublet strength (2.2.5); stability coefficient for curved flow (3.10.5); vertical decay exponent (5.10.4); bending over factor (10.3)
ν kinematic viscosity (1.3.28); vertical wave number (5.10.3)
Π impulsive pressure (1.1.11)
ρ density
σ Darcy porosity (2.6.7); concentration of pollution (8.9.11), of insects (12.6.1), frequency (2.11.1) (5.20.6) (Ch. 6) (7.11); convection fraction (9.4.1); dimensionless constant (8.12.13)
τ $1/\theta$(potential temperature)$^{-1}$ (1.6.5); trajectory torsion (3.1.7); shear stress (7.3); mixing time (7.9)
τ_{ij} stress tensor (4.11.7)
ϕ potential (1.1.6) (2.1.2) (6.1), angular displacement (1.8.9) (3.2.ii) (3.5.2), ridge inclination (5.15); entropy (11.2)
Φ buoyancy flux (8.11.2)

ψ	stream function (6.2); streamline inclination (6.3.2)
ξ	x-component of vorticity
ω	vorticity = **curl v** (1.2.3), = (ξ, η, ζ) (1.3.2)
ω_s	secondary vorticity = ω.**t** (3.1.18)
ω_n	normal component of vorticity = ω.**n** (3.3.1)
Ω	angular velocity of rotation (1.3.3) (2.5) (7.8) (7.11)
Ω	earth's angular velocity (Ch. 4)
ϖ	modified pressure (1.6.4)